Ecotoxicology and Environmental Pollution

Ecotoxicology and Environmental Pollution

Editor: Bruce Horak

R CALLISTO REFERENCE

www.callistoreference.com

Callisto Reference,
118-35 Queens Blvd., Suite 400,
Forest Hills, NY 11375, USA

Visit us on the World Wide Web at:
www.callistoreference.com

ISBN: 978-1-63239-819-2 (Hardback)

The publisher's policy is to use permanent paper from mills that operate a sustainable forestry policy. Furthermore, the publisher ensures that the text paper and cover boards used have met acceptable environmental accreditation standards.

Printed in the United States of America.

Cataloging-in-publication Data

Ecotoxicology and environmental pollution / edited by Bruce Horak.
    p. cm.
Includes bibliographical references and index.
ISBN 978-1-63239-819-2
1. Environmental toxicology. 2. Pollution--Environmental aspects. 3. Environmental health. I. Horak, Bruce.
RA1226 .E26 2017
615.902--dc23

Table of Contents

Preface

Ecotoxicology is the study of the effects of pollutants present in the atmosphere on population, wildlife and the environment. This book discusses the imminent dangers that are faced by our ecosystem due to long-term waste mismanagement and pollution. Environmental pollution is adversely affecting all forms of life. It has become a growing concern for researchers and scientists all around the world. Research in this area focuses on the impact of exotoxins on wildlife and resources and strategies of survival and adaptation. This book discusses the fundamentals as well as modern approaches of tackling environmental pollution. It consists of contributions made by international experts. The topics included herein are of utmost significance and bound to provide incredible insights to the readers. With state-of-the-art inputs by acclaimed experts of this field, this book targets students and professionals.

This book aims to highlight the current researches and provides a platform to further the scope of innovations in this area. This book is a product of the combined efforts of many researchers and scientists from different parts of the world. The objective of this book is to provide the readers with the latest information in the field.

I would like to express my sincere thanks to the authors for their dedicated efforts in the completion of this book. I acknowledge the efforts of the publisher for providing constant support. Lastly, I would like to thank my family for their support in all academic endeavors.

Editor

Comparison of the Effects of Air Pollution on Outpatient and Inpatient Visits for Asthma: A Population-Based Study in Taiwan

Hui-Hsien Pan[1,2,9], Chun-Tzu Chen[1,3,9], Hai-Lun Sun[1,2], Min-Sho Ku[1,2], Pei-Fen Liao[1], Ko-Hsiu Lu[2], Ji-Nan Sheu[1,2], Jing-Yang Huang[4], Jar-Yuan Pai[3], Ko-Huang Lue[1,2]*

1 Department of Pediatrics, Chung Shan Medical University Hospital, Taichung City, Taiwan R.O.C, 2 School of Medicine, Chung Shan Medical University, Taichung City, Taiwan R.O.C, 3 Department of Health Policy and Management, Chung Shan Medical University, Taichung City, Taiwan R.O.C, 4 Institute of Public Health, Department of Public Health, Chung Shan Medical University, Taichung City, Taiwan R.O.C

Abstract

Background: A nationwide asthma survey on the effects of air pollution is lacking in Taiwan. The purpose of this study was to evaluate the time trend and the relationship between air pollution and health care services for asthma in Taiwan.

Methods: Health care services for asthma and ambient air pollution data were obtained from the National Health Insurance Research database and Environmental Protection Administration from 2000 through 2009, respectively. Health care services, including those related to the outpatient and inpatient visits were compared according to the concentration of air pollutants.

Results: The number of asthma-patient visits to health-care facilities continue to increase in Taiwan. Relative to the respective lowest quartile of air pollutants, the adjusted relative risks (RRs) of the outpatient visits in the highest quartile were 1.10 (P-trend = 0.013) for carbon monoxide (CO), 1.10 (P-trend = 0.015) for nitrogen dioxide (NO$_2$), and 1.20 (P-trend < 0.0001) for particulate matter with an aerodynamic diameter $\leqq$10µm (PM$_{10}$) in the child group (aged 0–18). For adults aged 19–44, the RRs of outpatient visits were 1.13 (P-trend = 0.078) for CO, 1.17 (P-trend = 0.002) for NO$_2$, and 1.13 (P-trend < 0.0001) for PM$_{10}$. For adults aged 45–64, the RRs of outpatient visits were 1.15 (P-trend = 0.003) for CO, 1.19 (P-trend = 0.0002) for NO$_2$, and 1.10 (P-trend = 0.001) for PM$_{10}$. For the elderly (aged $\geqq$ 65), the RRs of outpatient visits in were 1.12 (P-trend = 0.003) for NO$_2$ and 1.10 (P-trend = 0.006) for PM$_{10}$. For inpatient visits, the RRs across quartiles of CO level were 1.00, 1.70, 1.92, and 1.86 (P-trend = 0.0001) in the child group. There were no significant linear associations between inpatient visits and air pollutants in other groups.

Conclusions: There were positive associations between CO levels and childhood inpatient visits as well as NO$_2$, CO and PM$_{10}$ and outpatient visits.

Editor: Stephania Ann Cormier, University of Tennessee Health Science Center, United States of America

Funding: The authors have no support or funding to report.

Competing Interests: The authors have declared that no competing interests exist.

* E-mail: cshy095@csh.org.tw

9 These authors contributed equally to this work.

Introduction

Asthma is a common chronic inflammatory respiratory disease that affects 300 million people of all ages and all ethnic backgrounds and accounts for about 1 in every 250 deaths worldwide [1]. In Taiwan, the prevalence of asthma increased from 5.07% in 1985 to 11.9% in 2007 [2,3]. The risk factors for asthma include many external determinants such as mites, dust, mold, indoor and outdoor air pollution, and season variations [4–6]. Although air pollution has not been shown as the sole cause of respiratory illnesses, there is evidence that air pollution episodes lead to respiratory irritation, increased use of asthma medications and hospitalizations [7,8]. Traffic and industry-related pollutants, nitrogen dioxide (NO$_2$) and carbon monoxide (CO), were associated with asthma hospitalizations and outpatient visits [9,10]. Elevated levels of ozone (O$_3$), sulfur dioxide (SO$_2$) and particulate matter with an aerodynamic diameter $\leqq$10µm (PM$_{10}$) were reported to be related with increased asthma emergency room visits and admissions [11–13]. It has been reported that the rise in air pollution has increased respiratory and cardiovascular complications leading to elevated risk of death [14].

If air pollution is responsible for the observed enhanced respiratory complications and mortality, one would also expect to see an impact on clinic visits, outpatient visits, emergency department (ED) visits and hospitalization rates for asthma. However, there are no data regarding a population based survey with seasonal and air pollutants in asthma outpatient and inpatient visits in Taiwan. The objective of this study was to assess asthma-related outpatient and inpatient patterns of visits in different age groups based on the National Health Insurance Research Database (NHIRD) in Taiwan, and to compare the association with utilization of health care services and concentrations of air pollutants.

Materials and Methods

Database

The data were obtained from the NHIRD released by the National Health Research Institute (NHRI) in Taiwan. The National Health Insurance Program finances compulsory universal health care for 99% of all of residents of Taiwan [15]. The database contains demographic data, all health-care encounters, expenditure and dates of enrollment and withdrawal. To facilitate research, the NHRI randomly sampled a representative database of one million subjects enrolled in the National Health Insurance program in the year 2005 by a systematic sampling method. This one-million sample was validated to be representative of the entire insured population as reported by the NHRI. The identification numbers and personal information of all individuals in the NHRID were erased to protect the privacy of the individuals. This study was approved by the Institutional Review Board of the Chung-Shan Medical University Hospital, Taiwan.

Study Population

Cases of asthma were ascertained by the service claim for either outpatient or inpatient visit with a primary diagnosis of asthma (ICD-9-CM code 493.xx). Daily counts of clinic visits, outpatient visits, ED visits and hospital admissions for asthma were extracted from the medical insurance file for the period of 2000–2009. The outpatient visit was defined as a patient visit to a physician's office, clinic, or hospital outpatient department. The inpatient visits include ED visits and admissions. The analyses covered 54 municipalities, each with its own air quality monitoring station. We identified 306628 men and 315178 women who lived in municipalities with air quality monitoring stations during the study period. There were 33934 men and 34527 women with asthma. The patient's exact addresses were not available from the database. Therefore, we assumed that the municipality where a patient visit occurred was regarded as the same area where the patient was most likely exposed to air pollutants. Each asthma outpatient or inpatient visit was matched with the municipality's seasonal average pollutant concentrations. Each occurrence, limited to municipality with air quality monitoring stations, was counted as one visit. The outpatient and inpatient visits were further analyzed by seasons (spring as February, March and April; summer as May, June and July; autumn as August, September and October; winter as November, December and January) and four age groups (0–18, 19–44, 45–64 and ≥65 years).

Outdoor Air Pollution Monitoring

There were 76 air quality monitoring stations in Taiwan, established by the Environmental Protection Administration. Data from background and industrial air quality monitoring stations were excluded to avoid extreme levels of air pollutants. Finally, data from 54 air quality monitoring stations were enrolled (24 municipalities in northern, 9 municipalities in central, 19 municipalities in southern and 2 municipalities in eastern Taiwan). Complete monitoring data for the air pollutants included PM_{10}, SO_2, CO, O_3, and NO_2. The average seasonal concentration of each air pollutant from the Taiwan Environmental Protection Administration's air quality monitoring stations was calculated for further analysis from 2000 to 2009.

Statistical Analysis

All analyses were done by using the SAS ver. 9.3 software package (SAS Institute, Cary, NC, USA). Multivariate Poisson Regression was made in order to determine the relative risk (RR) of asthma inpatient and outpatient visits by sex, age, year and seasons. For use of health services in relation to air pollutants, the municipalities were defined as the observed units, and the level of air pollutants and counts of patient visits were collected in each municipality for each season from 2000 to 2009. As multiple visits by the same patient in a season in the municipality who lived, the count of the visits was as one visit per season. Generalized estimating equations (PROC GENMOD with repeated statement by the SAS Institute) were used to analyze of levels of pollutants by stratifying municipalities and to compare whether the inpatient and outpatient visits in those municipalities correlated with levels of pollutants. A p value <0.05 was considered statistically significant.

Results

The descriptive statistics for outpatient and inpatient visits and corresponding period and season data are shown in Table 1. Compared with women, the RRs for inpatient and outpatient visits in men were 1.31 (p<0.0001) and 1.15 (p<0.0001), respectively. Furthermore, when divided into 4 age groups, the highest RRs for inpatient and outpatient visits were among the elderly group (i.e. ≥65 years) followed by the child group (0–18), the adults aged 45–64 group and the adult aged 19–44 group. For the period effect, there were increased outpatient visits (RR: 1.29, P<0.0001) since 2005 and inpatient visits (RR: 3.30, P<0.0001) since 2006 when compared with patient visits in 2000. The peak seasons for asthma inpatient and outpatient visits in Taiwan were spring (5.1/10,000 person-season) and winter (149.6/10,000 person-season), respectively.

Table 2 shows the RRs of outpatient health service use with respect to exposure to air pollutants in the four age groups following adjustments for sex, period and quartiles of air pollutants by the generalized estimating equations model. Relative to the respective lowest quartile of air pollutants, the adjusted RRs of the outpatient visits in the highest quartile were 1.10 (95% CI: 1.04, 1.16; P-trend = 0.013) for CO, 1.10 (95% CI: 1.01, 1.18; P-trend = 0.015) for NO_2, 0.94 (95% CI: 0.88, 0.99; P-trend = 0.016) for O_3, and 1.20 (95% CI:1.13, 1.27; P-trend <0.0001) for PM_{10} in the child group. For adults aged 19–44, the adjusted RRs of the outpatient visits were 1.13 (95% CI: 1.03, 1.23; P-trend = 0.078) for CO, 1.17 (95% CI: 1.05, 1.31; P-trend = 0.002) for NO_2, 0.88 (95% CI: 0.83, 0.94; P-trend<0.0001) for O_3, and 1.13 (95% CI: 1.05, 1.21; P-trend <0.0001) for PM_{10}. The adjusted RRs of the outpatient visits in the adults aged 45–64 were 1.15 (95% CI: 1.05, 1.26; P-trend = 0.003) for CO, 1.19 (95% CI: 1.08, 1.30; P-trend = 0.0002) for NO_2, 0.93 (95% CI: 0.89, 0.98; P-trend = 0.028) for O_3, and 1.10 (95% CI: 1.02, 1.18; P-trend = 0.001) for PM_{10}. The adjusted RRs for the elderly outpatient visits across quartiles of air pollutant level were 1.00, 1.02, 1.07, and 1.12 (P-trend = 0.003) for NO_2 and 1.00, 1.04, 1.06, 1.10 (P-trend = 0.006) for PM_{10}. There was no significant association between SO_2 and outpatient visits in any group.

The use of inpatient health services in relation to levels of air pollutants by different age groups is shown in Table 3. The adjusted RRs for the child inpatient visits across quartiles of CO level were 1.00, 1.70, 1.92, and 1.86 (P-trend = 0.0001) after multivariate adjustment by sex, year and other air pollutants. There were no significant linear associations between inpatient visits and air pollutants in other age groups.

Table 1. Inpatient and Outpatient Visits Stratified by Sex, Age, Year and Season in Taiwan.

	Person-seasons	Inpatient visits[a]				Outpatient visits[b]			
		Visits per season	Visit rate per 10000 person-season	RR	P value	Visits per season	Visit rate per 10000 person-season	RR	P value
Sex									
Female	8297518	2970	3.6	-		104358	125.7	-	
Male	7904684	3711	4.7	1.31	<0.0001	113915	144.0	1.15	<0.0001
Age									
0–18	3930291	2658	6.8	-		93580	237.9	-	
19–44	7373364	1083	1.5	0.22	<0.0001	40932	55.5	0.23	<0.0001
45–64	3674627	1317	3.6	0.53	<0.0001	46489	126.5	0.53	<0.0001
≥65	1223920	1623	13.3	1.96	<0.0001	37272	304.3	1.28	<0.0001
Year									
2000	1611185	369	2.3	-		18604	115.4	-	
2001	1611426	400	2.5	1.08	0.265	19932	123.6	1.07	<0.0001
2002	1628261	393	2.4	1.05	0.469	20624	126.6	1.10	<0.0001
2003	1635688	264	1.6	0.71	<0.0001	19501	119.2	1.03	0.002
2004	1692152	323	1.9	0.83	0.017	23061	136.2	1.18	<0.0001
2005	1719118	239	1.4	0.61	<0.0001	25529	148.4	1.29	<0.0001
2006	1699621	1284	7.6	3.30	<0.0001	24301	143.0	1.24	<0.0001
2007	1685165	1380	8.2	3.58	<0.0001	25208	149.6	1.30	<0.0001
2008	1672283	1212	7.3	3.17	<0.0001	23951	143.2	1.24	<0.0001
2009	1247303	817	6.6	2.86	<0.0001	17562	140.8	1.22	<0.0001
Season									
Spring (Feb-Apr)	4159829	2127	5.1	-		60648	145.7	-	
Summer(May-Jul)	4151745	1416	3.4	0.67	<0.0001	50591	121.8	0.84	<0.0001
Autumn(Aug-Oct)	4140026	1502	3.6	0.71	<0.0001	50887	122.9	0.84	<0.0001
Winter (Nov-Jan)	3750602	1636	4.4	0.85	<0.0001	56147	149.6	1.03	<0.0001

RR, relative risk.
[a]Inpatient visits include emergency department visits and hospitalizations.
[b]Outpatient visits include physician's office, clinic, or hospital outpatient department visits.

Table 2. Outpatient Visits in Relation to Air Pollutants by Different Age Groups.

Variables	Outpatient visits[a]							
	Age group							
	0–18		19–44		45–64		≥65	
	RR	95%CI	RR	95%CI	RR	95%CI	RR	95%CI
Men/women	1.43	1.37–1.48	0.84	0.79–0.90	0.77	0.72–0.82	1.20	1.12–1.29
Year	1.06	1.05–1.07	1.03	1.02–1.04	1.00	0.99–1.02	1.01	0.99–1.02
CO (ppm)								
0.19–0.43	–		–		–		–	
0.43–0.55	1.07	1.03–1.11	1.13	1.07–1.20	1.06	1.01–1.12	1.02	0.97–1.06
0.55–0.69	1.06	1.01–1.12	1.12	1.03–1.21	1.11	1.04–1.20	1.02	0.96–1.09
0.69–1.22	1.10	1.04–1.16	1.13	1.03–1.23	1.15	1.05–1.26	1.04	0.95–1.13
P-Trend	0.013		0.078		0.003		0.450	
NO₂ (ppm)								
4.45–14.01	–		–		–		–	
14.01–18.64	1.02	0.97–1.08	1.03	0.95–1.11	1.07	1.01–1.14	1.02	0.97–1.07
18.64–24.04	1.05	0.98–1.12	1.08	0.98–1.19	1.14	1.06–1.23	1.07	1.00–1.13
24.04–47.84	1.10	1.01–1.18	1.17	1.05–1.31	1.19	1.08–1.3	1.12	1.03–1.21
P-Trend	0.015		0.002		0.0002		0.003	
O₃ (ppm)								
14.23–23.83	–		–		–		–	
23.83–27.23	0.99	0.96–1.03	0.96	0.93–0.99	0.96	0.93–0.99	0.98	0.94–1.02
27.23–30.80	0.96	0.92–1.00	0.94	0.90–0.98	0.98	0.94–1.01	0.99	0.94–1.04
30.80–50.00	0.94	0.88–0.99	0.88	0.83–0.94	0.93	0.89–0.98	0.95	0.90–0.99
P-Trend	0.016		<0.0001		0.028		0.103	
PM₁₀ (µg/m³)								
23.00–43.67	–		–		–		–	
43.67–54.00	1.06	1.02–1.10	1.07	1.03–1.11	1.02	0.98–1.07	1.04	1.01–1.08
54.00–71.00	1.11	1.06–1.17	1.14	1.09–1.19	1.09	1.03–1.14	1.06	1.01–1.11
71.00–141.67	1.20	1.13–1.27	1.13	1.05–1.21	1.10	1.02–1.18	1.10	1.03–1.17
P-Trend	<0.0001		<0.0001		0.001		0.006	
SO₂ (ppm)								
0.20–2.80	–		–		–		–	
2.80–3.77	1.03	0.97–1.08	0.99	0.93–1.05	1.00	0.94–1.05	1.04	0.99–1.09
3.77–5.00	1.05	0.98–1.11	0.96	0.90–1.03	0.96	0.90–1.03	1.05	0.99–1.10
5.00–19.17	1.04	0.97–1.13	0.96	0.89–1.04	0.97	0.90–1.05	1.01	0.95–1.08
P-Trend	0.201		0.216		0.340		0.590	

Abbreviation: CO, carbon monoxide; NO_2, nitrogen dioxide; O_3, ozone; PM_{10}, particulate matter with an aerodynamic diameter $\leqq 10\mu m$; RR, relative risk; SO_2, sulfur dioxide.
[a]Outpatient visits include physician's office, clinic, or hospital outpatient department visits.

Table 3. Inpatient Visits in Relation to Air Pollutants by Different Age Groups.

	Inpatient visits[a]							
	Age group							
	0–18		19–44		45–64		≥65	
Variables	RR	95%CI	RR	95%CI	RR	95%CI	RR	95%CI
Men/women	1.86	1.67–2.07	0.98	0.83–1.16	0.82	0.66–1.01	1.14	0.96–1.34
Year	1.34	1.28–1.39	1.22	1.17–1.27	1.12	1.08–1.15	1.15	1.10–1.20
CO (ppm)								
0.19–0.43	–		–		–		–	
0.43–0.55	1.70	1.40–2.06	1.08	0.83–1.40	1.17	0.92–1.49	1.28	0.98–1.66
0.55–0.69	1.92	1.53–2.40	1.07	0.81–1.42	1.18	0.90–1.55	1.35	0.93–1.97
0.69–1.22	1.86	1.39–2.48	0.94	0.64–1.38	1.21	0.87–1.68	1.42	0.94–2.15
P-Trend	0.0001		0.856		0.343		0.106	
NO₂ (ppm)								
4.45–14.01	–		–		–		–	
14.01–18.64	0.85	0.70–1.03	0.86	0.65–1.13	0.85	0.68–1.07	1.09	0.85–1.40
18.64–24.04	0.85	0.66–1.08	1.08	0.75–1.55	1.09	0.83–1.44	1.22	0.85–1.76
24.04–47.84	0.90	0.67–1.20	1.33	0.92–1.94	1.08	0.76–1.53	1.17	0.77–1.76
P-Trend	0.615		0.068		0.447		0.434	
O₃ (ppm)								
14.23–23.83	–		–		–		–	
23.83–27.23	0.86	0.74–0.99	1.02	0.90–1.16	1.00	0.81–1.23	1.14	0.99–1.31
27.23–30.80	0.87	0.74–1.02	1.08	0.87–1.32	1.13	0.88–1.44	1.18	1.01–1.39
30.80–50.00	0.83	0.66–1.05	1.04	0.79–1.35	1.06	0.80–1.40	1.20	0.99–1.45
P-Trend	0.136		0.548		0.450		0.070	
PM₁₀ (µg/m³)								
23.00–43.67	–		–		–		–	
43.67–54.00	0.84	0.73–0.97	1.04	0.83–1.30	0.82	0.69–0.97	0.89	0.74–1.07
54.00–71.00	0.97	0.83–1.13	1.25	0.96–1.63	1.01	0.82–1.22	1.00	0.81–1.23
71.00–141.67	0.89	0.71–1.12	1.05	0.77–1.44	1.04	0.78–1.37	0.93	0.70–1.23
P-Trend	0.757		0.336		0.393		0.900	
SO₂ (ppm)								
0.20–2.80	–		–		–		–	
2.80–3.77	0.95	0.77–1.18	1.13	0.83–1.52	0.90	0.69–1.17	0.95	0.72–1.26
3.77–5.00	1.03	0.79–1.33	1.26	0.96–1.65	0.91	0.70–1.19	1.00	0.76–1.31
5.00–19.17	0.82	0.63–1.07	0.96	0.71–1.29	0.81	0.61–1.08	0.78	0.57–1.05
P-Trend	0.369		0.882		0.204		0.223	

Abbreviation: CO, carbon monoxide; NO₂, nitrogen dioxide; O₃, ozone; PM₁₀, particulate matter with an aerodynamic diameter ≦10μm; RR, relative risk; SO₂, sulfur dioxide.
[a]Inpatient visits include emergency department visits and hospitalizations.

The use of inpatient health services increased with time as shown in Fig 1. In general, there were trends of increasing asthma outpatient visits in the children's group (Fig. 1A & B). In 2003, there was an outbreak of Severe Acute Respiratory Syndrome (SARS) in Hong Kong and nearly became a pandemic event [16]. During this period, inpatient visits dramatically decreased in all age groups. After the SARS outbreak, the rates of inpatient visits have increased with time since 2006, especially in the child and the elderly groups (Fig 1C & D).

Discussion

Among patients with asthma, air pollutant exposure causes increased asthma morbidity. Little is known about changes over time in air pollutant exposure among patients with asthma in a national sample. During the study period, the inpatient and outpatient visits by men were higher than in women. The peak seasons of asthma inpatient and outpatient visits for the total population were spring and winter, respectively. The inpatient and outpatient visits of asthmatics have not reached a plateau and have continued to increase. Inpatient visits for asthma increased with increased levels of CO in children but not for any pollutants in adults in the present study. Our study found that CO, NO_2 and PM_{10} had significant estimated associations on outpatient visits due to asthma and children are more susceptible than other age groups.

Health care visits only illustrate a small percentage and most severe inpatients of the total impacts of air pollution. Our study used the count of health care visits from NHRI databases as the measure of morbidity in the population. Clinic visits, outpatient visits, ED visits and admissions are types of health care service, but also possessed of potentially important disparities [17,18]. Because asthma is a chronic disease, patients with asthma were taught to deal with their symptoms and discomfort of an asthma attack [19,20]. When the concentrations of air pollutants rise, patients with asthma may have treated symptoms by themselves or visited neighborhood clinics and hospital outpatient departments for medical treatment. Subsequently, if patients did not receive any treatment or if the condition was deteriorated or ineffective after an outpatient visit, they would then visit hospital emergency departments for assistance. This may explain why there was no increased risk for inpatient visits for adults with increasing levels of air pollutants.

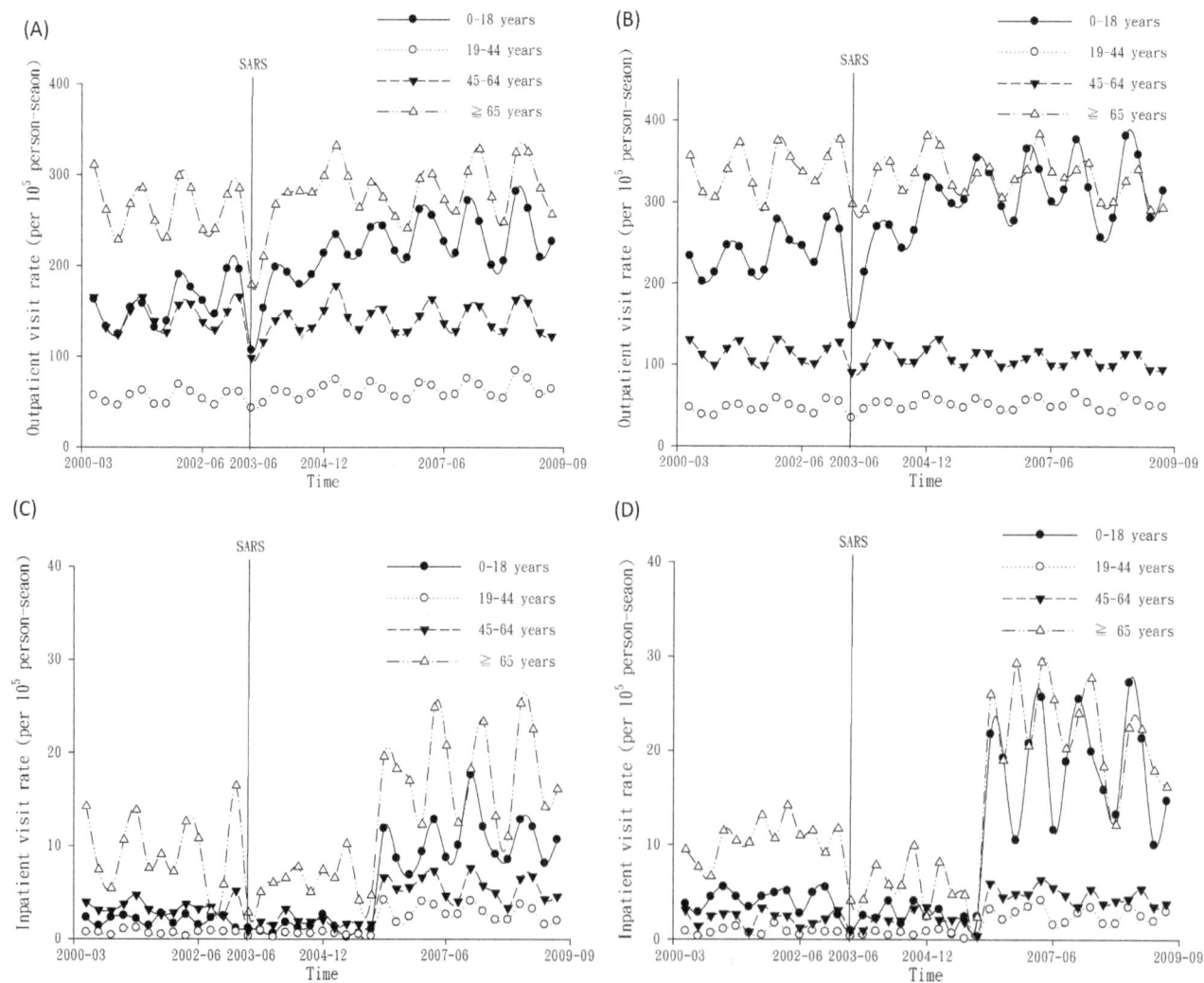

Figure 1. Outpatient visit rates for asthma in 4 age groups in women (A) and men (B), and inpatient visit rates in 4 age groups in women (C) and men (D) during 2000–2009. SARS, Severe Acute Respiratory Syndrome.

In response to the increasing prevalence, mortality rates, and medical cost of asthma, the Bureau of National Health Insurance initiated a Healthcare Quality Improvement Program for patients with asthma since November, 2001 [21]. The characteristics of the program were registry development, adherence to guidelines, patient education, and nursing care management. The Bureau offered financial incentives that motivated the nurses and physicians to change their practice patterns by following clinical guidelines; thereby, the asthma care team support from healthcare organizations was able to promote and enable patients to effectively self-manage their asthma with reduced healthcare resource utilization. During the SARS outbreak, people avoided hospital visits to prevent themselves from nasocomial infection and the inpatient visit rate for asthma decreased in all age groups [22]. However, there was an increase of inpatient and outpatient visits after the SARS outbreak. The exacerbation of air pollution probably plays a role in the rising rate of asthma visits due to growing populations, increased economic activity, rise in vehicular traffic, as well as the increasing intensity and occurrence of dust storms originating in Mongolia and China [23].

There have been associations between daily ambient air pollution levels and acute exacerbations of respiratory diseases in many time-series studies [11,12,24]. Urban air pollution constitutes a complex mixture of several compounds. The exposure of motor vehicle air pollutants, such as NO_2, CO, SO_2 and PM, increases the incidence and prevalence of asthma and bronchitic symptoms [25]. CO was reported to be associated with asthma admission and ED visits [9,26,27]. There were associations between short-term exposure to ambient CO and risk of cardiovascular disease hospitalizations, even at low ambient CO levels [28]. Sun et al conducted a one-year observation in central Taiwan that CO played a role in acute exacerbation of asthma in children and increased the number of childhood asthma ED visits, but not in adults [29]. Villeneuve et al reported that an increase in the interquartile range of the 5-day average for CO was associated with 48% increases in the risk of an asthma ED visit for children aged 2 to 4, but the associations were less pronounced in adult aged 15 to 44 [30]. In Rome, where air pollution comes mostly from combustion products of motor vehicles, CO was associated with most of the respiratory conditions in all ages, and it remained an independent predictor in multi-pollutant analysis for all respiratory admissions [31]. In London, there were significant associations between CO and daily consultations for asthma and other lower respiratory disease in children, whereas in adults the only consistent association was with PM_{10} [32]. In our study, CO was also significantly associated with asthma exacerbation and inpatient visits in children. However, a direct association between CO and asthma lacks a biologically plausible mechanism [33]. It is possible that CO might be a surrogate for other noxious incomplete combustion products [34]. Unlike with children, the other major confounders in adult asthmatic patients include occupational exposures, smoking, stress, emotional factors, and systemic diseases, which may also partly explain why outdoor air pollution was not associated with inpatient visits in the adults in our study.

Coal- and oil-fired power plants and diesel- and gasoline-powered motor vehicle engines are the main sources of ambient NOx emissions [35]. In a meta-analysis study, inhalation of NO_2 in the air significantly increased the development of childhood asthma and symptoms of wheezing [36]. A previous study in Taipei showed that most robust associations were found for NO_2 elevation and asthma admission rates [26]. NO_2 levels were associated with childhood asthma exacerbations and ED visits in Santa Clara, California [37]. In a spatiotemporal analysis of air pollution and asthma patient visits in Taipei, elevated levels of NO_2 had a positive association on outpatient visits [38]. In summary, there was a linear response in outpatient visits on days with elevated NO_2.

Previous studies have not yielded consistent results concerning associations between O_3 and SO_2 and asthma admissions. SO_2 was least frequently mentioned in the correlation with asthma hospitalization rate. Most previous studies have not shown a significant effect of SO_2 on asthma hospitalization rates supporting our findings [29,39]. O_3 is a highly reactive gas and induces bronchial inflammation, constriction of the airways and decreased lung function [40]. Long-term exposure to outdoor O_3 increases the prevalence of bronchitic symptoms among children [41]. O_3 levels have been previously reported to be associated with asthma admission and ED visits [11,24,38]. In contrast, asthma exacerbations were not associated with O_3 levels in North America [42]. Daily general-practice consultations for respiratory conditions were unrelated to O_3 in London [32] and Taiwan [43].

PM_{10} is a heterogeneous mixture of small solid or liquid particles with varying compositions in the atmosphere. There were no consistent results concerning associations between PM_{10} and asthma admissions. Some studies reported that PM_{10} was significantly associated with asthma admissions [8,39], other studies reported a lack of association between PM_{10} and asthma admission [27]. In our study, levels of PM_{10} were associated with outpatient visits for asthma, but not associated with admissions. Hwang and Chan used data obtained from clinic records and environmental monitoring stations in Taiwan and reported that PM_{10} had an impact on outpatient visits [43]. This was consistent with our findings.

Most of the previous studies are cross-sectional and have focused air pollutants on short-term, regional area and respiratory system diseases [8,29,38]. We conducted a nationwide asthma survey on the effects of air pollution in Taiwan and evaluated the association between different air pollutants and outpatient and inpatient visits. One of the strengths of the present study is the use of a computerized database, which is population-based and is highly representative. Because we enrolled all patients diagnosed with asthma from 2000 to 2009, we can rule out the possibility of selection bias. Since the data were obtained from a historical database, which collects all information, recall bias was avoided. Besides, we analyzed levels of pollutants at different municipalities and compared whether the inpatient and outpatient visits in those municipalities correlated with levels of pollutants. Thus this study directly associated the patient visits with the levels of pollutants in those municipalities. As multiple visits by the same patient would lead to misinterpretation of the data, we used the statistic method to reduce bias.

There were still several limitations of the present study. First, although we adjusted for several potential confounders in the statistical analysis, a number of possible confounding variables, including family history of atopy, dietary habits, physical activity, occupational exposures, smoking habits, stress and emotional factors, which are associated with asthma were not included in our database. Second, we were unable to ask patients for severity of asthma because of de-identifcation. Third, self-treatment with over-the-counter medications or alternative health services was not included in the database. These data also do not include the number of asthmatic subjects who had respiratory problems but did not search for health service. Therefore, the extent of the issue may have been considerably underestimated. Fourth, potentially inaccurate data in the records could lead to possible mis-classification.

In conclusion, the present study provides evidence that exposure to the outdoor air pollutant, CO, exerted adverse effects on health and increases in the child admission. Our study also showed a linear association between NO_2, CO, and PM_{10} and outpatient visits. It is an important public health policy to monitor air quality and warn the public about atmospheric factors that could be associated with increased risks of asthma.

Acknowledgments

This study is based on data from the National Health Insurance Research Database provided by the Bureau of National Health Insurance, Department of Health and managed by National Health Research Institutes. The descriptions or conclusions herein do not represent the viewpoints of the Bureau of National Health Insurance, Department of Health or National Health Research Institutes.

Author Contributions

Conceived and designed the experiments: K.H. Lue HHP CTC. Performed the experiments: JYH JYP. Analyzed the data: K.H. Lue HLS JNS. Wrote the paper: HHP CTC. Reviewed the manuscript and gave input to the final version: HLS MSK PFL K.H. Lu JNS.

References

1. Masoli M, Fabian D, Holt S, Beasley R (2004) The global burden of asthma: executive summary of the GINA Dissemination Committee report. Allergy 59: 469–478.

2. Hsieh KH, Shen JJ (1988) Prevalence of childhood asthma in Taipei, Taiwan, and other Asian Pacific countries. J Asthma 25: 73–82.

3. Hwang CY, Chen YJ, Lin MW, Chen TJ, Chu SY, et al. (2010) Prevalence of atopic dermatitis, allergic rhinitis and asthma in Taiwan: a national study 2000 to 2007. Acta Derm Venereol 90: 589–594.

4. Han YY, Lee YL, Guo YL (2009) Indoor environmental risk factors and seasonal variation of childhood asthma. Pediatr Allergy Immunol 20: 748–756.

5. Chiang CH, Wu KM, Wu CP, Yan HC, Perng WC (2005) Evaluation of risk factors for asthma in Taipei City. J Chin Med Assoc 68: 204–209.

6. Guo Y, Jiang F, Peng L, Zhang J, Geng F, et al. (2012) The association between cold spells and pediatric outpatient visits for asthma in Shanghai, China. PLoS One 7: e42232.

7. Yeh KW, Chang CJ, Huang JL (2011) The association of seasonal variations of asthma hospitalization with air pollution among children in Taiwan. Asian Pac J Allergy Immunol 29: 34–41.

8. Kuo HW, Lai JS, Lee MC, Tai RC, Lee MC (2002) Respiratory effects of air pollutants among asthmatics in central Taiwan. Arch Environ Health 57: 194–200.

9. Delamater PL, Finley AO, Banerjee S (2012) An analysis of asthma hospitalizations, air pollution, and weather conditions in Los Angeles County, California. Sci Total Environ 425: 110–118.

10. Wang KY, Chau TT (2013) An association between air pollution and daily outpatient visits for respiratory disease in a heavy industry area. PLoS One 8: e75220.

11. Wilson AM, Wake CP, Kelly T, Salloway JC (2005) Air pollution, weather, and respiratory emergency room visits in two northern New England cities: an ecological time-series study. Environ Res 97: 312–321.

12. Qiu H, Yu IT, Tian L, Wang X, Tse LA, et al. (2012) Effects of coarse particulate matter on emergency hospital admissions for respiratory diseases: a time-series analysis in Hong Kong. Environ Health Perspect 120: 572–576.

13. Cadelis G, Tourres R, Molinie J (2014) Short-term effects of the particulate pollutants contained in Saharan dust on the visits of children to the emergency department due to asthmatic conditions in Guadeloupe (French Archipelago of the Caribbean). PLoS One 9: e91136.

14. Wong TW, Tam WS, Yu TS, Wong AH (2002) Associations between daily mortalities from respiratory and cardiovascular diseases and air pollution in Hong Kong, China. Occup Environ Med 59: 30–35.

15. Lu JF, Hsiao WC (2003) Does universal health insurance make health care unaffordable? Lessons from Taiwan. Health Aff (Millwood) 22: 77–88.

16. Tsang KW, Ho PL, Ooi GC, Yee WK, Wang T, et al. (2003) A cluster of cases of severe acute respiratory syndrome in Hong Kong. N Engl J Med 348: 1977–1985.

17. Gold LS, Thompson P, Salvi S, Faruqi RA, Sullivan SD (2014) Level of asthma control and health care utilization in Asia-Pacific countries. Respir Med 108: 271–277.

18. Sun HL, Kao YH, Lu TH, Chou MC, Lue KH (2007) Health-care utilization and costs in Taiwanese pediatric patients with asthma. Pediatr Int 49: 48–52.

19. Centers for Disease Control and Prevention (2011) Vital signs: asthma prevalence, disease characteristics, and self-management education: United States, 2001–2009. MMWR Morb Mortal Wkly Rep 60: 547–552.

20. Weng HC (2005) Impacts of a government-sponsored outpatient-based disease management program for patients with asthma: a preliminary analysis of national data from Taiwan. Dis Manag 8: 48–58.

21. Bureau of National Health Insurance (2001) A national comprehensive disease management program. Taipei: Bureau of National Health Insurance.

22. Huang YT, Lee YC, Hsiao CJ (2009) Hospitalization for ambulatory-care-sensitive conditions in Taiwan following the SARS outbreak: a population-based interrupted time series study. J Formos Med Assoc 108: 386–394.

23. Bell ML, Levy JK, Lin Z (2008) The effect of sandstorms and air pollution on cause-specific hospital admissions in Taipei, Taiwan. Occup Environ Med 65: 104–111.

24. Winquist A, Klein M, Tolbert P, Flanders WD, Hess J, et al. (2012) Comparison of emergency department and hospital admissions data for air pollution time-series studies. Environ Health 11: 70.

25. Gasana J, Dillikar D, Mendy A, Forno E, Ramos Vieira E (2012) Motor vehicle air pollution and asthma in children: a meta-analysis. Environ Res 117: 36–45.

26. Yang CY, Chen CC, Chen CY, Kuo HW (2007) Air pollution and hospital admissions for asthma in a subtropical city: Taipei, Taiwan. J Toxicol Environ Health A 70: 111–117.

27. Slaughter JC, Kim E, Sheppard L, Sullivan JH, Larson TV, et al. (2005) Association between particulate matter and emergency room visits, hospital admissions and mortality in Spokane, Washington. J Expo Anal Environ Epidemiol 15: 153–159.

28. Bell ML, Peng RD, Dominici F, Samet JM (2009) Emergency hospital admissions for cardiovascular diseases and ambient levels of carbon monoxide: results for 126 United States urban counties, 1999–2005. Circulation 120: 949–955.

29. Sun HL, Chou MC, Lue KH (2006) The relationship of air pollution to ED visits for asthma differ between children and adults. Am J Emerg Med 24: 709–713.

30. Villeneuve PJ, Chen L, Rowe BH, Coates F (2007) Outdoor air pollution and emergency department visits for asthma among children and adults: a case-crossover study in northern Alberta, Canada. Environ Health 6: 40.

31. Fusco D, Forastiere F, Michelozzi P, Spadea T, Ostro B, et al. (2001) Air pollution and hospital admissions for respiratory conditions in Rome, Italy. Eur Respir J 17: 1143–1150.

32. Hajat S, Haines A, Goubet SA, Atkinson RW, Anderson HR (1999) Association of air pollution with daily GP consultations for asthma and other lower respiratory conditions in London. Thorax 54: 597–605.

33. American Thoracic Society (1996) Health effects of outdoor air pollution. Committee of the Environmental and Occupational Health Assembly of the American Thoracic Society. Am J Respir Crit Care Med 153: 3–50.

34. Norris G, YoungPong SN, Koenig JQ, Larson TV, Sheppard L, et al. (1999) An association between fine particles and asthma emergency department visits for children in Seattle. Environ Health Perspect 107: 489–493.

35. Trasande L, Thurston GD (2005) The role of air pollution in asthma and other pediatric morbidities. J Allergy Clin Immunol 115: 689–699.

36. Takenoue Y, Kaneko T, Miyamae T, Mori M, Yokota S (2012) Influence of outdoor NO2 exposure on asthma in childhood: meta-analysis. Pediatr Int 54: 762–769.

37. Lipsett M, Hurley S, Ostro B (1997) Air pollution and emergency room visits for asthma in Santa Clara County, California. Environ Health Perspect 105: 216–222.

38. Chan TC, Chen ML, Lin IF, Lee CH, Chiang PH, et al. (2009) Spatiotemporal analysis of air pollution and asthma patient visits in Taipei, Taiwan. Int J Health Geogr 8: 26.

39. Lin M, Stieb DM, Chen Y (2005) Coarse particulate matter and hospitalization for respiratory infections in children younger than 15 years in Toronto: a case-crossover analysis. Pediatrics 116: e235–240.

40. Khatri SB, Holguin FC, Ryan PB, Mannino D, Erzurum SC, et al. (2009) Association of ambient ozone exposure with airway inflammation and allergy in adults with asthma. J Asthma 46: 777–785.

41. Hwang BF, Lee YL (2010) Air pollution and prevalence of bronchitic symptoms among children in Taiwan. Chest 138: 956–964.

42. Schildcrout JS, Sheppard L, Lumley T, Slaughter JC, Koenig JQ, et al. (2006) Ambient air pollution and asthma exacerbations in children: an eight-city analysis. Am J Epidemiol 164: 505–517.

43. Hwang JS, Chan CC (2002) Effects of air pollution on daily clinic visits for lower respiratory tract illness. Am J Epidemiol 155: 1–10.

2

Azo Dye Biodecolorization Enhanced by *Echinodontium taxodii* Cultured with Lignin

Yuling Han[⑨], **Lili Shi**[⑨], **Jing Meng, Hongbo Yu, Xiaoyu Zhang***

College of Life Science and Technology, Huazhong University of Science and Technology, Wuhan, China

Abstract

Lignocellulose facilitates the fungal oxidization of recalcitrant organic pollutants through the extracellular ligninolytic enzymes induced by lignin in wood or other plant tissues. However, available information on this phenomenon is insufficient. Free radical chain reactions during lignin metabolism are important in xenobiotic removal. Thus, the effect of lignin on azo dye decolorization in vivo by *Echinodontium taxodii* was evaluated. In the presence of lignin, optimum decolorization percentages for Remazol Brilliant Violet 5R, Direct Red 5B, Direct Black 38, and Direct Black 22 were 91.75% (control, 65.96%), 76.89% (control, 43.78%), 43.44% (control, 17.02%), and 44.75% (control, 12.16%), respectively, in the submerged cultures. Laccase was the most important enzyme during biodecolorization. Aside from the stimulating of laccase activity, lignin might be degraded by E. taxodii, and then these degraded low-molecular-weight metabolites could act as redox mediators promoting decolorization of azo dyes. The relationship between laccase and lignin degradation was investigated through decolorization tests in vitro with purified enzyme and dozens of aromatics, which can be derivatives of lignin and can function as laccase mediators or inducers. Dyes were decolorized at triple or even higher rates in certain laccase–aromatic systems at chemical concentrations as low as 10 μM.

Editor: Ligia O Martins, Universidade Nova de Lisboa, Portugal

Funding: This work was supported by the National Natural Science Foundation of China (Grant No. 31170104/C010501), National High Technology Research and Development Program of China (863 Program) (No. 2012AA101805), and the Cooperation Project in Industry, Education and Research of Guangdong Province and Ministry of Education of P.R. China, 2008B090500217. The funders had no role in study design, data collection and analysis, decision to publish, or preparation of the manuscript.

Competing Interests: The authors have declared that no competing interests exist.

* Email: xyzhang.imer@gmail.com

⑨ These authors contributed equally to this work.

Introduction

More than 700,000 tonnes of dyes are produced annually [1]. Azo, anthraquinone, sulfur, indigoid, triphenylmethyl (trityl), and phthalocyanine are most commonly used in industries, but azo derivatives account for approximately 70% of all synthetic colorants [2]. Processing and daily washing release up to 15% of the used dyes into the aquatic environment [3]. This practice adversely affects the aesthetic value of water bodies and poisons aquatic and terrestrial organisms [4]. All synthetic dyestuffs have chemical structures resistant to light, water, chemicals (e.g., oxidizing and reducing agents), and biological corrosion, making effluents from these compounds refractory pollutants that notably affect conventional biological wastewater treatments [5,6]. Traditional biological treatments of synthetic dyes such as activated sludge and biofilm are inexpensive and green alternatives. However, the efficiency of these techniques in the treatment of textile effluents considerably varies. Thus, a more economical and effective approach to dispose of wastewater with an extensive variety of dyes should be explored and implemented.

White-rot fungi (WRF) have attracted increasing scientific attention in environmental pollutant removal because of their unique ligninolytic enzyme system. Those enzymes are mainly extracellular and exhibit broad substrate specificity to decompose the complex and random phenylpropanoic polymeric structure of lignin [7]. Bioprocessing of dyes using certain WRF strains cultured under optimal fermentation conditions programmed for higher ligninolytic enzyme production have been investigated. Given the importance of extracellular enzymes in biodegradation, decolorization using crude or pure enzymes has been extensively explored. However, whole cells are superior to crude enzymes and much better than purified enzymes in generating nontoxic, harmless, or even completely mineralized end products; thus, dye decolorization significantly depends on an integrated fungal working system [8]. Hence, whole-cell treatment systems are further investigated for better utilization. Stains were found to fade out more rapidly in WRF cultures with low-cost lignocellulose substrates, comprising a wide variety of wastes from agricultural, forest, and food industries. Lignin-degrading enzymes were generally observed to be induced by natural additives than by synthetic media [9–13]. However, the mechanism behind the simultaneous increase in decolorization and enzyme production in natural substrates has not been elucidated. The biodegradation enhancement has been perceived to have resulted from the added oxidoreductases caused by the complex carbon resources or from the phenolic extracts from lignocellulose. Ligninolytic enzymes are generally promoted by nutrient structure (especially nitrogen limitation) regardless of the existence of a pollutant [8]. Moreover, phenols in lignocellulose can easily be released and participate in the enzymatic reaction as cofactors or redox mediators [8,14] or in the activation and induction of ligninolytic enzyme. In addition, lignin extracted from effluents of pulping and papermaking industries was often used as a superior stimulator for ligninolytic enzyme production [15].

However, discussions on the improvement of decolorization using WRF grown on lignocelluloses, the mimetism of fungal primitive habitat, are speculations and ambiguous, without any records directly revealing the mechanism behind the lignin/lignocellulose induction of enzymes. Moreover, lignin metabolism during the secondary metabolic stage along with xenobiotic compound degradation is often neglected. Lignin is a highly stable natural heterogeneous polyphenolic biopolymer [16]. Thus, this compound can be selectively decomposed by fungal ligninolytic enzymes, such as low-redox potential laccase and high-redox potential ligninolytic peroxidases. These enzymes can nonspecifically attack the chemical bonds between phenylpropane units through free radical chain reactions and break down the structures of lignin as well as various xenobiotics, especially azo dyes, which share a similar aromatic skeleton with lignin [17,18]. The oxidative breakdown of lignin phenylpropanoid units releases low-molecular-weight products [19], which may act as stimulators for laccase or ligninolytic peroxidases [13]. Furthermore, humus and some quinone compounds, e.g., anthraquinone-2,6-disulfonate, which can be transformed from lignin by WRF [20], can function as redox mediator and significantly accelerate the biotransformation of azo dyes, nitroaromatics, or chlorinated aromatics [21]. Thus, when pollutants are degraded by WRF in the presence of lignocellulose, the metabolism of lignin (the second most abundant component in lignocellulose) positively participates in aromatic xenobiotic degradation. However, no direct proof for this phenomenon has been obtained, necessitating further investigations. The clarification of this phenomenon will facilitate the control of the bioprocessing of pollutant by WRF cultured with lignocelluloses.

This study was designed to evaluate the relationship between lignin decomposition and the efficiency of azo dye decolorization to determine the mechanism by which lignin metabolism affects aromatic pollutant degradation and provide guidance to the biological treatment of xenobiotics. The physicochemical properties and enzymatic activities of extracellular fluid were investigated in this study. Possible lignin metabolites and their function were determined by purifying and inoculating the key ligninolytic enzyme involved in biodegradation using dyes and various lignin-derived compounds.

Materials and Methods

Ethics statement

No specific permissions were required for Shen-nung-chia Nature Reserve (Hubei Province, P.R. China) and the location was a scenic spot. The field studies did not involve endangered or protected species and the fungus was common in this location. Echinodontium taxodii 2538 was reported in our previous studies [22–25] and the GenBank accession number of this fungus was EF422215.

Microorganism and growth conditions

Echinodontium taxodii (GenBank accession number, EF422215) was isolated from rotten wood in Shen-nung-chia Nature Reserve (a scenic spot in Hubei Province, P.R. China) and preserved by our laboratory [24]. This strain was chosen for the present study because of the following reason. In our previous researches on

biological pretreatment of lignocelluloses for energy production, the fungus exhibited great selective lignin-degrading ability [22–25]. Besides, *E. taxodii* was able to decolorize various dyestuffs in a pretest of environment pollutants removal. The organism was maintained on maize straw plate at 4°C with periodic transfer and activated on basic medium agar plate at 28°C for 7 days [26].

Dyestuffs and chemicals

Remazol Brilliant Violet 5R (RBV5), Direct Red 5B (DR5B), Direct Black 38 (DB38) and Direct Black 22 (DB22) were purchased from Aladdin. 2,2′-Azinobis,3-ethylbenzothiazoline-6-sulfonic acid (ABTS) and lignin (CAS Number: 8068-05-1, Product Number: 370959) were obtained from Sigma-Aldrich. Veratryl alcohol and other fine chemicals were bought from Sinopharm Chemical Reagent Co. Ltd. China.

Effect of lignin on azo dye decolorization by *Echinodontium taxodii*

Three agar plugs with 1 cm diameter, which were punched from the periphery of 7-day agar plates, were cultivated in a stationary 250-mL flask containing 30 mL medium at 28°C. Quadruplicate flasks were used for each group. The media components of the four groups for each dye set in this study are shown in Table 1. Dye groups were used as the control for lignin–dye groups. Media containing only lignin were set to examine the effect of lignin on *E. taxodii*, which was inoculated in basic media as the corresponding control.

The basic medium (per liter distilled water) contained 10.0 g glucose, 0.2 g ammonium tartrate, 2 g KH_2PO_4, 0.71 g $MgSO_4$, 0.1 g $CaCl_2$, and 20 mL trace element solution. The trace element solution had the following composition (per liter distilled water): 1 g NaCl, 0.5 g $MnSO_4 \cdot H_2O$, 0.1 g $CoCl_2$, 0.1 g $FeSO_4 \cdot 7H_2O$, 0.1 g $ZnSO_4 \cdot 7H_2O$, 0.1 g $CuSO_4 \cdot 5H_2O$, 0.01 g H_3BO_3, 0.01 g $Na_2MoO_4 \cdot 2H_2O$, and 0.01 g $KAl(SO_4)_2 \cdot 12H_2O$. The media were prepared as 1.2-fold-concentrated stock solutions and 25 mL of the media were split into 250-mL flasks. Azo dyes (RBV5, DR5B, DB38 and DB22, additional information on these dyes are shown in Table 2) were dissolved in distilled water with a final concentration of 600 mg/L. Then, distilled water, the media and dyes were autoclaved at 115°C for 20 min. After that, dye solutions or distilled water (5 mL) were added to each flask sterilely. Different concentrations of lignin (0.1, 0.3, and 0.5 g/L) were sterilized with ultraviolet light for 30 min before being added to the sterilized media. Considering the disoriented state of water-soluble components, lignin was washed thoroughly with distilled water and then freeze-dried to constant weight before use. The cultures of each group in quadruplicate flasks were taken on the day after inoculation. The fermentation liquor was obtained by filtering out the mixture of mycelia and lignin residues before centrifugation at 10,000×g for 10 min to remove water-insoluble lignin. The mycelia were oven-dried to a constant weight and were calculated as the biomass after the solid mixture was washed through a 100-mesh filter gauze to remove the lignin and media residues.

In order to evaluate the adsorption by lignin, different concentrations of lignin were added to media containing dyes sterilely, without inoculation. The samples were taken at the same time with cultures mentioned above, and centrifuged at 10,000×g for 10 min to remove water-insoluble lignin. The liquid supernatant was used to determine the decolorization of lignin adsorption.

Decolorization was monitored by measuring the absorbance of the fermentation broth at the maximum wavelength using a spectrophotometer. Decolorization percentage (%) was calculated as follows: decolorization percentage (%) = $[(A_0 - A)/A_0] \times 100$,

Table 1. The media used in this study.

contents	control	lignin	dye	Lignin-dye
basic medium	+	+	+	+
lignin	−	+	−	+
dye	−	−	+	+

+ With the content listed on the left added
− Without the content listed on the left added

where A_0 is the initial absorbance of the dye, and A is the absorbance of dye with time.

Laccase activity was determined by measuring the increase in A_{420} with 0.5 mM ABTS as the substrate in 50 mM sodium acetate buffer, pH 4.5 ($\varepsilon_{420} = 29,300$ $M^{-1} \cdot cm^{-1}$) [27]. Manganese peroxidase (MnP) activity was assayed using the method of Wariishi, Valli, Gold [28] and by measuring the increase in absorbance at 270 nm with 1 mM Mn^{2+} as the substrate in 50 mM malonic acid buffer, pH 4.5 ($\varepsilon_{270} = 11,590$ $M^{-1} \cdot cm^{-1}$). The reaction was started by adding 0.4 mM H_2O_2. Lignin peroxidase (LiP) activity was determined using the method of Tien, Kirk [29], with veratryl alcohol as a substrate in 0.1 M sodium tartrate, pH 3.0 ($\varepsilon_{310} = 9,300$ $M^{-1} \cdot cm^{-1}$). All reactions were performed at room temperature.

Effect of lignin-related compounds on azo dye decolorization by purified laccase of E. taxodii

Laccase was the only ligninolytic enzyme secreted by this strain during the whole dye removal period, and lignin might affect biodecolorization through this key oxidase. Laccase was purified by salting out with $(NH_4)_2SO_4$, hydrophobic interaction chromatography with Phenyl Sepharose 6 Fast Flow, and ion exchange chromatography with diethylethanolamine (DEAE)–sepharose Fast Flow.

The E. taxodii culture medium was passed through a filter paper to remove fungal mycelia. The supernatant was then subjected to a two-step ammonium sulfate precipitation scheme. First, the supernatant was brought to 45% saturation with ammonium sulfate, and the precipitated proteins were removed by centrifugation at 10,000×g for 15 min. Then, solid ammonium sulfate was added again, and the fraction precipitating at 45% to 85% saturation was collected by centrifugation. The precipitate was redissolved in 10 mL of 20 mM sodium acetate–acetic acid buffer (pH 6) containing 0.8 M ammonium sulfate. The supernatant was loaded onto a Phenyl Sepharose 6 Fast Flow column (1.6×20 cm) equilibrated with the same buffer. Unbound proteins were removed by washing the column at a flow rate of 1 mL·min^{-1}. Decreasing concentrations of $(NH_4)_2SO_4$ (0.8 M to 0 M) were used to elute the bound proteins, and the collected fractions (5 mL each) were subjected to enzymatic assay and protein measurement (A_{280}). The active fractions were pooled, dialyzed against 20 mM sodium acetate–acetic acid buffer (pH 6) overnight, and concentrated by PEG 20000. The concentrate was then applied to a DEAE–Sepharose Fast Flow column (1.6×20 cm) pre-equilibrated with the same buffer. Proteins were eluted with a linear gradient of NaCl (0 M to 0.8 M) at a flow rate of 1.5 mL·min^{-1}. Subsequently, fractions containing laccase activity were pooled, concentrated, and desalted. Enzyme purity was then confirmed by sodium dodecyl sulfate–polyacrylamide gel electrophoresis (SDS-PAGE).

Dyes were decolorized in a test tube using purified laccase at room temperature. The reaction mixture (3 mL) contained sodium acetate–acetic acid buffer (50 mM, pH 4.8), purified laccase (final concentration of 1 U/mL), and dyes (final concentration of 100 mg/L) in the presence or absence of lignin-related phenols. The reaction was initiated by adding the enzyme solution under mild shaking conditions. The duration of decolorization was determined by measuring the absorbance at the maximum wavelengths with a Mapada^{TM} UV-1600 spectrophotometer. Decolorization was calculated as described previously. A control test containing the same amount of heat-denatured laccase was conducted in parallel. All reactions were performed in triplicate.

To evaluate dye decolorization using different natural mediators, 12 lignin-related phenols as natural mediators were applied in the decolorization of dyes. These natural mediators (final concentration of 1 mM) were incubated in the buffer described previously at room temperature under mild shaking conditions. In order to evaluate the potential interference of the 12 lignin-related phenols as natural mediators on decolorization measurement, another control test without laccase and dyes were set. The results showed that these phenols had no absorption at those wave lengths, so decolorization was calculated as described previously. The decolorization of dyes was calculated after reaction for 2 h. Then, lower concentrations (10, 50, 100, 250 and 500 µM) of the efficient mediators were applied to the buffer under the same conditions mentioned above for further investigation of the

Table 2. Characteristics of the synthetic dyes studied in this work.

Dyes	CAS	Molecular Formula	λ_{max}(nm)	Number of azo bonds
RBV5	12226-38-9	$C_{20}H_{16}N_3Na_3O_{15}S_4$	560	1
DR5B	2610-11-9	$C_{29}H_{19}N_5Na_2O_8S_2$	510	2
DB38	1937-37-7	$C_{34}H_{25}N_9Na_2O_7S_2$	550	3
DB22	6473-13-8	$C_{44}H_{32}N_{13}Na_3O_{11}S_3$	482	4

efficiency of these mediators. The decolorization of dyes was detected at different time intervals (from 5 min to 240 min).

Results and Discussion

Effect of lignin on dye decolorization by *E. taxodii*

Azo reactive dyes RBV5, DR5B, DB38, and DB22 were chosen because they are extensively used in the textile industry worldwide. These azo dyes with 1 to 4 azo groups were tested to establish the relationship between the decolorization effect of *E. taxodii* (with/ without lignin) and the structures of azo dyes. *E. taxodii* could remove all four dyes (for as long as 2 wk; data not shown), and the total decolorization percentages of the RBV5, DR5B, DB38, and DB22 controls on day 6 were 65.96%, 43.78%, 53.02%, and 57.20%, respectively. The results showed that the decolorization effect differed from one dye to another, and there was no correlation between the difficulty degree of decolorization and the number of azo bonds. However, compared with the controls, lignin addition enhanced the biological treatment process, no matter how many azo linkages there was (Fig. 1).

In the case of RBV5, *E. taxodii* performed fairly in the first few days. Maximal differences in the decolorization percentage were observed between the control (65.96%) and lignin tests (75.54%, 100 mg/L; 80.64%, 300 mg/L; and 91.75%, 500 mg/L) on day

6, which were 15%, 22%, and 39% higher than that in non-lignin media. On day 7, lignin-free cultures still had dye residues at approximately 25 mg/L, whereas all the test groups had less than 10 mg/L residue. For DR5B, lignin resulted in a notable increase in the decolorization percentage from day 3, and the discrepancies were 5.69% (100 mg/L), 11.95% (300 mg/L), and 7.47% (500 mg/L). On the last day (day 7), control media still showed lower decolorization (75.99%) than the test groups by 16.29% (100 mg/L), 13.37% (300 mg/L), and 18.55% (500 mg/L). In addition, the discrepancies in the decolorization percentage first increased and then decreased along the degradation period. The maximal differences in the decolorization percentage between control and test groups varied, as follows: 100 mg/L, 16.77% on day 6; 300 mg/L, 31.67% on day 5; and 500 mg/L, 33.11% on day 6. For RBV5 and DR5B, the enhancement in decolorization seemed to have intensified with an increase in lignin dosage.

E. taxodii in lignin-containing media initially decolorized dye DB38 quickly and then the activity slowed down. However, the fungus worked at nearly constant speed in the blank culture. Different concentrations of lignin did not result in significant differences in dye degradation, but remarkable enhancements were observed when compared with lignin-free experiments in the first few days. Maximum differences in dye decolorization percentage compared with the blank were observed on day 2 or

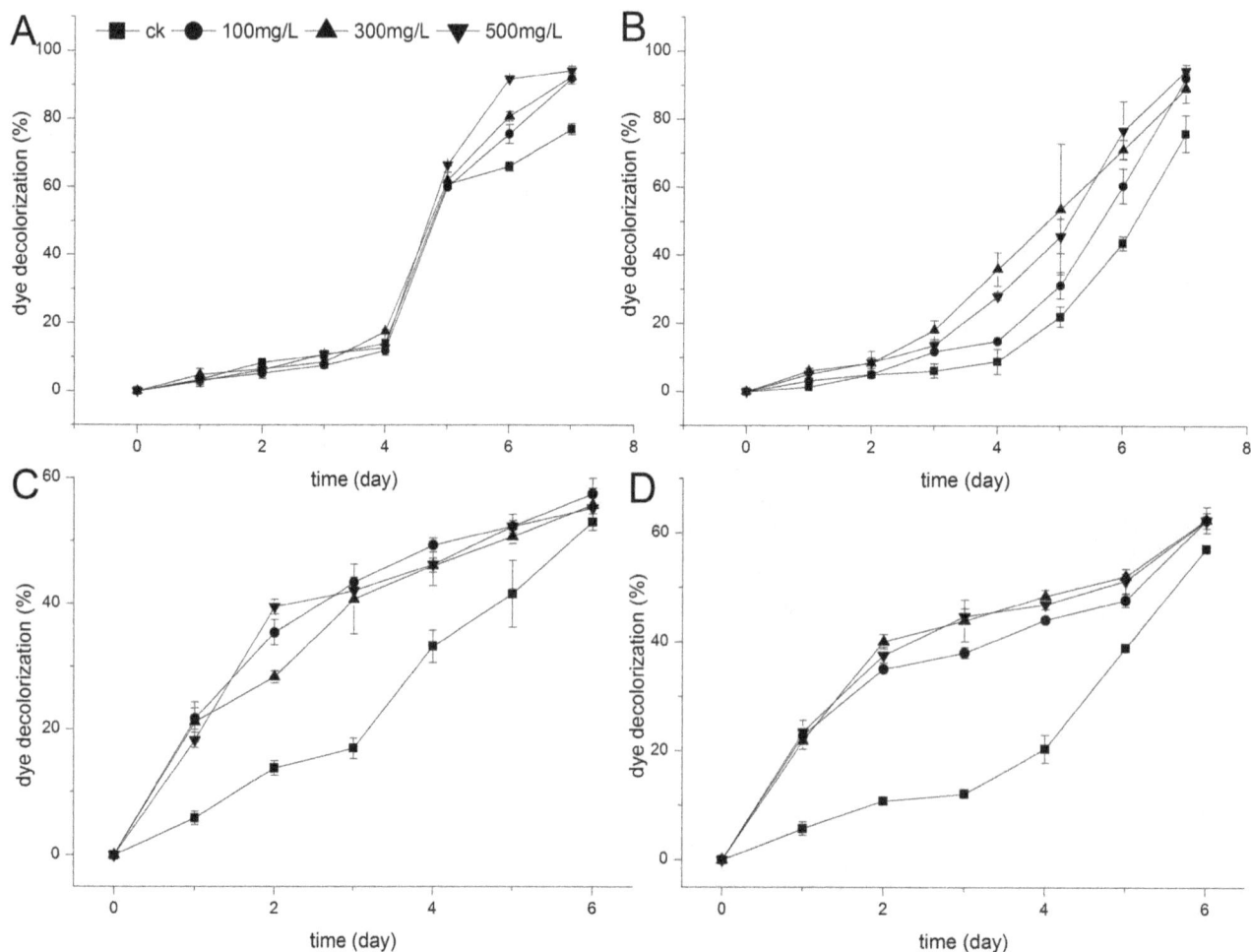

Figure 1. Effect of lignin on the decolorization of different dyes by *E. taxodii*. (A): RBV5. (B): DR5B. (C): DB38. (D): DB22. Dyes concentration: 100 mg/L. Lignin contents: ■, 0 mg/L; ●, 100 mg/L; ▲, 300 mg/L; ▼, 500 mg/L.

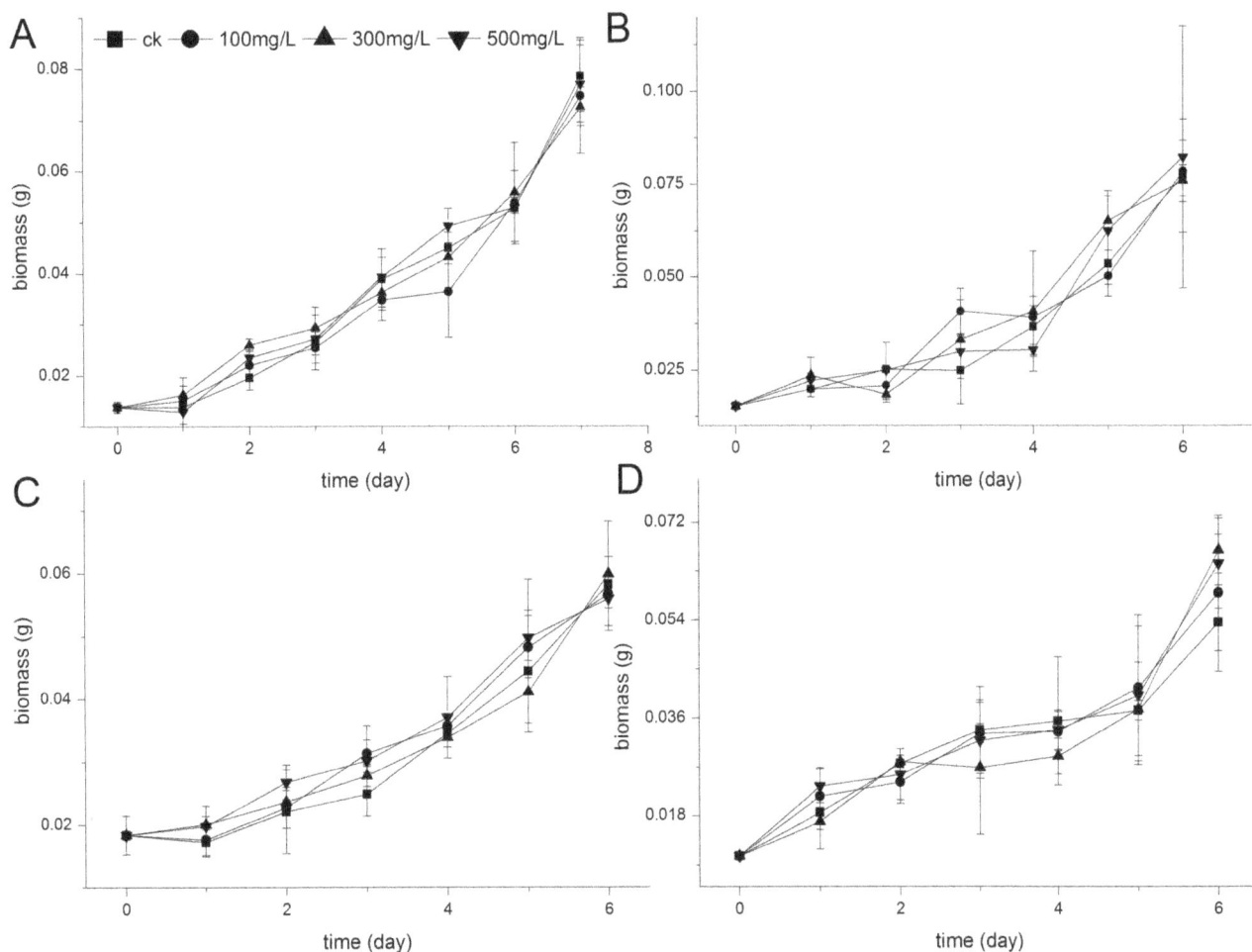

Figure 2. Effect of lignin on biomass during the decolorization of dyes by *E. taxodii.* (A): RBV5. (B): DR5B. (C): DB38. (D): DB22. Dyes concentration: 100 mg/L. Lignin contents: ■, 0 mg/L; ●, 100 mg/L; ▲, 300 mg/L; ▼, 500 mg/L.

3 after inoculation, as follows: 26.42% (100 mg/L on day 3), 23.74% (300 mg/L on day 3), and 25.67% (500 mg/L on day 2). Differences gradually decreased, and the control even performed fairly on day 6 (Fig. 1). Differences in decolorization curves caused by lignin indicated that *E. taxodii* treatment of DB38 may follow disparate pathways at first. Similar results were obtained when the fungus reacted with DB22, and more significant promotion than DB38 degradation was recorded. From days 1 to 6, the dye was eliminated much more rapidly along with lignin than the control, and the differences reached maximum values on day 3, as follows:

25.92% (100 mg/L), 31.81% (300 mg/L), and 32.59% (500 mg/L). Afterward, no significant differences were detected.

Dye bioadsorption using dead or living biomass of WRF has been extensively reviewed in several recent papers [30,31]. Hence, the absorption function of biomass and lignin was considered in determining the relationship between biological degradation and lignin metabolism. Stationary surface culture was adopted in this study. Results showed that lignin had different absorption patterns on the four dyes. Peak values were all reached on day 2 at the highest lignin concentration with less than 10% absorption as

Table 3. Effect of lignin on the adsorptions of azo dyes.

Adsorption Percentage of Azo dyes (%)[a]	Lignin content (mg/L)		
	100	**300**	**500**
RBV5	3.72(0.62)	4.31(0.53)	7.21(1.62)
DR5B	3.91(0.78)	6.75(2.09)	9.09(1.02)
DB38	None	None	None
DB22	3.62(0.78)	5.76(1.36)	8.96(1.02)

a Average Percentage with Standard Deviations in Parentheses; N = 3.

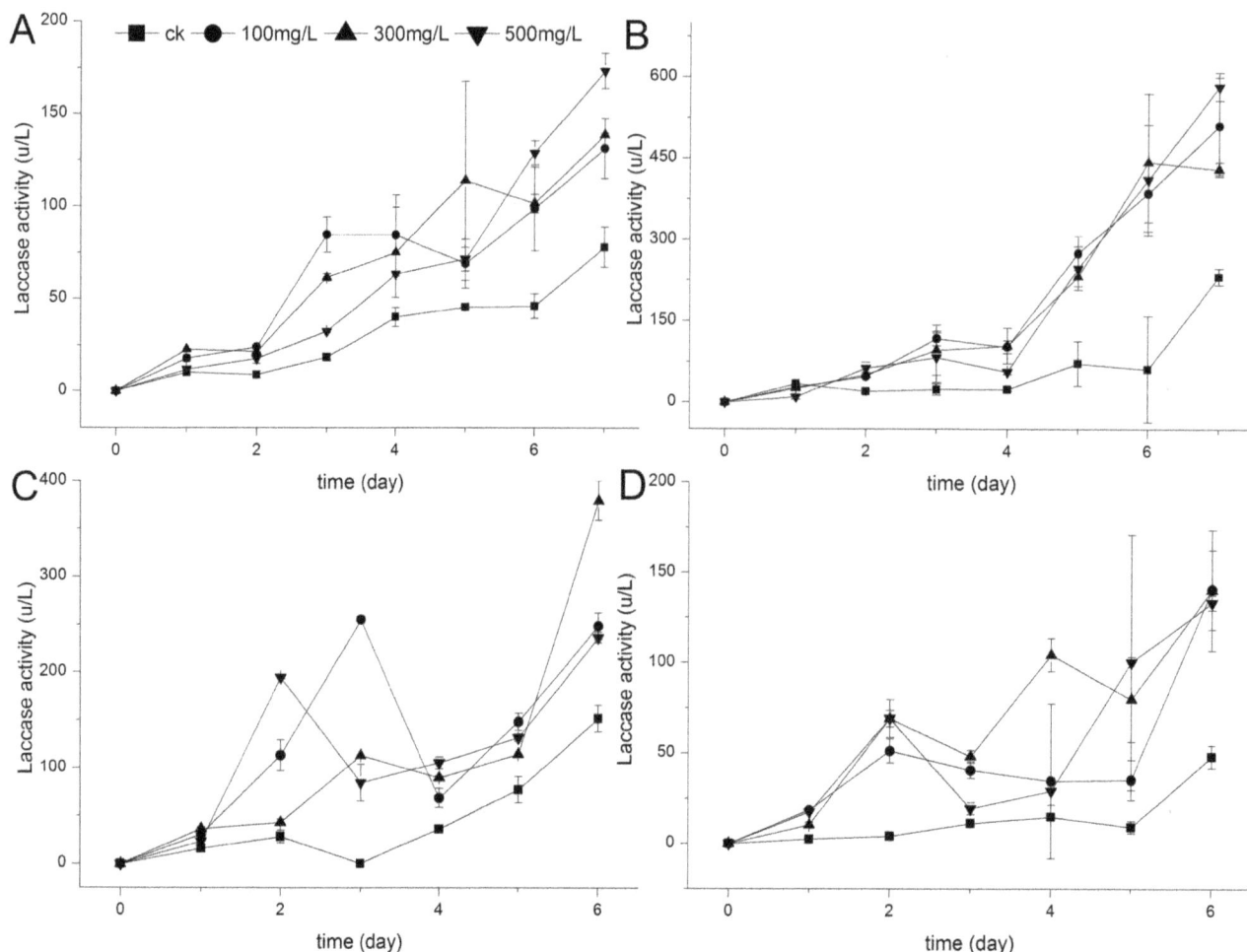

Figure 3. Effect of lignin on laccase activities during the decolorization of dyes by *E. taxodii*. (A): RBV5. (B): DR5B. (C): DB38. (D): DB22. Dyes concentration: 100 mg/L. Lignin contents: ■, 0 mg/L; ●, 100 mg/L; ▲, 300 mg/L; ▼, 500 mg/L.

shown in Table 3. In addition, RBV5 and DR5B showed no more than 10% adsorptive decolorization, whereas the actual dye removal was only approximately at 10% independent of the presence of lignin. We observed that lignin was not an ordinary adsorbent for RBV5 and DR5B when cooperating with living fungus. For DB38 and DB22, decolorization differences between lignin–dye groups and the dye controls (>20%) were larger than the highest adsorption values (<10%) on day 2 or 3 (Fig. 1). This result indicated that lignin might enhance dye elimination through some mechanism other than physical–chemistry adhesion. By contrast, biomass from culture inoculated with lignin did not grow faster than that from the control group. However, decolorization seemed to significantly speed up in the former, indicating the absence of difference in the biomass activity on absorption (Fig. 2). This result confirmed that enhanced decolorization was influenced by the metabolism of added lignin.

Effect of lignin on ligninolytic enzymes during the decolorization by *E. taxodii*

Fungi degrade lignin or azo dyes through oxidation, probably involving enzymes such as LiP, MnP, and laccase. Meanwhile, toxic aromatic compounds and lignin (lignosulfonate and synthetic lignin) were shown to stimulate the production of laccase, LiP, and MnP [32]. However, in our research, among the main ligninolytic

enzymes, only laccase was found in the liquid culture of *E. taxodii* throughout the biological decolorization (Fig. 3). Once lignin was present, significantly higher values of laccase activity were detected in the culture media, even when only lignin was present in the substrate with no dyes. Enzymatic activities were approximately twice to thrice higher than that in the lignin-dye-free blank (<5 IU/L). In addition, dyes alone can stimulate the yield of laccase which generally increased to different extents with the extension of incubation time, as follows: RBV5, 77.95 IU/L; DR5B, 231.92 IU/L; DB38, 53.03 IU/L; and DB22, 48.13 IU/L on the last day of degradation. The stimulation effects of dyes on laccase production during cultivation, which could be regarded as xenobiotic response to detoxification, have also been reported [33]. However, the fungus incubated with lignin produced more laccase and enhanced dye removal (Fig. 3). In the case of RBV5 and DR5B (Figs. 1A, 1B, 3A, and 3B), decolorization percentages were initially not enhanced along with the increase in laccase. Afterward, dyes fade out at higher rates when lignin was added in increasing amounts. However, laccase activities did not occur in similar patterns, i.e., no positive correlation was observed between the two phenomena. For DB38 and DB22 (Figs. 1C, 1D, 3C, and 3D), the decolorization curves were poorly consistent with enzymatic activity detection. All results showed that dyes decolorized in the presence of lignin were affected not only by laccase but also by lignin metabolism. Little information

KDa   line 1   line 2   line 3

116 ——

66.2 ——

45 ——

35 ——

25 ——

18.4 ——

14.4 ——

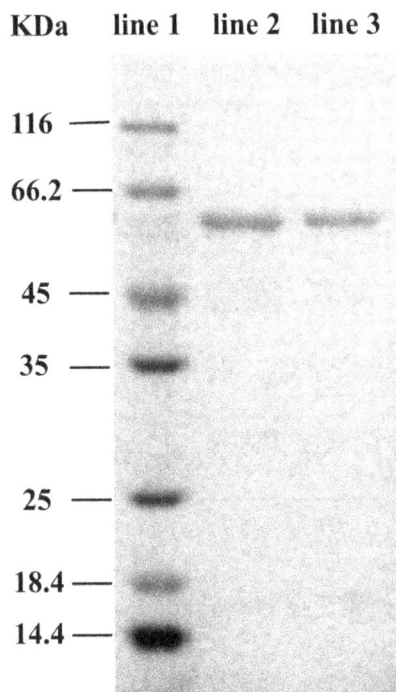

Figure 4. SDS-PAGE analyses of purified laccase of *E. taxodii.*
Lane 1, Molecular weight marker; Lane 2 and 3, purified laccase.

on the combined degradation of pollutants and lignin or the resolution of the inner mechanism is available. According to Tychanowicz, Zilly, de Souza, Peralta [34], structurally different dyes were decolorized because of the high level of laccase generated in response to the presence of soluble phenolic compounds in the glucose/ammonium tartrate–corncob solid-state medium. The fungi produced laccase and removed color at the same time with the supplementation of lignocellulose, providing an alternative approach to deal with textile wastewater and increasing the economic value added by producing laccase [9]. The underlying promoting mechanism should be uncovered to provide a theoretical guideline with strong feasibility in depleting diverse dyes or pollutants.

Thus, Figs. 1 to 3 show that lignin can influence dye removal through laccase, the only oxidase during biotreatment, in some pathway different from the common improvement of enzymatic activity. According to papers published worldwide, WRF secrete a large amount of volatile aromatic compounds, giving rise to natural fragrances, providing a promising source of flavors and aromas. Gutierrez, Caramelo, Prieto, Martínez, Martinez [35] reported that simple aromatic compounds could be generated by fungi with or even without aromatic origins. However, the products varied in quantities and kinds as stimulators were added. Low-molecular-weight aromatic acids have been identified during the long-term degradation of lignin in spruce wood by *Phanerochaete chrysosporium* by high-performance liquid chromatography and after acetylation and methylation by gas chromatography/mass spectrometry (GC/MS) [19]. Additionally, some of these chemicals were low-molecular-weight phenol derivatives and considered potential laccase substrates, which could also work as natural mediators for the oxidation reactions by laccase. Johannes, Majcherczyk [36] tested different kinds of mediators for laccase in

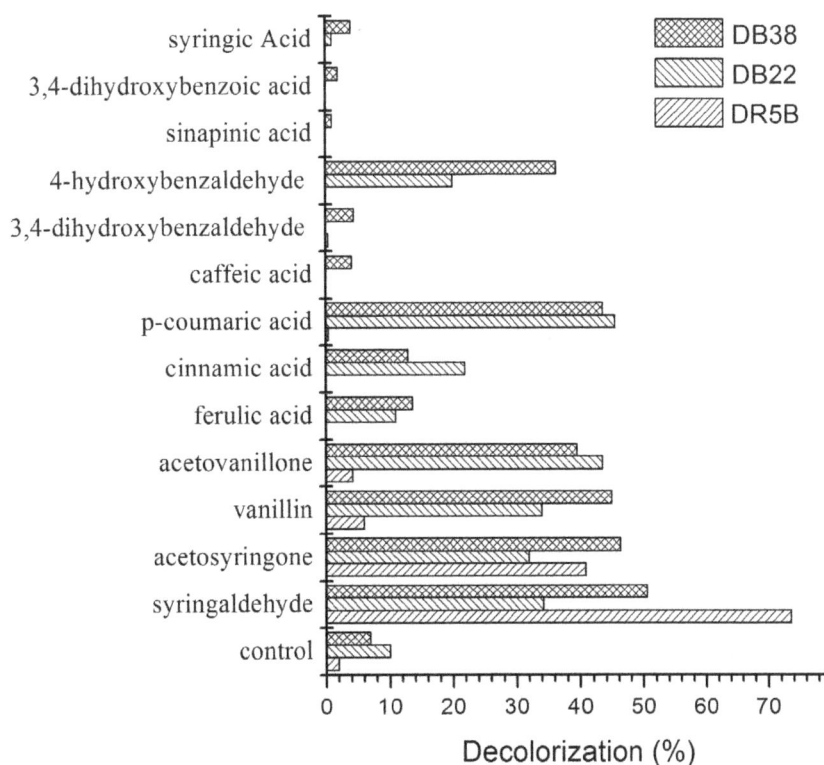

Figure 5. Evaluating for nature mediators based on decolorization of DR5B, DB22 and DB38 (100 mg/L) in presence of mediator (1 mM) for 2 h.

the oxidation of polycyclic aromatic hydrocarbons (PAH) and determined that natural compounds such as phenol, aniline, 4-hydroxybenzoic acid, and 4-hydroxybenzyl alcohol are efficient as synthetic chemicals ABTS and 1-hydroxybenzotriazole. Moreover, p-hydroxycinnamic acids, namely, p-coumaric (4-hydroxycinnamic) acid, ferulic (3-methoxy-4-hydroxycinnamic) acid, and sinapic (3,5-dimethoxy-4-hydroxycinnamic) acid, were used as natural mediators for laccase oxidation of recalcitrant dyes and PAH [37]. Thus, we infer that the metabolites of lignin may act as redox mediators in the oxidation of dyes by laccase, the main oxidative enzyme in decolorization.

In this study, lignin metabolite detection was also conducted. However, we did not observe similar results on the chromatogram map from GC/MS (data not shown) [19]. We infer that lignin degradation in large quantities started during the secondary metabolism of WRF and was a long-term process. Thus, low-molecular-weight lignin derivates were likely in trace amount and were difficult to identify during the short-term decolorization. Moreover, the metabolic intermediates were probably initiated by the increase in laccase during the active growth phase (based on biomass accumulation) when the fungus did not function on lignin degradation. Hence, the metabolic intermediates were difficult to accumulate and often existed briefly. In addition, laccase can graft aromatic substrates back onto lignin [38,39]. Thus, various lignin-derived aromatics were studied in vitro in mediating degradation by purified laccase.

Effect of lignin related compounds on the azo dyes decolorization by purified laccase of E. taxodii

Laccase is the most potent oxidase involved in the degradation of lignin and dyes. Laccase was purified from *E. taxodii* cultures to homogeneity by salt fractionation with ammonium sulfate, hydrophobic interaction, and ion exchange chromatography. Approximately 23.3-fold purification was achieved, with an overall yield of 38.8% (data not shown). The purified laccase showed a single protein band on SDS-PAGE (Fig. 4).

The purified laccase of *E. taxodii* slowly decolorized RBV5, DR5B, DB22, and DB38 in the absence of mediators, with corresponding decolorization percentages of 1%, 2%, 10%, and 7% after 2 h reaction. However, the dye decolorization percentages by laccase were enhanced in the presence of natural mediators, except that of RBV5, which did not disappear more rapidly with mediators present. Among the 13 aromatics (phenolic aldehydes, ketones, and acids), acetosyringone and syringaldehyde significantly promoted the decolorization of DR5B, DB22, and DB38. Decolorization of the last two dyes was also accelerated by p-coumaric acid, acetovanillone, and vanillin (Fig. 5). However, we did not obtain any suitable redox mediators in the candidates for RBV5. Among the most efficient mediators, syringaldehyde provided the highest decolorization percentages for DR5B and DB38. The decolorization percentages increased 37-fold and 3.7-fold, respectively, from 2% to 74% for DR5B and from 7% to 51% for DB38. Compared with syringaldehyde, p-coumaric acid was more effective in the decolorization of DB22. The decolorization percentage increased 4.5-fold from 10% to 45%. Camarero, Ibarra, Martínez, Martínez [40] determined that syringaldehyde, acetosyringone, vanillin, acetovanillone, methyl vanillate, and p-coumaric acid, which were ubiquitous laccase producers of basidiomycetes in decayed wood and forest soils, significantly promoted the oxidation of recalcitrant dyes by laccase.

The influences of natural mediator concentrations on the decolorization of DR5B, DB22, and DB38 (4 h treatment) are shown in Fig. 6. The decolorization percentages of the three dyes were correlated with the concentration of mediators. Laccase in the presence of p-coumaric acid produced a high decolorization

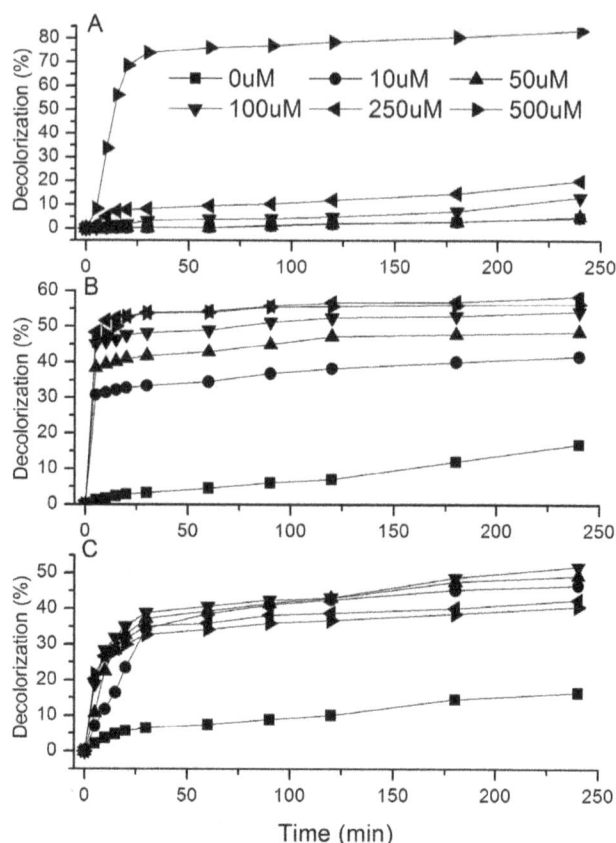

Figure 6. Effects of mediator concentration on dye decolorization (%) by purified laccase (4-h treatment of 100 mg/l DR5B, DB22 and DB38). (A): DR5B + syringaldehyde. (B): DB22 + p-coumaric acid. (C): DB38 + syringaldehyde. Mediator concentration: ■, 0 μM; ●, 10 μM; ▲, 50 μM; ▼, 100 μM; ◄, 250 μM; ►, 500 μM.

percentage for DB22, whereas the presence of syringaldehyde promoted the decolorization percentage for DB38 at a low concentration (10 μM). By contrast, the decolorization percentage for DR5B was directly proportional to syringaldehyde concentration. The addition of syringaldehyde of 1 mM concentration enhanced the decolorization percentage from 0.2% to 74.0% after reaction for 30 min, whereas the decolorization percentage did not change at 10 μM concentration of syringaldehyde.

Although different mediators apparently influence the efficiency of laccase to decolorize the dyes in vitro [40], actual decolorization by *E. taxodii* in the presence of lignin might be affected by the various previously tested lignin-derived compounds. Meanwhile, the contents of each chemical vary from those obtained in this study and were possibly much lower. The result showed that aromatics like p-coumaric acid and syringaldehyde enhanced decolorization by purified laccase, and the enhancement was positively related with the concentration of aromatics. When phenols at a low concentration, they did not increase the decolorization as effectively as they were at a concentration of 1 mM. It was presumed that lignin decomposition product a small quantity of aromatics, and they slowly boosted decolorization in the cultures. By contrast, Chen, Chang, Kirk [19] studied the degradation products of lignin in decayed wood. Monomers and dimers were detected, in which dimers, e.g., dihydroxydivanillic acid and 2'-hydroxy-2,3'-dimethoxydiphenylether, were also produced from monomer polymerization by laccase [41]. Thus, dimers could be detected when aromatics mediated lignin or dye oxidation by laccase. However, natural mediators in the fungal system are not always at a detectable concentration (they are difficult to accumulate and often existed only briefly, but usually performing better than synthetic ones), and the derivatives are in low quantity to be identified, contrary to that in vitro which is set to a much higher level in lignin or xenobiotic compound degradation. Thus, E. taxodii degraded lignin and yielded hundreds of metabolites, among which some chemicals worked as redox mediators in laccase oxidation of dyes.

Conclusions

In the current work E. taxodii was observed to degrade four azo dyes more efficiently when lignin was present in the media. Laccase, the dominant enzyme in the biodegradation, was found to be induced by lignin. Experiments in vitro with purified laccase demonstrated that contribution of lignin-related aromatics as laccase mediators might be the explanation for the improvement in decolorization.

Supporting Information

File S1 Data of Figure 1. Effect of lignin on the decolorization of different dyes by E. taxodii. Table S1, RBV5. Table S2, DR5B. Table S3, DB38. Table S4, DB22.

File S2 Data of Figure 2. Effect of lignin on biomass during the decolorization of dyes by E. taxodii. Table S5, RBV5. Table S6, DR5B. Table S7, DB38. Table S8, DB22.

File S3 Data of Figure 3. Effect of lignin on laccase activities during the decolorization of dyes by E. taxodii. Table S9, RBV5. Table S10, DR5B. Table S11, DB38. Table S12, DB22.

File S4 Data of Figure 5. Evaluating for nature mediators based on decolorization of DR5B, DB22 and DB38 (100 mg/L) in presence of mediator (1 mM) for 2 h. Table S13.

File S5 Data of Figure 6. Effects of mediator concentration on dye decolorization (%) by purified laccase (4-h treatment of 100 mg/l DR5B, DB22 and DB38). Table S14, DR5B. Table S15, DB38. Table S16, DB22.

Author Contributions

Conceived and designed the experiments: YLH LLS HBY XYZ. Performed the experiments: YLH LLS JM. Analyzed the data: YLH LLS XYZ HBY. Contributed reagents/materials/analysis tools: YLH LLS JM HBY. Wrote the paper: YLH LLS XYZ.

References

1. Burkinshaw S, Salihu G (2013) The wash-off of dyeings using interstitial water Part 4: disperse and reactive dyes on polyester/cotton fabric. Dyes Pigment 99: 548–560.
2. Erkurt EA, Erkurt HA, Unyayar A (2010) Decolorization of azo dyes by white rot fungi. Biodegradation of Azo Dyes. Springer. pp. 157–167.
3. Parshetti G, Telke A, Kalyani D, Govindwar S (2010) Decolorization and detoxification of sulfonated azo dye methyl orange by Kocuria rosea MTCC 1532. J Hazard Mater 176: 503–509.
4. Ulson de Souza SMAG, Forgiarini E, Ulson de Souza AA (2007) Toxicity of textile dyes and their degradation by the enzyme horseradish peroxidase (HRP). J Hazard Mater 147: 1073–1078.
5. Işik M, Sponza DT (2007) Fate and toxicity of azo dye metabolites under batch long-term anaerobic incubations. Enzyme Microb Technol 40: 934–939.
6. de Souza SMdAGU, Bonilla KAS, de Souza AAU (2010) Removal of COD and color from hydrolyzed textile azo dye by combined ozonation and biological treatment. J Hazard Mater 179: 35–42.
7. Gao D, Du L, Yang J, Wu W-M, Liang H (2010) A critical review of the application of white rot fungus to environmental pollution control. Crit Rev Biotechnol 30: 70–77.
8. Kaushik P, Malik A (2009) Fungal dye decolourization: recent advances and future potential. Environ Int 35: 127–141.
9. Liu J, Cai Y, Liao X, Huang Q, Hao Z, et al. (2012) Simultaneous Laccase Production and Color Removal by Culturing Fungus Pycnoporus sp. SYBC-L3 in a Textile Wastewater Effluent Supplemented with a Lignocellulosic Waste Phragmites australis. Bull Environ Contam Toxicol 89: 269–273.
10. Ozmen N, Yeşilada O (2012) Valorization and biodecolorization of dye adsorbed on lignocellulosics using white rot fungi. BioResources 7: 1656–1665.
11. Keliang Y, Wang H, Zhang X, Yu H (2009) Bioprocess of triphenylmethane dyes decolorization by Pleurotus ostreatus BP under solid-state cultivation. J Microbiol Biotechnol 19: 1421–1430.
12. Yan K, Wang H, Zhang X (2009) Biodegradation of Crystal Violet by low molecular mass fraction secreted by fungus. J Biosci Bioeng 108: 421–424.
13. Verma P, Madamwar D (2002) Production of ligninolytic enzymes for dye decolorization by cocultivation of white-rot fungi Pleurotus ostreatus and Phanerochaete chrysosporium under solid-state fermentation. Appl Biochem Biotechnol 102: 109–118.
14. Cañas AI, Camarero S (2010) Laccases and their natural mediators: biotechnological tools for sustainable eco-friendly processes. Biotechnol Adv 28: 694–705.
15. Tinoco R, Acevedo A, Galindo E, Serrano-Carreón L (2011) Increasing Pleurotus ostreatus laccase production by culture medium optimization and copper/lignin synergistic induction. J Ind Microbiol Biotechnol 38: 531–540.
16. Ruiz-Dueñas FJ, Martínez ÁT (2009) Microbial degradation of lignin: how a bulky recalcitrant polymer is efficiently recycled in nature and how we can take advantage of this. Microb Biotechnol 2: 164–177.
17. Bugg TD, Ahmad M, Hardiman EM, Rahmanpour R (2011) Pathways for degradation of lignin in bacteria and fungi. Nat Prod Rep 28: 1883–1896.
18. Asgher M, Bhatti HN, Ashraf M, Legge RL (2008) Recent developments in biodegradation of industrial pollutants by white rot fungi and their enzyme system. Biodegradation 19: 771–783.
19. Chen C-L, Chang H-M, Kirk TK (1983) Carboxylic acids produced through oxidative cleavage of aromatic rings during degradation of lignin in spruce wood by Phanerochaete chrysosporium. J Wood Chem Technol 3: 35–57.
20. Piccolo A (2002) The supramolecular structure of humic substances: a novel understanding of humus chemistry and implications in soil science. Adv Agron. pp. 57–134.
21. Kudlich M, Keck A, Klein J, Stolz A (1997) Localization of the enzyme system involved in anaerobic reduction of azo dyes by Sphingomonas sp. Strain BN6 and effect of artificial redox mediators on the rate of azo dye reduction. Appl Environ Microbiol 63: 3691–3694.
22. Yu H, Zhang X, Song L, Ke J, Xu C, et al. (2010) Evaluation of white-rot fungi-assisted alkaline/oxidative pretreatment of corn straw undergoing enzymatic hydrolysis by cellulase. J Biosci Bioeng 110: 660–664.
23. Ma F, Yang N, Xu C, Yu H, Wu J, et al. (2010) Combination of biological pretreatment with mild acid pretreatment for enzymatic hydrolysis and ethanol production from water hyacinth. Bioresour Technol 101: 9600–9604.
24. Zhang X, Yu H, Huang H, Liu Y (2007) Evaluation of biological pretreatment with white rot fungi for the enzymatic hydrolysis of bamboo culms. Int Biodeterior Biodegrad 60: 159–164.
25. Yu H, Guo G, Zhang X, Yan K, Xu C (2009) The effect of biological pretreatment with the selective white-rot fungus Echinodontium taxodii on enzymatic hydrolysis of softwoods and hardwoods. Bioresour Technol 100: 5170–5175.
26. Yang X, Zeng Y, Ma F, Zhang X, Yu H (2010) Effect of biopretreatment on thermogravimetric and chemical characteristics of corn stover by different white-rot fungi. Bioresour Technol 101: 5475–5479.
27. Wolfenden BS, Willson RL (1982) Radical-cations as reference chromogens in kinetic studies of one-electron transfer reactions: pulse radiolysis studies of 2, 2′-azinobis-(3-ethylbenzthiazoline-6-sulphonate). J Chem Soc Perkin Trans 2: 805–812.
28. Wariishi H, Valli K, Gold MH (1992) Manganese (II) oxidation by manganese peroxidase from the basidiomycete Phanerochaete chrysosporium. Kinetic mechanism and role of chelators. J Biol Chem 267: 23688–23695.
29. Tien M, Kirk TK (1988) Lignin peroxidase of Phanerochaete chrysosporium. Methods Enzymol 161: 238–249.
30. Binupriya A, Sathishkumar M, Swaminathan K, Kuz C, Yun S (2008) Comparative studies on removal of Congo red by native and modified mycelial pellets of Trametes versicolor in various reactor modes. Bioresour Technol 99: 1080–1088.

31. Lin Y, He X, Han G, Tian Q, Hu W (2011) Removal of Crystal Violet from aqueous solution using powdered mycelial biomass of *Ceriporia lacerata* P2. J Environ Sci 23: 2055–2062.

32. Rogalski J, Lundell TK, Leonowicz A, Hatakka AI (1991) Influence of aromatic compounds and lignin on production of ligninolytic enzymes by *Phlebia radiata*. Phytochemistry 30: 2869–2872.

33. Puvaneswari N, Muthukrishnan J, Gunasekaran P (2006) Toxicity assessment and microbial degradation of azo dyes. Indian J Exp Biol 44: 618.

34. Tychanowicz GK, Zilly A, de Souza CGM, Peralta RM (2004) Decolourisation of industrial dyes by solid-state cultures of *Pleurotus pulmonarius* Process Biochem 39: 855–859.

35. Gutierrez A, Caramelo L, Prieto A, Martínez MJ, Martinez AT (1994) Anisaldehyde production and aryl-alcohol oxidase and dehydrogenase activities in ligninolytic fungi of the genus *Pleurotus*. Appl Environ Microbiol 60: 1783–1788.

36. Johannes C, Majcherczyk A (2000) Natural mediators in the oxidation of polycyclic aromatic hydrocarbons by laccase mediator systems. Appl Environ Microbiol 66: 524–528.

37. Camarero S, CaÑas AI, Nousiainen P, Record E, Lomascolo A, et al. (2008) p-Hydroxycinnamic acids as natural mediators for laccase oxidation of recalcitrant compounds. Environ Sci Technol 42: 6703–6709.

38. Moilanen U, Kellock M, Galkin S, Viikari L (2011) The laccase-catalyzed modification of lignin for enzymatic hydrolysis. Enzyme Microb Technol 49: 492–498.

39. Rencoret J, Aracri E, Gutiérrez A, del Rio JC, Torres AL, et al. (2014) Structural insights on laccase biografting of ferulic acid onto lignocellulosic fibres. Biochem Eng J 86: 16–23.

40. Camarero S, Ibarra D, Martínez MJ, Martínez ÁT (2005) Lignin-derived compounds as efficient laccase mediators for decolorization of different types of recalcitrant dyes. Appl Environ Microbiol 71: 1775–1784.

41. Ibrahim V, Volkova N, Pyo S-H, Mamo G, Hatti-Kaul R (2013) Laccase catalysed modification of lignin subunits and coupling to *p*-aminobenzoic acid. J Mol Catal B: Enzym 97: 45–53.

3

Promoted Relationship of Cardiovascular Morbidity with Air Pollutants in a Typical Chinese Urban Area

Ling Tong[1,3], Kai Li[2], Qixing Zhou[3]*

1 School of Environmental Science and Engineering, Tianjin University, Tianjin, China, 2 Department of Industrial Engineering, Nankai University, Tianjin, China, 3 Key Laboratory of Pollution Process and Environmental Criteria (Ministry of Education), College of Environmental Science and Engineering, Nankai University, Tianjin, China

Abstract

Background: A large number of studies about effects of air pollutants on cardiovascular mortality have been conducted; however, those investigating association between air pollutants and cardiovascular morbidity are limited, especially in developing countries.

Methods: A time-series analysis on the short-term association between outdoor air pollutants including particulate matter (PM) with diameters of 10 μm or less (PM_{10}), sulfur dioxide (SO_2) and nitrogen dioxide (NO_2) and cardiovascular morbidity was conducted in Tianjin, China based on 4 years of daily data (2008–2011). The morbidity data were stratified by sex and age. The effects of air pollutants during the warm season and the cool season were also analyzed separately.

Results: Each increase in PM_{10}, SO_2, and NO_2 by increments of 10 μg/m^3 in a 2-day average concentration was associated with increases in the cardiovascular morbidity of 0.19% with 95 percent confidence interval (95% CI) of 0.08–0.31, 0.43% with 95% CI of 0.03–0.84, and 0.52% with 95% CI of −0.09–1.13, respectively. The effects of air pollutants were more evident in the cool season than those in the warm season, females and the elderly were more vulnerable to outdoor air pollution.

Conclusions: All estimated coefficients of PM_{10}, SO_2 and NO_2 are positive but only the effect of SO_2 implied statistical significance at the 5% level. Moreover, season, sex and age might modify health effects of outdoor air pollutants. This work may bring inspirations for formulating local air pollutant standards and social policy regarding cardiovascular health of residents.

Editor: Qinghua Sun, The Ohio State University, United States of America

Funding: This research was financially supported by the Ministry of Science and Technology, People's Republic of China as a 973 project (grant No. 2011CB503802), and by the National Natural Science Foundation as a key project (grant No. 21037002). The funders had no role in study design, data collection and analysis, decision to publish, or preparation of the manuscript.

Competing Interests: The authors have declared that no competing interests exist.

* Email: zhouqx@nankai.edu.cn

Introduction

As the largest harbor in northern China, Tianjin is a fast-growing and economically developed city. It has an area of approximately 11,919 km^2 and a population of 10 million. A continental monsoonal climate featuring hot and humid summer, and dry and cold winter is typical here. The mean annual temperature and precipitation is 13.1°C and 389.4 mm, respectively. It is wetter in summer than in winter. The domestic heating season is generally between November and March. The average wind speeds range from 16 to 24 km/h. A comprehensive industrial system including integrated machinery, electronics, petroleum and chemicals, metallurgy, textiles and vehicles has been built in this city. The industrial prosperity brings about severe air pollution problems. From 2008 to 2011, both daily concentrations of PM_{10} and SO_2 (80.2 μg/m^3 and 148.1 μg/m^3, respectively) exceeded the Class I levels of the Chinese national standards (daily average: 50 μg/m^3 for both PM_{10} and SO_2) [1] and the standards of the World Health Organization (WHO) (daily average: 50 μg/m^3 and 24 μg/m^3 for PM_{10} and SO_2,

respectively) [2]. Air pollution exerts tremendous burdens to public health, ranking as the 13th leading cause of mortality [3–5].

There are numerous studies of similar questions that have been conducted in North America and Western Europe. Several years ago, the Health Effects Institute sponsored some studies under the acronym PAPA (http://pubs.healtheffects.org/types.php?type=1 for published reports), including "Public Health and Air Pollution in Asia (PAPA): Coordinated Studies of Short-Term Exposure to Air Pollution and Daily Mortality in Two Indian Cities" and "Public Health and Air Pollution in Asia (PAPA): Coordinated Studies of Short-Term Exposure to Air Pollution and Daily Mortality in Four Cities". Pathophysiologic mechanisms of air pollutant-induced cardiovascular morbidity and mortality are also being studied widely [6–8], besides, some studies on the associations between air pollutants and cardiovascular diseases have been conducted in terms of epidemiology [9–13]. However, investigations on the relationship between air pollutants and cardiovascular morbidity are scarce at present [14], especially in Asian countries where, arguably, both living conditions and health indicators may be quite different.

This study presents an investigation of this issue in a typical Chinese city to increase our understanding of cardiovascular health risks associated with polluted air and provide some scientific basis for establishment of public health and environmental protection policies [15].

Materials and Methods

2.1. Subject Data

Daily air pollution data on PM_{10}, SO_2 and NO_2 were obtained from the website of the Tianjin Environmental Monitoring Centre. Daily mean temperature and relative humidity were obtained from the China Meteorological Data Sharing Service System (Data S1).

Daily cardiovascular morbidity data from 1 January 2008 to 31 November 2011 was collected from the Centers for Disease Control and Prevention of Urban Districts in Tianjin, China (Nankai, Heping, Hexi, Hedong and Hongqiao districts), covering around 77000 local residents (SI). Information of patients including those under the age of 18 was anonymized and were validated each year by China CDC. They were coded according to the ICD-10 (the 10th revision of International Classification of Diseases) and classified into cardiovascular causes including cerebral infarction, primary diagnosed hypertension, cerebral hemorrhage, acute myocardial infarction and subarachnoid hemorrhage [16]. They were also stratified by sex and age (0-18, 18-44, 45-64, and ≥65 years) [17]. The ethical committee of the coordinating center of five urban CDCs in Tianjin approved the study (Full name is "CDC biomedical ethics council").

Air pollutant concentrations and meteorological measures including temperature and humidity are shown in Table 1 and 2.

2.2. Statistical methods

All analyses were conducted with statistical software package SAS version 9.1. Time-series analysis was utilized to explore modification effects of season, age and sex on the association between air pollutants and morbidity in Tianjin from 2008 to 2011. In detail, the generalized linear model with natural splines (ns) functions of time, weather conditions accommodate nonlinear and non-monotonic relationships of morbidity with time, temperature and relative humidity were utilized for the analysis [17]. In the basic model, morbidity outcomes were included without air pollution variables. Residuals of the basic models were examined to determine whether there were discernible patterns and autocorrelation by means of residual plots and partial autocorrelation function (PACF) plots. Day of the week (DOW) was considered as a dummy variable in the basic modes. After the establishment of the basic model, PM_{10} and covariates (temperature, humidity, and SO_2 and NO_2 concentrations) were introduced into it and their effects were analyzed. The selection of df (degrees of freedom) for time trends was done with the PACF [17,18]. 4 df per year for time trends were used in our basic models for cardiovascular morbidity. In addition, we used 3 df

(during whole period of study) for temperature and humidity. This modeling procedure was carried out for each series studied and the core models were assessed with plots of model residuals and fitted values as well as plots of the estimated partial autocorrelation functions.

The estimated pollution log-relative rate β is obtained through fitting of the following log-linear generalized linear model:

$$logE(Yt) = βZt + DOW + ns(time, df) + ns(temperature/humidity, 3) + intercept (1)$$

Where E(Yt) and β represent the expected morbidity numbered at day t and the log-relative rate of morbidity corresponding to a unit increase of air pollutants, respectively; Zt indicates pollutant concentrations at day t; DOW is dummy variable for day of the week; ns (time, df) is the ns function of calendar time with 4 df to adjust for seasonality and other time-varying influences on admissions (e.g. influenza epidemics and longer-term trends); and ns (temperature/humidity, 3) is the ns function for temperature and humidity with 3 df.

The data were stratified by sex and age. We analyzed the associations for the warm season (April-September), the cool season (October-March) and the entire year, respectively (Kan et al., 2008b). The basic models of seasonal analyses were different from those of the whole-period in terms of df for time trends. The effects were quantified on the basis of the percentage change in risk per 10 mg/m^3 increase in the concentration of each pollutant. The statistical significance was defined as $p<0.05$.

Results

3.1. Time series plot of the morbidity, exposure-response relationships, ACF and PACF of morbidity

The time series plot of cardiovascular morbidity from 2008 to 2011 in Tianjin is shown in Fig. 1. Fig. 2 demonstrates the exposure–response relationships for PM_{10}, SO_2 and NO_2 (2-day moving average of air pollutant concentration (lag 01)) with cardiovascular mortality during 2008–2011 in Tianjin. In Fig. 3, ACF and PACF of cardiovascular morbidity are depicted as original data (log). ACF and PACF of cardiovascular morbidity as original data (log+Difference) are shown in Fig. 4. Fig. 5 gives ACF and PACF of residual after modeling (basic model).

3.2. Effects by sexes and ages

As shown in Table 3, the percent increases associated with air pollutant concentrations varied by sex or age groups. The effect estimates of pollutants among females were greater than those among males, especially the estimate for NO_2 was thrice as much as that among males, although their between-sex differences were insignificant statistically. The effects among those ≥65 were greater than those in the other two groups.

3.3. Effects by seasons

The daily average morbidity in the warm and cool season was 51.8±0.5 and 50.9±0.6, respectively. For the entire period, the

Table 1. Description of air pollutants and meteorological parameters during 2008–2011 in Tianjin.

Season	PM_{10}	SO_2	NO_2	Temperature (°C)	Humidity (%)
Warm season (n = 732)	79.1±1.3	11.8±1.3	6.2±0.4	22.5±0.2	61.2±0.6
Cool season (n = 698)	78.7±1.6	142.5±4.0	21.2±0.8	3.5±0.3	52.0±0.7
Entire period (n = 1430)	78.9±1.1	75.6±2.9	13.5±0.5	13.2±0.3	56.7±0.5

Table 2. Quartile values of three pollutants.

Quartile	PM$_{10}$ (µg/m^3)	SO$_2$ (µg/m^3)	NO$_2$ (µg/m^3)
5th	29.4	11.3	0.1
25th	58.0	62.4	5.9
50th	73.6	119.1	13.7
75th	97.0	215.5	25.3
95th	149.8	368.5	50.5

average value was 51.4 ± 0.4 and an increase of 10 µg/m^3 of 2-day average concentration of PM$_{10}$, SO$_2$ and NO$_2$ correspond to 0.19% (95% CI, 0.08–0.31), 0.43% (95% CI, 0.03–0.84) and 0.52% (95% CI, −0.09–1.13) increases, respectively (Table 4).

Sensitivity analyses results for PM$_{10}$ as shown in Table 5 and Table 6. They showed that the results were fairly robust for various concentrations, specifications for temperature, methods of aggregating daily data, df used in the smoothers, and alternative spline models.

Discussion

PM$_{10}$ in developed countries are mainly discharged from various automobiles including large amount of secondary organic aerosols. In Tianjin, ground dust, vehicle, cement dust and incineration are the primary PM$_{10}$ sources [19]. Coal and crude oil are responsible for 66% and 30% of energy supply, respectively [20]. Crude oil dominated by PM with fine fractions carries more toxic substance and is easier to penetrate into the circulation system of humans than coal. The main components of other PM sources such as raise dust and sand dust are less toxic inorganic minerals.

Generally speaking, the smoking habit exerted much more oxidative and inflammatory influences on males than air pollution [21]. In China, smoking is much more prevalent among men than women [22], this may have affected male health to a greater degree than environmental factors, therefore we suspect females were more sensitive to smoking than males. As for age, older people ≥65 were more vulnerable to air pollution (especially SO$_2$) than people in the other two age groups. In terms of their occupation, percent increase of myocardial infarction exposed to welding and soldering fumes in the Copenhagen male study

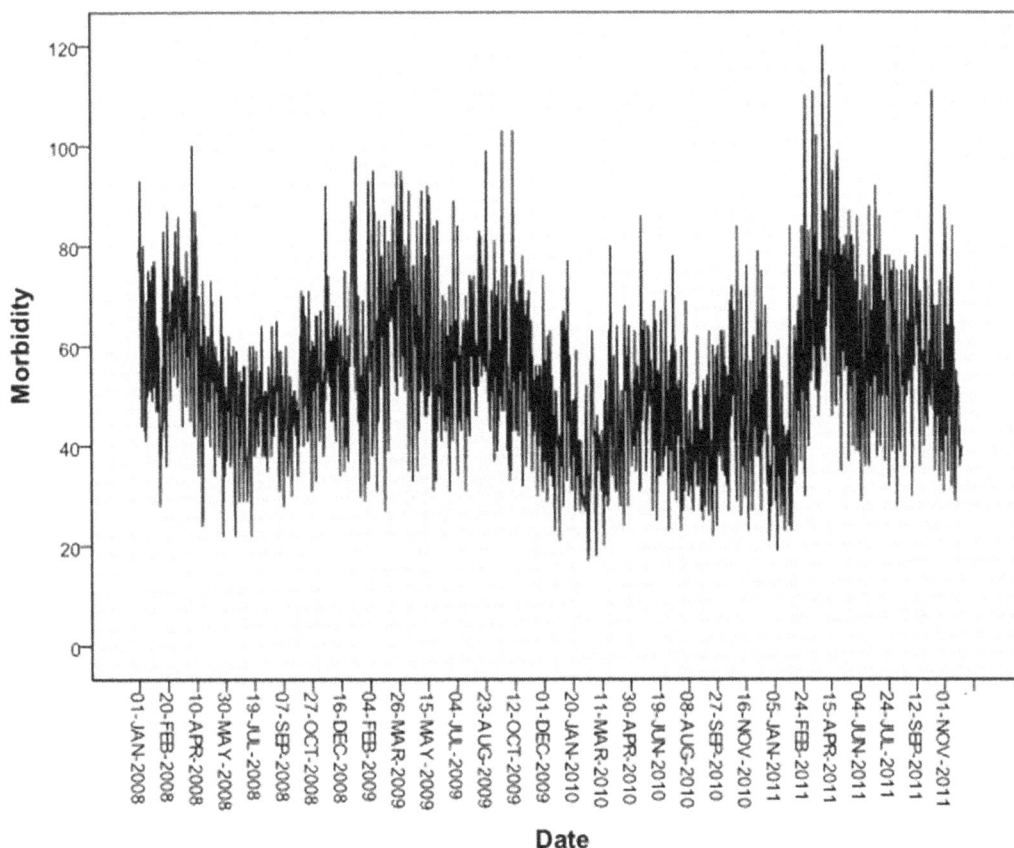

Figure 1. Time series of cardiovascular morbidity from 2008 to 2011 in Tianjin.

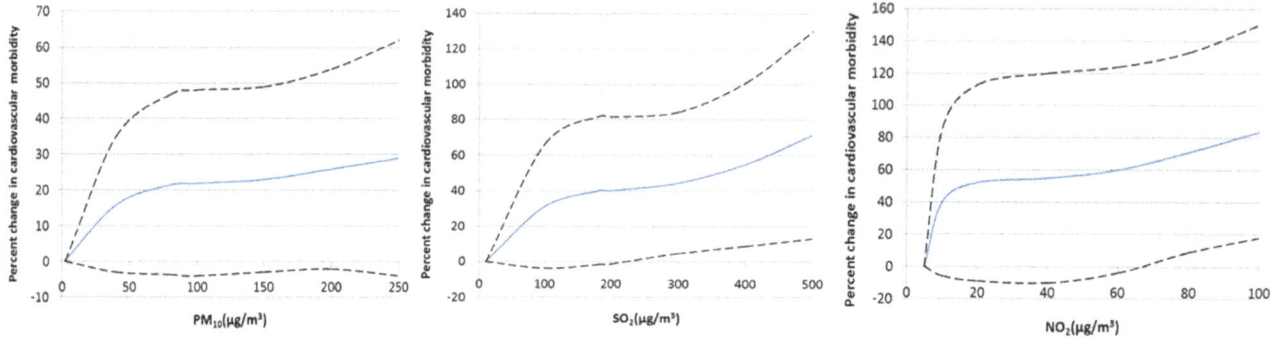

Figure 2. Exposure-response relationships (smoothing plots) of air pollutants against cardiovascular morbidity. The x-axis is the pollutant concentrations; the y-axis is the estimated percent change in cardiovascular mortality; the solid blue lines indicate the estimated mean percent change in daily mortality outcomes with the dashed lines representing the 95% CI.

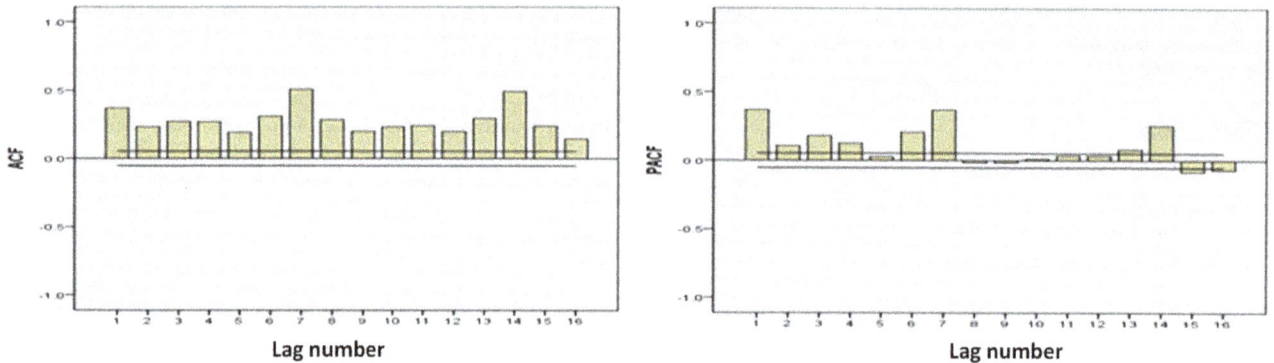

Figure 3. ACF and PACF of cardiovascular morbidity (log) and (log+difference). The x-axis is the lag number.

population was 1.1 (95%CI, 0.6–2.2) [23], much higher than the effect estimates on males in this study (Table 3). This is mainly attributed to the higher concentration PM exposure with more toxicity in the workplace than public environment.

Though not measured in this study, other factors such as underlying diet could also have affected our observed results. Because nutrients with natural chelating properties, including

antioxidants, herbs, minerals, essential amino acids and fiber, can detoxify human bodies, they can protect humans from oxidative stress of free radicals derived from air pollutants to some extent. Vitamins C, E and A in majority of plant and fish foods favored by residents in Tianjin as an important coastal city can interfere with or scavenge reactive oxygen species within cells [24–26]. To better understand the modification effects of living habits, socioeconomic

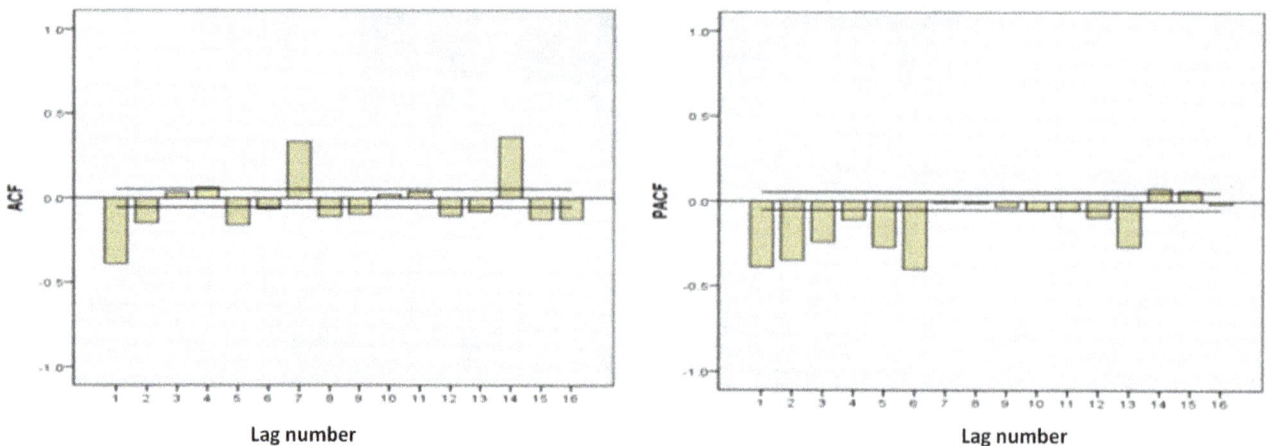

Figure 4. ACF and PACF of residual after modeling (basic model). The x-axis is the lag number.

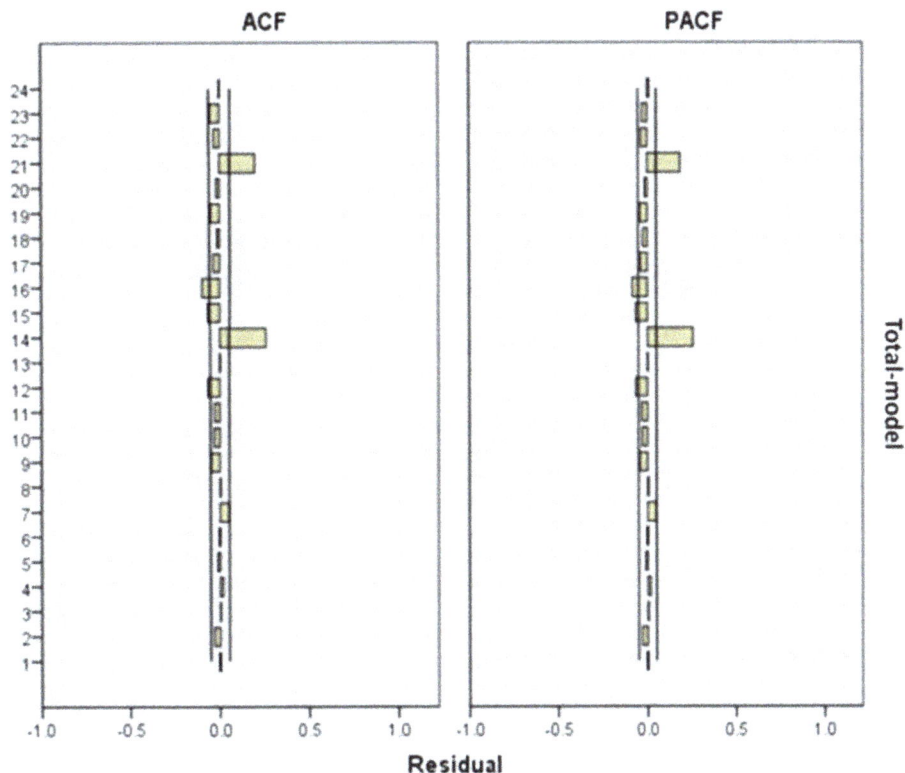

Figure 5. ACF and PACF of residual after modeling (basic model).

and demographic factors on the associations, more individual features (smoking and eating habit, occupation, education attainment levels, physical activity, socioeconomic status, etc.) should be taken into consideration in the future.

As for seasonal effects, some studies indicated that high temperatures in the warm season and low temperatures in the cold season are associated with increased cardiovascular mortality [17,27–29]. It is likely that they can also affect cardiovascular morbidity. Therefore, morbidity data were stratified by the warm and cool season defined as April-September and the cool season defined as October-March, respectively. Therefore, this research also demonstrated that the effect estimates of air pollutants in Tianjin might be modified by seasons. Exposure patterns may

contribute to this season-specific observation. Personal exposure was reduced by decreased time spent outdoors caused by heavy rain. In contrast, people are more likely to go outdoors and open windows in the cool season with the drier and less variable in Tianjin. The effect estimates of SO_2 and NO_2 in the cool season were much higher than those in the warm season. Significant associations were observed for PM_{10} and SO_2 in the cool season (0.24 (0.20, 0.28) and 0.66 (0.35, 0.97), respectively), while the effect of SO_2 in the warm season was insignificant. It may be caused by coal combustion accounting for the largest proportion of heat supply in the cool season. Besides, the significant Pearson correlation coefficient between PM_{10} and SO_2 (0.197, $p<0.01$) might partially explain this phenomenon. NO_2 was mainly

Table 3. Mean daily morbidity and percent increase (95% CI) in cardiovascular morbidity of Tianjin residents associated with a 10-$\mu g/m^3$ increase in air pollutant concentrations by sex and age[a].

	Mean daily morbidity	Pollutant		
		PM_{10}	SO_2	NO_2
Sex				
Female	21.86±0.23	0.25 (0.10, 0.39) <0.001	0.47 (0.10, 0.85) 0.019	0.57 (0.07, 1.06) 0.022
Male	31.52±0.29	0.13 (0.01, 0.25) 0.042	0.39 (0.21, 0.58) <0.001	0.17 (−0.16, 0.50) 0.892
Age				
5–44	1.94±0.04	0.07 (−0.05, 0.21) 0.687	0.11 (−0.02, 0.24) 0.566	0.42 (−0.19, 1.03) 0.601
45–64	20.27±0.19	0.12 (−0.16, 0.39) 0.844	0.21 (−0.10, 0.53) 0.947	0.26 (0.02, 0.51) 0.009
≥65	31.17±0.27	0.20 (0.09, 0.30) <0.001	0.47 (0.09, 0.85) 0.001	0.53 (0.31, 0.76) <0.001

[a] Current day temperature and humidity (lag 0) and 2-day moving average of air pollutant concentration (lag 01) with 3df of temperature and humidity were used.

Table 4. Mean percent increase (95% CI) of cardiovascular morbidity outcomes associated with 10-$\mu g/m^3$ increase in air pollutant concentrations by season, 2008–2011[a].

Pollutant	Warm season	Cool season	Entire period
PM_{10}	0.15 (−0.19, 0.50) $p=0.869$	0.24 (0.20, 0.28)* $p<0.001$	0.19 (0.08, 0.31) $p=0.016$
SO_2	0.24 (0.01, 0.47) $p=0.046$	0.66 (0.35, 0.97)* $p=0.001$	0.43 (0.03, 0.84) $p=0.037$
NO_2	0.21 (−0.20, 0.63) $p=0.667$	0.71 (0.15, 1.27) $p=0.033$	0.52 (−0.09, 1.13) $p=0.424$

[a] We used current day temperature and humidity (lag 0) and 2-day moving average of air pollutant concentration (lag 01), and applied 3df to temperature and humidity.
Significantly different from the warm season (p<0.05).

Table 5. Sensitivity analyses for varying degrees of freedom for time trend and weather conditions: df = 4–7 per year for time trend (A), df = 3–6 for current day temperature/relative humidity (B)[a].

	A			
	4	**5**	**6**	**7**
Warm season	0.153 (−0.191, 0.498)	0.154 (−0.190, 0.499)	0.156 (−0.189, 0.502)	0.158 (−0.187, 0.504)
Cool season	0.242 (0.203, 0.280)	0.242 (0.203, 0.281)	0.244 (0.204, 0.283)	0.245 (0.205, 0.285)
Entire period	0.194 (0.076, 0.311)	0.195 (0.078, 0.313)	0.197 (0.078, 0.315)	0.199 (0.080, 0.318)
	B			
	3	**4**	**5**	**6**
Warm season	0.153 (−0.191, 0.498)	0.153 (−0.191, 0.498)	0.153 (−0.192, 0.498)	0.152 (−0.192, 0.497)
Cool season	0.242 (0.203, 0.281)	0.242 (0.203, 0.281)	0.242 (0.203, 0.281)	0.242 (0.202, 0.281)
Entire period	0.194 (0.076, 0.311)	0.194 (0.076, 0.311)	0.194 (0.076, 0.311)	0.193 (0.076, 0.310)

[a] We used 2-day moving average (lag 01) of PM_{10} concentration, and current day temperature and humidity (lag0).

released from vehicle emission and partially from coal combustion, which may account for the insignificant relationship between NO_2 and morbidity. The constituents of the complex mix of PM_{10} may vary by seasons, different from the gaseous pollutants (SO_2 and NO_2).

There are some limitations in our analyses. Ozone was not included in our studies due to a lack of monitoring. Our ability to separate the independent effect of each pollutant was limited by high correlations between PM and gaseous pollutants including SO_2 and NO_2 in Tianjin. Moreover, the role of each specific component of air pollution should be examined to determine the combination of particles responsible for the increases in environment-induced health concerns. These investigations are paramount for policy makers to carry out improved interventions that will impact health hazards related with air pollutants, especially on the increased risks of cardiovascular morbidity.

Conclusions

In summary, we found that PM_{10} and SO_2 were significantly associated with cardiovascular morbidity in the cool season in Tianjin in this time-series analysis. The mean percent increases (95% CI) of cardiovascular morbidity outcomes per 10 $\mu g/m^3$ increase in PM_{10} and SO_2 concentrations by season were 0.24 (0.20, 0.28) and 0.66 (0.35, 0.97), respectively. Besides modification effect of a season, individual features (sexes, ages) might also interfere with the effect estimates of air pollutants. Females were more sensitive than males to air pollutants. The effects of air pollutants on elder people (≥65) were greater than those <65. These data can enable policy makers to enforce or improve existing legislation that control air pollution after weighing the disadvantages of potentially slowing rapid economic development.

Table 6. Sensitivity analyses for two different spline models[a].

	Quadratic spline	Cubic nature spline
Warm season	0.151 (−0.194, 0.495)	0.153 (−0.191, 0.498)
Cool season	0.251 (0.190, 0.311)	0.242 (0.203, 0.281)
Entire period	0.196 (0.078, 0.315)	0.194 (0.076, 0.311)

[a] Current day temperature and humidity (lag 0) and 2-day moving average of PM_{10} (lag 1) are used, 3 df for temperature and humidity is applied.

Acknowledgments

We cordially thank the Centers for Disease Control and Prevention of Urban Districts in Tianjin for assisting us to collect medical data.

Author Contributions

Performed the experiments: LT. Analyzed the data: KL LT QZ. Wrote the paper: LT QZ KL.

References

1. GB/T 3095-2012 (2012) Ambient air quality standards. Chinese Standard Press, Beijing.
2. WHO (2005) Air quality guidelines global update for particulate matter, ozone, nitrogen dioxide and sulfur dioxide. Geneva: World Health Organization.
3. Roberts S (2011) Can mortality displacement mask thresholds in the concentration-response relation between particulate matter air pollution and mortality? Atmo Environ 45: 4728–4734.
4. Kan HD, Chen RJ, Tong SL (2012) Ambient air pollution, climate change, and population health in China. Environ Int 42: 10–19.
5. Cheng Z, Jiang JK, Fajardo O, Wang SX, Hao JM (2013) Characteristics and health impacts of particulate matter pollution in China (2001–2011). Atmos Environ 65: 186–194.
6. Brook DR (2008) Cardiovascular effects of air pollution. Clin Sci 115: 175–187.
7. Koulova A, Frishman HW (2014) Air Pollution Exposure as a Risk Factor for Cardiovascular. Cardiology 22: 30–36.
8. Shrey K, Suchit A, Deepika D, Shruti K, Vibha R (2011) Air pollutants: The key stages in the pathway towards the development of cardiovascular disorders. Environ Toxicol Phar 31: 1–9.
9. Wong CM, Atkinson WR, Anderson RH, Hedley AJ, Ma S, et al. (2002) A tale of two cities: effects of Air Pollution on Hospital Admissions in Hong Kong and London Compared. Environ Health Persp 110: 67–77.
10. Moolgavkar HS (2000) Air pollution and hospital admissions for diseases of the circulatory system in three U.S. metropolitan areas. Air & Waste Manage Assoc 50: 1199–1206.
11. Roberts S (2003) Combining data from multiple monitors in air pollution mortality time series studies. Atmo Environ 37: 3317–3322.
12. Guo YM, Barnett AG, Zhang YS, Tong SL, Yu WW, et al. (2010) The short-term effect of air pollution on cardiovascular mortality in Tianjin, China: Comparison of time series and case-crossover analyses. Sci Total Environ 409: 300–306.
13. Tsai DH, Wang JL, Chuang KJ, Chan CC (2010) Traffic-related air pollution and cardiovascular mortality in central Taiwan. Sci Total Environ 408: 1818–1823.
14. Brook DR, Rajagopalan S, Pope AC III, Brook RJ, Bhatnagar A, et al. (2010) Particulate Matter Air Pollution and Cardiovascular Disease: An Update to the Scientific Statement from the American Heart Association. Circulation 1: 2331–2378.
15. Hu XG, Zhou QX (2013) Health and ecosystem risks of graphene. Chem Rev 113: 3815–3835.
16. WHO (1993) International Classification of Diseases, Tenth Revision. Geneva: World Health Organization.
17. Kan HD, London JS, Chen GH, Zhang YH, Song GX, et al. (2008) Season, sex, age, and education as modifiers of the effects of outdoor air pollution on daily mortality in Shanghai, China: The Public Health and Air Pollution in Asia (PAPA) study. Environ Health Persp 116: 1183–1188.
18. Klea K, Giota T, Evangelia S, Alexandros G, Alain LT, et al. (2001) Confounding and effect modification in the short-term effects of ambient particles on total mortality: results from 29 European cities within the APHEA2 Project. Epidemiology 12: 521–531.
19. Xiao ZM, Wu JH, Han SQ, Zhang YF, Xu H, et al. (2012) Vertical characteristics and source identification of PM10 in Tianjin. J Environ Sci-China 24: 112–115.
20. Li WF, Bai ZP, Liu AX, Chen J, Li C (2009) Characteristics of major $PM_{2.5}$ components during winter in Tianjin, China. Aerosol Air Qual Res 9: 105–119.
21. Grundtvig M, Hagen PT, Amrud SE, Reikvam Å (2013) Reduced life expectancy after an incident hospital diagnosis of acute myocardial infarction-Effects of smoking in women and men. Int J Cardiol 167: 2792–2797.
22. Johnson AC, Palmer HP, Chou CP, Pang ZC, Zhou Dunjin, et al. (2006) Tobacco use among youth and adults in Mainland China: The China Seven Cities Study. Public Health 120: 1156–1169.
23. Fang CS, Cassidy A, Christiani CD (2010) A Systematic Review of Occupational Exposure to Particulate Matter and Cardiovascular Disease. Inter J Environmental Research Pub Heal 7: 1773–1806.
24. Kampa M, Castanas E (2008) Human health effects of air pollution. Environ Pollu 151(2): 362–367.
25. Kelly JF (2004) Dietary antioxidants and environmental stress. P Nutr Soc 63: 579–585.
26. Chalamaiah M, Dinesh kumar B, Hemalatha R, Jyothirmayi T (2012) Fish protein hydrolysates: Proximate composition, amino acid composition, antioxidant activities and applications: A review. Food Chem, 135: 3020–3038.
27. Curriero FC, Heiner KS, Samet JM, Zeger LS, Strug L, et al. (2002) Temperature and mortality in 11 cities of the eastern United States. Am J Epidemiol 155: 80–87.
28. Kan HD, Heiss G, Rose MK, Whitsel AE, Lurmann F, et al. (2008) Prospective analysis of traffic exposure as a risk factor for incident coronary heart disease: The Atherosclerosis Risk in Communities (ARIC) Study. Environ Health Persp 116: 1463–1468.
29. Zanobetti A, Schwartz J (2008) Temperature and mortality in nine US cities. Epidemiology 19: 563–570.

4

Increased Risk of Dementia in Patients Exposed to Nitrogen Dioxide and Carbon Monoxide: A Population-Based Retrospective Cohort Study

Kuang-Hsi Chang[1,2], Mei-Yin Chang[3], Chih-Hsin Muo[4], Trong-Neng Wu[1], Chiu-Ying Chen[1], Chia-Hung Kao[5,6]*

1 Department of Public Health, China Medical University, Taichung, Taiwan, 2 Department of Medical Research, Taichung Veterans General Hospital, Taichung, Taiwan, 3 Department of Medical Laboratory Science and Biotechnology, School of Medical and Health Sciences, Fooyin University, Kaohsiung, Taiwan, 4 Management Office for Health Data, China Medical University Hospital, Taichung, Taiwan, 5 Graduate Institute of Clinical Medical Science, College of Medicine, China Medical University, Taiwan, 6 Department of Nuclear Medicine and PET Center, China Medical University Hospital, Taichung, Taiwan

Abstract

Background: The air pollution caused by vehicular emissions is associated with cognitive decline. However, the associations between the levels of nitrogen dioxide (NO_2) and carbon monoxide (CO) exposure and dementia remain poorly defined and have been addressed in only a few previous studies.

Materials and Methods: In this study, we obtained data on 29547 people from the National Health Insurance Research Database (NHIRD) of Taiwan, including data on 1720 patients diagnosed with dementia between 2000 and 2010, and we evaluated the risk of dementia among four levels of air pollutant. Detailed data on daily air pollution were available from January 1, 1998 to December 31, 2010. Yearly average concentrations of pollutants were calculated from the baseline to the date of dementia occurrence, withdrawal of patients, or the end of the study, and these data were categorized into quartiles, with Q1 being the lowest level and Q4 being the highest.

Results: In the case of NO_2, the adjusted hazard ratios (HRs) of dementia for all participants in Q2, Q3, and Q4 compared to Q1 were 1.10 (95% confidence interval (CI), 0.96–1.26), 1.01 (95% CI, 0.87–1.17), and 1.54 (95% CI, 1.34–1.77), and in the case of CO, the adjusted HRs were 1.07 (95% CI, 0.92–1.25), 1.37 (95% CI, 1.19–1.58), and 1.61 (95% CI, 1.39–1.85).

Conclusion: The results of this large retrospective, population-based study indicate that exposure to NO_2 and CO is associated with an increased risk of dementia in the Taiwanese population.

Editor: Gianluigi Forloni, "Mario Negri" Institute for Pharmacological Research, Italy

Funding: The study was supported in part by China Medical University (CMU102-BC-2), Taiwan Ministry of Health and Welfare Clinical Trial and Research Center of Excellence (MOHW103-TDU-B-212-113002), Taiwan Ministry of Health and Welfare Cancer Research Center for Excellence (MOHW103-TD-B-111-03), and International Research-Intensive Centers of Excellence in Taiwan (I-RiCE) (NSC101-2911-I-002-303). The funders had no role in study design, data collection and analysis, decision to publish, or preparation of the manuscript.

Competing Interests: The authors have declared that no competing interests exist.

* Email: d10040@mail.cmuh.org.tw

Introduction

Ambient air pollution includes solid and gaseous pollutants [1,2]. Most of the studies that have investigated the effects of pollutants on cognitive functions have examined the influence of solid pollutants [3–8]. However, exposure to ambient gaseous pollutants such as nitrogen dioxide (NO_2) is known to increase the risk of cerebrovascular and neurodegenerative diseases and ischemic stroke [9–12]. Cerebrovascular disease is the principal contributor to dementia [13,14], and Alzheimer's disease (AD) is the most common neurodegenerative disease. Moreover, a population-base study reported that dementia often developed after the occurrence of an ischemic stroke [15]. Several previous studies have suggested negative associations between NO_2 exposure and cognitive development in children, including preschool children [16–18], and animal studies have indicated that NO_2 exposure inhibits the recovery of nerve function after a stroke [19,20]. In addition, one animal study reported that nitration can induce beta-amyloid aggregation and plaque formation [21]; beta-amyloid aggregation is a pathologic hallmark of AD. However, a literature search indicated that only a few studies have been conducted to address the link between NO_2 exposure and cognitive function in adults. In a recent study conducted on 1496 middle-aged people living in Los Angeles, no statistically significant correlation was detected between the level of NO_2 exposure and cognitive functions [22]. Therefore, we conducted a retrospective cohort study to determine the association between NO_2 and dementia risk. Furthermore, in this study, we evaluated the influence of carbon monoxide (CO), because acute CO poisoning may cause headache, nausea, malaise, and fatigue [23], and chronic CO exposure has been linked to depression, confusion, memory loss, and cognitive decline [24,25]. Comparison between this study with other environmental study of Taiwan NHRID, the main difference is the residential area definition. In previous studies, the residential area is as the insurance area [26]. In present study, we defined the residential areas as the location of clinics which subjects sought treatment for acute upper respiratory infections.

Materials and Methods

Data sources and study population

In March 1995, the Taiwan National Health Insurance (NHI) program, which is a single-payer, compulsory social insurance system that has provided insurance coverage to almost every citizen in Taiwan, was established. The NHI covered approximately 99% of the 22.96 million citizens in Taiwan at the end of 2007 [27]. To protect patient privacy, the data on patient identities are encrypted in the National Health Insurance Research Database (NHIRD), and the database is accessible to researchers and the public in Taiwan. In this study, we used a subset of the NHIRD data containing comprehensive health-care data, including files on ambulatory care claims, inpatient claims, and prescriptions received by 1000000 people who were randomly selected from all insured beneficiaries. These data files can be linked through an encrypted but unique personal identification number and, thus, provide a longitudinal medical history of each patient. The health status of each person was identified according to the International Classification of Disease, Ninth Revision, Clinical Modification (ICD-9-CM).

Exposure assessment

Across Taiwan, 74 ambient air quality monitoring stations are located based on population density. Air quality data are maintained by Taiwan Environmental Protection Administration. [28]. A database containing daily NO_2 and CO concentrations measured at the monitoring stations was available for the period from January 1, 1998 to December 31, 2010. The people included in this study were assigned pollutant-exposure values based on the data obtained from the monitoring station present in the residential district in which the clinic where the people most frequently sought treatment for acute upper respiratory infection was located (ICD-9-CM Code 460). Yearly average concentrations of pollutants were calculated from the baseline to the date of dementia occurrence, the withdrawal of patients, or the end of the study period, and the data were categorized into quartiles.

Study patients

We identified 29547 people who were aged 50 years or older and for whom estimable air pollution data were available, but who did not present a history of head injury (ICD-9-CM Codes 800.804, 850.854.1, 310.2, and 959.01), stroke (ICD-9-CM Codes 430–438), or dementia (ICD-9-CM Codes 290.0–290.4, 294.1, and 331.0) before 2000.

Data Availability Statement

All data and related metadata are deposited in an appropriate public repository: The study population's data were from Taiwan NHIRD (http://w3.nhri.org.tw/nhird//date_01.html) are maintained by Taiwan National Health Research Institutes (http://nhird.nhri.org.tw/) [27]. The National Health Research Institutes (NHRI) is a non-profit foundation established by the government. Air quality data were from Taiwan Air Quality Monitoring Network (http://taqm.epa.gov.tw/taqm/en/PsiMap.aspx) in Taiwan Environmental Protection Administration (http://www.epa.gov.tw/) [28].

Ethics statement

Because identification numbers of patients had been encrypted, patient consent was not required for this study. This study was approved by the Research Ethic Committee at China Medical University (CMU-REC-101-012). The committee waived the requirement for consent.

Statistical analysis

We used x^2 tests to examine the distributions of sex, monthly income (New Taiwan Dollar<14 400, 14 400–18 300, 18 301–21 000, and >21 000), diabetes (DM, ICD-9-CM Code 250), ischemic heart disease (IHD, ICD-9-CM Codes 410–414), hypertension (HT, ICD-9-CM Codes 401–405),chronic obstructive pulmonary disease(COPD, ICD-9-CM Codes 490–496), alcoholism (ICD-9-CM Codes 303.305.0andV113),and the quartiles of NO_2 concentration (ppb; <6652.3, 6652.3–8349.0, 8349.1–9825.5,>9825.5) and CO concentration (ppm; <196.2, 196.2–241.6, 241.7–296.9, >296.9). A one-way analysis of variance (ANOVA) was performed to compare the age among the quartiles of NO_2 and CO concentrations. We calculated the incidence density rates of dementia in person-years in each quarter stratified according to sex. The incidence rate ratio (IRR) was estimated using a Poisson regression. Univariate and multivariate Cox proportional hazard regression analyses were performed to calculate the hazard ratios (HRs) and 95% confidence intervals (CIs) of the risk of dementia in association with pollutant levels. Multiple models were tested by controlling for age, sex, monthly income, DM, HT, IHD, COPD, alcoholism, and urbanization. Plots of the Kaplan-Meier analysis were used to determine the probability of people remaining without dementia, and the log-rank test was used to evaluate the differences among quartiles of pollutant concentrations. All analyses were performed using SAS 9.2 software (SAS Institute Inc., Cary, NC, USA), and the Kaplan-Meier survival curve was plotted using the Statistical Package for the Social Sciences (Version 15.1; SPSS Inc, Chicago, IL, USA). All tests were considered statistically significant when two-tailed P values were <.05.

Results

We obtained a total of 29547 and 29537 data on daily NO_2 and CO exposure, respectively. Dementia was not present at the baseline (2000), and 1720 people developed dementia after follow-up (yearly CO data were available for 1718 people). We categorized the NO_2 and CO levels into quartiles, with Q1 being the lowest level and Q4 being the highest. The people included in this study had a mean age of 61.4 years (SD 8.5 y). In both the NO_2 and CO groups, the highest level of the quartiles was associated with the people being slightly younger, more frequently earning a high monthly income, and living in a highly urbanized residential area, but less frequently exhibiting IHD and COPD compared with other quartiles (Tables 1 and 2). Table 3 shows the associations between the gaseous pollutant levels and the risk of dementia. Among the quartiles Q1, Q2, Q3, and Q4 of NO_2 in all patients, the IRRs in Q2, Q3, and Q4 compared with that in Q1 were 1.05, 0.90, and 1.35, and the adjusted HRs of dementia were 1.10 (95% CI, 0.96–1.26), 1.01 (95% CI, 0.87–1.17), and 1.54 (95% CI, 1.34–1.77), respectively. Among men, we determined that the IRRs in Q2, Q3, and Q4 compared with that in Q1 were 1.08, 0.79, and 1.28, and the adjusted HRs were 1.16 (95% CI, 0.95–1.43), 0.89 (95% CI, 0.71–1.11), and 1.52 (95% CI, 1.23–1.88), respectively. Among women, the IRRs in Q2, Q3, and Q4 compared with that in Q1 were 1.05, 1.11, and 1.56, and the adjusted HRs were 1.05 (95% CI, 0.87–1.27), 1.11 (95% CI, 0.92–1.35), and 1.56 (95% CI, 1.29–1.87), respectively. When the data on sex were stratified or merged for analysis, statistically significant correlations of IRRs and adjusted HRs were measured in Q4 compared with those in Q1.

Among the quartiles of CO concentration, the IRRs in Q2, Q3, and Q4 compared with that in Q1 were 0.96, 1.23, and 1.36, and the adjusted HRs were 1.07 (95% CI, 0.92–1.25), 1.37 (95% CI,1.19–1.58),and 1.61 (95% CI, 1.39–1.85), respectively, in all people included in the study. Among men, the IRRs in Q2, Q3, and Q4 compared with that in Q1 were 0.97, 1.18, and 1.28, and the adjusted HRs were 1.16 (95% CI, 0.93–1.45), 1.28 (95% CI, 1.04–1.58), and 1.57 (95% CI, 1.26–1.94), respectively. Among women, the IRRs in Q2, Q3, and Q4 compared with that in Q1 were 0.95, 1.28, and 1.43, and the adjusted HRs were 1.01 (95% CI, 0.82–1.24), 1.46 (95% CI, 1.21–1.77), and 1.64 (95% CI, 1.36–1.98), respectively. A clear trend that was detected was an increase in the risk of dementia as CO exposure increased. Figures 1 and 2 show the Kaplan-Meier curves of freedom that were calculated for dementia and are separated according to pollutant level. Statistically significant differences in the occurrence of dementia were observed among the quartiles of NO_2 and CO concentrations (log-rank test, $P<.001$).

Discussion

The major finding of previous animal study was that nitration was highly correlated with beta-amyloid aggregation and plaque formation, and beta-amyloid aggregation is a pathologic hallmark of AD [21]. Another animal study indicated that NO_2 expose can exacerbate the ultra structural impairment of synapses in stroke rats, and induce neuronal damage in healthy rats [29]. The apolipoprotein E (APOE) e4 allele was a well know genetic risk factor or AD, and a randomized clinical trial has found CO poisoning can induce APOE e4 carriers suffer greater morbidity [30].

The major finding of our study was that increased exposure to NO_2 (Q4) is associated with an enhanced risk of dementia in men and women. The probability of dementia occurrence was increased by 52%–56% in Q4 compared with Q1. A similar trend was observed in the CO group, and the results collectively showed that increasing levels of the 2 pollutants increased the risk of dementia in a dose-dependent manner.

This study was a national population-based investigation on ambient air pollution and dementia. Therefore, collecting individual exposure data was not feasible. To obtain exposure data associated with study patients, previous studies have identified the residential areas of patients by employing a GIS-based system. To protect the privacy of patients, the NHIRD does not provide patients' addresses. Therefore, we identified the residential areas of the patients based on the location of the clinic at which the patients most frequently sought treatment for acute upper respiratory tract infection. In the United States, upper respiratory

tract infections are the most common type of infectious disease, and each adult experiences approximately 3 respiratory infections annually [31]. Identifying residential areas in the accessible medical resources, as we did in this study, is more accurate than listing patients according to insurance area [32,33].

Previous studies have suggested that smoking and drinking alcohol are highly correlated with the risk of AD [34–40]. Because of the limitations of the NHIRD, we could not obtain data on the smoking or drinking status of the patients. Therefore, we performed multivariate analysis with COPD and alcoholism adjusted in accordance with previous studies that indicated that smoking is a major causative factor in the development of COPD, and in which alcoholism was diagnosed based on drinking patterns and the attitudes of patients [41–43]. In Taiwan, women are not encouraged to smoke or drink alcohol, as reflected in the low prevalence of these behaviors among women (3% and 1%, respectively) [44,45]. We were able to overcome this limitation by stratifying and adjusting the data according to sex [46].

We adjusted for urbanization in the multivariate analysis. The level of urbanization was determined according to population density (number of people/km^2), the population ratio of people with a college-level education or higher, the population ratio of people aged over 65 years, the population ratio of agricultural workers, and the number of physicians per 100000 people [47]. The 359 communities in Taiwan were classified into 7groups: highly urbanized area, moderately urbanized area, boomtown, general town, aging town, agricultural town, and remote town. This classification method has been used in several studies [48–50].

In addition, we obtained results contrasting those related to dementia, as shown in Tables 1 and 2: the frequency of IHD and COPD were low at the highest level of the pollutants. These results agree with the explanation provided by previous studies suggesting that patients who are highly educated and earn a high monthly income live in areas where the level of air pollutants is high [6,22].

The strengths of this study are the following. First, this study was based on a long follow-up period, which allowed the possible occurrence of dementia to be assessed. Second, Taiwan launched a national health insurance (NHI) in 1995, operated by a single-buyer, the government. All insurance claims should be scrutinized by medical reimbursement specialists and peer review. The diagnoses of dementia were based on the ICD-9 code determined by qualified clinical neurology physicians under strict audit in the reimbursement process. Therefore the diagnoses and codes for dementia should be accurate and reliable. Third, this study was conducted using a large population derived from the NHIRD. In Taiwan, the government is the only compulsory social insurance provider; approximately 99% of the 23.74 million citizens of Taiwan are enrolled in the NHI program. Because this was a nationwide study, we considered urbanized towns throughout Taiwan. Lastly, in this study, cerebrovascular and cardiovascular diseases were considered and the association between pollutants and dementia was evaluated. We excluded subjects with cardiovascular before the index date in this study because cardiovascular was a widely known predictor for dementia. IHD increased 27% risk for dementia in both model 1 and model 2. (Table S1).

Certain limitations of this study should be considered. First, the evidence derived from a retrospective cohort study is generally lower in statistical quality than that obtained from randomized trials because, in such retrospective studies, potential biases exist that are related to the adjustment of confounding variables. Despite our meticulous study design and the measures adopted to control for confounding factors, bias resulting from unknown confounders may have affected our results. Second, all data in the

Increased Risk of Dementia in Patients Exposed to Nitrogen Dioxide and Carbon Monoxide: A Population-Based...

29

Table 1. Comparison of Baseline Characteristics among quartiles of NO$_2$ yearly average.

| | Quartiles of NO$_2$ yearly average | | | | | | | | p | Total (n = 29547) | |
	Q1 (n = 7349)		Q2 (n = 7425)		Q3 (n = 7572)		Q4 (n = 7201)				
Dementia	406	5.5	425	5.7	374	4.9	515	7.2	<0.001	1720	5.8
Age (mean, SD)	61.8	8.4	61.4	8.5	61.0	8.4	61.4	8.8	<0.001†	61.4	8.5
Male	3365	45.8	3469	46.7	3474	45.9	3298	45.8	0.611	13606	46.0
Monthly income									<0.001		
<14400	1481	20.2	1814	24.4	2004	26.5	1991	27.7		7290	24.7
14400–18300	1054	14.3	1324	17.8	1511	20.0	1480	20.6		5369	18.2
18301–21000	3255	44.3	2399	32.3	1992	26.3	1785	24.8		9431	31.9
>21000	1559	21.2	1887	25.4	2062	27.2	1944	27.0		7452	25.2
DM	845	11.5	837	11.3	916	12.1	850	11.8	0.421	3448	11.7
IHD	1347	18.3	1354	18.2	1295	17.1	1222	17.0	0.047	5218	17.7
HT	2899	39.4	2906	39.1	2889	38.2	2785	38.7	0.391	11479	38.8
COPD	2612	35.5	2608	35.1	2579	34.1	2376	33.0	0.005	10175	34.4
Alcoholism	19	0.3	19	0.3	22	0.3	10	0.1	0.250	70	0.2
Urbanization									<0.001		
Highly urbanization	1330	18.1	1668	22.5	2503	33.1	3720	51.7		9221	31.2
Moderate urbanization	2157	29.4	2782	37.5	2908	38.4	1828	25.4		9675	32.7
Boomtown	907	12.3	986	13.3	1485	19.6	1126	15.6		4504	15.2
General town	1692	23.0	1160	15.6	412	5.4	298	4.1		3562	12.1
Aging town	304	4.1	56	0.8	68	0.9	72	1.0		500	1.7
Agricultural town	658	9.0	321	4.3	111	1.5	88	1.2		1178	4.0
Remote town	301	4.1	452	6.1	85	1.1	69	1.0		907	3.1

Chi-square test;
†T-test;

Table 2. Comparison of Baseline Characteristics among quartiles of CO yearly average.

	Quartiles of CO yearly average									p	Total (n = 29537)	
	Q1 (n = 7565)		Q2 (n = 6428)		Q3 (n = 7681)		Q4 (n = 7863)					
Dementia	391	5.2	321	5.0	476	6.2	530	6.7		<0.001	1718	5.8
Age (mean, SD)	61.8	8.3	61.1	8.3	61.4	8.6	61.3	8.8		<0.001†	61.4	8.5
Male	3532	46.7	2882	44.8	3587	46.7	3597	45.7		0.084	13598	46.0
Monthly income										<0.001		
<14400	1477	19.5	1477	23.0	2190	28.5	2144	27.3			7288	24.7
14400–18300	1074	14.2	1189	18.5	1513	19.7	1591	20.2			5367	18.2
18301–21000	3401	45.0	2095	32.6	1887	24.6	2046	26.0			9429	31.9
>21000	1613	21.3	1667	25.9	2088	27.2	2080	26.5			7448	25.2
DM	862	11.4	712	11.1	918	12.0	954	12.1		0.173	3446	11.7
IHD	1430	18.9	1054	16.4	1394	18.1	1339	17.0		<0.001	5217	17.7
HT	2980	39.4	2455	38.2	3021	39.3	3017	38.4		0.306	11473	38.8
COPD	2785	36.8	2189	34.1	2607	33.9	2587	32.9		<0.001	10168	34.4
Alcoholism	19	0.3	15	0.2	24	0.3	12	0.2		0.232	70	0.2
Urbanization										<0.001		
Highly urbanization	912	12.1	1697	26.4	2694	35.1	3918	49.8			9221	31.2
Moderate urbanization	2388	31.6	2615	40.7	2323	30.2	2346	29.8			9672	32.7
Boomtown	1084	14.3	819	12.7	1576	20.5	1024	13.0			4503	15.2
General town	1684	22.3	772	12.0	781	10.2	322	4.1			3559	12.0
Aging town	336	4.4	22	0.3	65	0.8	74	0.9			497	1.7
Agricultural town	699	9.2	253	3.9	120	1.6	106	1.3			1178	4.0
Remote town	462	6.1	250	3.9	122	1.6	73	0.9			907	3.1

Chi-square test;
†T-test;

Table 3. Comparisons of difference dementia incidences and associated hazard ratios among four levels of air pollutants by gender stratification.

		Dementia	PY	Incidence rate#	IRR*	95%CI	aHR†	95%CI
NO₂	Total							
	Q1	406	75461.4	5.38	1.00		1.00	
	Q2	425	75246.1	5.65	1.05	0.92, 1.20	1.10	0.96, 1.26
	Q3	374	77576.5	4.82	0.90	0.78, 1.03	1.01	0.87, 1.17
	Q4	515	71461.0	7.21	1.35	1.18, 1.54	1.54	1.34, 1.77
	Male							
	Q1	186	33853.8	5.49	1.00		1.00	
	Q2	206	34587.2	5.96	1.08	0.89, 1.32	1.16	0.95, 1.43
	Q3	152	34973.3	4.35	0.79	0.64, 0.98	0.89	0.71, 1.11
	Q4	224	31976.0	7.01	1.28	1.05, 1.56	1.52	1.23, 1.88
	Female							
	Q1	220	41607.6	5.29	1.00		1.00	
	Q2	219	40658.9	5.39	1.02	0.85, 1.23	1.05	0.87, 1.27
	Q3	222	42603.2	5.21	0.99	0.82, 1.19	1.11	0.92, 1.35
	Q4	291	39485.0	7.37	1.41	1.18, 1.67	1.56	1.29, 1.87
CO	Total							
	Q1	391	77816.4	5.02	1.00		1.00	
	Q2	321	66509.7	4.83	0.96	0.83, 1.11	1.07	0.92, 1.25
	Q3	476	77215.4	6.16	1.23	1.08, 1.41	1.37	1.19, 1.58
	Q4	530	78172.7	6.78	1.36	1.19, 1.55	1.61	1.39, 1.85
	Male							
	Q1	182	35681.8	5.10	1.00		1.00	
	Q2	145	29334.5	4.94	0.97	0.78, 1.20	1.16	0.93, 1.45
	Q3	212	35371.7	5.99	1.18	0.97, 1.44	1.28	1.04, 1.58
	Q4	227	34977.3	6.49	1.28	1.05, 1.55	1.57	1.26, 1.94
	Female							
	Q1	209	42134.6	4.96	1.00		1.00	
	Q2	176	37175.2	4.73	0.95	0.78, 1.16	1.01	0.82, 1.24
	Q3	264	41843.8	6.31	1.28	1.07, 1.54	1.46	1.21, 1.77
	Q4	303	43195.4	7.01	1.43	1.20, 1.70	1.64	1.36, 1.98

Incidence rate#: per 1,000 person-years;
IRR*: incidence rate ratio;
Adjusted HR†: multiple analysis including age, sex, monthly income, DM, IHD, HT, COPD, alcoholism and urbanization.

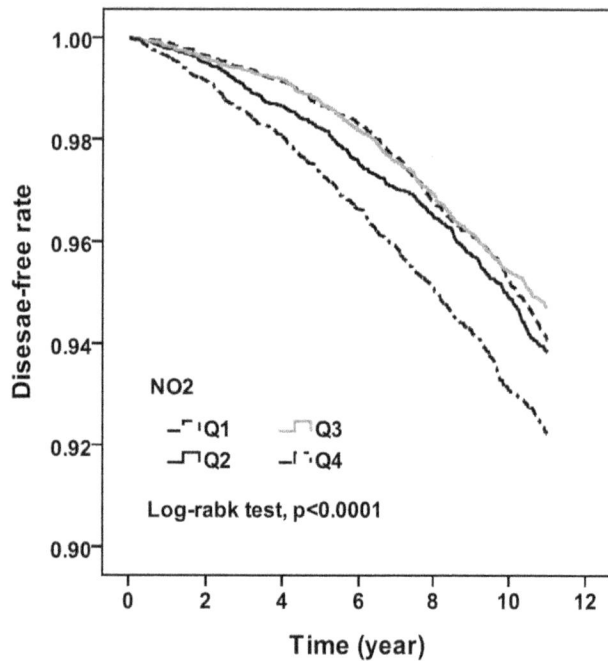

Figure 1. Probability free of dementia among quartiles of yearly average concentration in NO₂.

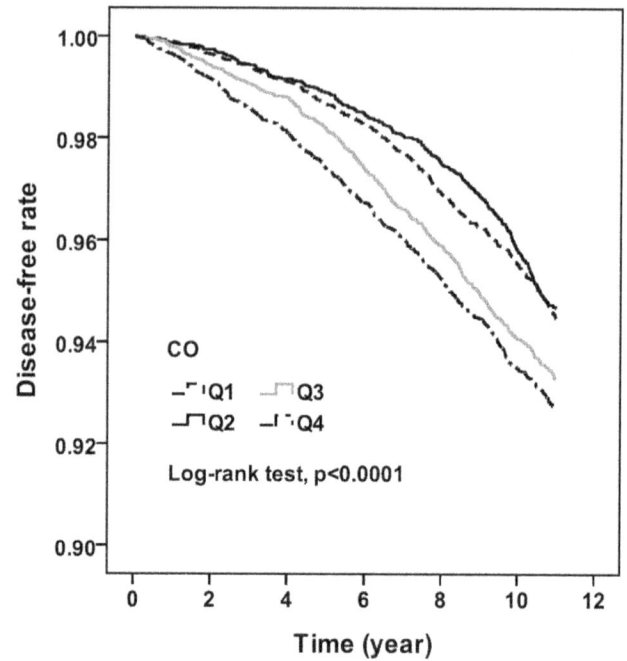

Figure 2. Probability free of dementia among quartiles of yearly average concentration in CO.

NHIRD are anonymous. Thus, relevant clinical variables, such as imaging results and pathology findings, were unavailable for the patient cases included in this study. Third, the participants were assigned to residential districts based on the clinic where they most frequently sought treatment for acute upper respiratory infection. Therefore, the resident who has no acute upper respiratory infection during study period had being excluded in this study. In our opinion, the resident without respiratory infection related medical record exposed to low level air pollutants. It might under the estimated risk of dementia. Nevertheless, the data on air pollutants and dementia diagnoses were reliable.

Conclusions

Understanding the regional distribution of human health statuses can facilitate the investigation of the spread of diseases and the related risk factors as well as the assessment of medical resources and the planning of the use of these resources. In future

research, animal studies can be conducted to further examine the association between air pollutants and neurological disorders.

Author Contributions

Study concept and design: KHC CHK. Acquisition of data: KHC MYC CHM TNW CYC CHK. Analysis and interpretation of data: KHC CHM CHK. Drafting of the manuscript: KHC MYC CHM TNW CYC CHK. Critical revision of the manuscript for important intellectual content: KHC CHM CHK. Statistical analysis: CHM. Obtained funding: CHK. Administrative, technical, or material support: KHC MYC CHM TNW CYC CHK. Study supervision: CHK.

References

1. Dickey JH, Part VII (2000) Air pollution: overview of sources and health effects. Dis Mon 46:566–89.
2. Lewtas J (2007) Air pollution combustion emissions: characterization of causative agents and mechanisms associated with cancer, reproductive, and cardiovascular effects. Mutat Res 636:95–133.
3. Weuve J, Puett RC, Schwartz J, Yanosky JD, Laden F, et al. (2012) Exposure to particulate air pollution and cognitive decline in older women. Arch Intern Med 172:219–27.
4. Srám RJ, Benes I, Binková B, Dejmek J, Horstman D, et al. (1996) Teplice program–the impact of air pollution on human health. Environ Health Perspect 104 Suppl 4:699–714.
5. Suglia SF, Gryparis A, Wright RO, Schwartz J, Wright RJ (2008) Association of black carbon with cognition among children in a prospective birth cohort study. Am J Epidemiol 167:280–6.
6. Chen JC, Schwartz J (2009) Neurobehavioral effects of ambient air pollution on cognitive performance in US adults. Neurotoxicology 30:231–9.
7. Ranft U, Schikowski T, Sugiri D, Krutmann J, Krämer U (2009) Long-term exposure to traffic-related particulate matter impairs cognitive function in the elderly. Environ Res 109:1004–11.

8. Power MC, Weisskopf MG, Alexeeff SE, Coull BA, Spiro A 3rd, et al. (2011) Traffic-related air pollution and cognitive function in a cohort of older men. Environ Health Perspect 119:682–7.
9. Lisabeth LD, Escobar JD, Dvonch JT, Sánchez BN, Majersik JJ, et al. (2008) Ambient air pollution and risk for ischemic stroke and transient ischemic attack. Ann Neurol 2008;64:53–9.
10. Migliore L, Coppedè F (2009) Environmental-induced oxidative stress in neurodegenerative disorders and aging. Mutat Res 674:73–84.
11. Turin TC, Kita Y, Rumana N, Nakamura Y, Ueda K, et al (2012) Ambient air pollutants and acute case-fatality of cerebro-cardiovascular events: Takashima Stroke and AMI Registry, Japan (1988–2004). Cerebrovasc Dis 34(2):130–9.
12. Andersen ZJ, Kristiansen LC, Andersen KK, Olsen TS, Hvidberg M, et al. (2012) Stroke and long-term exposure to outdoor air pollution from nitrogen dioxide: a cohort study. Stroke 43:320–5.
13. Knopman DS (2007) Cerebrovascular disease and dementia. Br J Radiol 80 :S121–7.
14. O'Brien JT (2006) Vascular cognitive impairment. Am J Geriatr Psychiatry 14:724–33.

15. Kokmen E, Whisnant JP, O'Fallon WM, Chu CP, Beard CM (1996) Dementia after ischemic stroke: a population-based study in Rochester, Minnesota (1960–1984). Neurology 46:154–9.

16. Morales E, Julvez J, Torrent M, de Cid R, Guxens M, et al. (2009) Association of early-life exposure to household gas appliances and indoor nitrogen dioxide with cognition and attention behavior in preschoolers. Am J Epidemiol 169:1327–36.

17. Freire C, Ramos R, Puertas R, Lopez-Espinosa MJ, Julvez J, et al. (2010) Association of traffic-related air pollution with cognitive development in children. J Epidemiol Community Health 64:223–8.

18. Clark C, Crombie R, Head J, van Kamp I, van Kempen E, et al. (2012) Does traffic-related air pollution explain associations of aircraft and road traffic noise exposure on children's health and cognition? A secondary analysis of the United Kingdom sample from the RANCH project. Am J Epidemiol 176:327–37.

19. Zhu N, Li H, Han M, Guo L, Chen L, et al. (2012) Environmental nitrogen dioxide (NO2) exposure influences development and progression of ischemic stroke. Toxicol Lett 214:120–30.

20. Li H, Xin X (2013) Nitrogen dioxide (NO(2)) pollution as a potential risk factor for developing vascular dementia and its synaptic mechanisms. Chemosphere 92:52–8.

21. Kummer MP, Hermes M, Delekarte A, Hammerschmidt T, Kumar S, et al. (2011) Nitration of tyrosine 10 critically enhances amyloid β aggregation and plaque formation. Neuron 71:833–44.

22. Gatto NM, Henderson VW, Hodis HN, St John JA, Lurmann F, et al. (2013) Components of air pollution and cognitive function in middle-aged and older adults in Los Angeles. Neurotoxicology 40C:1–7.

23. Blanco F, Alkorta I, Solimannejad M, Elguero J (2009) Theoretical study of the 1:1 complexes between carbon monoxide and hypohalous acids. J Phys Chem A 113:3237–44.

24. Roberts GP, Youn H, Kerby RL (2004) CO-sensing mechanisms. Microbiol Mol Biol Rev 68:453–73, table of contents.

25. Chen HL, Chen PC, Lu CH, Hsu NW, Chou KH, et al. (2013) Structural and cognitive deficits in chronic carbon monoxide intoxication: a voxel-based morphometry study. BMC Neurol 13:129.

26. Jung CR, Lin YT, Hwang BF (2013) Air pollution and newly diagnostic autism spectrum disorders: a population-based cohort study in Taiwan. PLoS One 8:e75510.

27. National Health Insurance Research Database (NHIRD): Introduction to the National Health Insurance Research Database (NHIRD), Taiwan (2010) Available: http://w3.nhri.org.tw/nhird//date_01.html

28. Taiwan Air Quality Monitoring Network in Taiwan Environmental Protection Administration. Available: http://taqm.epa.gov.tw/taqm/en/PsiMap.aspx

29. Li H, Xin X (2013) Nitrogen dioxide (NO (2)) pollution as a potential risk factor for developing vascular dementia and its synaptic mechanisms. Chemosphere 92:52–8

30. Hopkins RO, Weaver LK, Valentine KJ, Mower C, Churchill S, et al. (2007) Apolipoprotein E genotype and response of carbon monoxide poisoning to hyperbaric oxygen treatment. Am J Respir Crit Care Med 176: 1001–6.

31. Garibaldi RA (1985) Epidemiology of community-acquired respiratory tract infections in adults. Incidence, etiology, and impact. Am J Med 78:32–7.

32. Kuo SS, Chang RE (2010) Geographical analysis of ESRD incidence and environment [Dissertation]. Taipei: Graduate Institute of Health Care Organization Administration, National Taiwan University. [In Chinese: English abstract]

33. Ministry of the Interior, R.O.C. (Taiwan). Monthly bulletin of interior statistics. Available at: http://sowf.moi.gov.tw/stat/month/list.htm. Accessed 2011 March 3. [In Chinese: English abstract]

34. Cataldo JK, Prochaska JJ, Glantz SA (2010) Cigarette smoking is a risk factor for Alzheimer's Disease: an analysis controlling for tobacco industry affiliation. J Alzheimers Dis 19:465–80.

35. Deng J, Shen C, Wang YJ, Zhang M, Li J, et al. (2010) Nicotine exacerbates tau phosphorylation and cognitive impairment induced by amyloid-beta 25–35 in rats. Eur J Pharmacol 637:83–8.

36. Oddo S, Caccamo A, Green KN, Liang K, Tran L, et al. (2005) Chronic nicotine administration exacerbates tau pathology in a transgenic model of Alzheimer's disease. Proc Natl Acad Sci U S A 102:3046–51.

37. Juan D, Zhou DH, Li J, Wang JY, Gao C, et al. (2004) A 2-year follow-up study of cigarette smoking and risk of dementia. Eur J Neurol 11:277–82.

38. Peters R, Peters J, Warner J, Beckett N, Bulpitt C (2008) Alcohol, dementia and cognitive decline in the elderly: a systematic review. Age Ageing 37:505–12.

39. Deng J, Zhou DH, Li J, Wang YJ, Gao C, et al. (2006) A 2-year follow-up study of alcohol consumption and risk of dementia. Clin Neurol Neurosurg 108:378–83.

40. Anstey KJ, Mack HA, Cherbuin N (2009) Alcohol consumption as a risk factor for dementia and cognitive decline: meta-analysis of prospective studies. Am J Geriatr Psychiatry 17:542–55.

41. Pauwels RA, Rabe KF (2004) Burden and clinical features of chronic obstructive pulmonary disease (COPD). Lancet 364:613–20.

42. Patel BD, Loo WJ, Tasker AD, Screaton NJ, Burrows NP, et al. (2006) Smoking related COPD and facial wrinkling: is there a common susceptibility? Thorax 61:568–571.

43. Enoch MA, Goldman D (2002) Problem drinking and alcoholism: diagnosis and treatment. Am Fam Physician 65:441–8.

44. Liang CY, Chou TM, Ho PS, Shieh TY, Yang YH (2004) Prevalence Rates of Alcohol Drinking in Taiwan. Taiwan Journal of Oral Medicine & Health Sciences 20:91–104

45. Chuang YC, Chuang KY (2008) Gender differences in relationships between social capital and individual smoking and drinking behavior in Taiwan. Soc Sci Med 67:1321–30.

46. Chang KH, Chung CJ, Lin CL, Sung FC, Wu TN, et al. (2014) Increased risk of dementia in patients with osteoporosis: a population-based retrospective cohort analysis. Age (Dordr) 36:967–75.

47. Liu C, Hung Y, Chuang Y, Chen Y, Weng W, et al. (2006) Incorporating development stratification of Taiwan townships into sampling design of large scale health interview survey. Journal of Health Management 4:1–22 [in Chinese].

48. Chiang PH, Chang YC, Lin JD, Tung HJ, Lin LP, et al. (2013) Healthcare utilization and expenditure analysis between individuals with intellectual disabilities and the general population in Taiwan: a population-based nationwide child and adolescent study. Res Dev Disabil 34:2485–92.

49. Lin YJ, Tian WH, Chen CC (2011) Urbanization and the utilization of outpatient services under National Health Insurance in Taiwan. Health Policy 103:236–43.

50. Lin HC, Lin YJ, Liu TC, Chen CS, Lin CC (2007) Urbanization and place of death for the elderly: a 10-year population-based study. Palliat Med 21:705–11.

5

New Insights into Handling Missing Values in Environmental Epidemiological Studies

Célina Roda[1], Ioannis Nicolis[1], Isabelle Momas[1,2]*, Chantal Guihenneuc[1]

1 Laboratoire Santé Publique et Environnement, EA 4064, Faculté de Pharmacie, Université Paris Descartes, Sorbonne Paris Cité, Paris, France, 2 Mairie de Paris, Direction de l'Action Sociale de l'Enfance et de la Santé, Cellule Cohorte, Paris, France

Abstract

Missing data are unavoidable in environmental epidemiologic surveys. The aim of this study was to compare methods for handling large amounts of missing values: omission of missing values, single and multiple imputations (through linear regression or partial least squares regression), and a fully Bayesian approach. These methods were applied to the PARIS birth cohort, where indoor domestic pollutant measurements were performed in a random sample of babies' dwellings. A simulation study was conducted to assess performances of different approaches with a high proportion of missing values (from 50% to 95%). Different simulation scenarios were carried out, controlling the true value of the association (odds ratio of 1.0, 1.2, and 1.4), and varying the health outcome prevalence. When a large amount of data is missing, omitting these missing data reduced statistical power and inflated standard errors, which affected the significance of the association. Single imputation underestimated the variability, and considerably increased risk of type I error. All approaches were conservative, except the Bayesian joint model. In the case of a common health outcome, the fully Bayesian approach is the most efficient approach (low root mean square error, reasonable type I error, and high statistical power). Nevertheless for a less prevalent event, the type I error is increased and the statistical power is reduced. The estimated posterior distribution of the OR is useful to refine the conclusion. Among the methods handling missing values, no approach is absolutely the best but when usual approaches (e.g. single imputation) are not sufficient, joint modelling approach of missing process and health association is more efficient when large amounts of data are missing.

Editor: Guy Brock, University of Louisville, United States of America

Funding: The PARIS cohort is supported by the Paris council, within its Social, Childhood and Health Direction (DASES). The environmental investigation of the cohort received financial support from the French National Agency for Food, Environment, and Occupational health Safety (Anses) and the French Institute for Public Health Surveillance (InVS). Célina Roda received a doctoral grant from the French Environment and Energy Management Agency (ADEME) and the French Scientific and Technical Construction Center (CSTB). The funders had no role in study design, data collection and analysis, decision to publish, or preparation of the manuscript.

Competing Interests: The authors have declared that no competing interests exist.

* Email: isabelle.momas@parisdescartes.fr

Introduction

In epidemiological studies, accurate estimate of exposure is very important for assessment of health risk. However missing data are often unavoidable, resulting from loss to follow-up in longitudinal studies, or non-responses in questionnaires. In large scale studies, certain strategies have been developed reducing the high cost of environmental measurements, e.g. collecting exposure surrogates for all subjects generally by questionnaires, and performing exact personal or environmental measurements only in subsamples of population [1]; most of the time these subsamples are small due to economic and logistic reasons (high cost, noise, bulk samplers…). For instance, in the National health and nutrition examination survey (NHANES), personal exposure to volatile organic compounds was assessed in a subsample of 851 adults, i.e. 8.5% of the population study [2]. Furthermore, even these specific studies can suffer from missing data due to measuring instrument failure, routine maintenance of monitors, and human error. Whatever the reasons for incomplete data, it can be a significant obstacle for researchers.

Most statistical software omits records with missing values by default, and analysis is conducted on a subset with the available data. This approach is commonly used for handling missing data, but can lead to loss of statistical power which can be problematic in environmental surveys where associations between environmental factors and health outcome are generally weak. Furthermore, the results of the complete cases analyses are imprecise, given that part of the data is not considered.

An alternative is to use measurements issued from subsamples to build predictive models and then, apply them to the whole population. Among these imputation techniques, single imputation approach is the most common and easily conducted as standard methodology [3], it involves a single estimated value for each missing value. It can be applied directly without loss of power due to the sample size being brought back to its original size. However, single imputation ignores the uncertainty of estimation due to the imputation.

Consequently, in 1987, Rubin proposed multiple imputation [4], where each missing value is imputed by multiple simulated data leading to multiple "completed datasets". Each generated dataset is analyzed by standard methodology and the results combined, enabling the uncertainty attached to missing data to be assessed. Whilst several authors have declared multiple imputation their method of choice [5,6], a recent review has suggested that its use is still quite rare: less than 2% of papers published in epidemiology journals have used multiple imputation and often omitting important details of the methodology used [7].

Another more recent alternative for dealing with missing data is to jointly model through a Bayesian approach the missing process and the association between health outcome and covariates [8,9]. This kind of models is a part of more general approaches referred to hierarchical Bayesian modelling [10] which combines several sub-models. In our case, there are two sub-models where the first one connects missing exposure and predictive factors, and the second one assesses the association between health outcome and covariates including missing exposure. These two sub-models could be implemented separately but the estimated exposure by the first sub-model would be used in the second sub-model of disease as if it had been measured without uncertainty. Through the Bayesian modelling, the sub-models are integrated together allowing to take into account all uncertainty.

The aim of this paper is to examine performances of several approaches for handling large amounts of missing data in environmental epidemiological surveys when the data are missing completely at random (MCAR) using both a case study and a simulation study. Omitting missing values, imputation techniques (single and multiple imputation) and fully Bayesian approach are considered.

Materials and Methods

In order to compare results from methods for handling missing values, a real dataset was used. In the PARIS (Pollution and Asthma Risk: an Infant Study) cohort, measurements were performed and pollutant levels, such as formaldehyde levels, are available in a subset of the population, but missing for infants not involved in the environmental investigation [11].

Data

The cohort enrolled 3 840 full-term healthy babies, recruited between February 2003 and June 2006. The study protocol is described elsewhere [12]. At birth, an interview with the mother was conducted to collect data about the history of allergic conditions in both parents. Gender, parity, anthropometric parameters of the child, maternal history of pregnancy and delivery were also registered from newborn's and mother's medical records. Parents regularly documented health outcomes in mailed questionnaires. The health outcomes of interest were defined by the occurrence of lower respiratory infection (LRI) and a dry night cough (DNC) during the first twelve months of life [11,13].

Concerning environmental and lifestyle data, a trained interviewer interviewed parents by phone during maternity leave to describe in detail home characteristics (construction date, number of occupants, home surface area, heating and cooking appliances, presence of mechanical ventilation and double glazing, wall and floor coverings and signs of dampness) and family living conditions (duration of breastfeeding, information on day-care attendance, keeping of pets, aeration, smoking, use of air fresheners, do-it-yourself (DIY)). Any changes were assessed by mailed questionnaires at each time points.

Aldehyde air sampling measurements were performed in a random sample of 196 babies' dwellings using a passive sampler [14]. Predictive factors of formaldehyde levels were previously identified: sources (presence and age of wall coating, wood-pressed products for flooring or varnished parquet floor, and particle board furniture), parameters of aeration and air stuffiness (length of window opening, presence of mechanical ventilation and double glazing), and home characteristics (construction date, housing area, and number of occupants) [11].

This study was approved by the National Ethics Committee (permissions 031153 and 051289), and parents of participating infants gave their written informed consent. Data are stored in the Paris council, within its Social, Childhood and Health Direction (DASES). All data were anonymized before statistical analysis.

Association between formaldehyde, including missing values, and health outcomes in the PARIS cohort

In the context of the study of formaldehyde exposure impact (variable including missing values) on LRI (a relative common health outcome in infancy) or DNC (a less prevalent event), results from methods for handling missing values were compared. All infants who move during the first year of life and those with no data on health outcome were excluded from the analyses.

Unmeasured formaldehyde values are assumed MCAR [15] since families where measurements were carried out were selected at random, values are missing by design. The methods for dealing with missing data that we considered were: omitting missing data, imputation methodologies, and a fully Bayesian approach. For the first one, as its name implies, only cases with available information are considered, with cases with missing data being discarded. Concerning the imputation approach, missing data are imputed from the available information. Whatever the choice between single or multiple, an imputation model has to be established. Two approaches imputing missing formaldehyde values were examined, the linear regression model (LM), and the partial least squares (PLS) model. PLS regression is particularly suited when there are more predictors variables than observations, and contrary to LM, it allows multicolinearity between variables [16]. PLS method is based on the reduction of predictor variables dimension by using techniques near principal component analysis. As recommended in the literature, the imputation model includes variables that are used in subsequent analyses such as the outcome [17,18]. Therefore, the imputation model included formaldehyde predictive factors and LRI or DNC. The predicted formaldehyde mean was used in the single imputation. Note that other approaches exist for single imputation where, for instance, the missing value is replaced by local or adaptative estimate [19,20,21] and not by the global mean. Whatever the chosen technique, the missing value is always replaced by a single value underestimating the variability due to this estimation. As the health outcomes are binary variables, the association between formaldehyde levels and LRI or DNC was then examined using logistic regression whatever the imputed model approach.

In the multiple imputation approach, several imputations are generated for given missing data. As previously described by Little and Rubin [22], "m completed datasets" were firstly created by filling in the missing values through the imputation model: missing formaldehyde values were imputed randomly from an approximate predictive distribution based on the fitted regression. For example in the case of LM, regression coefficients were sampled from their multivariate Gaussian distribution obtained on observed data and then missing formaldehyde values were replaced by their corresponding predicted values. This procedure was repeated m times. Here 10 000 imputed datasets were fitted. The completed datasets were analyzed separately, the association between formaldehyde levels (observed and imputed) and health was examined using logistic regression, and the results of all datasets were then combined, applying Rubin's rules, to yield final inference on the parameters of interest. The variance for the combined parameter estimates included between and within imputation variation.

Finally, a fully Bayesian model was implemented, as suggested by Carpenter and Kenward [18]. Two sub-models were fitted jointly using Markov Chain Monte Carlo methods [17]. The first one modelled the association between health indicator and exposure, and the second one modelled the relation between missing and observed exposure measurements including covariates supposed to be linked to exposure. Such joint modelling has advantages as mutually enhanced estimates precision of two sub-models parameters, extending multiple imputation methodology. The algorithm was run for 10 000 iterations with 1000 iterations discarded for burn-in. Inspection of posterior time series plots for the parameters as well as autocorrelation plots indicated that the model mixed well. For each model, posterior mean of OR with 95% credibility interval (95% Cr) is shown for the formaldehyde exposure. Note that since the PLS approach is not based on a probability model, Bayesian modelling cannot be used.

Simulation study

Facing missing data, the choice of approach is crucial in terms of the conclusion, particularly when there is a high proportion of missing data (near 94% in our case), and weak associations. Comparisons between approaches have to be based on the quality of estimates, and on the ability to conclude or not a significant association. Simulation studies with characteristics near those of real data but controlling the true value of OR without omitting the case of no association (OR = 1.0) were therefore conducted. Two cases were considered: one frequent outcome similar to LRI (named "event 1"), and a second case close to DNC ("event 2"). Sample sizes are similar to those in real data sets (n = 2 551 for event 1 and n = 2 342 for event 2).

Datasets were simulated from the following steps: $ln\ E \sim N(X \varphi;\ \sigma^2\ Id)$ and $Y_i \sim B(\pi_i)$ with $logit(\pi_i) = \alpha + \beta\ E_i + \gamma\ Z_i$, $\beta = ln\ OR_{true}$ and $i = 1, ..., n$, and where formaldehyde factors are denoted by X, exposure variable by E, covariates by Z and health indicator by Y. Exposure variable (corresponding to formaldehyde in real dataset) was simulated on a logarithmic scale from a linear model with formaldehyde predictors obtained from real data [11], and coefficients φ were equal to those estimated in this study. Then, the health indicator was simulated from a logistic model with the resulting formaldehyde levels and covariates from real data (coefficients γ associated with covariates were those estimated on real data as well as the residual variance σ^2). Formaldehyde predictors and covariates are given in Table S1.

Three different OR_{true} (and then three $\beta_{true} = ln\ OR_{true}$) between formaldehyde and event 1 or event 2 were considered: 1.0, 1.2, and 1.4. Missing values for formaldehyde were assigned at random. A case of 95% missing values was considered. To assess the robustness of our conclusions, simulations with different missing values percentages (85%, 75% and 50%) were also conducted. For each scenario, a total of 100 datasets were generated. This number of simulations is required to obtain an estimate of the regression coefficient associated with formaldehyde exposure within 10% of its true value when missing values are omitted in the real data. Indeed, the equation given by Burton et al. [23] for the number of simulations B is $B = \left(\frac{Z_{1-\frac{\alpha}{2}}}{\delta}\right)^2$ here δ is the specified level of accuracy of the estimate of interest "accepted", σ, the standard error for the parameter of interest and, $Z_{1-\alpha/2}$, the $(1-\alpha/2)$ quantile of the standard normal distribution. When $B = 100$, $\alpha = 0.05$ and $\sigma^2 = 0.26$ (estimated variance on real data set), δ is equal to 10%.

The quality of estimates for the different approaches was assessed by the root mean square error of beta coefficients (RMSE,

$RMSE = \sqrt{\frac{1}{m}\sum_{i=1}^{m}(\widehat{\beta}_i - \beta_{true})^2}$ where $\widehat{\beta}_i$ is estimated β on i^{th} dataset

$(i=1,...,m)$ and $CI_{95\%}(RMSE) = \left[\sqrt{MSE_{inf}};\ \sqrt{MSE_{sup}}\right]$ where $MSE_{inf} = MSE - 1.96\frac{\widehat{sd(MSE)}}{\sqrt{m}}$ and $MSE_{sup} = MSE + 1.96 \frac{\widehat{sd(MSE)}}{\sqrt{m}}$ and $MSE = RMSE^2$). The proportion of "significant" associations (PS, $PS = \frac{1}{m}\sum_{i=1}^{m} 1_{1\notin CI(\widehat{OR}_i)}$ where $1_{1\notin CI(\widehat{OR}_i)} = 1$, if $1 \notin CI(\widehat{OR}_i) = 1$ and $1_{1\notin CI(\widehat{OR}_i)} = 0$, otherwise) i.e. of confidence or credibility interval of OR excluding 1 was also calculated. This criterion assesses risk of type I error when $OR_{true} = 1$, and statistical power when OR_{true} was different from 1.

Confidence intervals for RMSE were based on a Gaussian approximation using the empirical standard deviation of the MSE, and confidence intervals for the PS were based on the Clopper-Pearson "exact" confidence intervals [24] avoiding the Gaussian approximation. As the question is basically if formaldehyde exposure increased LRI and DNC risk according to previous studies in literature, the one-sided approach was chosen. Statistical analyses were conducted using R 2.14.0 [25] and WinBUGS software [26].

Results

Results on real data

In this study, 2 551 infants of the PARIS birth cohort were completely observed, independently of health outcomes and formaldehyde levels, and pollutant levels were available for 142 of them. Most of infants lived in apartments. Nearly 30% of buildings were built after 1975, and two-thirds were equipped with double glazing. Around half of babies had wood-pressed products for flooring or varnished parquet floor in their bedroom. Recent (less than one year old) particle board furniture was present in 49.8% dwellings. About half of infants (46.9%) had at least one LRI during their first year of life and the prevalence of DNC was 14.9%.

OR estimates between LRI or DNC occurrence and formaldehyde exposure levels are given in Table 1. The single imputation techniques (LM or PLS) clearly induced high estimates of association compared to all other approaches. The estimated OR with fully Bayesian approach was similar to that obtained with multiple imputation particularly for LRI, but lower than that with PLS imputation. However, intervals were different leading to a significant association with Bayesian modelling for LRI and nearly significant for DNC but not significant with both multiple imputation.

Simulations results

From the 100 simulated datasets, the resulting mean prevalence of event 1 was 29.6% (range: 18.4%–41.3%) and of event 2 was 12.3% (range: 6.8%–18.5%). Table 2 shows RMSE and PS assessed on simulated data when no data are missing. Results were quasi similar for all approaches. As RMSE depends on OR_{true} and on prevalence of event, it increased between events 1 and 2 and slightly with OR_{true}. Concerning event 1, frequentist and Bayesian approaches always concluded a significant association when $OR_{true} = 1.4$ while when $OR_{true} = 1.2$, statistic power was near 60%. For an infrequent event as event 2, statistical power decreased being near 85% when $OR_{true} = 1.4$ and near 40% when $OR_{true} = 1.2$. When $OR_{true} = 1.0$, PS had to be equal to 5%, which was the case for the two approaches even if there was a slight increase for the less frequent event. These results can be considered as reference results as no data are missing.

Table 1. Associations between environmental risk factor[a] including missing values, and health outcomes, by different methods handling missing values (OR [95% CI or 95% Cr]).

		LRI[b]	DNC[c]
Na omitted		1.11 [0.55, +∞)	1.31 [0.45, +∞)
Single imputation	LM	1.91 [1.53, +∞)	5.63 [3.69, +∞)
	PLS[d]	3.27 [1.61, +∞)	3.69 [2.57, +∞)
Multiple imputation	LM	1.28 [0.91, +∞)	1.35 [0.14, +∞)
	PLS[d]	2.81 [0.35, +∞)	2.69 [0.39, +∞)
Bayesian approach		1.27 [1.10, +∞)	1.16 [0.95, +∞)

Abbreviations: CI, 95% confidence interval; Cr, 95% credibility interval, LM, linear regression model; OR, odds ratio; PLS, partial least squares.
[a]: Environmental factor: formaldehyde exposure, expressed in $\mu g/m^3$, (for one unit increase in the logarithmic scale), and health outcome: lower respiratory infection (LRI) or dry nigh cough (DNC).
[b]: Association was adjusted for gender, socio-economic status, siblings, parental history of asthma, breastfeeding, daycare attendance, pre/postnatal tobacco smoke exposure, sign(s) of dampness, and presence of pets at home.
[c]: Association was adjusted for gender, socio-economic status, parental history of allergy, breastfeeding, pre/postnatal tobacco smoke exposure, gas heating, cockroaches, infant's mattress age, family events, and number of episodes of lower respiratory infections.
[d]: PLS imputation with two components.

Table 3 provides results of the RMSE on replicates with 95% of missing values. RMSE values ranged from 0.18 to 0.83 for event 1, and from 0.09 to 1.06 for event 2. For each event, RMSE slightly increased with OR. RMSE values were always lower in multiple imputation than in single imputation, whatever the imputation model (LM or PLS). Single imputation led to huge RMSE reflecting poor qualities of estimates. All results were confirmed with proportion of missing values of 75% and 85% (Table S2).

Figure S1 shows boxplots of beta estimates obtained from Bayesian approach, from 100 simulations for the three different values of OR_{true}. When OR_{true} is equal to 1.0, the range of estimates increased with the proportion of missing values and with the decrease of event prevalence. These results were also observed when $OR_{true} = 1.2$ and $OR_{true} = 1.4$. Moreover, an underestimation of beta (and so of OR) was obtained when $OR_{true} = 1.4$.

When OR_{true} is equal to 1, real risk of type I error is assessed by PS while the theoretical one was fixed at 5% (Table S3). For event 1 with 95% of missing data, single imputation led to very high risks (23% [15.2, 32.5] and 21% [13.5, 30.3] for LM and PLS, respectively). If missing proportion is 75% or 85%, huge risks of type I error were again found for single imputation. For event 2, PS considerably increased for single imputation, 35% [25.7, 45.2] and 42% [32.2, 52.3] for LM and PLS, respectively. This increase was confirmed with 75% and 85% of missing values. Multiple imputation led to always conservative results explained by large confidence intervals of OR obtained with this approach whether for event 1 or 2. Bayesian approach led to increase risk of type I error. For event 1, this increase seems reasonable because 5% is always in the confidence interval (7% [2.9, 13.9] with 95% missing values). For the infrequent event 2, risk of type I error increases and excluding 5% from confidence intervals when 85% and 95% of data are missing (e.g. 16% [9.4, 24.7] with 95% missing values).

Figure 1 presents PS when OR_{true} is greater than 1 and 95% of missing values for two events and for all approaches, excluding single imputation which had given a weak quality on estimates and overestimated risk of type I error. As expected, statistical power increased with OR_{true} and decreased for infrequent event. Weak performances were obtained for multiple linear and PLS imputation. This figure clearly shows highest PS for Bayesian model near reference power on complete data (Table 1) especially for event 1, all other approaches giving a null or quasi null power for $OR_{true} = 1.2$ and $OR_{true} = 1.4$, respectively. Figure S2 presents

PS when OR_{true} is greater than 1 for 85% and 75% of missing values. Highest PS were clearly obtained for Bayesian model.

And with 50% of missing data, statistical power increased remaining the best for Bayesian approach especially for infrequent event (near 80% by Bayesian approach against 53% by LM multiple imputation when $OR_{true} = 1.4$).

Discussion

This paper addresses the crucial question of how to handle large amounts of missing data in environmental epidemiological surveys. Till now, as far as we know, very few teams have compared performances of approaches handling missing values with a proportion of missingness above 75% [27,28]. To solve this question, we used both real and simulated data to determine the most appropriate approach when there is a large amount of missing data. Results on RMSE and PS showed poor performances with single imputation. Fully Bayesian approach seems better, followed by imputation approaches, which in turn gave better results than omitting missing observations.

As expected, we found that omitting missing values is less efficient than single and multiple imputations. Even if it is the easiest approach for handling missing values, it should be used only in presence of less than 5% of missing values [29] because it should induce a significant loss of statistical power: unrealistic when the health outcome is infrequent and problematic in environmental studies where pollutants have often a weak impact.

Another common approach for handling missing values is to impute them before any analysis which is commonly used in environmental epidemiological studies to estimate exposure levels for all study members. An increase in power is the substantial benefit of this alternative over omitting missing values. Nevertheless, the specification of the imputation model is an important step. As previously demonstrated, if the imputation model is not properly specified the imputation approach could introduce a bias which is not present in omitting missing values when missingness is MCAR [28]. It has been observed that including many relevant variables in the imputation model tends to make the missing at random assumption more plausible [30], even if computational problems could occur such as multicollinearity and a large number of predictors might provide instable estimates. Previous authors showed the importance of including the health outcome because

Table 2. Root mean square error, and proportion of "significant" association with 95% confidence interval or credibility interval, on 100 replicates with no missing values.

		OR = 1.0		OR = 1.2		OR = 1.4	
		event 1	event 2	event 1	event 2	event 1	event 2
RMSE	Frequentist	0.10 [0.09, 0.11]	0.17 [0.15, 0.19]	0.10 [0.09, 0.11]	0.12 [0.11, 0.13]	0.09 [0.08, 0.10]	0.12 [0.11, 0.13]
	Bayesian	0.10 [0.10, 0.11]	0.17 [0.15, 0.19]	0.10 [0.08, 0.11]	0.12 [0.10, 0.14]	0.09 [0.08, 0.11]	0.12 [0.11, 0.14]
PS	Frequentist	5 [1.6, 11.3]	8 [3.5, 15.2]	59 [48.7, 68.7]	40 [30.3, 50.2]	100	83 [74.2, 89.8]
	Bayesian	4 [1.1, 9.9]	7 [2.9, 13.9]	61 [50.7, 70.6]	39 [29.4, 49.3]	100	86 [77.6, 92.1]

Abbreviations: RMSE, root mean square error; OR, odds ratio; PS, proportion of "significant" association.
Sample size for each simulated dataset: event 1 N= 2 551/event 2 N= 2 342.

Table 3. Root mean square error of beta coefficients with 95% confidence interval based on 100 replicates with 95% of missing values.

		OR = 1.0		OR = 1.2		OR = 1.4	
		event 1	event 2	event 1	event 2	event 1	event 2
Na omitted		0.33 [0.29, 0.36]	0.11 [0.10, 0.12]	0.22 [0.20, 0.24]	0.09 [0.08, 0.10]	0.18 [0.17, 0.20]	0.17 [0.15, 0.18]
Single imputation	LM	0.46 [0.42, 0.50]	1.06 [0.95, 1.16]	0.57 [0.52, 0.61]	0.96 [0.89, 1.03]	0.70 [0.66, 0.74]	1.02 [0.95, 1.09]
	PLS	0.58 [0.47, 0.67]	0.89 [0.75, 1.01]	0.63 [0.52, 0.73]	0.81 [0.70, 0.91]	0.83 [0.71, 0.94]	0.99 [0.82, 1.13]
Multiple imputation	LM	0.30 [0.27, 0.32]	0.33 [0.30, 0.36]	0.29 [0.26, 0.32]	0.29 [0.26, 0.32]	0.30 [0.27, 0.33]	0.29 [0.26, 0.31]
	PLS	0.48 [0.40, 0.56]	0.75 [0.62, 0.85]	0.49 [0.41, 0.56]	0.66 [0.57, 0.74]	0.65 [0.57, 0.73]	0.76 [0.65, 0.86]
Bayesian approach		0.18 [0.16, 0.20]	0.26 [0.23, 0.29]	0.18 [0.16, 0.20]	0.24 [0.20, 0.27]	0.19 [0.17, 0.22]	0.24 [0.21, 0.26]

Abbreviations: LM, linear model; OR, odds ratio; PLS, partial least squares.
Sample size for each simulated dataset: event 1 N= 2 551/event 2 N= 2 342.

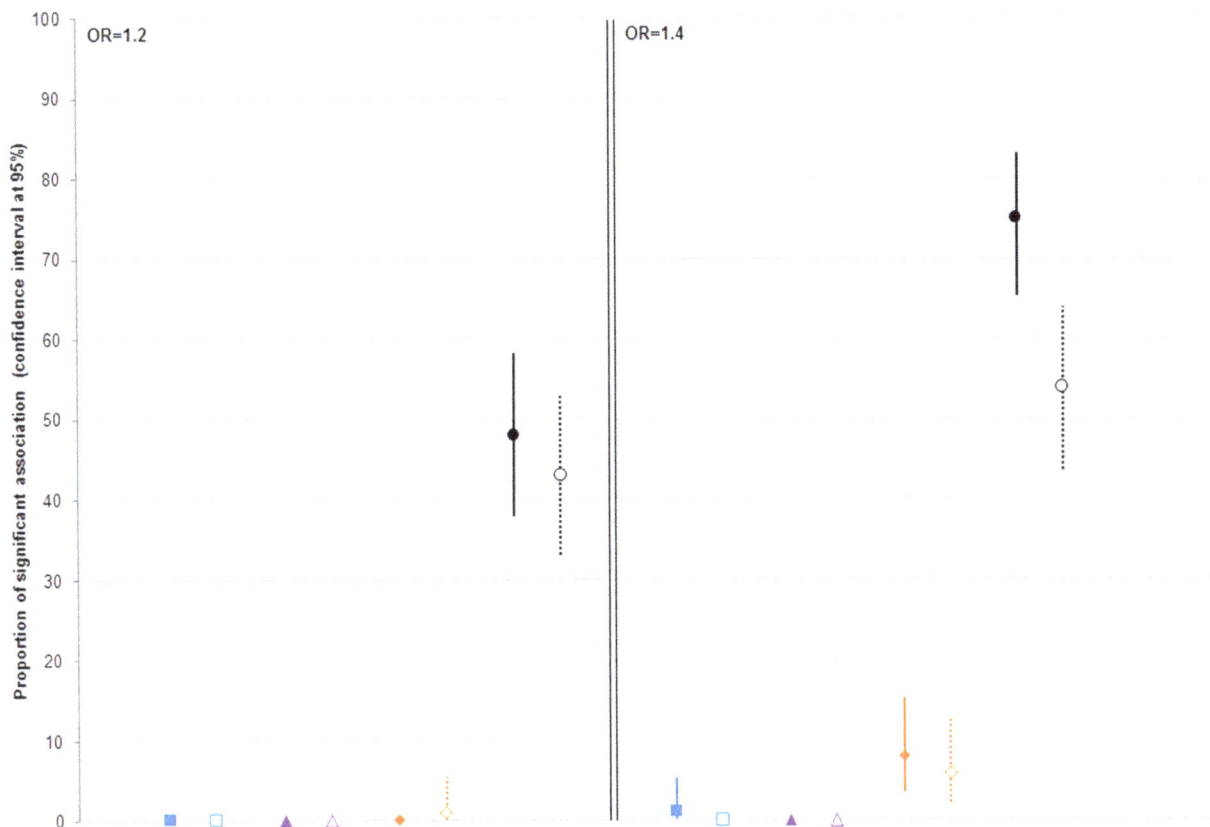

Figure 1. Proportions of significant associations based on 100 replicates, for each approach dealing with 95% of missing values

Solid line: event 1 / dotted line: event 2

Figure 1. Proportions of significant associations based on 100 replicates, for each approach dealing with 95% missing values.

regression coefficients came close to the truth [31,32]. Note that inclusion of the health outcome implies that the imputation procedure has to be renewed for each new health outcome.

Single imputation is easy to implement, but the major disadvantage is the overestimation of association between exposure and health outcome increasing with the strength of the association. As explained by Rubin [4], this overstatement is certainly due to that one imputed value cannot itself represent uncertainty about imputed value.

Conversely, multiple imputation takes into account the uncertainty and thus, does not underestimate the variance of estimates. But this approach is conservative. In fact, as previously described in the literature [31], large imprecision of OR estimates was observed, thus yielding to no significant associations. Moreover, concerning the choice of the number of imputations, it has been suggested [4], that less than 10 imputed datasets are useful compared with an infinite number of imputations. However, when the percentage of missing values is huge, more than 10 imputations may be needed and the number of imputed data should approximate the number of observations with missing data, as previously suggested [33]. Although 2 000 imputations could have been sufficient in this study, we fitted 10 000 imputed datasets as a precaution even if it was time consuming. Bayesian joint modelling appears to be less conservative with a statistical power near 75% for event 1 (near 55% for the infrequent event 2) when

$OR_{true} = 1.4$ with 95% of missing data. Nevertheless, it is important to notice that risk of type I error tends to increase, slightly for event 1 (near 7%) and much more for the infrequent event 2 (near 16%). Even if the RMSE is always smaller for the Bayesian approach compared to the other approaches and indicates a better performance in terms of estimates, boxplots of the beta estimates from 100 simulations under $OR_{true} = 1.0$ clearly show that the range of estimates increases with decrease of event prevalence and with increase in the missing values proportion. This result is confirmed when OR is greater than one but a bias appears when OR_{true} increases. Thus, caution should be taken when interpreting results for an infrequent event with high proportion of missing values. In addition, the use of empirical approaches (e.g. bootstrap, Monte Carlo study) could be useful to assess the real risk of type I error.

For infrequent events (for instance with prevalence less than 10%) and high proportion of missing values, statistical power remains too weak. Bayesian approach offers the possibility to obtain easily estimated posterior distribution of OR which could be a useful tool to refine conclusions. Posterior probability of OR being smaller than 1 can indeed be deduced and such probability between 5% and 10% could be considered weak enough to conclude an "almost significant" association. This strategy would lead to three possible conclusions: "not significant", "significant" or "almost significant". Thus, for event 2, the classical approach yields 46 non significant associations when $OR_{true} = 1.4$, but among them, 7 would be declared as "almost significant". It is noteworthy that the proportion of non significant associations now labeled "almost significant" increases with OR_{true} from 8.8% ($OR_{true} = 1.2$) to 15.2% ($OR_{true} = 1.4$). If the type I error is assessed only among associations not declared "almost significant", it remains stable (16% and 17.8% for classical and this new strategy, respectively).

In conclusion, among the methods dealing with missing data, no approach is absolutely better than the others in all circumstances. In the presence of high proportions of missing values, using only complete data yields to a significant loss of statistical power. Single imputation underestimates the variance, thus overestimating the association between environmental factor and health outcome. Multiple imputation, due to overcoverage, is too conservative and unable to show significant associations. When the health outcome is frequent, joint modelling seems to be more efficient than other approaches, combining low RMSE, limited increase of risk of type I error, and high statistical power. The simulation study is useful for explaining the disparity of associations found in the real data, for example for LRI (Table 1) corresponding to a frequent event. Indeed, the characteristics of each method highlighted by the simulation study are found in the real case, i.e. bias using simple imputation, lack of power using multiple imputation, and significant association using Bayesian approach. The conclusion of a significant association between formaldehyde exposure and LRI is strengthened. With regards to the infrequent event, DNC, only a tendency of an association is observed. No approach gives completely satisfactory results when the health outcome is infrequent and the proportion of missing values is high. Though the Bayesian modelling has the best power and precision of estimates, this comes at a cost of inflated risk of type I error. However, estimated posterior distribution of OR would be helpful to refine the conclusion by introducing a new category of "almost significant association" when probability of OR less than 1 is between 5% and 10%. Concerning inflation of risk of type I error, correcting methodology as bootstrap could be implemented. This would lead certainly to very huge computer time as MCMC iterative algorithm would be used on each bootstrapped sample. An alternative approach to MCMC in repeated Bayesian estimations such as "Importance Sampling" could be envisaged [34].

Supporting Information

Figure S1 Boxplots of β ($\beta = ln\ OR$) estimates under Bayesian approach from 100 simulated datasets for the three different values of true OR dealing with no missing values, 75%, 85% and 95% of missing values.

Figure S2 Proportions of significant associations based on 100 replicates, for each approach dealing with 85%, and 75% of missing values.

Table S1 Predictive factors of formaldehyde exposure and covariates used for adjustment in the model relating formaldehyde exposure and health indicator.

Table S2 Root mean square error of beta coefficients with 95% confidence interval based on 100 replicates with 85%, and 75% of missing values.

Table S3 Proportions of significant associations with 95% confidence interval based on 100 replicates with 95%, 85% and 75% of missing values when $OR_{true} = 1$.

Acknowledgments

The authors wish to thank the families for their participation and the administrative staff for their involvement in the PARIS study. The authors would like to thank the editor and the referees for helpful suggestions that lead to significant improvements.

Author Contributions

Conceived and designed the experiments: CR IM CG. Analyzed the data: CR IN CG. Contributed reagents/materials/analysis tools: CR IN IM CG. Wrote the paper: CR IN IM CG.

References

1. Vach W, Blettner M (2005) Missing data in epidemiologic studies. Encyclopedia of Biostatistics: John Wiley & Sons.
2. Jia C, D'Souza J, Batterman S (2008) Distributions of personal VOC exposures: a population-based analysis. Environ Int 34: 922–931.
3. Fusco D, Forastiere F, Michelozzi P, Spadea T, Ostro B, et al. (2001) Air pollution and hospital admissions for respiratory conditions in Rome, Italy. Eur Respir J 17: 1143–1150.
4. Rubin DB (2004) Multiple imputation for nonresponse in surveys: Wiley.
5. Barzi F, Woodward M (2004) Imputations of missing values in practice: results from imputations of serum cholesterol in 28 cohort studies. Am J Epidemiol 160: 34–45.
6. Schafer J (1997) Analysis of incomplete multivariate data. Monographs on statistics and applied probability.
7. Klebanoff MA, Cole SR (2008) Use of multiple imputation in the epidemiologic literature. Am J Epidemiol 168: 355–357.
8. Carrigan G, Barnett AG, Dobson AJ, Mishra G (2007) Compensating for Missing Data from Longitudinal Studies Using WinBUGS. J Stat Softw 19: 1–17.
9. Gryparis A, Paciorek CJ, Zeka A, Schwartz J, Coull BA (2009) Measurement error caused by spatial misalignment in environmental epidemiology. Biostatistics 10: 258–274.
10. Gelman A, Carlin JB, Stern HS, Rubin DB (2003) Bayesian Data Analysis. 2nd Edition. CRC Press.
11. Roda C, Kousignian I, Guihenneuc-Jouyaux C, Dassonville C, Nicolis I, et al. (2011) Formaldehyde Exposure and Lower Respiratory Infections in Infants: Findings from the PARIS Cohort Study. Environ Health Perspect 119: 1653–1658.

12. Clarisse B, Nikasinovic L, Poinsard R, Just J, Momas I (2007) The Paris prospective birth cohort study: which design and who participates? Eur J Epidemiol 22: 203–210.
13. Roda C, Guihenneuc-Jouyaux C, Momas I (2013) Environmental triggers of nocturnal dry cough in infancy: New insights about chronic domestic exposure to formaldehyde in the PARIS birth cohort. Environ Res 123: 46–51.
14. Dassonville C, Demattei C, Laurent AM, Le Moullec Y, Seta N, et al. (2009) Assessment and predictor determination of indoor aldehyde levels in Paris newborn babies' homes. Indoor Air 19: 314–323.
15. Rubin DB (1976) Inference and missing data. Biometrika 63: 581.
16. Tenenhaus M (1998) [La régression PLS: théorie et pratique]: Technip.
17. Gilks WR, Richardson S, Spiegelhalter DJ (1996) Markov Chain Monte Carlo in Practice. London, UK: Chapman & Hall.
18. Carpenter J, Kenward M (2005) Example Analyses Using WinBUGS 1.4. Available: http://www.missingdata.org.uk/. Accessed 2011 Feb 2.
19. Bo TH, Dysvik B, Jonassen I (2004) LSimpute: accurate estimation of missing values in microarray data with least squares methods. Nucleic Acids Res 32: e34.
20. Brock GN, Shaffer JR, Blakesley RE, Lotz MJ, Tseng GC (2008) Which missing value imputation method to use in expression profiles: a comparative study and two selection schemes. BMC Bioinformatics 9: 12.
21. Kim H, Golub GH, Park H (2005) Missing value estimation for DNA microarray gene expression data: local least squares imputation. Bioinformatics 21: 187–198.
22. Little RJA, Rubin DB (2002) Statistical analysis with missing data, 2nd edition. New York: John Wiley.
23. Burton A, Altman DG, Royston P, Holder RL (2006) The design of simulation studies in medical statistics. Stat Med 25: 4279–4292.

24. Clopper C, Pearson E (1934) The use of confidence or fiducial limits illustrated in the case of the binomial. Biometrika 26: 404–413.
25. R Development Core Team (2011) R: A language and environment for statistical computing. In: R Foundation for Statistical Computing, editor. Vienna, Austria.
26. Lunn DJ, Thomas A, Best N, Spiegelhalter D (2000) WinBUGS-A Bayesian modelling framework: Concepts, structure, and extensibility. Stat Comput 10: 325–337.
27. Scheffer J (2002) Dealing with Missing Data. Res Lett Inf Math Sci: 153–160.
28. Lee KJ, Carlin JB (2012) Recovery of information from multiple imputation: a simulation study. Emerg Themes Epidemiol 9: 3.
29. Rubin DB (1987) Multiple imputation for nonresponse in surveys. New York, NY: John Wiley and Sons, Inc.
30. Steyerberg EW (2009) Clinical prediction models: A practical approach to development, validation, and updating: Springer.
31. Moons KG, Donders RA, Stijnen T, Harrell FE Jr (2006) Using the outcome for imputation of missing predictor values was preferred. J Clin Epidemiol 59: 1092–1101.
32. Sterne JA, White IR, Carlin JB, Spratt M, Royston P, et al. (2009) Multiple imputation for missing data in epidemiological and clinical research: potential and pitfalls. BMJ 338: b2393.
33. Bodner TE (2008) What improves with increased missing data imputations? Struct Equ Modeling-a Multidisciplinary Journal 15: 651–675.
34. Gajda D, Guihenneuc-Jouyaux C, Rousseau J, Mengersen K, Nur D (2010) Use in practice of importance sampling for repeated MCMC for Poisson models. Electron J Stat 4: 361–383.

6

Inhibitory Action of Benzo[α]pyrene on Hepatic Lipoprotein Receptors *In Vitro* and on Liver Lipid Homeostasis in Mice

Hamed Layeghkhavidaki[1,2], **Marie-Claire Lanhers**[1,2], **Samina Akbar**[1,2], **Lynn Gregory-Pauron**[1,2], **Thierry Oster**[1,2], **Nathalie Grova**[3], **Brice Appenzeller**[3], **Jordane Jasniewski**[4], **Cyril Feidt**[1,2], **Catherine Corbier**[1,2], **Frances T. Yen**[1,2,5]*

1 Unité de Recherche Animal et Fonctionnalités des Produits Animaux EA3998, Université de Lorraine, Vandœuvre-lès-Nancy, France, 2 Institut National de Recherche Agronomique USC 0340, Vandœuvre-lès-Nancy, France, 3 Laboratory of Analytical Human Biomonitoring, Centre de Recherche Public de la Santé, Luxembourg, Luxembourg, 4 Laboratoire d'Ingénérie des Biomolécules, Université de Lorraine, Vandœuvre-lès-Nancy, France, 5 Institut National de la Santé et de la Recherche Médicale, Vandœuvre-lès-Nancy, France

Abstract

Background: Dyslipidemia associated with obesity often manifests as increased plasma LDL and triglyceride-rich lipoprotein levels suggesting changes in hepatic lipoprotein receptor status. Persistent organic pollutants have been recently postulated to contribute to the obesity etiology by increasing adipogenesis, but little information is available on their potential effect on hepatic lipoprotein metabolism.

Objective: The objective of this study was to investigate the effect of the common environmental pollutant, benzo[α]pyrene (B[α]P) on two lipoprotein receptors, the LDL-receptor and the lipolysis-stimulated lipoprotein receptor (LSR) as well as the ATP-binding cassette transporter A1 (ABCA1) using cell and animal models.

Results: LSR, LDL-receptor as well as ABCA1 protein levels were significantly decreased by 26–48% in Hepa1-6 cells incubated (<2 h) in the presence of B[α]P (≤1 μM). Real-time PCR analysis and lactacystin studies revealed that this effect was due primarily to increased proteasome, and not lysosomal-mediated degradation rather than decreased transcription. Furthermore, ligand blots revealed that lipoproteins exposed to 1 or 5 μM B[α]P displayed markedly decreased (42–86%) binding to LSR or LDL-receptor. B[α]P-treated (0.5 mg/kg/48 h, *i.p.* 15 days) C57BL/6J mice displayed higher weight gain, associated with significant increases in plasma cholesterol, triglycerides, and liver cholesterol content, and decreased hepatic LDL-receptor and ABCA1 levels. Furthermore, correlational analysis revealed that B[α]P abolished the positive association observed in control mice between the LSR and LDL-receptor. Interestingly, levels of other proteins involved in liver cholesterol metabolism, ATP-binding cassette transporter G1 and scavenger receptor-BI, were decreased, while those of acyl-CoA:cholesterol acyltransferase 1 and 2 were increased in B[α]P-treated mice.

Conclusions: B[α]P demonstrates inhibitory action on LSR and LDL-R, as well as ABCA1, which we propose leads to modified lipid status in B[α]P-treated mice, thus providing new insight into mechanisms underlying the involvement of pollutants in the disruption of lipid homeostasis, potentially contributing to dyslipidemia associated with obesity.

Editor: Alberico Catapano, University of Milan, Italy

Funding: These studies were supported by grants from the French Ministry of Higher Education and Research, and from the University of Lorraine SF (Structure Fédérative) EFABA. HL and SA were supported by thesis fellowships from the Avicenna Research Institute, Iran, and Pakistan Ministry of Research, respectively. The authors hereby state that the funders had no role in study design, data collection and analysis, decision to publish, or preparation of the manuscript.

Competing Interests: The authors have declared that no competing interests exist.

* Email: frances.yen-potin@univ-lorraine.fr

Introduction

Obesity has become an increasing problem of public health in the industrialized world [1,2], and represents a significant risk factor for many pathologies including cardiovascular disease and diabetes. Its origin is complex and multifactorial, where numerous genetic and environmental factors have been shown to contribute to the etiology of obesity. While the adipocyte represents the site of fat accumulation in obesity, the role of the liver as a central station for the reception and delivery of lipids in the form of lipoproteins to the different peripheral tissues should also be considered in investigations of the underlying causes of disrupted lipid status. Indeed, obesity is often associated with dyslipidemia, demonstrated by changes in plasma lipoproteins, including increased levels of triglyceride (TG)-rich lipoproteins (TGRL), and low-density lipoprotein (LDL), and sometimes decreased levels of high density lipoprotein (HDL) [3]. Increased TGRL in obesity has been attributed to overproduction of VLDL [4], but elevated postprandial lipemia in obese subjects suggests decreased ability to remove intestinally-derived chylomicrons. The removal of lipoproteins from the circulation is mediated by hepatic lipoprotein receptors, including the LDL-receptor (LDL-R), well-known for its key role in the regulation of cholesterol metabolism [5], but which also participates in the removal of TGRL. The lipolysis stimulated lipoprotein receptor (LSR) has recently been shown by our laboratory and others to play an important role in the removal of TGRL during the postprandial phase [6–8]. Interestingly, if these receptors are deficient ($LSR^{+/-}$) or lacking ($LDL-R^{-/-}$) in mice, the animals exhibit not only changes in plasma lipid levels, but also significantly higher weight gain when placed under high-fat diets as compared to wild-type littermates [9,10]. $LSR^{+/-}$ mice shown to have 50% reduced LSR expression and to display elevated postprandial lipemia and reduced lipid clearance also develop obesity with age, even if maintained on a standard diet [10], therefore suggesting that changes in lipoprotein receptor status leading to dyslipidemia may also increase disposition towards excess weight gain, and thus increased propensity towards developing obesity.

Recent attention has focused on persistent organic pollutants (POP) as potential environmental factors involved in the etiology of obesity, due in part to epidemiological studies demonstrating a correlation between exposure to pollutants and increased risk or incidence of obesity [11,12] as well as studies demonstrating their role as endocrine disruptors leading to increased adipogenesis [13,14]. Among them includes the polycyclic aromatic hydrocarbon (PAH) benzo[α]pyrene (B[α]P), which is a common pollutant that is produced during incomplete combustion of organic material including wood, coal, diesel, tobacco [15-17], and is also present in foods due to cooking processes including frying, smoking and grilling [18]. Levels of B[α]P in the circulation have been shown to be correlated to body mass index (BMI) in a population study [19]. We have shown that B[α]P can exert inhibitory effects on epinephrine-induced lipolysis in the adipocyte, and reported that mice exposed to this pollutant exhibit higher weight gain as compared to control mice [20]. Also, the carcinogenic properties of this environmental pollutant has already been demonstrated where reactive dihydrodiol epoxide derivatives of B[α]P metabolites derived from the cytochrome P450 system bind covalently to DNA, leading to adduct formation and subsequent tumors [21–23]. Further, B[α]P serves as ligand with the aryl hydrocarbon receptor (AHR) [24,25], shown to be rate-limiting step in B[α]P-induced tumor formation [26].

While the carcinogenic aspect of this pollutant has been well-documented, little is known on its potential effect on lipid and lipoprotein metabolism. This pollutant circulates in the blood associated with lipoproteins [27,28], can penetrate cells *via* biological membranes [29,30] and subsequently accumulate not only in the lipophilic adipose tissue and mammary fat, but also in other tissues including the liver and kidney [31]. This led us to question if B[α]P could disrupt lipid homeostasis *via* an effect on lipoprotein metabolism in the liver. Also, although B[α]P may be involved in regulating expression of multidrug resistance transporters [32], no information is available on its potential effect on ATP-binding cassette (ABC) proteins including ABCA1 which is involved in cholesterol efflux from the liver. In this study, we investigate and show evidence for the effect of B[α]P on hepatic lipoprotein receptors LDL-R and LSR, and on ABCA1, using both cell and animal models.

Materials and Methods

Materials, antibodies and standards

Chemicals and reagents were purchased from Sigma-Aldrich (St Quentin Fallavier, France) unless otherwise indicated. Cell culture media and supplements were obtained from Invitrogen (Alfortville, France). Rabbit anti-LDL-R and anti-LSR antibodies were obtained from Abcam (Paris, France) and Sigma-Aldrich, respectively. Mouse anti-ABCA1 antibodies were purchased from Millipore (Darmstadt, Germany), and goat anti-ACAT1 and anti-ACAT2 antibodies from Santa Cruz Biotechnology (Heidelberg, Germany), respectively. Antibodies recognizing SR-BI or ABCG1 were obtained from Novus Biologicals (Cambridge, United Kingdom). Secondary anti-rabbit and anti-mouse peroxide-conjugated IgG were acquired from Cell Signaling Technology (Boston, MA). Anti-apolipoprotein (apo)B and anti-apoE antibodies, as well as rabbit peroxidase-conjugated anti-goat IgG were obtained from Santa Cruz Biotechnology. B[α]P-d_{12}, 1-OH-benz[α]anthracene-$^{13}C_6$ and all B[α]P metabolites (including 1-, 2-, 3-, 4-, 5-, 6-, 7-, 8-, 9-, 10-, 11-, 12-OH-B[α]P, 4,5-diol- B[α]P cis, 7,8-diol-B[α]P cis, 7,8-diol-B[α]P trans, 9,10-diol-B[α]P trans) investigated in this study were purchased in powder form from MRI Global (Kansas City, MO, USA). B[α]P was purchased in powder form from Sigma-Aldrich. B[α]P standard, internal standards and OH-B[α]P standard stock solutions were prepared in acetonitrile (10 mg/L). Working solutions were prepared in acetonitrile by successive ten-fold dilutions at concentration ranging from 1000 μg/L to 10 μg/L and were stored at $-20°C$.

Cell culture studies

The mouse Hepa1–6 liver cell line (DSMZ, Brunswick, Germany) was maintained in Dulbecco's modified Eagle Medium containing 10% fetal bovine serum and 1 mM glutamine [10]. Cells were seeded in 12-well or 24-well plates and used after 48 h at 80–90% confluence. On the day of the experiment, cells were washed in phosphate-buffered saline (PBS), and then incubated at 37°C in 95% air, 5% CO_2 environment for the indicated times with 0.1 or 1 μM B[α]P in 1% (v/v) DMSO; control cells were treated with vehicle (1% (v/v) DMSO). The proteasome inhibitor lactacystin was prepared in DMSO and incubated with cells, with 1% (v/v) DMSO used as control. In experiments using chloroquine, this lysosomal inhibitor (25 μM) was first pre-incubated with cells 2 h at 37°C, before a second incubation of 1 h with 0.1 μM B[α]P or 1% (v/v) DMSO. At the end of the incubation periods, cells were then placed on ice and washed 2 times with ice-cold PBS. Cell lysates were recovered using ice-cold radioimmunoprecipitation assay buffer containing anti-proteases [10], followed by centrifugation at 13,000 x g for 30 min at 4°C. Protein was measured in the recovered supernatants using BCA assay (Thermo Scientific, Courtaboeuf, France).

Cell viability and metabolic activity assays. In preliminary studies, viability of Hepa1-6 cells was determined using the MTT [3-(4,5-dimethylthiazol-2-yl)2,5-diphenyl tetrazolium bromide] assay [33]. Briefly, Hepa1-6 cells were incubated with filtered (0.22 μM) MTT (67 μg/mL, final concentration) for 20 min at 37°C, followed by incubation at room temperature with DMSO (50% v/v, final concentration) with gentle agitation for 10 min to dissolve the formazan crystals; absorbance was measured at 570 nm. A second method was used to assess cell viability and metabolic activity [34]. Hepa1-6 cells were incubated for 30 min at room temperature with 2 μM calcein-AM. After washing with PBS, cells were incubated for 15 min at room temperature with 1% (v/v) Triton X-100 with gentle agitation.

Fluorescence emission intensity was measured (λemission = 530 nm and λexcitation = 485 nm).

Cell viability was also measured using the trypan blue exclusion assay. Aliquots of cells in serum-free medium were diluted with equal volumes of 0.4% (w/v) trypan blue. After 3 min incubation at room temperature, unstained and stained cells were counted in a hemacytometer.

Immunoblot analysis. Identical amounts of membrane protein or cell lysate (15-20 µg) were separated on 10% SDS-PAGE gels, followed by transfer to nitrocellulose membranes. Loading was systematically verified using Red Ponceau staining. Immunoblotting was performed as previously described using the different antibodies as indicated [8,10]. For cell lysates, β-tubulin was used as loading control. Protein bands were revealed by chemiluminescence (GE Healthcare, Orsay, France) using a peroxidase-conjugated secondary antibody and a chemiluminescence kit (GE Healthcare), followed by imaging on a Bio-Rad Fusion FX5 (Vilber Lourmat, France). Densitometric analysis was performed using ImageJ software.

In vivo study

Adult 10 week-old male C57Bl/6J mice weighing 20–22 g were obtained from Charles River Laboratories (L'Arbresle, France) and housed in temperature-regulated (20°C), ventilated cabinets with a 12 h light, 12 h dark cycle (8AM to 8PM) in a certified animal facility. Animals were acclimated in this controlled environment for 1 week prior to the study with a normal rodent chow diet and water *ad libitum* in a room with a mean temperature of 21–22°C and relative humidity of 50 ± 20%.

The study protocols for animal handling and experiments were authorized by the Department for the Protection of Populations (DDPP, Meurthe et Moselle, authorization N° 54-547-24) and in accordance with the European Communities Council Directive of 2010/63/EU. All efforts were made to minimize suffering.

B[α]P was solubilized in physiological saline solution containing 5% DMSO and 1% methyl carboxy cellulose, and injected *i.p.* into mice at 8AM every 48 h at a dose of 0.5 mg/kg [20]. The first and last injections were on day 1 and day 15, respectively. The 11-week old animals were selected randomly for each group. Control animals received vehicle (n = 9 per group). Body weight and food intake of individually housed animals were measured at day 0, before every injection, and on day 16. On day 0 and day 16, animals were fasted for 4 h, and lightly anesthetized with isoflurane before blood sampling by submandibular bleeding. Blood (100 µL) was directly placed into tubes containing EDTA on ice, and plasma was obtained by centrifugation at 13,000 x g at 4°C for 10 min. Samples were stored for analysis at −20°C. On day 16, animals were exsanguinated by cardiac puncture. Liver, epididymal fat pads of adipose tissue, and gastrocnemius muscle were rapidly dissected, rinsed in physiological saline, and snap frozen in liquid N_2 for storage at −80°C. Liver total membranes were prepared as described previously [35].

Analysis of lipoprotein profiles. Plasma samples of 4 mice from each group were pooled (210 µL total) and then added to 290 µL of 30 mM phosphate buffer containing 150 mM NaCl, 1 mM EDTA, and 0.02% sodium azide, pH 7.4. This was applied (0.2 mL/min) to a Superose 6 10–300 GL column (GE Healthcare) equilibrated with the same buffer. Fractions of 500 µL were collected and then analyzed for total cholesterol (TC) and triglyceride (TG) content using the enzymatic kits as described previously [8].

Biochemical determinations. Lipids [(TG, TC, phospholipids (PL)] of plasma, tissue lipid extracts, and fractions from lipoprotein profiles were analyzed as previously described [8] using

colorimetric enzymatic kits (Biomerieux, Craponne, France) according to the manufacturer's instructions. A serum control (Unitrol) was included with each assay performed.

Real-time PCR analysis. Cell pellets ($1–2 \times 10^6$ cells) or frozen liver samples (40-60 mg) were homogenized in QIAzol Lysis reagent (Qiagen, Courtaboeuf, France), according to manufacturer's instructions. Total RNA was extracted using RNeasy lipid tissue minikit (Qiagen); the integrity of the RNA was verified by the presence of 28S and 18S bands on agarose gels. Ten micrograms of total RNA was used for RT from which 500 ng was used for real-time PCR, as described previously [8]. For LSR and LDL-R, reactions were prepared using the Applied Biosystems (Foster City, CA, USA) SYBR Green PCR Master Mix and then performed on the StepOnePlus real-time PCR system (Applied Biosystems). Real-time PCR analysis for mouse ABCA1 was performed using a validated Taqman assay (Mm00442646_m1) obtained from Applied Biosystems. Relative expression calculations and statistical analyses were performed using the Relative Expression Software Tool (REST) 2009.

Ligand blot studies

Rat liver plasma membranes were prepared as previously described [36]. LDL (1.025< density (d) <1.055 g/mL) and VLDL (d<1.006 g/mL) were isolated from pooled human plasma [8]. Ligand blots on solubilized protein from rat liver plasma membranes were performed as previously described for LSR [35] and LDL-R [37]. Solubilized membrane protein from rat hepatocyte were separated on a 10% SDS-PAGE gel, followed by transfer to nitrocellulose. After blocking with 3% BSA, nitrocellulose strips were incubated for 30 min at 37°C with 0.8 mM oleate in the presence of 0.1 M phosphate buffer, 350 mM NaCl and 2 mM EDTA (pH 8.0) as described previously for optimal binding of lipoprotein to LSR [35]. For the LDL-R, the binding buffer used was 50 mM Tris-HCl, pH 8 containing 2 mM $CaCl_2$, 50 mg/mL of BSA and 90 mM NaCl [37] since binding of LDL-R to its ligand is Ca^{2+}-dependent, unlike LSR. VLDL and LDL were preincubated at room temperature for 30 min in the absence or presence of 0, 1, or 5 µM B[α]P. The nitrocellulose strips were then incubated for 1 h at 37°C with 20 µg/mL B[α]P-VLDL or B[α]P-LDL protein, maintaining the same concentrations of B[α]P. Following washes with PBS containing 0.5% Triton X-100 for LSR, or 50 mM Tris-HCl, pH 8 containing 2 mM $CaCl_2$, 5 mg/mL of BSA and 90 mM NaCl for LDL-R, strips were incubated with rabbit anti-apoB and apoE IgG, followed by secondary conjugated antibodies to detect bound lipoprotein.

GC-MS/MS Analysis of B[α]P and its metabolites in blood

Samples (100 µL) were supplemented with 20 µL of 1-OHbenz[α]anthracene-$^{13}C_6$, (at 1 mg/mL) and 10 µL of B[α]P-d_{12} (at 1 mg/mL) as internal standards and adjusted to pH 5.7 with 200 µL of 1 M sodium acetate buffer. The hydrolysis, extraction and purification procedures were carried out in accordance with the analytical method recently described by Peiffer *et al.* [38]. Plasma extracts were reconstituted in 25 µL of MSTFA (N-methyl-N-(trimethylsilyl) trifluoroacetamide, Derivatization of target analytes was completed after 30 min at 60°C, and 2 µL of the extract were injected into the GC–MS/MS system. Analyses were carried out with an Agilent 7890A gas chromatograph equipped with a HP-5MS capillary column (30 m, 0.25 mm i.d., 0.25 µm film thickness), coupled with an Agilent 7000B triple quadrupole mass spectrometer operating in electron impact ionization mode and an Agilent CTC PAL autosampler. Analytical conditions used for chromatography and MS/MS detection were as previously described [39]. Calibration curves were performed using plasma specimens supplemented with increased concentration levels of B[α]P and of their metabolites from 0 to 125 ng/mL of plasma. Limits of quantification (LOQs) were evaluated at 0.25 ng/mL of plasma for B[α]P, ranged from 0.1 to 1.0 ng/mL for monohydroxylated- and dihydroxylated-forms of B[α]P.

Dynamic light scattering (DLS) and Zeta potential measurements

DLS and Zeta potential measurements were performed with a Zetasizer Nano ZS (Malvern, England) equipped with a 532 nm frequency doubled DPSS laser, a measurement cell, a photo-multiplier and a correlator. Scattering intensity was measured at a scattering angle of 173° relative to the source using an avalanche of photodiode detector. The software used to collect and analyse the data was the Zetasizer Software version 6.34 from Malvern.

One mL of LDL or VLDL (refractive index 1.47; concentration of 20 μg protein/mL in Tris 0.2 mM) was measured in a clear disposable Zeta cell (DTS1060 Malvern). The measurements were made at a position of 5.50 mm and 2.00 mm from the cuvette wall for the DLS or Zeta potential measurements, respectively, with an automatic attenuator selection and at a controlled temperature of 25°C. This setup allows considerable reduction of the signal due to multiple scattering events and enables working in slightly turbid media. Intensity autocorrelation functions were analysed by CONTIN algorithm in order to determine the distribution of translational z-averaged diffusion coefficient of particles, D_T (m². s⁻¹). The D_T parameter is related to the hydrodynamic radius (R_h) of particles through the Stokes-Einstein relationship: $D_T = k_B T / 6\pi\eta R_h$, where η is the solvent viscosity (Pa.s), k_B is the Boltzmann constant (1.38×10^{-23} N.m.K⁻¹), T is the absolute temperature (°K) and R_h (m) is the equivalent hydrodynamic radius of a sphere having the same diffusion coefficient than the particles. For each sample, 15 runs of 70 s were performed with three repetitions. Approximate Zeta Potential measurements were obtained from the Smoluchowski equation and the electophoretic mobility values determined using an automatic voltage selection. For each sample, 20 runs were performed with three repetitions.

Statistical analyses

All results are shown as mean ± SEM, unless otherwise indicated. Statistical differences were tested using one-way or two-way ANOVA, or Student's t test as indicated; statistical significance was considered as $P<0.05$. Correlations were evaluated using Pearson or Spearman rank correlation coefficients.

Results

Effect of B[α]P on protein levels and mRNA expression of LDL-R, LSR and ABCA1 in mouse Hepa1-6 cells

We first sought to determine the effect of B[α]P on LSR, LDL-R in mouse Hepa1-6 cells, which have been shown previously to express both of these lipoprotein receptors [8]. In preliminary studies in which cells were incubated 1 h at 37°C with increasing concentrations of B[α]P from 0.1-100 μM, cell viability, mito-chondrial and metabolic activities were not significantly affected, using 3-(4,5-dimethylthiazol-2-yl)-2,5-diphenyltetrazolium bromide (MTT), calcein, and trypan blue as read-outs (Figure S1). Since we had previously observed that this pollutant displayed inhibitory action on epinephrine-induced lipolysis in adipocytes at concentrations <1 μM [20], experiments were performed using similar concentrations. Immunoblots of the lysates from Hepa 1-6 cells incubated at 37°C with 0.1 and 1 μM of B[α]P revealed that

B[α]P induced a significant decrease in the levels of LSR and LDL-R in a dose-dependent manner (Fig. 1A, left and middle panels). For LSR (Fig. 1A, left panel), two LSR bands are typically observed, corresponding to the α (or α') and the β LSR subunits [40]; both subunits were decreased after 1 h incubation with B[α]P (Fig. 1A, left panel), while for LDL-R a 2-h incubation was necessary to observe a significant B[α]P-induced decrease of this receptor (Fig. 1A, middle panel). Similarly to LSR, ABCA1 protein levels were significantly decreased after 1 h incubation of cells with B[α]P (Fig. 1A, right panel). No significant change was observed in the β-tubulin loading control, indicating that these effects did not appear to be related to the general state of the cells.

A second set of cells were treated similarly, from which mRNA was isolated to perform qPCR analysis. Results presented in Figure 1B revealed no significant change in LSR expression relative to the housekeeping gene HPRT, nor for that of LDL-R and ABCA1 in cells incubated in the presence of 0.1 μM B[α]P. A small but significant increase in relative expression of LDL-R and ABCA1 was observed when B[α]P concentrations were increased to 1 μM (Fig. 1B, middle and right panels) contrary to the decreased protein levels (Fig. 1A). Nonetheless, the results showed that the decrease in protein levels of LSR, LDL-R and ABCA1 was not due to decreased transcription.

Inhibition of B[α]P effect by lactacystin in cells

The lack of significant decreases of mRNA levels in Hepa1-6 cells led us to question if the B[α]P-induced effect could have occurred by increasing catabolism, rather than by inhibition of the synthesis of the proteins affected. Ubiquitin-mediated proteolysis in the proteasome is intimately involved in the removal of numerous cellular proteins, and little is known regarding PAH's potential involvement in the proteasome pathway. Cells were exposed to the proteasome inhibitor lactacystin for up to 24 h, and immunoblots revealed that LSR, LDL-R and ABCA1 protein levels in Hepa1-6 cells were all significantly increased after 2 h incubation in the presence of lactacystin (Figure S2). Another set of cells were then treated 1 h with 10 μM lactacystin, followed by addition of 0.1 μM B[α]P and an additional incubation of 1 h. Results of the immunoblots confirmed the B[α]P-induced decrease of LSR, LDL-R and ABCA1, as well as their increased protein levels in the presence of lactacystin (Fig. 2). In cells pre-incubated with lactacystin, protein levels of LSR and LDL-R were slightly lower when B[α]P was present, but did not reach statistical significance as compared to cells pre-incubated with lactacystin in absence of B[α]P. The B[α]P-induced decrease of ABCA1 in the presence of lactacystin was significant as compared to cells incubated with lactacystin alone. However, this decrease was 2-fold lower as compared to the effect of B[α]P on cells in the absence of lactacystin. These data suggest therefore that lactacystin prevented the B[α]P-induced decrease of LSR, LDL-R and ABCA1 protein levels, thus suggesting that the effect of this pollutant was mediated through increased degradation in the proteasome.

Once internalized by endocytosis, ligand-bound receptors can be eventually degraded in the lysosome. We performed experiments similar to those described for lactacystin, using chloroquine instead to inhibit lysosomal activity. Results of immunoblots revealed a small but significant increase in LDL-R, but not LSR in cells incubated with chloroquine alone (Fig. 3). B[α]P-induced decrease of both LSR and LDL-R remained identical, either in the absence or presence of chloroquine (Fig. 3), suggesting that the observed B[α]P-mediated effect was not due to increased degradation in the lysosome.

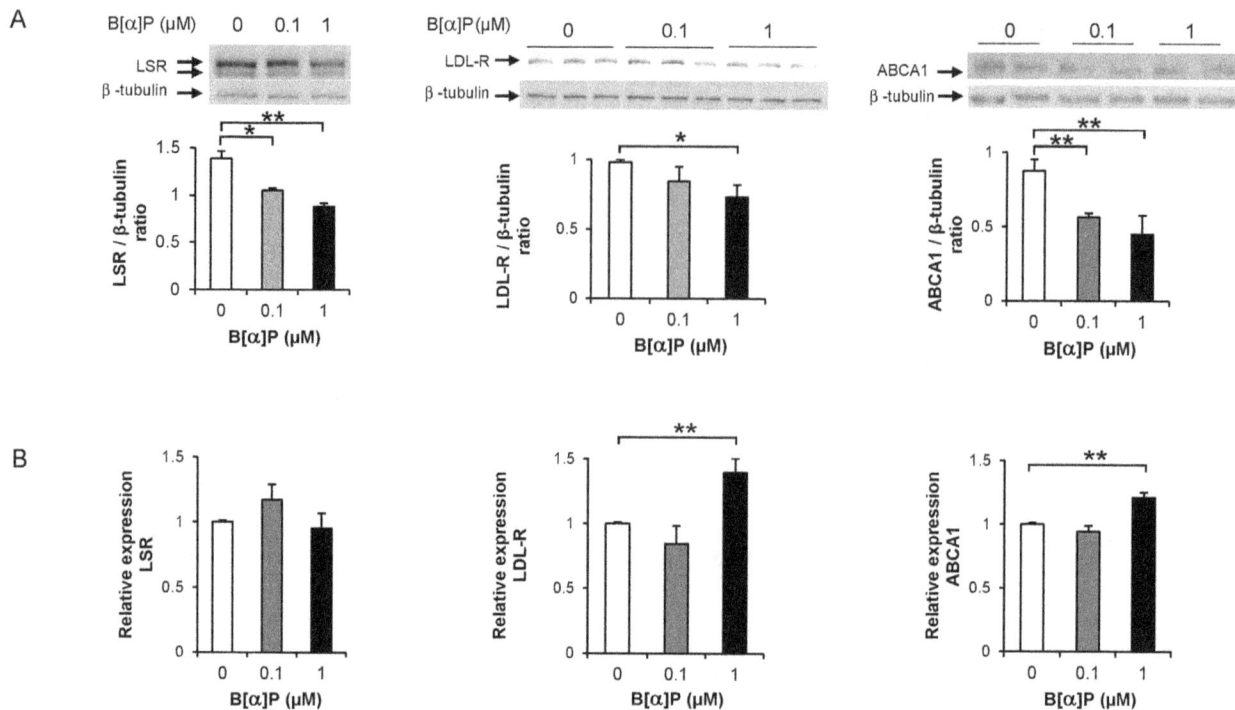

Figure 1. Effect of B[α]P on LSR, LDL-R and ABCA1 protein and mRNA levels in Hepa1-6 cells. A) Hepa1-6 cells were incubated at 37°C for up to 2 h in the absence or presence of the indicated concentrations of B[α]P, after which cell lysates and were prepared as described in Methods. Representative immunoblots (for LSR and ABCA1, results of 1-h incubation are shown; for LDL-R, results for 2-h incubation are shown) and corresponding densitometric analyses using β-tubulin as internal control are shown (one-way ANOVA, *$P<0.05$, **$P<0.01$, as compared to 0 μM B[α]P, n = 3 different cell preparations using duplicate or triplicate wells). **B)** mRNA levels of LSR, LDLR and ABCA1 were determined in cells incubated with the indicated concentrations of B[α]P using real-time PCR, as described in Materials and Methods. Results are shown (n = 4 in triplicate) of LSR, LDLR and ABCA1 mRNA expression relative to HPRT, used as reference housekeeping gene. It should be noted that there was no significant changes in HPRT expression under the different conditions (**$P<0.01$ as compared to 0 μM B[α]P).

In view of these changes, we next sought to determine the effect of B[α]P *in vivo*. We had previously established that mice treated with B[α]P displayed significantly higher weight gain as compared to controls [20]. The same experimental conditions were therefore used to determine the effect of B[α]P on hepatic lipid and lipoprotein metabolism.

Effect of B[α]P on plasma and tissue lipids

Eleven-week old C57BL/6J male mice on a standard diet were injected *i.p.* with 0.5 mg B[α]P/kg/48 h (mice were used one week after delivery following a 1 week quarantine). As reported previously [20], body weight increased in both groups continuously during the treatment period (Fig. 4A). At the end of the experimental period, B[α]P-treated mice exhibited 26% greater weight gain as compared to controls ($P<0.03$). Monitoring of food intake revealed no detectable difference between the control and experimental groups (Fig. 4B). Plasma levels of B[α]P were measured on day 16, and found to be significantly higher in B[α]P-treated animals as compared to controls (0.38 ± 0.04 ng/mL and 0.69 ± 0.09 ng/mL for control and B[α]P-treated animals, respectively, $P<0.05$). With regards to B[α]P metabolites, none were detected in plasma of both control and B[α]P-treated mice.

Analysis of plasma after a 4-h fasting period revealed that both plasma TC and TG significantly increased (32 and 51%, respectively) between day 0 and day 16 in animals treated with B[α]P, whereas there was no significant difference in mice treated with the vehicle (Fig. 4C). Consequently, the levels of plasma TC

and TG were higher (22 and 31%, respectively), in the B[α]P-treated mice as compared to controls on day 16. The analysis of lipoprotein profiles from pooled plasma samples using gel filtration chromatography revealed a slight increase of cholesterol levels in the fraction containing TGRL (VLDL and chylomicrons) (Fig. 4D, upper panel), as well as that containing HDL. Since TG is the major lipid component of VLDL lipoproteins, we also measured TG in the fractions and observed an increase primarily in the fractions containing TGRL as compared to those containing LDL or HDL (Fig. 4D, lower panel). This suggests that the observed increase in plasma TG was most likely due to an increase in the TGRL fraction in plasma of B[α]P-treated animals.

Mice were sacrificed on day 16 and liver, adipose tissue and skeletal muscle were removed for lipid composition analysis. A small, but significant increase in liver TC content was observed in the B[α]P group compared with controls (Table 1). Since liver PL also increased, although not significantly, this may explain the similar TC/PL ratios in both groups. A significant increase (29%) in TG content of adipose tissue of B[α]P-treated mice was observed, as compared to those of control animals (Table 1), consistent with the increased fat mass observed in B[α]P-treated animals [20]. TC content of adipose tissue, although a minor component in adipose tissue compared to TG, was also significantly increased in mice treated with B[α]P. This was similarly the case if expressed as ratios relative to PL in adipose tissue. If it is assumed that PL as a major cell membrane constituent represents an estimate of cell number, this would

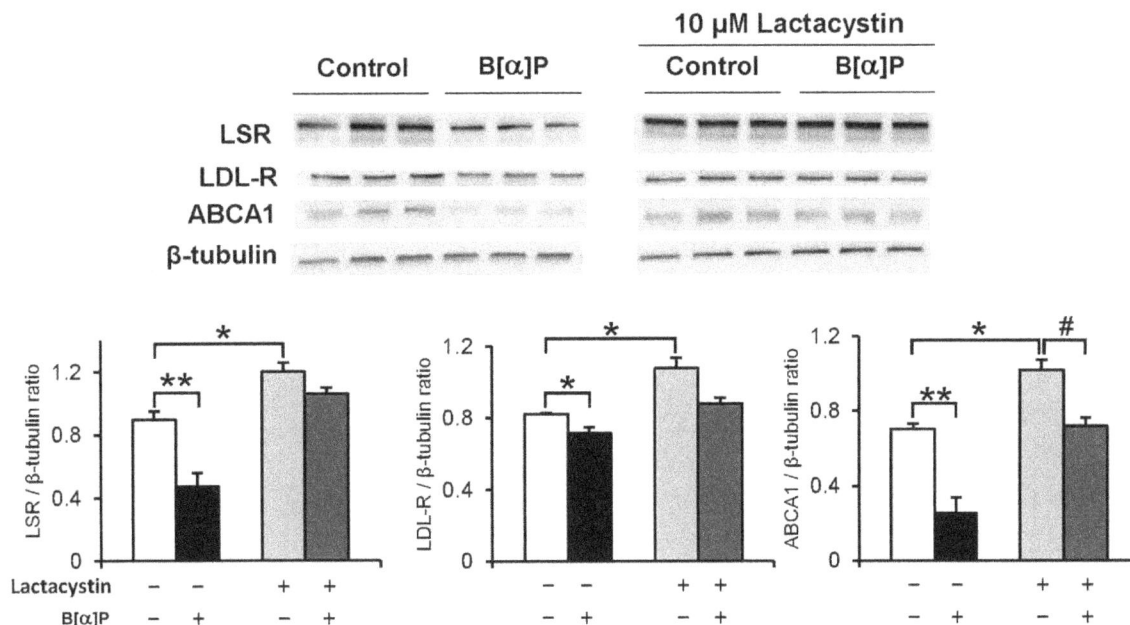

Figure 2. Lactacystin effect on B[α]P-induced decrease of LSR, LDL-R and ABCA1 protein levels in Hepa1-6 cells. Hepa1-6 cells were preincubated for 1 h at 37°C with 10 μM lactacystin, followed by 1-h incubation with 0.1 μM B[α]P with lactacystin still maintained in the cell medium. Immunoblots were performed on cell lysates, to detect LSR, LDL-R and ABCA1, and are shown with corresponding densitometric analyses (two-way ANOVA, *$P<0.05$, **$P<0.01$, vs cells incubated in absence of lactacystin and B[α]P; # $P<0.05$ vs cells incubated with lactacystin alone, n = 3 different wells per treatment).

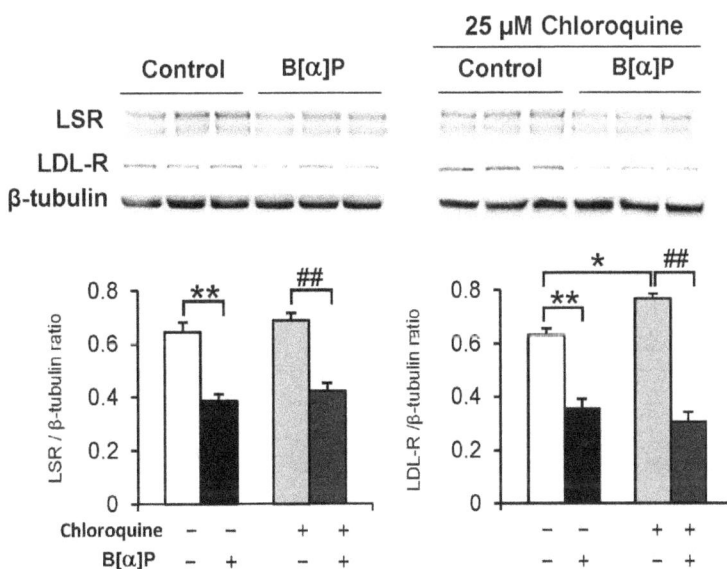

Figure 3. Chloroquine effect on B[α]P-induced decrease of LSR, LDL-R protein levels in Hepa1-6 cells. Hepa1-6 cells were preincubated for 2 h at 37°C with 25 μM chloroquine, followed by 1-h incubation with 0.1 μM B[α]P with chloroquine still maintained in the cell medium. Immunoblots were performed on cell lysates, to detect LSR and LDL-R, and are shown with corresponding densitometric analyses (two-way ANOVA, *$P<0.05$, **$P<0.01$, vs cells incubated in absence of chloroquine and B[α]P; ## $P<0.01$ vs cells incubated with chloroquine alone, n = 3 different wells per treatment).

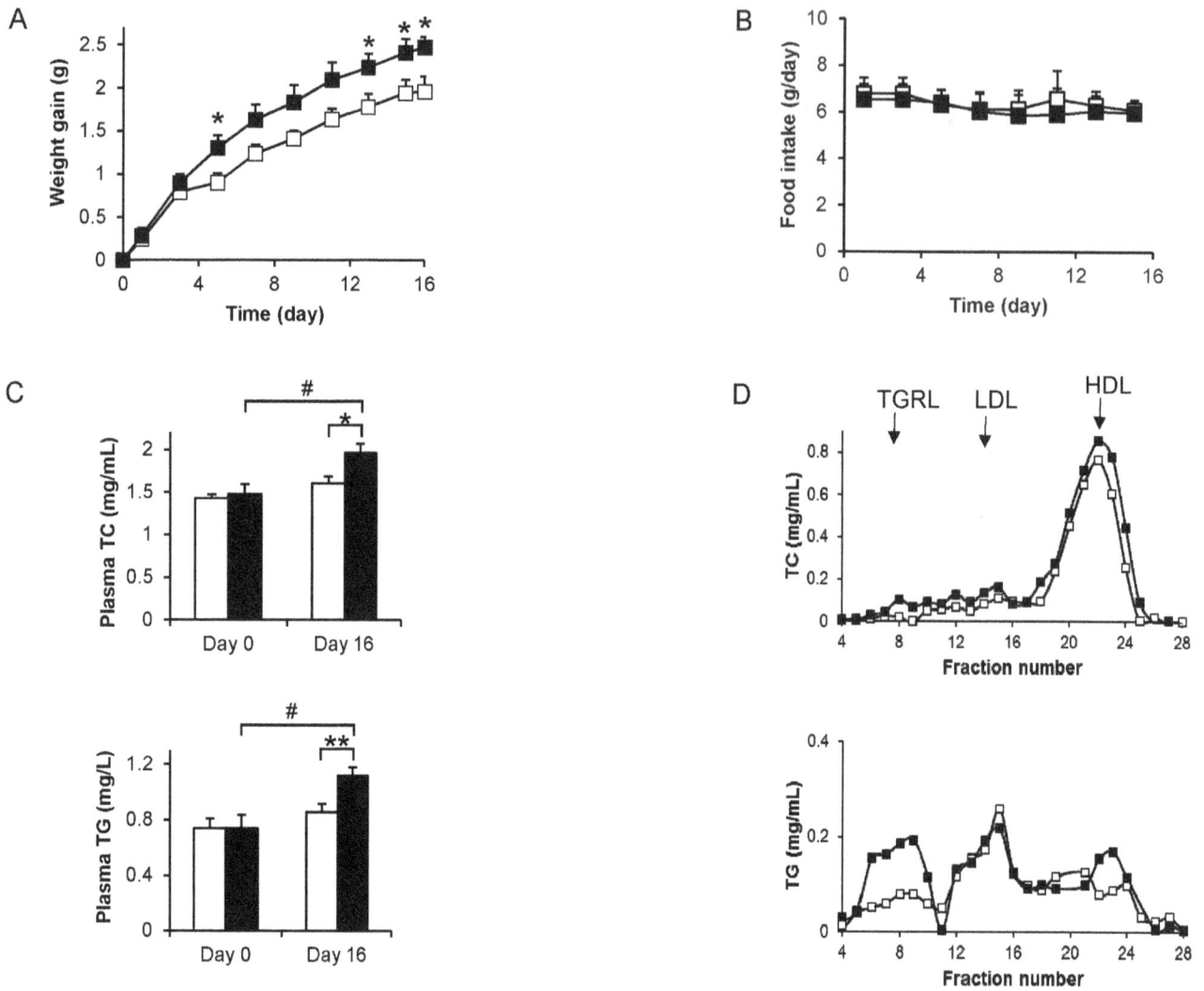

Figure 4. B[α]P effect on (A) body weight gain, (B) food intake, (C) plasma lipids, (D) lipoprotein profile in mice. Male 11 week-old C57BL/6J mice were *i.p.* injected every 48 h with vehicle alone (□, n = 9) or vehicle containing 0.5 mg/kg B[α]P (■, n = 9) from day 1 to day 15 and monitored for (**A**) weight gain (two-way ANOVA, *$P<0.05$ B[α]P group as compared to control group), and (**B**) food intake. **C**) Plasma total cholesterol (TC) and triglycerides (TG) were measured on Day 0 and Day 16 in plasma of 4-h fasted animals (two-way ANOVA*$P<0.05$, **$P <0.01$ B[α]P Day 16 vs control Day 16, # $P<0.05$; B[α]P Day 16 vs B[α]P Day 0). **D**) Lipoprotein profiles were obtained from pooled plasma samples obtained on day 16 using gel filtration chromatography as described in Materials and Methods. Fractions were analyzed for TC (upper panel) and TG (lower panel) content. The elution of TG-rich lipoproteins (TGRL), LDL and HDL are indicated by arrows.

Table 1. Tissue lipid content of control and B[α]P-treated mice.

	Liver		Adipose Tissue		Skeletal muscle	
	Control[a]	B[α]P[a]	Control[a]	B[α]P[a]	Control[a]	B[α]P[a]
	(μg/mg dry weight)					
TC	4.4±0.14	5.1±0.21*	0.3±0.02	0.5±0.07*	1.8±0.16	1.9±0.07
TG	9.3±0.51	9.6±0.69	3.5±0.22	4.5±0.49*	4.1±0.26	4.4±0.23
PL	18.2±0.61	19.9±0.96	2.6±0.30	2.2±0.30*	16.9±0.52	15.6±0.85
	(ratio)					
TC/PL	0.24±0.01	0.26±0.01	0.14±0.02	0.24±0.02**	0.10±0.01	0.11±0.004
TG/PL	0.52±0.04	0.50±0.06	1.50±0.20	2.23±0.25**	0.24±0.01	0.26±0.01

[a]mean ± SEM (n = 9 for each group).
*$P<0.05$, **$P<0.01$, as compared to control group.

suggest increased lipid content relative to the number of adipocytes. No detectable difference in lipid content was observed in skeletal muscle.

Immunoblots on protein solubilized from liver total membranes revealed that protein levels of both LDL-R and ABCA1 were significantly decreased (34% and 22%, respectively) in B[α]P-treated mouse liver membranes as compared to those of controls (Fig. 5A, middle and right panels), qPCR analysis revealed that mRNA levels were modified in a similar manner (Fig. 5B, middle and right panels), although the difference was only significant for ABCA1. The increase in hepatic TC also led us to examine other proteins involved in cholesterol uptake and transport in the liver. Hepatic protein levels of ABCG1 were decreased as well as that of the scavenger receptor SR-BI, which is involved in HDL uptake (Fig. S3A and S3B). Interestingly, both ACAT1 and ACAT2 which are enzymes involved in cholesterol esterification were elevated in mice treated with B[α]P (Fig. S3C and S3D).

LSR protein levels did not appear significantly different (Fig. 5A, left panel), while LSR mRNA was slightly but significantly higher in B[α]P-treated mice (Fig. 5B, left panel) As we had previously suggested a potential cooperativity between both receptors in the removal of TG-rich lipoproteins [8], and in view of the inter-individual variation of LSR protein observed in B[α]P-treated mice (Fig. 5A, left panel), protein levels for both LSR and LDL-R in each animal were compared on a scatter plot (Fig. 6).

Interestingly, a significant positive correlation between hepatic LDL-R and LSR protein levels in control animals was observed. However, this correlation was no longer present in animals exposed to B[α]P, which led us to question if other potential correlations could be detected between plasma and tissue lipid parameters. Spearman rank correlation analysis was thus performed, and confirmed the positive correlation (Fig. 6) between LSR and LDL-R in control, but not B[α]P-treated animals (Table 2). Also, in B[α]P-treated mice, but not control animals, a positive correlation was detected between LDL-R and plasma TC, and negative correlations were observed between LDL-R and liver TC/PL ratio, as well as between LSR and liver TG/PL. ABCA1 protein levels were significantly correlated positively with plasma TG and negatively with liver TG/PL and TC/PL ratios, but again, only in B[α]P-exposed animals, and not in controls (Table 2). B[α]P exposure therefore appears to significantly modify relationships between LSR, LDL-R, or ABCA1 and various plasma and liver lipid levels, which would support the notion that exposure to B[α]P leads to dysfunction of the regulation of lipid homeostasis in these mice.

Effect of B[α]P on lipoprotein binding to receptors

Since studies have shown that B[α]P in the circulation is found associated with lipoproteins [27,28], the question arose as to whether B[α]P could directly affect the ability of lipoproteins to

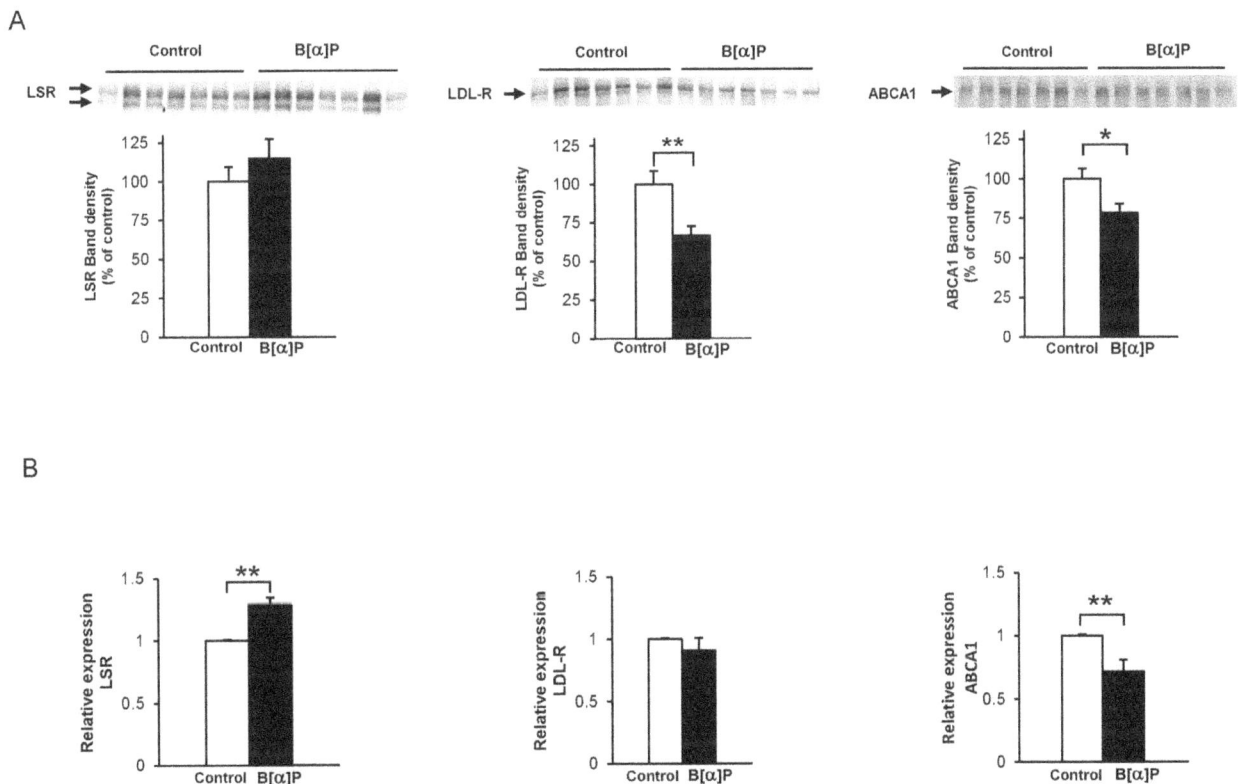

Figure 5. Effect of B[α]P on mouse hepatic LDL-R, LSR, and ABCA1 protein and mRNA levels. A) Immunoblots to detect LDL-R, LSR, and ABCA1 were performed on protein solubilized from liver total membranes isolated from control and B[α]P-treated mice. Blots are shown with corresponding densitometric analysis (Student's t-test, *P<0.05, **P<0.01 B[α]P as compared to control groups, n = 7 for each group). **B)** mRNA levels of liver LSR, LDLR and ABCA1 isolated from control and B[α]P-treated mice were determined using real-time PCR, as described in Materials and Methods. Results are shown for LSR, LDLR and ABCA1 mRNA expression relative to HPRT, used as reference housekeeping gene. There was no significant changes in HPRT expression under the different conditions (**P<0.01 as compared to control mice; triplicate determinations of n = 4 per group).

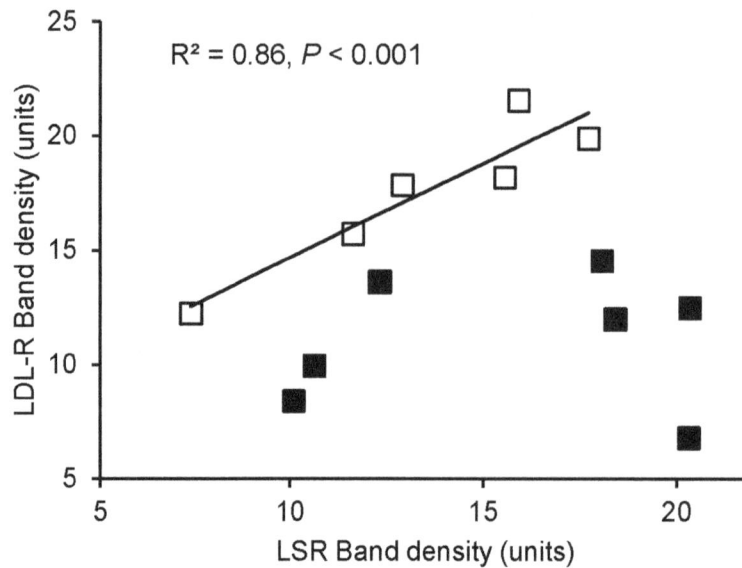

Figure 6. Scatter plot between hepatic LSR and LDL-R in mice. Distribution of individual values for liver membrane LSR and LDL-R protein levels are shown for mice from control (□) and B[α]P-treated (■) groups.

bind to LSR and LDL-R. To address this, we chose to use *in vitro* ligand blots, where a constant amount of receptor protein could be maintained, while exposing lipoproteins to different concentrations of the pollutant. The Ca^{2+}-dependent (LDL-R) and -independent (LSR) binding properties, as well as apparent molecular weight of these two receptors allowed us to clearly distinguish the two receptors. Furthermore, LSR binds apoB- and apoE-containing lipoproteins (VLDL and LDL) only in the presence of oleate [7,41], while this free fatty acid inhibits binding of LDL to its receptor [42]. Ligand blots were conducted using VLDL and LDL that had been pre-incubated 30 min at room temperature with 0, 1, and 5 μM B[α]P. LDL bound to the LDL-R or LSR was then detected by anti-apoB antibodies, while VLDL bound to LSR was detected using either anti-apoB or anti-apoE antibodies. The densities of bands revealing apoB or apoE bound to LSR (Fig. 7A) or LDL-R (Fig. 7B) became less pronounced with increasing concentrations of B[α]P, indicating a decrease in binding between the lipoprotein and LSR or LDL-R. Therefore, these results demonstrate that B[α]P can also directly interfere with the binding

of the lipoprotein ligand to its receptor. The addition of 1 μM B[α]P on LDL and VLDL did not have any effect on their hydrodynamic radius (Table S1), nor on protein content, suggesting that this PAH did not change the structure of these lipoproteins. Similarly, incubation with 1 μM of B[α]P did not significantly change their approximate Zeta potential value (Table S1).

Discussion

The objective of this study was to determine the effect of the common PAH pollutant B[α]P on lipid homeostasis in the liver. *In vitro* cell and ligand blot studies showed that B[α]P induced proteasome-mediated degradation of LSR, LDL-R and ABCA1, and also inhibited lipoprotein ligand binding to LSR and LDL-R, all of which could contribute to the changes in lipid status observed in mice exposed to this pollutant. Indeed, repeated treatment of B[α]P *in vivo* resulted in dyslipidemia in the form of increased plasma TC, TG, and TGRL, as well as increased liver TC

Table 2. Correlational analysis of lipid parameters measured in control and B[α]P-treated mice.

Parameter	LSR		LDL-R		ABCA1	
	Control[a]	B[α]P[a]	Control[a]	B[α]P[a]	Control[a]	B[α]P[a]
	Spearman correlation (ρ)					
Plasma TC	−0.42	−0.39	−0.31	**0.82***	0.25	0.42
Plasma TG	0.10	0.54	0.60	0.71	0.00	**0.80***
Liver TC/PL	−0.32	0.14	−0.50	**−0.93****	0.64	**−0.79***
Liver TG/PL	0.04	**−0.75***	0.32	−0.61	−0.43	**−0.82***
LSR	—	—	**0.94****	0.10	0.21	0.57
LDL-R	—	—	—	—	−0.21	0.67
ABCA1	—	—	—	—	—	—

[a]n = 7 for each group.
* P<0.05, ** P<0.01, compared to control group.

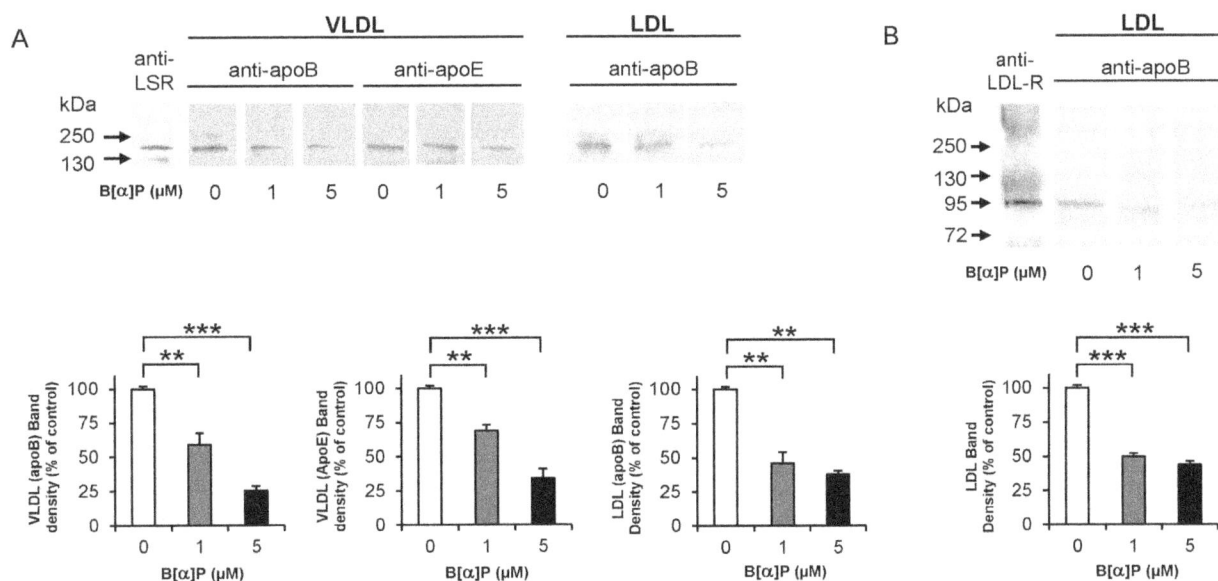

Figure 7. Effect of B[α]P on lipoprotein binding to (A) LSR and (B) LDL-R. Solubilized proteins from rat liver plasma membranes were separated on SDS-PAGE under non-reducing conditions and immobilized onto nitrocellulose membrane. For LSR, nitrocellulose membranes were pre-incubated with 0.8 mM oleate in order to activate the LSR complex. Strips were incubated at 37°C for 1 h with VLDL or LDL pre-incubated with 0, 1, or 5 μM B[α]P Strips were then washed and immunoblots were performed using anti-apoB or anti-apoE antibodies to identify LDL or VLDL, as indicated. Immunoblots using anti-LSR or anti-LDL-R antibodies were performed to verify the presence of the LSR complex or LDL-R (left strip for A and B, respectively). Densitometric analysis and representative blots are shown here of ligand blots performed on two different preparations of lipoproteins.

content, and adipose tissue TG levels, the latter being compatible with the previously reported higher weight gain and fat mass.

Our results show that brief exposure of Hepa1-6 cells to <1 μM concentrations of B[α]P significantly diminished protein levels of LSR, LDL-R, as well as ABCA1. This effect was not related to a general cell loss since no significant changes were observed in the β-tubulin loading control, nor was cell death observed with the concentrations of the pollutant used (Supplementary Figure S1). Other PAH's including pyrene and phenanthrene did not display any effect on protein levels of LSR, LDL-R or ABCA1 when added to Hepa1-6 cells at similar concentrations, despite similar lipophilicities based on their octanol/water partition coefficients (log Kow for pyrene, 4.50-5.52 and for phenanthrene, 4.28-4.67) to that of B[α]P (log Kow 5.85-6.78) (H. Layeghkhavidaki,*et al*, unpublished data). This would therefore suggest that the B[α]P-induced decrease of these 3 proteins is specific to this pollutant. B[α]P has been shown to induce DNA adduct formation, however, only after prolonged incubations of at least 24 h [43,44], while here, we used incubation times for ≤2 h, and with concentrations as low as 0.1 μM, 20-fold lower than those used for studies related to its carcinogenic properties [43,44]. Furthermore, in the *in vivo* study, mice were exposed to concentrations of the pollutant that were 100-fold lower than those used to induce a tumorigenic response [45]. It is difficult to know precisely the levels of B[α]P to which individuals are exposed. This is a common pollutant formed by incomplete combustion of organic material, and can be inhaled, absorbed through the skin, and ingested. In the general population, different PAH levels range between 0.001 to 10 ng/mL [46]. In this study, the concentrations of B[α]P measured in plasma at the time of tissue sampling was measured at 0.69 ng/mL; this value was in the similar range as those found previously in

plasma (0.04-1.62 ng/mL) from individuals in which this pollutant was found to be correlated to body mass index [19].

With regard to B[α]P metabolites, their toxico-kinetics in animal system are nowadays well established [47]. For instance, relative short half-lives (7.6 h to 9.2 h) have been described for 3-OH-B[α]P (which is often used as biomarker for assessing B[α]P exposure) following a single intravenous administration of B[α]P at 0.01 mg/kg and 0.05 mg/kg in rats [48]. The low level of rat exposure (0.5 mg/kg body weight of B[α]P every 48 h for 15 days) associated with the relative short half-lives of its metabolites may therefore explain why the concentration levels of all the metabolites analyzed 24 h after the last B[α]P administration were below the limit of quantification of the method used. Further studies would be needed to determine if metabolites of B[α]P and other PAH could exert similar effects.

Cell studies performed here showed that the proteasome was involved in the catabolism of LSR, LDL-R and ABCA1 in Hepa1-6 cells. This is actually the first evidence for proteasome degradation of LSR; proteasome involvement in the LDL-R and ABCA1 removal is consistent with previous work showing that the proteasome pathway can participate in part to the removal LDL-R and ABCA1 in HepG2 and macrophages, respectively [49,50]. The more recently characterized PCSK9-mediated removal of LDL-R has also been shown to rely on the proteasome, but this may be cell-type dependent [51,52]. Indeed, the lactacystin-induced increase of LDL-R appeared somewhat less pronounced as compared to that of LSR and ABCA1. B[α]P-induced reduction of the levels of these three proteins was practically abolished in the presence of the proteasome inhibitor, lactacystin, but not in the presence of the lysosomal inhibitor chloroquine, pointing towards proteasome-mediated removal as the mechanism underlying the effect of B[α]P. A previous observation showing that PAHs including B[α]P increase ubiquitylation of p21 protein in the A549 lung cancer cell line [53] lead us to suggest ubiquitylation as a potential mechanism by which this pollutant increased proteasomal degradation of the three proteins studied here. Additional investigation is required to determine if this represents increased degradation of newly-synthesized LSR or LDL-R, or vesicular LSR or LDL-R as a result of endocytosis from the plasma membrane.

Ligand blots using constant amounts of lipoprotein receptor revealed that the ability of lipoproteins exposed to B[α]P to bind LSR or LDL-R was significantly diminished. In a study in which B[α]P was added to plasma, apoB was found to be the major carrier of this pollutant [54]. LDL contains one apoB per particle, which is the apolipoprotein that is recognized by LDL-R or LSR. The presence of B[α]P on this apoB-containing lipoprotein could therefore interfere with the ability of LDL-R or LSR to recognize the binding epitope(s) of apoB, which may explain the observed decrease in binding. A similar effect could also explain the lower binding of VLDL to LSR. Binding of the apoE component of VLDL to LSR was also affected by B[α]P treatment, which may be due to direct interference in the binding of apoE to LSR, or simply to the indirect effect of reduced apoB-mediated binding of VLDL to LSR, since VLDL particles are large TGRL that contain both apoB and apoE. However, incubation of VLDL or LDL with 1 μM B[α]P did not lead to significant changes of the hydrodynamic radius and the approximate Zeta potential values obtained by dynamic light scattering and electrophoretic mobility measurements (Table S1), suggesting that B[α]P was adsorbed by hydrophobic interaction with the lipid component of the lipoproteins. Further biochemical studies are needed using purified receptor to determine the mechanisms by which this pollutant interferes with binding of the lipoproteins to these receptors.

We are unable to explain the differences in results for mRNA levels in the cell culture and animal studies. Despite this, the results based on cell culture and ligand blot studies obtained *in vitro* could provide a possible explanation for the changes in hepatic lipid metabolism reported in the *in vivo* experiments. In the liver of B[α]P-treated animals, LDL-R protein levels were significantly decreased, which may be due in part to a direct effect of the pollutant based on the cell culture studies, and/or as a result of the increase, albeit modest, in hepatic cholesterol content. Indeed, cholesterol acts as a sensor in the liver through the sterol regulatory element-binding protein pathway, which regulates hepatic needs for exogenous cholesterol by regulating LDL-R expression [55]. Since the absence of LDL-R expression is associated with increased plasma cholesterol and LDL [56,57], the changes in LDL-R protein levels would be consistent with the increased plasma cholesterol measured. Direct interference of B[α]P with LDL binding to the LDL-R, as demonstrated in the ligand blots, also would contribute to increased plasma cholesterol. However, an increase in the plasma LDL fraction was not detectable in lipoprotein profiles obtained after separation by gel filtration, which may be due to the fact that LDL-R protein levels were reduced, and not completely absent.

ABCA1 was significantly diminished in the presence of B[α]P, in both the cell and animal studies. Indeed, this protein belongs to the family of multi-resistance drug transporters, which have been shown to be modulated through xenobiotic effects of B[α]P [32]. A previous study reported that B[α]P induced expression of ABCG1 after a 3-day exposure of Caco2 cells [58], however they used 50-fold higher concentrations than those used in this study with Hepa1-6 cells. Since ABCA1 plays a role in cholesterol efflux by providing cholesterol to lipid-poor apoA-I for secretion in the form of HDL, its reduced levels in B[α]P-treated mice may contribute to

the observed increase in hepatic TC content. Indeed, significant negative correlations were observed between liver TC/PL ratio and ABCA1, but only in animals treated with B[α]P, and not in controls. A similar negative correlation was also observed between liver TG/PL ratio and ABCA1. A previous study showing that low ABCA1 levels are associated with non-alcoholic steatohepatitis and that siRNA-mediated knock-down of hepatic ABCA1 can lead to TG accumulation in the liver [59] would therefore be consistent with the negative correlations observed between liver lipid levels and ABCA1 in the B[α]P-treated mice.

The *in vitro* study suggests that B[α]P increased proteasome-mediated degradation of LSR. However, no significant changes were observed in hepatic LSR protein levels of B[α]P-treated mice. The fact that decreases in LDL-R and ABCA1 were nevertheless observed in B[α]P-treated mice suggests that LSR turnover rates may be different as compared to these other proteins; however, this remains to be verified. Ligand blots did demonstrate that B[α]P interferes with binding of VLDL and LDL to LSR. Since hepatic LSR is a receptor that plays a significant role in the removal of TGRL during the postprandial phase [6,8], and taking into account that mice were injected with B[α]P during the postprandial period, this could explain the observed increase in plasma TG and TGRL. Indeed, radioactively-labelled B[α]P associated with chylomicrons is transported directly to the liver to be removed from the circulation [28]. We would therefore postulate that B[α]P-mediated inhibition of binding of TGRL to LSR at the site of the liver could contribute to the observed increase of TGRL in the circulation of B[α]P-treated mice.

Correlational studies revealed a strong positive correlation between LSR and the LDL-R, but only in the control group, which is consistent with a potential cooperativity between LSR and LDL-R regarding the removal of TGRL [8]. Interestingly, this correlation was no longer evident in B[α]P-treated mice. Furthermore, rather strong negative correlations were revealed between LSR and liver TG/PL, and between LDL-R and liver TC/PL, but only in the B[α]P group of mice. While this type of analysis cannot reveal causal relationships, it nevertheless supports the notion that B[α]P exposure can disrupt normal lipid homeostasis.

This study represents the first report on the direct effect of B[α]P on the hepatic lipoprotein receptors LDL-R and LSR. Increased plasma lipids have been shown in LSR$^{+/-}$ mice [8] as well as in siRNA-mediated liver-specific inactivation of LSR mRNA [6], and in LDL-R$^{-/-}$ mice [56,57]. Interestingly, LSR$^{+/-}$ exhibited higher weight gain as compared to controls with age, or if placed on a high-fat diet [10]. In addition, LDL-R$^{-/-}$ mice placed on a high-fat and high-carbohydrate diet gained more weight as compared to controls [9]. This leads us to suggest that in a similar manner, B[α]P exposure *in vivo* leads to reduced functional LSR and LDL-R, which contribute to the changes in lipid status, as well as to increased susceptibility towards increased weight gain. Interestingly, we observed that levels of other hepatic proteins involved in cholesterol transport, SR-BI and ABCG1 were decreased in B[α]P –treated mice. On the contrary, protein levels of hepatic ACAT1 and ACAT2 involved in the synthesis of cholesteryl esters were actually increased, which may have contributed to the increased hepatic TC content observed in the mice exposed to this pollutant. These results would support the idea that besides LSR, LDL-R and ABCA1, other proteins involved in hepatic lipid and lipoprotein metabolism may also be affected by B[α]P and other pollutants. Additional studies are needed to explore the underlying molecular mechanisms. We have previously observed that B[α]P directly affects the ability of the adipocyte to release fatty acids after stimulation by adrenaline

[20]. We have now shown in this study that this pollutant also contributes towards disrupting lipid homeostasis in the liver leading to dyslipidemia, thus revealing potential mechanisms that could explain part of the contribution of pollutants such as B[α]P to the etiology of dyslipidemia-linked obesity.

Supporting Information

Figure S1   Effect of B[α]P on cell viability. Hepa1-6 cells were incubated at 37°C for 1 h with the indicated concentrations of B[α]P. Cell viability and metabolic activity tests were performed using (A) MTT, (B) calcein, and (C) trypan blue exclusion assays, as described in Materials and Methods. Results are shown as mean ± SEM of triplicate determinations.

Figure S2   Time course effect of lactacystin on LSR, LDL-R and ABCA1 protein levels in Hepa1-6 cells. Hepa1-6 cells were incubated at 37°C with 10 μM lactacystin for the indicated times. Immunoblots and densitometric analyses of the signal as a ratio to that of β-tubulin are shown for LSR, LDL-R and ABCA1 (*$P<0.03$, **$P<0.01$, compared to time 0, n = 3 per treatment).

Figure S3   Effect of B[α]P treatment on hepatic protein levels of ABCG1, SR-BI, ACAT1 and ACAT2. Immunoblots were performed to detect A) ABCG1, B) SR-BI, C) ACAT1 and D) ACAT2 protein levels in membrane (SR-BI) or cytosolic (ABCG1, ACAT1, ACAT2) fractions prepared from liver homogenates from mice treated with or without B[α]P. Densitometric analyses were performed, and are shown here as means ± SEM of n = 7 per group. Student's t-test was used to determine statistical difference (* $P<0.05$, ** $P<0.01$) as compared to control values.

Table S1   Effect of B[α]P on the hydrodynamic radius and on Zeta potential of LDL and VLDL.

Acknowledgments

The authors thank Dr. Henri Schroeder for providing the rats and Dr. Nazir Ahmad for the preparation of rat liver plasma membranes.

Author Contributions

Conceived and designed the experiments: HL MCL CF CC FTY. Performed the experiments: HL MCL SA NG. Analyzed the data: HL MCL SA TO NG BA LGP CC FTY JJ. Contributed reagents/materials/analysis tools: JJ. Wrote the paper: HL SA TO NG BA CC FTY.

References

1. Caballero B (2007) The global epidemic of obesity: an overview. Epidemiol Rev 29: 1–5.
2. Cunningham E (2010) Where can I find obesity statistics? J Am Diet Assoc 110: 656.
3. Howard BV, Ruotolo G, Robbins DC (2003) Obesity and dyslipidemia. Endocrinol Metab Clin North Am 32: 855–867.
4. Eckel RH (2011) The complex metabolic mechanisms relating obesity to hypertriglyceridemia. Arterioscler Thromb Vasc Biol 31: 1946–1948.
5. Brown MS, Goldstein JL (1986) A receptor-mediated pathway for cholesterol homeostasis. Science 232: 34–47.
6. Narvekar P, Berriel Diaz M, Krones-Herzig A, Hardeland U, Strzoda D, et al. (2009) Liver-specific loss of lipolysis-stimulated lipoprotein receptor triggers systemic hyperlipidemia in mice. Diabetes 58: 1040–1049.
7. Stenger C, Corbier C, Yen FT (2012) Structure and function of the lipolysis stimulated lipoprotein receptor. In: D Ekinci, editor editors. Chemical Biology. Rijeka: InTech. pp. 267–292.
8. Yen FT, Roitel O, Bonnard L, Notet V, Pratte D, et al. (2008) Lipolysis stimulated lipoprotein receptor: a novel molecular link between hyperlipidemia, weight gain, and atherosclerosis in mice. J Biol Chem 283: 25650–25659.
9. Schreyer SA, Vick C, Lystig TC, Mystkowski P, LeBoeuf RC (2002) LDL receptor but not apolipoprotein E deficiency increases diet-induced obesity and diabetes in mice. Am J Physiol Endocrinol Metab 282: E207–214.
10. Stenger C, Hanse M, Pratte D, Mbala ML, Akbar S, et al. (2010) Up-regulation of hepatic lipolysis stimulated lipoprotein receptor by leptin: a potential lever for controlling lipid clearance during the postprandial phase. Faseb J 24: 4218–4228.
11. Heindel JJ, vom Saal FS (2009) Role of nutrition and environmental endocrine disrupting chemicals during the perinatal period on the aetiology of obesity. Mol Cell Endocrinol 304: 90–96.
12. Schug TT, Janesick A, Blumberg B, Heindel JJ (2011) Endocrine disrupting chemicals and disease susceptibility. J Steroid Biochem Mol Biol 127: 204–215.
13. Hao C, Cheng X, Xia H, Ma X (2012) The endocrine disruptor mono-(2-ethylhexyl) phthalate promotes adipocyte differentiation and induces obesity in mice. Biosci Rep 32: 619–629.
14. Sargis RM, Johnson DN, Choudhury RA, Brady MJ (2010) Environmental endocrine disruptors promote adipogenesis in the 3T3-L1 cell line through glucocorticoid receptor activation. Obesity (Silver Spring) 18: 1283–1288.
15. Bostrom CE, Gerde P, Hanberg A, Jernstrom B, Johansson C, et al. (2002) Cancer risk assessment, indicators, and guidelines for polycyclic aromatic hydrocarbons in the ambient air. Environ Health Perspect 110 Suppl 3: 451–488.
16. Brevik A, Lindeman B, Rusnakova V, Olsen AK, Brunborg G, et al. (2012) Paternal benzo[α]pyrene exposure affects gene expression in the early developing mouse embryo. Toxicol Sci 129: 157–165.
17. Palanikumar L, Kumaraguru AK, Ramakritinan CM, Anand M (2011) Biochemical response of anthracene and benzo [α] pyrene in milkfish Chanos chanos. Ecotoxicol Environ Saf 75: 187–197.
18. EFSA (2008) Scientific opinion of the panel on contaminants in the food chain on a request from the European Commission on polycyclic aromatic hydrocarbons in food. The EFSA Journal 724: 1–114.
19. Hutcheon DE, Kantrowitz J, Van Gelder RN, Flynn E (1983) Factors affecting plasma benzo[α]pyrene levels in environmental studies. Environ Res 32: 104–110.
20. Irigaray P, Ogier V, Jacquenet S, Notet V, Sibille P, et al. (2006) Benzo[α]pyrene impairs beta-adrenergic stimulation of adipose tissue lipolysis and causes weight gain in mice. A novel molecular mechanism of toxicity for a common food pollutant. Febs J 273: 1362–1372.
21. Boysen G, Hecht SS (2003) Analysis of DNA and protein adducts of benzo[α]pyrene in human tissues using structure-specific methods. Mutat Res 543: 17–30.
22. Ekstrom G, von Bahr C, Glaumann H, Ingelman-Sundberg M (1982) Interindividual variation in benzo(a)pyrene metabolism and composition of isoenzymes of cytochrome P-450 as revealed by SDS-gel electrophoresis of human liver microsomal fractions. Acta Pharmacol Toxicol (Copenh) 50: 251–260.
23. Mitchell CE, Fischer JP, Dahl AR (1987) Differential induction of cytochrome P-450 catalyzed activities by polychlorinated biphenyls and benzo [α]pyrene in B6C3F1 mouse liver and lung. Toxicology 43: 315–323.
24. Fujisawa-Sehara A, Yamane M, Fujii-Kuriyama Y (1988) A DNA-binding factor specific for xenobiotic responsive elements of P-450c gene exists as a cryptic form in cytoplasm: its possible translocation to nucleus. Proc Natl Acad Sci U S A 85: 5859–5863.
25. Schmidt JV, Bradfield CA (1996) Ah receptor signaling pathways. Annu Rev Cell Dev Biol 12: 55–89.
26. Shimizu Y, Nakatsuru Y, Ichinose M, Takahashi Y, Kume H, et al. (2000) Benzo[α]pyrene carcinogenicity is lost in mice lacking the aryl hydrocarbon receptor. Proc Natl Acad Sci U S A 97: 779–782.
27. Busbee DL, Norman JO, Ziprin RL (1990) Comparative uptake, vascular transport, and cellular internalization of aflatoxin-B1 and benzo(a)pyrene. Arch Toxicol 64: 285–290.
28. Vauhkonen M, Kuusi T, Kinnunen PK (1980) Serum and tissue distribution of benzo[α]pyrene from intravenously injected chylomicrons in rat in vivo. Cancer Lett 11: 113–119.
29. Sureau F, Chinsky L, Duquesne M, Laigle A, Turpin PY, et al. (1990) Microspectrofluorimetric study of the kinetics of cellular uptake and metabolization of benzo(a)pyrene in human T 47D mammary tumor cells: evidence for cytochrome P1450 induction. Eur Biophys J 18: 301–307.
30. Verma N, Pink M, Petrat F, Rettenmeier AW, Schmitz-Spanke S (2012) Exposure of primary porcine urothelial cells to benzo(a)pyrene: in vitro uptake, intracellular concentration, and biological response. Arch Toxicol 86: 1861–1871.
31. Kelman BJ, Springer DL (1982) Movements of benzo[α]pyrene across the hemochorial placenta of the guinea pig. Proc Soc Exp Biol Med 169: 58–62.

32. Sarkadi B, Homolya L, Szakacs G, Varadi A (2006) Human multidrug resistance ABCB and ABCG transporters: participation in a chemoimmunity defense system. Physiol Rev 86: 1179–1236.

33. Sieuwerts AM, Klijn JG, Peters HA, Foekens JA (1995) The MTT tetrazolium salt assay scrutinized: how to use this assay reliably to measure metabolic activity of cell cultures in vitro for the assessment of growth characteristics, IC50-values and cell survival. Eur J Clin Chem Clin Biochem 33: 813–823.

34. Bratosin D, Mitrofan L, Palii C, Estaquier J, Montreuil J (2005) Novel fluorescence assay using calcein-AM for the determination of human erythrocyte viability and aging. Cytometry A 66: 78–84.

35. Mann CJ, Khallou J, Chevreuil O, Troussard AA, Guermani LM, et al. (1995) Mechanism of activation and functional significance of the lipolysis-stimulated receptor. Evidence for a role as chylomicron remnant receptor. Biochemistry 34: 10421–10431.

36. Ahmad N, Girardet JM, Akbar S, Lanhers MC, Paris C, et al. (2012) Lactoferrin and its hydrolysate bind directly to the oleate-activated form of the lipolysis stimulated lipoprotein receptor. Febs J 279: 4361–4373.

37. Daniel TO, Schneider WJ, Goldstein JL, Brown MS (1983) Visualization of lipoprotein receptors by ligand blotting. J Biol Chem 258: 4606–4611.

38. Peiffer J, Cosnier F, Grova N, Nunge H, Salquèbre G, et al. (2013) Neurobehavioral toxicity of a repeated exposure (14 days) to the airborne polycyclic aromatic hydrocarbon fluorene in adult Wistar male rats. PLoS One 8: e71413.

39. Grova N, Salquèbre G, Schroeder H, Appenzeller BM (2011) Determination of PAHs and OH-PAHs in rat brain by gas chromatography tandem (triple quadrupole) mass spectrometry. Chem Res Toxicol 24: 1653–1667.

40. Yen FT, Masson M, Clossais-Besnard N, Andre P, Grosset JM, et al. (1999) Molecular cloning of a lipolysis-stimulated remnant receptor expressed in the liver. J Biol Chem 274: 13390–13398.

41. Bihain BE, Yen FT (1992) Free fatty acids activate a high-affinity saturable pathway for degradation of low-density lipoproteins in fibroblasts from a subject homozygous for familial hypercholesterolemia. Biochemistry 31: 4628–4636.

42. Bihain BE, Deckelbaum RJ, Yen FT, Gleeson AM, Carpentier YA, et al. (1989) Unesterified fatty acids inhibit the binding of low density lipoproteins to the human fibroblast low density lipoprotein receptor. J Biol Chem 264: 17316–17321.

43. Huang M, Blair IA, Penning TM (2013) Identification of stable benzo[a]pyrene-7,8-dione-DNA adducts in human lung cells. Chem Res Toxicol 26: 685–692.

44. Pruess-Schwartz D, Baird WM, Nikbakht A, Merrick BA, Selkirk JK (1986) Benzo(a)pyrene:DNA adduct formation in normal human mammary epithelial cell cultures and the human mammary carcinoma T47D cell line. Cancer Res 46: 2697–2702.

45. Wu K, Shan YJ, Zhao Y, Yu JW, Liu BH (2001) Inhibitory effects of RRR-alpha-tocopheryl succinate on benzo(a)pyrene (B(a)P)-induced forestomach carcinogenesis in female mice. World J Gastroenterol 7: 60–65.

46. Pleil JD, Stiegel MA, Sobus JR, Tabucchi S, Ghio AJ, et al. (2010) Cumulative exposure assessment for trace-level polycyclic aromatic hydrocarbons (PAHs) using human blood and plasma analysis. J Chromatogr B Analyt Technol Biomed Life Sci 878: 1753–1760.

47. Chien YC, Yeh CT (2012) Excretion kinetics of urinary 3-hydroxybenzo[a]pyrene following dietary exposure to benzo[a]pyrene in humans. Arch Toxicol 86: 45–53.

48. Payan JP, Lafontaine M, Simon P, Marquet F, Champmargin-Gendre C, et al. (2009) 3-Hydroxybenzo(a)pyrene as a biomarker of dermal exposure to benzo(a)pyrene. Arch Toxicol 83: 873–883.

49. Miura H, Tomoda H, Miura K, Takishima K, Omura S (1996) Lactacystin increases LDL receptor level on HepG2 cells. Biochem Biophys Res Commun 227: 684–687.

50. Ogura M, Ayaori M, Terao Y, Hisada T, Iizuka M, et al. (2011) Proteasomal inhibition promotes ATP-binding cassette transporter A1 (ABCA1) and ABCG1 expression and cholesterol efflux from macrophages in vitro and in vivo. Arterioscler Thromb Vasc Biol 31: 1980–1987.

51. Chen Y, Wang H, Yu L, Yu X, Qian YW, et al. (2011) Role of ubiquitination in PCSK9-mediated low-density lipoprotein receptor degradation. Biochem Biophys Res Commun 415: 515–518.

52. Wang Y, Huang Y, Hobbs HH, Cohen JC (2012) Molecular characterization of proprotein convertase subtilisin/kexin type 9-mediated degradation of the LDLR. J Lipid Res 53: 1932–1943.

53. Nakanishi Y, Pei XH, Takayama K, Bai F, Izumi M, et al. (2000) Polycyclic aromatic hydrocarbon carcinogens increase ubiquitination of p21 protein after the stabilization of p53 and the expression of p21. Am J Respir Cell Mol Biol 22: 747–754.

54. Polyakov LM, Chasovskikh MI, Panin LE (1996) Binding and transport of benzo[a]pyrene by blood plasma lipoproteins: the possible role of apolipoprotein B in this process. Bioconjug Chem 7: 396–400.

55. Brown MS, Goldstein JL (1997) The SREBP pathway: regulation of cholesterol metabolism by proteolysis of a membrane-bound transcription factor. Cell 89: 331–340.

56. de Faria E, Fong LG, Komaromy M, Cooper AD (1996) Relative roles of the LDL receptor, the LDL receptor-like protein, and hepatic lipase in chylomicron remnant removal by the liver. J Lipid Res 37: 197–209.

57. Ishibashi S, Perrey S, Chen Z, Osuga J, Shimada M, et al. (1996) Role of the low density lipoprotein (LDL) receptor pathway in the metabolism of chylomicron remnants. A quantitative study in knockout mice lacking the LDL receptor, apolipoprotein E, or both. J Biol Chem 271: 22422–22427.

58. Ebert B, Seidel A, Lampen A (2005) Identification of BCRP as transporter of benzo[a]pyrene conjugates metabolically formed in Caco-2 cells and its induction by Ah-receptor agonists. Carcinogenesis 26: 1754–1763.

59. Yang Y, Jiang Y, Wang Y, An W (2010) Suppression of ABCA1 by unsaturated fatty acids leads to lipid accumulation in HepG2 cells. Biochimie 92: 958–963.

7

Holding Thermal Receipt Paper and Eating Food after Using Hand Sanitizer Results in High Serum Bioactive and Urine Total Levels of Bisphenol A (BPA)

Annette M. Hormann[1], Frederick S. vom Saal[1], Susan C. Nagel[2], Richard W. Stahlhut[1], Carol L. Moyer[1], Mark R. Ellersieck[3], Wade V. Welshons[4], Pierre-Louis Toutain[5,6], Julia A. Taylor[1]*

1 Division of Biological Sciences, University of Missouri, Columbia, Missouri, United States of America, 2 Department of Obstetrics, Gynecology and Women's Health, University of Missouri, Columbia, Missouri, United States of America, 3 Department of Statistics, University of Missouri, Columbia, Missouri, United States of America, 4 Department of Biomedical Sciences, University of Missouri, Columbia, Missouri, United States of America, 5 Université de Toulouse, INPT, ENVT, UPS, UMR1331, F- 31062 Toulouse, France, 6 INRA, UMR1331, Toxalim, Research Centre in Food Toxicology, F-31027 Toulouse, France

Abstract

Bisphenol A (BPA) is an endocrine disrupting environmental contaminant used in a wide variety of products, and BPA metabolites are found in almost everyone's urine, suggesting widespread exposure from multiple sources. Regulatory agencies estimate that virtually all BPA exposure is from food and beverage packaging. However, free BPA is applied to the outer layer of thermal receipt paper present in very high (~20 mg BPA/g paper) quantities as a print developer. Not taken into account when considering thermal paper as a source of BPA exposure is that some commonly used hand sanitizers, as well as other skin care products, contain mixtures of dermal penetration enhancing chemicals that can increase by up to 100 fold the dermal absorption of lipophilic compounds such as BPA. We found that when men and women held thermal receipt paper immediately after using a hand sanitizer with penetration enhancing chemicals, significant free BPA was transferred to their hands and then to French fries that were eaten, and the combination of dermal and oral BPA absorption led to a rapid and dramatic average maximum increase (Cmax) in unconjugated (bioactive) BPA of ~7 ng/mL in serum and ~20 µg total BPA/g creatinine in urine within 90 min. The default method used by regulatory agencies to test for hazards posed by chemicals is intra-gastric gavage. For BPA this approach results in less than 1% of the administered dose being bioavailable in blood. It also ignores dermal absorption as well as sublingual absorption in the mouth that both bypass first-pass liver metabolism. The elevated levels of BPA that we observed due to holding thermal paper after using a product containing dermal penetration enhancing chemicals have been related to an increased risk for a wide range of developmental abnormalities as well as diseases in adults.

Editor: David O. Carpenter, Institute for Health & the Environment, United States of America

Funding: This work was supported by a grant from The Passport Foundation (no URL available) and by NIEHS grant ES018764 to FvS (NIEHS website http://www.niehs.nih.gov). The funders had no role in study design, data collection and analysis, decision to publish, or preparation of the manuscript.

Competing Interests: The authors have declared that no competing interests exist.

* Email: TaylorJA@missouri.edu

Introduction

Bisphenol A [BPA; bis(4-hydroxyphenyl)propane; CAS #80-05-7] is one of the highest volume chemicals in commerce with 15-billion pounds produced per year [1], and based on the presence of BPA metabolites in urine, it can be concluded that virtually everyone is exposed [2,3]. BPA has estrogenic and other endocrine disrupting activities [4,5]. BPA molecules are polymerized to make polycarbonate plastic used for food and beverage containers, epoxy resins used to line cans, and dental composites and sealants, but free (unpolymerized) BPA is also used as an additive (plasticizer), such as in polyvinyl chloride (PVC) products. Our interest is in the use of BPA in thermal paper, which is used for airline ticket, gas, ATM, cash register and other types of receipts (Figure 1). The print surface of thermal paper is coated with milligrams of free BPA per gram paper as a heat-activated print developer [6], and it appears that free BPA is readily transferred to other materials that the thermal paper contacts [7].

While small lipophilic compounds such as BPA (logP = 3.4; molecular weight 228 Da) can pass through skin [8,9], regulatory agencies have assumed that this route of human BPA exposure should not be significant in spite of the lack of data and acknowledged "significant uncertainties" around the issue of human exposure to BPA from thermal paper [10]. However, a factor that has not been considered in estimating transdermal exposure to BPA from thermal paper is that hand sanitizers are now commonly used, particularly in fast-food restaurants where people may handle thermal receipts before eating or ordering food. Hand sanitizer and other skin care products may also be used by cashiers while working. Exposure to BPA from thermal paper goes beyond just transdermal exposure and consumption of food that is picked up and eaten with a BPA-contaminated hand.

Thermal head

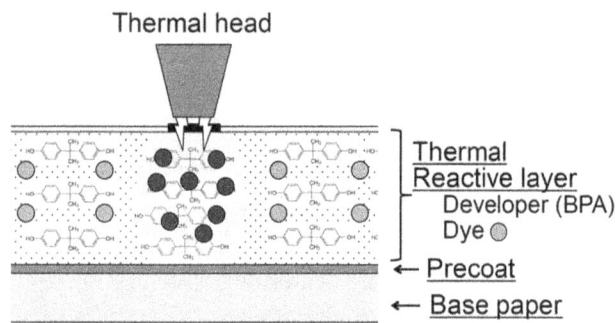

Figure 1. Schematic diagram of thermal receipt paper identifying the thermal reactive layer that contains BPA as a developer and a leuco dye, as well as stabilizers and binders (not shown).

The transfer of a chemical directly from hand-to-mouth (mouthing behavior) has been proposed to be an important variable for estimating total chemical exposure in humans [11], particularly in young children [12].

The use of hand sanitizers and other skin-care products, including soaps, lotions and sunscreens, is significant because some contain mixtures of chemicals that are also used as dermal penetration enhancers to increase the transdermal delivery of drugs. Drugs and chemicals that are suitable for transdermal delivery and are impacted by dermal penetration enhancers have a LogP>1.5 and a molecular weight <500 Da [9]. There are many factors that impact the ability of compounds to pass through skin in addition to molecular weight and lipophilicity, including differences arising from the location of skin on the body, gender and age [13]. Mixtures of dermal penetration enhancing chemicals can act synergistically to increase by up to 100 fold the dermal penetration of small lipophilic molecules such as estradiol [8,9], with which BPA shares physical-chemical and biological properties [4]. For example, Purell hand sanitizer (Gojo Industries), which we used in the current study, contains a number of dermal penetration enhancers, such as isopropyl myristate and propylene glycol, and is (63% w/w) ethanol. The use of hand sanitizers has increased in recent years and is now about a 200 million dollar a year industry just in the USA [14]. The impact of the use of personal care products such as moisturizing lotions that contain dermal penetration enhancing chemicals on exposure to environmental chemicals has been identified as a concern [15].

To assess the relevance of this research to real-world behavior, we conducted a preliminary observational study in fast-food restaurants, food courts and shopping malls in Columbia Missouri. Receipt contact time varied widely, but was sometimes substantial. In one restaurant, we found that receipt contact time ranged up to 65 sec for people purchasing food that was eaten in the restaurant; the 75th percentile for time holding the receipt was >12 sec, and the 90th percentile >32 sec. In a fast-food restaurant that is part of an international chain, take-out food was placed into a bag and the top of the bag was folded, then the thermal receipt was stapled to the top of the bag; the result was that the print surface of the receipt (coated with BPA) was grabbed when the bag was picked up. The contact time between the hand and thermal receipt was thus considerably longer than would be the case for food eaten in the restaurant. In a food court we observed that some fast-food restaurants had hand sanitizer dispensers available for use by customers next to the cash register, and customers were observed using the hand sanitizer before handling the thermal receipt. The estimate is that 50 million people eat in a fast-food establishment

every day in the USA [16]. Finally, our experiments here are also relevant to occupational exposures, because we observed in a national chain big-box store that all cash registers had a hand sanitizer dispenser next to them for use by the cashiers.

Our objectives were to examine the impact of having dry hands vs. wet hands due to using a popular hand sanitizer that contains dermal penetration enhancing chemicals on extraction of BPA from the surface of thermal receipt paper coated with BPA. We also measured (using a LC/MSMS assay) unconjugated, bioactive BPA (uBPA) and its conjugated metabolites, BPA-glucuronide (BPA-G) and BPA-monosulfate (BPA-S), in serum and urine in adult male and female subjects after holding a thermal receipt. To determine the proportion of thermal receipts that contained BPA, we examined receipt papers for the presence and amount of BPA. We also examined receipts for the most commonly used BPA replacement chemical, bisphenol S [bis(4-hydroxyphenyl)sulfone; BPS; CAS #80-09-1].

Methods

Ethics statement

The University of Missouri School of Medicine Institutional Review Board approved all procedures involving human subjects, and sample collection was conducted by licensed personnel in the Clinical Research Center (CRC) within the University of Missouri School of Medicine. Subjects were informed of the procedures, and provided written consent. The signed consent forms were retained. The University IRB approved the consent procedure.

Subjects

Participants for the different experiments in this study were recruited through a weekly University of Missouri campus-wide email newsletter. Candidates (men and women) were pre-screened by age, height, weight, and health status. Participants selected were 20–40 years old (average 27.0 yrs), and an attempt was made to select those with average height, weight and normal-range body-mass index. Participants selected were not taking any prescription or non-prescription medication other than oral contraceptives; the type of oral contraceptive used was recorded. To ensure that pregnant women were excluded from the study, all women were administered a pregnancy test when they arrived at the CRC.

For all studies participants were asked to refrain from touching thermal paper receipts, consuming food or beverages stored in polycarbonate or other types of plastic containers as well as canned food and beverages during the 48 hr prior to participating in the study, in order to reduce background BPA levels in body fluids as much as possible. The participants also filled out a questionnaire concerning their activities during the prior 48 hr (see Section S3 in File S1 for questionnaire).

For experiments in which there was hand contact with thermal receipt paper, subjects were required to wash their hands with soap and water, rinse thoroughly, and then dry using Kimwipes (Kimberly-Clark, Irving, TX). A number of soaps were screened for BPA content and/or chromatographic interference prior to the start of the study, and the soap chosen was Softsoap "Aquarium series" (Colgate Palmolive Company, Manhattan, NY), which showed no detectable BPA or chromatographic interference with the assay of BPA. Standard brown laboratory paper towels were tested and found to contain BPA at around 6 μg/towel. Because of this, Kimwipes, which tested negative for BPA, were used throughout for drying hands. Water from faucets used in the CRC was tested and found to be below the limit of detection (LOD) for BPA content (detection limit was 10 pg/mL by HPLC with CoulArray detection based on C-18 extraction of 250 ml of water).

Sample analysis

Analysis of BPA in extracted samples occurred within an accredited facility (Veterinary Medicine Diagnostic Laboratory) within the College of Veterinary Medicine at the University of Missouri.

Reagents. Solvents (methanol, acetonitrile) and water were HPLC grade, and were obtained from Fisher Scientific. BPA, bisphenol S (BPS), and BPA monosulfate (BPA-S) were obtained from Sigma-Aldrich (St. Louis MO; purity >99%, 98% and 95% respectively). C^{13}-BPA was obtained from Cambridge Isotope Laboratories Inc. (Andover, MA; purity 99%), and both BPA-G (purity 98%) and BPA D-glucuronide (BPA-DG; purity >99%) were provided by the National Institute of Environmental Health Sciences (NIEHS), Research Triangle Park, NC. Ethanol (200 proof) used for hand swipes was obtained from Decon Labs, Inc. (King of Prussia, PA).

Total receipt BPA and BPS content. Weighed samples of each receipt (3×3 cm) were incubated overnight in methanol at room temperature. The methanol extracts were diluted in methanol, typically to a final dilution of 1/10,000, and BPA content was analyzed by HPLC with CoulArray detection (see Section S1 in File S1 for details). We also analyzed the same receipt sample extracts for BPS using LC/MSMS (see Section S1 in File S1 for details).

BPA levels in Kimwipe hand swipes. Kimwipe swipes were incubated in methanol at room temperature overnight, and aliquots were taken from the methanol extract for analysis. BPA in the methanol extract was determined by HPLC with CoulArray detection.

BPA levels in French fries. French fries were incubated individually in methanol overnight. The fries were then removed, and the samples centrifuged briefly to separate any solid and/or oily matter, and a sample of the clear methanol extract was assayed. Equal volumes from the 10 extracts from the 10 French fries touched by each participant were pooled, and a single measurement was made for each participant. Quantitation was made by HPLC with CoulArray detection.

Serum sample collection and extraction. Multiple-point blood samples were collected via IV catheter into 10 mL syringes, and the syringes were emptied into the same uncoated vacutainer tubes (for details and catalog numbers of collection materials Section S1 in File S1). Single point blood samples were collected by venipuncture into uncoated glass vacutainer tubes (Becton Dickinson, Franklin Lakes, NJ). All blood samples were allowed to clot at room temperature for 15–30 min and then refrigerated until centrifugation at 4°C for 15 min. The serum was transferred with glass Pasteur pipets into 15 mL centrifuge tubes and then frozen at −20°C. Samples were extracted using C18 SPE as previously described [17]; see Section S1 in File S1. Procedural blanks were also run alongside the samples to monitor for reagent contamination or interference. Serum extracts were analyzed by LC/MSMS.

Urine sample collection and extraction. All urine samples were collected directly into Samco polypropylene specimen cups (Fisher Scientific, Waltham, MA) and were immediately refrigerated (4°C for 2–5 hours) until they could be transferred to the research laboratory, at which point they were frozen at −20°C. The total BPA concentration (representing a combined measure of unconjugated and conjugated BPA) was measured by LC/MSMS (see Section S1 in File S1).

Assay of creatinine in urine. To calculate creatinine-corrected urine BPA concentrations, urine creatinine was measured using an ELISA kit (R&D Systems Inc., Minneapolis, MN), according to manufacturer's instructions. Sensitivity of this assay is 0.02 mg/dL.

Field blanks. The possibility of BPA leaching from each piece of equipment used in the collection or processing of samples identified above was determined by passing BPA-free water through all collection equipment, which was then handled and assayed for BPA as described below for the actual samples. All equipment and sample handling was determined to not leach detectable BPA before any sample collections occurred.

Statistical methods and calculation of pharmacokinetic parameters

For both uBPA and BPA-G, the area under the concentration-time curve (AUC) up to the last measured serum concentration above the LOQ, i.e. AUC (0–90 min), was calculated by using the linear trapezoidal rule. The average AUC (0–90 min) (ng/mL) was calculated by dividing AUC (0–90 min ng/mL)/90 min. Time (Tmax) of maximal plasma BPA concentration (Cmax) was directly obtained from the raw data. Comparisons of men and women were conducted using the Mann-Whitney U test or ANOVA. Statistical significance was set a $P<0.05$, two-tailed test. All data are presented as mean±SEM.

Experiment 1: Measurement of BPA and BPS in 50 used thermal receipt papers

The objective of this experiment was to determine the amount of BPA and BPS in thermal receipt paper and to determine the proportion of receipts that contained BPA or BPS, which is the most commonly used BPA replacement chemical. Thermal paper sales receipts were obtained by purchasing items from 41 different vendors in Columbia, MO and from a further 9 vendors in Southern Missouri (50 receipts total). Weighed portions of each paper were extracted and assayed for BPA by HPLC with CoulArray detection and for BPS by LC/MSMS. After screening, an unused roll was obtained from a vendor from which a BPA-positive receipt had been identified. The BPA content of paper from this roll was confirmed prior to being used for testing with human subjects in Experiments 2, 3 and 4.

Experiment 2: BPA transferred to a hand with and without using hand sanitizer due to holding a thermal receipt for different lengths of time

The objective of this experiment was to determine the amount of BPA extracted by a hand from a standard piece of thermal receipt paper immediately after using Purell hand sanitizer (Experiment 2-A) or with dry hands (Experiment 2-B). Subjects in both experiments cleaned and dried their hands prior to the experiment and between each trial. For Experiment 2-A the subjects (2 men and one woman) each held the thermal paper for different lengths of time: 2, 15, 30, 45, 60 or 240 sec (in 6 separate trials for each subject). Both hands were wetted by applying three "squirts" of Purell to each hand, and the hands were then briefly rubbed together to distribute the hand sanitizer evenly across both palms and fingers, but the sanitizer was not allowed to dry prior to holding the receipt paper. In experiment 2-B the subjects (2 men and 2 women) held the receipt with dry hands for 60 or 240 sec (2 separate trials for each subject). In both experiments an 8×12 cm portion of thermal paper cut from an unused receipt roll that was obtained from a local merchant (previously identified as containing 27.2 mg BPA/g paper) was placed BPA-coated (print surface) side down into the right hand. The hand was swiped 3 times each with 3 ethanol-soaked Kimwipes, and BPA was extracted from the Kimwipes with methanol and measured by HPLC with CoulArray detection.

Experiment 3: Serum and urine BPA in men and women before and after transdermal and oral exposure to BPA from thermal receipt paper after using hand sanitizer

The objective of this experiment was to measure the transfer of BPA from thermal paper receipts to hands, and the amount of BPA remaining on the surface of a hand 90-min later, after using Purell hand sanitizer (as described in the prior experiment) in 5 male and 5 female subjects. In addition, we measured the amount of BPA transferred from a BPA-contaminated hand to 10 French fries, and measured blood and urine concentrations of uBPA, BPA-G and BPA-S before and after ingestion of the French fries and BPA absorption through skin. The design of the study is shown in Figure 2.

The background level of BPA on the dominant hand was determined when the subjects first arrived at the CRC. The dominant hand was swiped 3 times with 3 separate Kimwipes soaked with ethanol, from which we extracted BPA for analysis by HPLC with CoulArray detection, and the hands were then cleaned. The subjects' weight and height were determined, after which they provided a baseline urine specimen, an IV port was inserted into the cubital vein, and a baseline blood sample was collected. Purell hand sanitizer was applied to the hands as described in Experiment 2. An 8×12 cm piece of thermal paper cut from an unused receipt roll (used in Experiment 2) was then placed BPA-coated side down into each hand with the hands still wet. The subjects held the receipt papers for 4 min in each hand. The dominant arm of each subject was determined based on whether the person was right or left handed, and in this experiment the non-dominant hand remained contaminated with BPA for the duration of the experiment. Blood was collected from the cubital vein in the contaminated arm of one set of subjects (N = 7) and from the cubital vein in the non-contaminated arm of other subjects (N = 3). We note that the phlebotomist did not handle the thermal paper for either Experiment 3 or Experiment 4. The study coordinator who did handle the paper wore gloves to do so and did not touch the blood tubes of other equipment. A separate person swiped the subjects' hands after thermal paper exposure and wore fresh gloves for each swipe session and discarded them immediately afterwards.

French fries that had been purchased from a local fast food restaurant and had been found to not contain detectable BPA were briefly warmed in a toaster oven. Immediately after holding the thermal receipts in each hand, the subjects picked up a French fry in each hand, and held both fries for 10 sec. The fry held in the dominant hand was placed into a labeled glass tube, and the fry that was held in the non-dominant hand was eaten. A total of 10 French fries was handled by each hand and either placed in a test tube or eaten using this same procedure. Approximately 4 min elapsed between removal of the receipt paper from the hand and consumption of the last French fry. Thus, it took about 8 min from the time that the thermal receipt paper was first touched and consumption of the last French fry.

After the last French fry was consumed, the subject's dominant hand was swiped with 3 ethanol-soaked Kimwipes to clean BPA off the hand and for determination (by extracting BPA from the Kimwipes) of the amount of BPA remaining on the hand immediately after holding the 10 French fries that were placed into test tubes. The non-dominant hand was not cleaned after holding the receipt paper and eating French fries, and thus was a

continuing source of transdermal BPA exposure over the following 90-min period of blood collection.

Blood samples were collected from the cubital vein from the still contaminated arm of 7 subjects, 4 males and 3 females, and from the uncontaminated arm of 3 subjects, one male and 2 females. The blood collected from the BPA-contaminated arm provided direct information about BPA absorbed from the hand on which BPA remained for 90 min, since the cubital vein is one of the major veins draining the hand; this blood is not subject to first-pass liver metabolism prior to going to the heart and being transported in the arterial circulation to tissues. The blood collected from the uncontaminated arm provided information about BPA in the systemic (mixed) circulation.

Blood was collected from the IV port before holding the thermal paper (baseline) and at 15, 30, 60 and 90 min after consumption of the last French fry. The non-dominant contaminated hand (from which the French fries were eaten) was not allowed to touch anything during the 90-min after holding the receipt paper and then picking up the 10 French fries; this hand was swiped with 3 ethanol-soaked Kimwipes after the final 90-min blood collection at the end of the study. After these swipes were obtained, both hands were thoroughly cleaned and a second urine sample was collected.

Experiment 4: Serum and urine BPA in men and women before and after transdermal exposure to BPA from thermal receipt paper with dry hands

The objective of this study was to examine the amount of BPA transferred to a clean dry hand and then present in serum and urine without using hand sanitizer. In this study we examined 12 adult men and 12 adult women subjects. The subjects washed and dried their hands and provided a baseline blood and urine sample as described in Experiment 3. The non-dominant hand was swiped 3 times each with 3 ethanol-soaked Kimwipes to obtain a baseline measure of BPA on the hand prior to holding a thermal receipt. After the hand was dry, subjects held an 8×12 cm piece of thermal receipt paper (from the roll used in Experiment 1) with the non-dominant dry hand for 4 min. Thirty minutes later a second blood sample was collected from the contaminated arm, after which the BPA was swiped from the contaminated hand with ethanol-soaked Kimwipes as described previously. As above, the contaminated hand was not allowed to touch anything during the 30-min period prior to the second blood collection. The hands were washed, and a second urine sample was collected 60 min after holding the receipt paper.

Results

Experiment 1: Measurement of BPA and BPS in thermal receipt paper

Thermal receipts were collected at stores, bars and restaurants in mid-Missouri. Of the 50 receipts, 22 (44%) contained high levels of BPA (Table 1). High levels of the BPA replacement chemical BPS were found in 26 (52%) of the receipts, and 2 receipts contained an unidentified chemical as the print developer [18]; see Section S2 in File S1 for individual values. These findings suggest that BPS is now as commonly used as BPA as a developer in thermal receipt paper. Note that these receipts had been obtained with purchases, while the receipt paper used in Experiments 2, 3 and 4 came from an unused roll of thermal paper that we determine contained BPA as the developer.

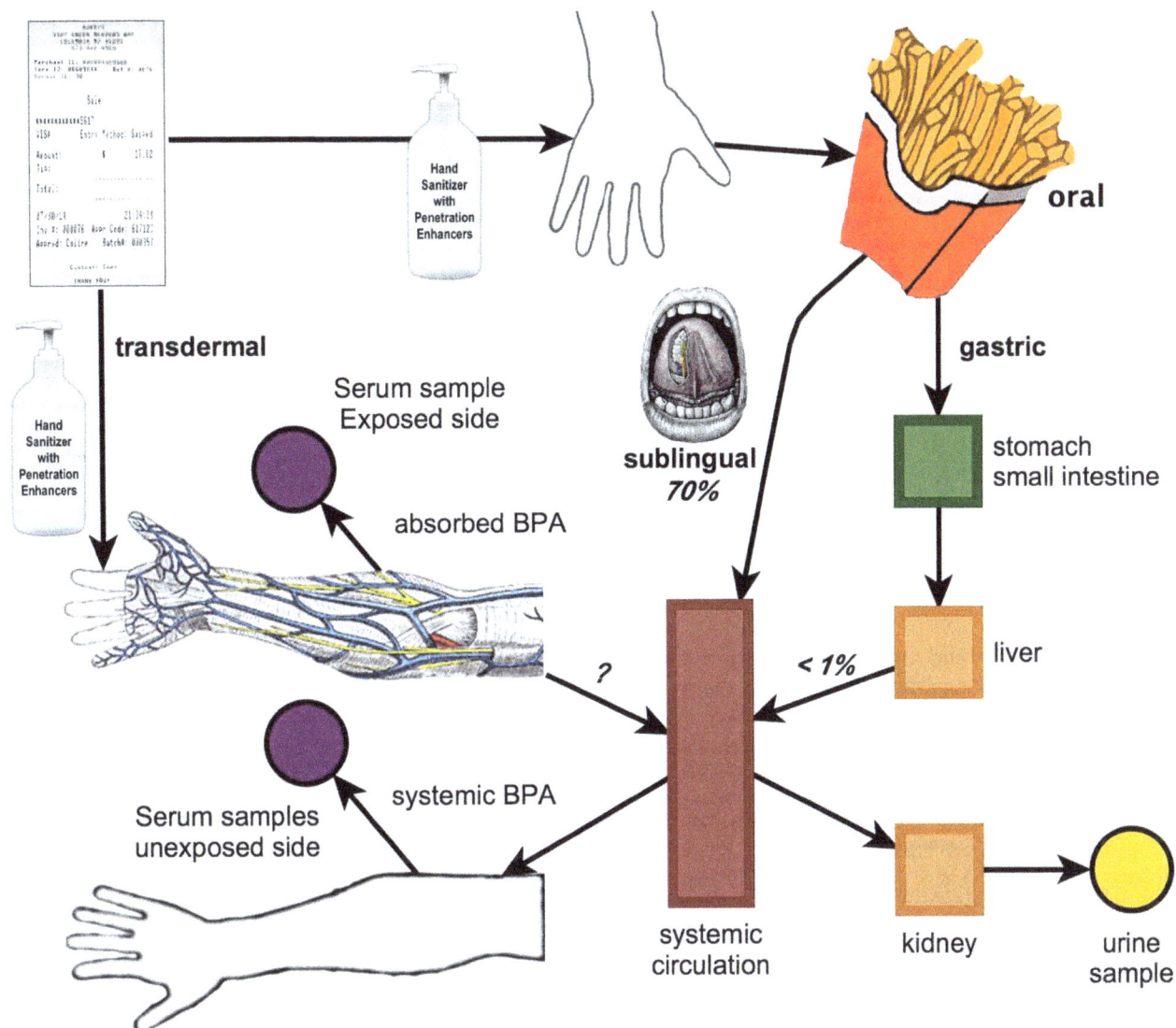

Figure 2. Schematic diagram of the protocol for Experiment 3 in which thermal receipt paper containing BPA was held with a hand wet from using Purell hand sanitizer, after which the subjects picked up 10 French fries and ate them, resulting in both oral and transdermal routes of exposure. Of the 5 male and 5 female subjects, 7 subjects had serum collected from the cubital vein in the arm with a contaminated hand that contained the BPA from holding thermal paper. Three subjects had blood collected from the cubital vein in the unexposed arm that did not have BPA on the hand throughout the 90-min test period during which blood was collected. Urine samples were obtained before and at the end of the test period.

Table 1. BPA and BPS concentrations in 50 thermal paper receipt samples.

Chemical in paper	mg/g receipt	mg/8×12 cm receipt
BPA-positive (44%)	19.6±1.0	9.0±0.4
	(11.5–26.3)	(6.1–11.3)
BPS-positive (52%)	23.5±0.7	10.8±0.3
	(15.2–30.1)	(7.1–13.2)

Two (4%) of 50 papers tested did not contain either BPA or BPS and did not show any estrogenic activity in a MCF-7 breast cancer cell proliferation assay (data not shown). Values are mean±SEM, with the range of measured values given in parentheses. See Section S2 in File S1 for individual receipt data.

Experiment 2: BPA transferred to a hand with and without using hand sanitizer due to holding a thermal receipt for different lengths of time

The data shown in Figure 3-A reveal that after using Purell hand sanitizer with the hand still wet, the maximum amount of BPA swiped from the palm and fingers of the hand (581 µg BPA) occurred after holding a receipt for 45 sec. After holding a receipt for 2 sec, 40% (235 µg BPA) of maximum was recovered from the hand, and within 15 sec 58% (339 µg BPA) of maximum was recovered from the hand. The decrease in BPA swiped from the hand between 45 sec and 4 min to 73% of maximum (425 µg BPA) may have been due to absorption into skin occurring at a greater rate than transfer to the skin from the thermal receipt.

The data in Figure 3-B show that holding a thermal receipt with dry hands resulted in dramatically lower amounts of BPA being extracted from the receipt relative to the amounts extracted immediately after using hand sanitizer. The ratio of the extracted BPA swiped from the wet vs. dry hand was higher at 60 sec (ratio = 185) than at 240 sec (ratio = 51), reflecting the fact that while the amount of BPA swiped from a wet hand decreased between 60 and 240 sec, the levels increased over this time when the hand was dry, likely due to a reduced rate of absorption with dry relative to wet hands.

Experiment 3: Serum and urine BPA before and after transdermal and oral exposure to BPA from thermal receipt paper after using hand sanitizer

We measured the amount of BPA swiped from the dominant hand after using hand sanitizer, holding a receipt and then eating 10 French fries, which took 8 min. BPA levels were not significantly different for the 5 males (mean±SEM: 126±19 µg) and the 5 females (mean±SEM: 128±10 µg). These levels measured at about 8-min after first touching the thermal receipt paper were lower than levels measured at 45 sec and 4 min in Experiment 2 (Figure 3-A), which likely reflects rapid transdermal absorption of BPA due to the use of hand sanitizer as well as some of the BPA having been transferred to the French fries. Importantly, females transferred significantly more (58±19 µg)

BPA from their dominant hand to the 10 French fries than males (15±3 µg; Mann-Whitney U; P<0.05), resulting in females having a significantly higher oral BPA dose than males between 4–8 min after applying the hand sanitizer.

Since the participants had been instructed to avoid known sources of BPA, such as canned products, and instructed not to touch thermal paper, 9 of the 10 subjects had undetectable BPA on their dominant hand prior to washing their hands when they first arrived at the Clinical Research Center; none of the subjects was a cashier. However, one female had 0.9 µg of BPA extracted from her hand upon arriving at the CRC, and she was also found to have a very high background concentration of serum uBPA (14.3 ng/mL) prior to holding the thermal receipt paper (subject #3; Figure 4-B). This was the only female subject who was menstruating and thus using products to control menstrual flow, and she also indicated use of hand and body lotion 7–9 times in the prior 48 hr, which was more than any other female or male subject (see Section S3 in File S1). However, even though female #3 (Figure 4-B) had very high background serum uBPA, she showed a dramatic 9.5 ng/mL increase relative to baseline in serum uBPA after holding the thermal receipt and eating 10 contaminated fries at the 15 min blood collection time (15 min after consuming the last French fry). The increase relative to baseline in serum uBPA for female #3 was thus virtually identical to the maximum increase (relative to baseline) found for the other 2 females who had low baseline serum uBPA levels and that were tested in the same way (blood was collected from the BPA contaminated arm; Figure 4-A; Table 1).

Experiment 3-A: Collection of blood from the cubital vein in the contaminated arm with BPA remaining on the hand throughout the 90-min test period

The data for female subject #3 are not included in the pharmacokinetic data (Table 2) calculated for the remaining 6 subjects that had blood collected from their contaminated arm but who had undetectable baseline levels of BPA on their hands when they first entered the CRC and also had very low baseline uBPA in serum (0.23±0.15 ng/mL; N = 6). These 6 subjects showed a dramatic increase in serum uBPA after holding the thermal receipt

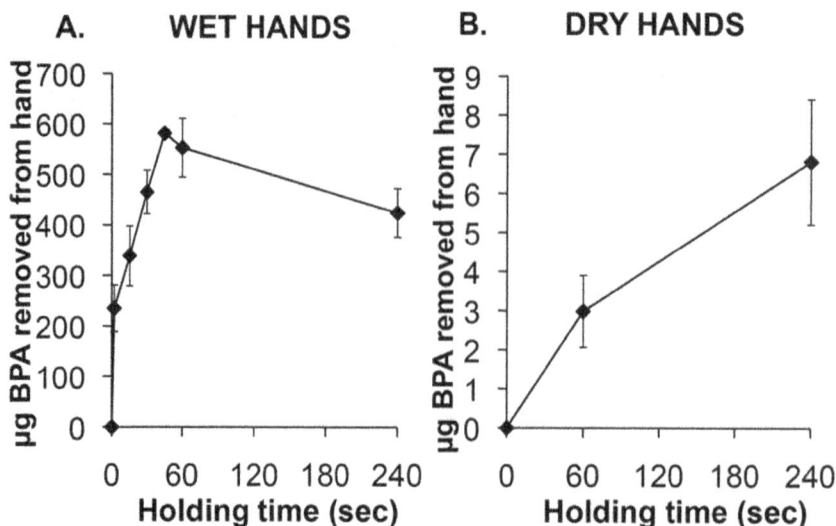

Figure 3. Effect of length of time holding an 8×12 cm thermal receipt on the amount of BPA (µg) swiped from the hand, when hands were pre-wetted with hand sanitizer (Panel A) or left dry (Panel B). The hand was swiped with KimWipes wetted with ethanol to remove the BPA from the surface of the palm and fingers.

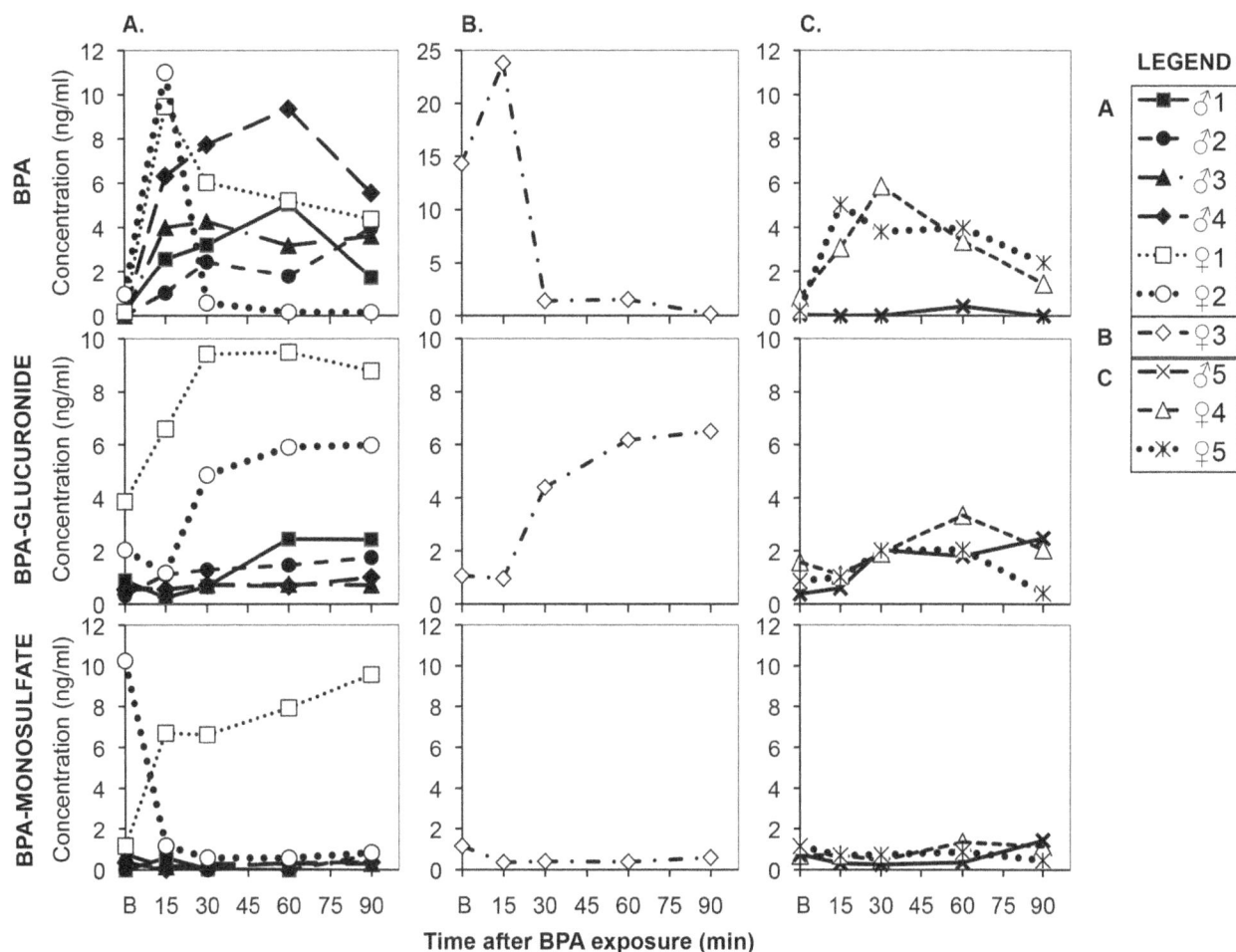

Figure 4. Individual serum profiles of BPA, BPA-G and BPA-MS in men and women prior to (B = baseline levels) and after holding BPA-containing receipt paper for 4 min followed by picking up and eating 10 French fries over about 4 min with a BPA-contaminated hand. The BPA then remained on the contaminated hand throughout the following 90-min period of blood collection (blood was collected between 15–90 min after eating the last French fry). Panel A: data for serum BPA collected from the contaminated arm with BPA remaining on the hand for 4 males and 2 females that had very low baseline serum uBPA. Panel B: serum BPA data collected from the contaminated arm from Female #3 who had a high baseline serum concentration of uBPA. Panel C: serum BPA data for one male and 2 females who had systemic blood collected from the uncontaminated arm.

and eating 10 contaminated fries. Females had a greater Cmax and maximum increase relative to baseline in serum uBPA and BPA-G than males after holding the thermal paper, while males reached peak levels of uBPA (Tmax) later than females. The average uBPA value, based on the area under the concentration-time curve [AUC (0–90 min)] did not differ between males and females, while for BPA-G, the AUC (0–90 min) was greater for females than males. The ratio of BPA-G/uBPA based on the average AUC (0–90 min) was very low (0.35±0.12 for males and 1.82±0.30 for females), consistent with routes of absorption of BPA (dermal and sublingual) that bypass first pass metabolism [19,20].

Only female #1 (Figure 4-A) showed a marked increase in serum BPA-S relative to baseline, revealing that while BPA-G is the major conjugated metabolite of BPA in most men and non-pregnant women, some individuals do form significant amounts of BPA-S. Urine total BPA (unconjugated and conjugated) increased dramatically between baseline and 90 min after handling thermal paper, although unlike the serum data, there was no difference between males and females (Table 2).

Experiment 3-B: Collection of blood from the cubital vein in the uncontaminated arm to measure BPA in mixed systemic blood throughout the 90-min test period

We also obtained data from 3 subjects who had the same procedures described above except that they had blood collected from the cubital vein in the opposite uncontaminated arm that did not have BPA remaining on the hand during the 90-min period of blood collection. The 3 subjects consisted of one male and 2 females (age 22.3±0.9 yrs, BMI 26.0±0.9). While the baseline serum uBPA levels were very low (Figure 4-C; Table 3), the average serum uBPA AUC (0–90 min) for the two female subjects (Table 3) was similar to the data from the other 7 subjects discussed in Experiment 3-A (Table 2). Even for the male subject with low serum uBPA after holding thermal receipt paper, BPA-G in serum increased between baseline and 90 min (Figure 4-C), and urine total BPA increased dramatically over the 90-min test, similar to the increase in total urine BPA in the women (Table 3). These findings show that high levels of uBPA could be detected in the systemic circulation of subjects after holding thermal receipt paper and eating 10 French fries. Levels of total BPA in urine at

Table 2. Unconjugated BPA (uBPA) and glucuronidated BPA (BPA-G) pharmacokinetic parameters for 4 male and 2 female subjects who held thermal receipt paper and ate French fries after using hand sanitizer (shown in Figure 5-A).

Analyte	Parameter	Male	Female	All
Serum uBPA	Baseline (ng/ml)	0.06±0.04	0.57±0.28	0.23±0.15
	Cmax (ng/ml)	5.66±1.25	10.24±0.77	7.19±1.26
	Maximum increase (ng/ml)	5.60±1.24	9.66±0.37	6.95±1.17
	Range of increase (ng/ml)	3.96–9.28	9.29–10.03	3.96–10.03
	Tmax (min)	60.00±12.25	15.00±0.00	45.00±12.25
	Range of Tmax	30.00–90.00	15.00	15.00–90.00
	Average AUC (0–90 min) (ng/ml)	3.92±1.08	3.85±0.04	3.90±0.75
Serum BPA-G	Baseline (ng/ml)	0.60±0.10	2.95±0.91	1.38±0.55
	Cmax (ng/ml)	1.49±0.39	7.74±1.75	3.57±1.42
	Maximum increase (ng/ml)	0.89±0.36	4.79±0.85	2.19±0.88
	Range of increase (ng/ml)	0.09–1.58	3.94–5.63	0.09–5.63
	Tmax (min)	75.00±8.66	75.00±15.00	75.00±6.71
	Range of Tmax	60.00–90.00	60.00–90.00	60.00–90.00
	Average AUC (0–90 min) (ng/ml)	1.05±0.21	6.48±1.93	2.86±1.25
	BPA-G/uBPA AUC (0–90 min) (ng/mL)	0.35±0.12	1.82±0.30	0.84±0.33
Urine Total BPA	Baseline (ng/ml)	0.15±0.04	1.10±0.58	0.46±0.24
	Baseline (μg/g creatinine)	0.20±0.09	1.22±0.24	0.54±0.19
	90 min (ng/ml)	23.36±6.66	10.62±3.16	19.11±4.32
	90 min (μg/g creatinine)	18.20±5.33	40.93±22.56	25.77±8.56

For these subjects blood was collected from the arm draining the hand that remained contaminated with BPA throughout the 90-min period of blood collection. Urine total BPA at baseline and at 90 min are presented as both actual concentration (ng/ml) and creatinine adjusted (μg BPA/g creatinine) values.
For the one female (Female #3; Figure 5-B) who had very high baseline serum uBPA (data not included in this table), the average serum AUC (0–90) values for uBPA and BPA-G were 6.05 and 4.49 ng/ml, respectively, and urine total BPA levels at baseline and at 90 min were 0.41 and 41.41-μg/g creatinine, respectively.

baseline or 90 min in these 3 subjects (Table 3) were similar to levels measured in the other 7 subjects (Table 2).

Experiment 4: Serum and urine BPA in men and women before and after transdermal exposure to BPA from thermal receipt paper with dry hands

We conducted this study with 12 male (age 27.7±1.6 yrs, BMI 26.9±0.9) and 12 female (age 25.8±1.6 yrs, BMI 25.2±1.5) subjects. Male and females subjects held a single 8×12 cm piece of thermal receipt paper in the non-dominant hand for 4 min, but unlike the prior experiment, no hand sanitizer was used prior to holding the thermal paper with a dry hand. No BPA was detected on the hands of any subject after washing and drying the hands prior to holding the thermal receipt paper. We did not determine the amount of BPA transferred to the hand immediately after holding the thermal receipt paper, since we had previously examined this (Figure 3-B). However, 30 min after holding the thermal paper BPA was swiped from the surface of the hand: for men (5.5±1.7 μg; range: 0.8–22.5 μg) and women (6.1±0.8 μg; range: 1.3–10.6 μg). There was no difference between males and females in the urine total BPA concentration at baseline or at 60 min, and there was also no difference between total urine BPA at baseline vs. 60 min for either males or females (Figure 5-A). There was a tendency (based on 2-tailed t-tests) for males to have higher serum uBPA at baseline (P=0.08) and after 30 min (P=0.06) relative to females (Figure 5-B). While there was no sex difference in conjugated BPA (cBPA), consisting of both BPA-G and BPA-S, at the baseline blood collection (Figure 5-C), at

30 min after holding the thermal receipt males had significantly higher serum conjugated BPA than females (ANOVA; P<0.001).

Discussion

Our data provide the first evidence that the use of very large amounts of free BPA as a developer on the print surface of thermal paper (~20 mg BPA/g paper) could be an important factor in accounting for the high levels of bioactive serum uBPA and urine total excreted BPA reported previously in various human populations [21]. We conducted this study to mimic aspects of the behavior of people in a fast-food restaurant where we have observed people using hand sanitizer and handling a thermal receipt for variable periods of time prior to picking up and eating food with their hands. In Figure 3 we show that holding a receipt for 45 sec immediately after using hand sanitizer containing dermal penetration enhancing chemicals resulted in the maximum amount of BPA that was swiped from the palm and fingers (581 μg BPA). After holding the receipt for 2 sec 40% of maximum was recovered from the hand, and within 15 sec 58% of maximum was recovered. Between 45 sec and 4 min, the amount of BPA recovered from the surface of the hand decreased, which may have been due to absorption into skin occurring at a greater rate than transfer to the skin from the thermal receipt. These findings show that a very large amount of BPA is transferred from thermal paper to a hand as a result of holding a thermal receipt for only a few seconds immediately after using a product with dermal penetration enhancing chemicals. The data in Figure 3-A also suggest that transdermal BPA absorption is very rapid due to the penetration enhancing chemicals in the hand sanitizer that we used, and thus

Table 3. Unconjugated BPA (uBPA) and glucuronidated BPA (BPA-G) pharmacokinetic parameters for one male and two female subjects who held thermal receipt paper and ate French fries after using hand sanitizer (shown in Figure 5-C).

Analyte	Parameter	Male	Female	All
Serum uBPA	Baseline (ng/ml)	0.06	0.53±0.30	0.37±0.24
	Cmax (ng/ml)	0.42	5.44±0.42	3.77±1.69
	Maximum increase (ng/ml)	0.37	4.91±0.13	3.40±1.53
	Range of increase (ng/ml)	0.37	4.80–5.03	0.37–5.03
	Tmax (min)	60.00	22.50±7.50	35.00±13.23
	Range of Tmax	60.00	15.00–30.00	15.00–60.00
	Average AUC (0–90 min) (ng/ml)	0.15	3.47±0.06	2.36±1.11
Serum BPA-G	Baseline (ng/ml)	0.38	1.22±0.36	0.94±0.35
	Cmax (ng/ml)	2.47	2.69±0.65	2.62±0.39
	Maximum increase (ng/ml)	2.09	1.47±0.30	1.68±0.27
	Range of increase (ng/ml)	2.09	1.18–1.77	1.18–2.09
	Tmax (min)	90.00	60.00	70.00±10.00
	Range of Tmax	90.00	60.00	60.00–90.00
	Average AUC (0–90 min) (ng/ml)	1.65	1.87±0.38	1.80±0.23
	BPA-G/uBPA AUC (0–90 min) (ng/mL)	11.00	0.54±0.12	4.03±3.49
Urine Total BPA	Baseline (ng/ml)	1.92	0.38±0.15	0.89±0.52
	Baseline (µg/g creatinine)	1.17	0.21±0.08	0.53±0.40
	90 min (ng/ml)	27.86	29.34±0.35	28.85±0.54
	90 min (µg/g creatinine)	27.35	14.11±1.16	18.53±4.46

For these subjects mixed systemic blood was collected from the uncontaminated arm over the 90 min after BPA exposure. Urine total BPA at baseline and at 90 min are presented as both actual concentration (ng/ml) and creatinine adjusted (µg/g creatinine) values.

measurement of BPA swiped from the surface of the hand likely underestimates the actual amount of free BPA transferred from the print surface of thermal paper. We note that since the thermal receipt paper is sold in rolls, the non-print surface has BPA transferred to it from the print surface (Figure 1). By swiping the two surfaces with ethanol on Kimwipes, the print surface was found to contain an 8.7-fold greater amount of BPA relative to the non-print surface of the thermal receipt paper roll used in these experiments (data not shown).

In both men and women there was a dramatic increase in serum uBPA after using hand sanitizer with dermal penetration enhancing chemicals and then holding thermal receipt paper and eating French fries with the BPA-contaminated hand (Figure 4-A and 4-B). While the sample size was small, our data suggest higher maximum serum levels (Cmax) for females and a greater maximum increase relative to baseline for both uBPA and BPA-G (the primary conjugated BPA metabolite) for females relative to males (Table 2). This finding was related to a greater transfer of BPA from the hand to the French fries and thus a greater oral dose (by about 4 fold) in females relative to males. However, we cannot rule out that the skin of females also allows greater transdermal transport of BPA relative to males due to sex differences in skin permeability [13]. In fact, our data are consistent with the hypothesis that a combination of both transdermal and buccal/sublingual absorption (Figure 2) resulted in the dramatic increase in both serum uBPA (Figure 4-A and 4-B) and total BPA excreted in urine (Table 2). The profile of serum uBPA suggests that females absorbed BPA more rapidly than

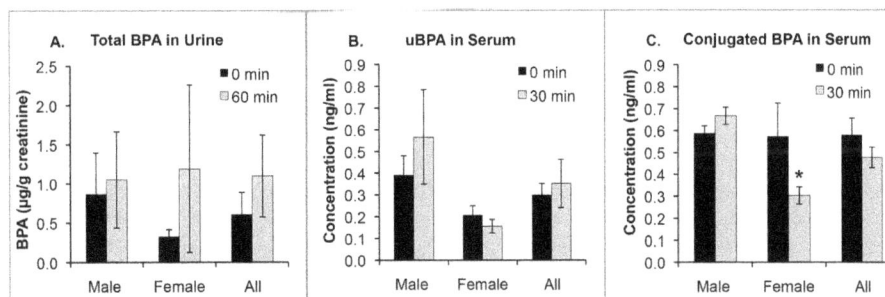

Figure 5. BPA in urine and serum of 12 men and 12 women who held thermal receipt paper with dry hands for 4 min, Panel A: the total concentration of BPA in urine (expressed relative to creatinine) at baseline and 60-min after holding the thermal receipt. Panel B: unconjugated BPA (uBPA) in serum at baseline and 30 min after holding the thermal receipt. Panel C: conjugated BPA (BPA-G and BPA-S) in serum at baseline and 30 min after holding the thermal receipt paper. * = significant difference between males and females (P<0.001).

males (Figure 4-A), consistent with females having a shorter time to reach the maximum serum level (Tmax) of uBPA (Table 2). The later Tmax in men than in women is consistent with men having a thicker stratum corneum (the outermost layer of the epidermis) relative to women [22,23]. In addition to skin thickness, another possible explanation for the sex differences we observed (Table 2) would be a greater use of skin moisturizers in females than in males, which could impact both the transfer of BPA to the hand from the surface of thermal paper as well as transdermal absorption of BPA. Our finding that males tended to have higher serum uBPA and had significantly higher serum conjugated BPA than women at 30 min after holding a receipt with a dry hand (Figure 5-B and 5-C) requires further study, since the more rapid absorption of BPA found for women after using hand sanitizer (Figure 4-A and 4-B) would have been missed at the 30-min blood collection time.

For the 3 females that had blood collected from the cubital vein in the same arm with the BPA contaminated hand (Figure 4-A and 4-B; Table 2), the maximum increase in serum uBPA relative to baseline (~10 ng/mL) was about two-times greater than the maximum increase found in the other two females whose blood was collected from the opposite uncontaminated arm (~5 ng/mL; Table 3; Figure 4-C). This difference between blood collected from a vein draining the contaminated hand vs. blood from the opposite uncontaminated arm (reflecting uBPA in the systemic circulation) suggests that a substantial amount of uBPA in blood from the contaminated arm was due to the BPA that was transdermally absorbed before its mixing in the general circulation. Our data from the cubital vein draining the contaminated hand that had BPA remaining on it for 90 min thus support the hypothesis of a higher arterial than mixed venous blood BPA concentration during the dermal absorption phase in the framework of a physiologically based pharmacokinetic (PBPK) model [24]. The hypothesis that served as the basis for collecting blood from the same arm in which BPA was being absorbed is that a portion of the BPA contaminated blood would be transported from the contaminated hand through the cubital vein and then to the heart. Subsequently, the contaminated blood would enter the arterial circulation and be transported to the tissues in the body, including endocrine target tissues. This leads to the prediction that during the dermal absorption of BPA, the BPA concentration in arterial blood is likely more relevant to consider in terms of exposure than the BPA concentration in the mixed venous blood, because blood in a vein draining the contaminated hand is not subjected to clearance by enzymes in the liver prior to reaching endocrine target tissues in arterial blood.

When examining all of our data for serum unconjugated and conjugated BPA, we show significant inter-subject variability in the absorption and clearance of BPA (Figure 4), which was also previously found for the estrogenic drug used in oral contraceptives, ethinylestradiol [25]. A particular concern is that there are individuals who have limited capacity to excrete BPA or other estrogenic compounds; one population at risk is patients with early-stage or advanced kidney disease [26,27].

The default method of administration of chemicals to animals by regulatory agencies is by intra-gastric gavage, regardless of how the chemical is used or whether there are known non-oral routes of exposure [28]. It is thus not surprising that US and European regulatory agencies [29,30] have modeled human exposure to BPA based on results from intra-gastric gavage administration of BPA to animals, which results in direct transport of BPA to the liver via the mesenteric vessels and extensive first-pass metabolism (detoxification) in the liver (Figure 2); the result is less than 1% of the gavage administered dose being bioavailable in blood [19,31].

However, Gayrard et al. [19] found high absorption and bioavailability (~70%) of BPA following sublingual administration that was dramatically different than the much lower bioavailability (<1%) of BPA following gavage administration in a parallel experiment. These findings directly challenge predictions that it is not possible to find the high blood levels of biologically active uBPA that have actually been measured in numerous human biomonitoring studies [21,32] but are currently being rejected for use in risk assessments by the US-Food and Drug Administration (US-FDA) as not plausible [31].

In contrast to the extremely high ratio of BPA-G to uBPA (> 100:1) predicted by gavage exposure studies due to rapid phase 2 metabolism in the liver, the average ratio of BPA-G to uBPA in our Experiment 3 was 0.84 ± 0.33 based on the average AUC (0-90 min) for the 6 subjects with blood collected from the arm with the contaminated hand (Figure 5-A; Table 2); this ratio was also low for the subjects with systemic blood collected from the uncontaminated arm (Table 3). These findings indicate that the primary route of BPA exposure was not via gastrointestinal absorption after eating the BPA-contaminated French fries, since this ratio would be predicted to exceed 100:1 [31,33]. One reason may have been that, in addition to transdermal absorption, the BPA transferred to the French fries would have been on the surface of the fries and thus easily absorbed by the highly vascularized epithelium in the mouth [19].

In the present study we measured total BPA in urine to be able to relate our findings to a very large epidemiological literature showing BPA in urine to be correlated with abnormal development and diseases in children and adults [5,34]. The geometric mean for adults in the 95th percentile for total BPA in urine reported in NHANES 2003/4 was about 11 μg/g creatinine [2] and the range of values at the 95th percentile include values we measured here. Periodically, BPA levels exceeding those we found are reported in studies that measured BPA in urine [21], and it is possible that those assays are of people who had very recently been exposed to BPA in a manner similar to our experiment.

Our findings thus provide evidence regarding how some people could be found to have very high urine levels of BPA. Importantly, the amount of total BPA in urine was ~20 μg BPA/g creatinine (~20 ng/mL urine uncorrected for creatinine). This high level of urine total BPA collected 90 min after using hand sanitizer and holding a thermal receipt (Table 2 and Table 3) has been associated with a significant increase in the likelihood of developing cardiovascular disease and type 2 diabetes [35,36]. BPA levels in human urine have also been related to a wide range of other diseases in over 60 human epidemiological studies [5,34]. Published findings include: reproductive effects in women (polycystic ovary syndrome, altered ovarian response to hormones, reduced fertilization success, implantation failure, endometrial disorders, reduced embryo quality, miscarriage, premature delivery and breast cancer), reproductive effects in men (reduced libido, sperm quality, altered sex hormone concentrations and embryo quality), altered thyroid hormone concentrations, obesity, impaired liver function, impaired immune and kidney function, inflammation, and neurobehavioral deficits such as aggressiveness, hyperactivity and impaired learning [5,34]. The estimate of the costs per year of additional cases of just cardiovascular disease in the USA attributable to BPA is 1.5 billion dollars [37].

In a study that involved handling thermal receipts (without using hand sanitizer) continuously for 2 hr, which would be relevant for a cashier, there was a significant increase in urine total BPA relative to baseline [38]. This finding is consistent with prior data that cashiers have higher levels of BPA in urine than the general public [39]. Blood concentrations of BPA were not determined in these studies, and they also did not take into account that perhaps 50% of the receipts handled may have contained BPS rather than BPA (Table 1). The results of these studies indicate that with repeated handling of thermal receipts in an occupational setting, even without the use of hand sanitizer, there is a significant increase in BPA exposure. Future studies involving handling of thermal paper need to include analysis of the thermal paper to determine if the developer used is BPA or some other chemical. Related to the issue of occupational exposure to BPA is our observation that at least one big-box store in Columbia, Missouri provides hand sanitizer dispensers for use by all cashiers, and our data suggest this can not only markedly increase transfer of BPA from the thermal paper to hands but also increase transdermal absorption of BPA.

Our findings that thermal receipt paper is a potential source of high exposure to BPA are supported by data showing that BPA readily leaches from thermal receipts and thus likely contaminates anything that a receipt contacts. Thus, environmental contamination caused by the use of unpolymerized (free) BPA in thermal paper is widespread [7]. The dermal penetration enhancing chemicals present in personal care products as well as hand sanitizers cause a breakdown of the dermal barrier leading to an increase in transdermal absorption [8,9]. While BPA was reported to be absorbed through pig and human skin *in vitro* [40], our data show after holding a receipt for 60 sec, there was 185-times more BPA transferred to a wet hand due to holding thermal receipt paper immediately after using hand sanitizer with penetration enhancing chemicals as opposed to when the hands were dry (Figure 3). The specific mixtures of chemicals used in products will impact transdermal exposure to environmental chemicals such as BPA [8,9], and additional research is needed to determine the degree to which alcohols and other chemicals impact exposures. This is important because when soap and water are not available, hand sanitizers are recommended to reduce infectious disease transmission [http://www.cdc.gov/handwashing/when-how-handwashing.html]. It is also important to determine the length of time after using skin-care products with dermal penetration enhancing chemicals that there is an impact on absorption of environmental contaminants.

The issue of assay performance is obviously very important and was examined using a round robin validation process in Europe for a number of chemicals, including BPA, which identified that some laboratories were able to accurately assay uBPA and other chemicals without contamination, while other laboratories were unable to assay uBPA or other chemicals accurately [41]. Importantly, our findings reported here are based on measurement of uBPA with a sensitive, validated, contamination-free LC/MSMS assay. Specifically, our laboratory is one of three in the U.S. that recently successfully completed a NIH-sponsored round-robin measuring uBPA in human serum [17]. In addition, in the present experiment, the time course of blood uBPA concentrations after a controlled BPA exposure in our university Clinical Research Center (Figure 4) is not consistent with any spurious contamination; this conclusion was also supported by the use of field blanks. It is thus clear that the prediction that any finding of uBPA in human serum must be due to sample contamination [31] is not valid, even though a few investigators report being unable to control BPA contamination in their assays [42,43]. Supporting our conclusion is a report from the CDC in which sources of contamination were identified and systematically eliminated during the successful development of assays for BPA and three

other chemicals [44]. The issue of the potential for assay contamination is thus not unique to BPA and simply requires the use of standard assay procedures and appropriate controls that should be routinely employed.

Conclusions

Thermal paper requires a chemical in the surface coating as a print developer. The current preferred developers, BPA and BPS, have both been shown to have estrogenic activity [45,46]. This is leading to widespread exposure to both of these endocrine-disrupting chemicals [7,47], and BPS is more persistent in the environment relative to BPA and is thus an unacceptable replacement for BPA [18,48]. A recent EPA report examined 19 alternative chemicals, including BPS, that could potentially replace BPA as a developer in thermal paper and concluded that "No clearly safer alternatives to BPA were identified in this report; most alternatives have Moderate or High hazard designations for human health or aquatic toxicity endpoints" [18]. The report identified that "decision makers may wish to consider alternative printing systems". Two of the papers screened for our current study employed a developer other than BPA or BPS that was not estrogenic in a MCF-7 human breast cancer cell proliferation assay (data not shown), but lack of estrogenic activity does not imply safety, as indicated in the EPA report.

Thermal paper is a major source of BPA contamination in recycled paper, and its use results in the widespread contamination of other products and the environment [49] due to the presence of large amounts of free, unpolymerized BPA in the surface coating of thermal paper (Figure 1). Further, our findings are consistent with other data reporting that BPA can be transferred from the surface of thermal paper to items it contacts. Because no safe alternatives to the use of BPA or its primary replacement chemical BPS in thermal paper have been identified, our findings provide support for the EPA's recommendation that thermal paper should be replaced with other safer technologies [18].

Our study provides the first data that thermal paper may be a significant factor in accounting for high levels of bioactive BPA in human serum and total BPA in urine that have been associated with diseases that are increasing in frequency in human populations [21,34]. Our findings also suggest that the impact of the use of dermal penetration enhancing chemicals in skin care products on transdermal absorption of environmental contaminants should be taken into consideration in risk assessments and should be a priority for future research.

Supporting Information

File S1 Section S1: Sample handling, extraction and assay methods. Section S2: Table of individual BPA and BPS values for 50 thermal receipt papers. Section S3: List of questions asked each subject.

Author Contributions

Conceived and designed the experiments: FVS AMH JAT. Performed the experiments: AMH JAT CLM. Analyzed the data: JAT PLT RWS MRE FVS. Contributed reagents/materials/analysis tools: FVS. Contributed to the writing of the manuscript: FVS AMH CLM JAT SCN WVW PLT RWS.

References

1. GrandViewResearch (2014) Global bisphenol A (BPA) market by appliation (appliances, automotive, consumer, construction, electrical & electronics) expected to reach USD 20.03 billion by 2020. http://www.digitaljournal.com/pr/2009287, Accessed June 24, 2014. http://www.digitaljournal.com/pr/2009287, Accessed June 24, 2014.

2. Calafat AM, Ye X, Wong LY, Reidy JA, Needham LL (2008) Exposure of the U.S. population to bisphenol A and 4-tertiary-octylphenol: 2003–2004. Environ Health Perspect 116: 39–44.

3. Liao C, Kannan K (2012a) Determination of free and conjugated forms of bisphenol a in human urine and serum by liquid chromatography-tandem mass spectrometry. Environ Sci Technol 46: 5003–5009.

4. Welshons WV, Thayer KA, Judy BM, Taylor JA, Curran EM, et al. (2003) Large effects from small exposures. I. Mechanisms for endocrine-disrupting chemicals with estrogenic activity. Environ Health Perspect 111: 994–1006.

5. Vandenberg LN, Ehrlich S, Belcher SM, Ben-Jonathan N, Dolinoy DC, et al. (2013a) Low Dose Effects of Bisphenol A: An Integrated Review of In Vitro, Laboratory Animal and Epidemiology Studies. Endocrine Disruption 1: E1–E20.

6. Mendum T, Stoler E, VanBenschoten H, Warner JC (2011) Concentration of bisphenol A in thermal paper. Green Chemistry Letters and Reviews 4: 81–86.

7. Liao C, Kannan K (2011) Widespread Occurrence of Bisphenol A in Paper and Paper Products: Implications for Human Exposure. Environ Sci Technol 45: 9372–9739.

8. Funke AP, Schiller R, Motzkus HW, Gunther C, Muller RH, et al. (2002) Transdermal delivery of highly lipophilic drugs: in vitro fluxes of antiestrogens, permeation enhancers, and solvents from liquid formulations. Pharm Res 19: 661–668.

9. Karande P, Mitragotri S (2009) Enhancement of transdermal drug delivery via synergistic action of chemicals. Biochim Biophys Acta 1788: 2362–2373.

10. EFSA (2013) Draft scientific opinion on the risks to public health related to the presence of bisphenol A (BPA) in foodstuffs – exposure assessment; EFSA Panel on Food Contact Materials, Enzymes, Flavourings and Processing Aids; European Food Safety Authority (EFSA), Parma, Italy, July 1.

11. Xue J, Zartarian V, Moya J, Freeman N, Beamer P, et al. (2007) A meta-analysis of children's hand-to-mouth frequency data for estimating nondietary ingestion exposure. Risk Anal 27: 411–420.

12. Heffernan AL, Aylward LL, Samidurai AJ, Davies PS, Toms LM, et al. (2014) Short term variability in urinary bisphenol A in Australian children. Environ Int 68C: 139–143.

13. Singh I, Morris AP (2011) Performance of transdermal therapeutic systems: Effects of biological factors. Int J Pharm Investig 1: 4–9.

14. IBISWorld (2012) Hand sanitizer manufacturing in the US: Market research report. Feb 2012. http://www.ibisworld.com/industry/hand-sanitizer-manufacturing.html.

15. Brand RM, Charron AR, Sandler VL, Jendrzejewski JL (2007) Moisturizing lotions can increase transdermal absorption of the herbicide 2,4-dichlorophenoxacetic acid across hairless mouse skin. Cutan Ocul Toxicol 26: 15–23.

16. StatisticsBrain (2014) Fast food statistics, http://www.statisticbrain.com/fast-food-statistics/, Accessed March 26.Statistic Brain.

17. Vandenberg LN, Gerona RR, Kannan K, Taylor JA, van Breemen RB, et al. (2014a) A round robin approach to the analysis of bisphenol a (BPA) in human blood samples. Environ Health 13: 25.

18. EPA (2014) Bisphenol A alternatives in thermal paper. United States Environmental Protection Agency, Final document, January 2014. http://www.epa.gov/dfe/pubs/projects/bpa/bpa-report-complete.pdf. Accessed 9/9/14.

19. Gayrard V, Lacroix MZ, Collet SH, Viguie C, Bousquet-Melou A, et al. (2013) High bioavailability of bisphenol a from sublingual exposure. Environ Health Perspect 121: 951–956.

20. vom Saal FS, Vandevoort CA, Taylor JA, Welshons WV, Toutain PL, et al. (2014) Bisphenol A (BPA) pharmacokinetics with daily oral bolus or continuous exposure via silastic capsules in pregnant rhesus monkeys: Relevance for human exposures. Reprod Toxicol 45: 105–116.

21. Vandenberg LN, Chahoud I, Heindel JJ, Padmanabhan V, Paumgartten FJ, et al. (2010a) Urinary, circulating, and tissue biomonitoring studies indicate widespread exposure to bisphenol A. Environ Health Perspect 118: 1055–1070.

22. Fitzmaurice S, Maibach HI (2010) Gender differences in skin. In: M.A F, Miller KM, Maibach HI, editors. Textbook of Aging Skin. New York: Springer. 999–1017.

23. Polak S, Ghobadi C, Mishra H, Ahamadi M, Patel N, et al. (2012) Prediction of concentration-time profile and its inter-individual variability following the dermal drug absorption. J Pharm Sci 101: 2584–2595.

24. Levitt DG (2004) Physiologically based pharmacokinetic modeling of arterial - antecubital vein concentration difference. BMC Clin Pharmacol 4: 2–7.

25. Goldzieher JW, Stanczyk FZ (2008) Oral contraceptives and individual variability of circulating levels of ethinyl estradiol and progestins. Contraception 78: 4–9.

26. You L, Zhu X, Shrubsole MJ, Fan H, Chen J, et al. (2011) Renal function, Bisphenol A, and Alkylphenols: Results from the National Health and Nutrition Examination Survey (NHANES 2003–2006). Environ Health Perspect 119: 527–533.

27. Krieter DH, Canaud B, Lemke HD, Rodriguez A, Morgenroth A, et al. (2013) Bisphenol A in chronic kidney disease. Artif Organs 37: 283–290.

28. Vandenberg LN, Welshons WV, Vom Saal FS, Toutain PL, Myers JP (2014b) Should oral gavage be abandoned in toxicity testing of endocrine disruptors? Environ Health 13: 46.

29. EFSA (2008) Toxicokinetics of Bisphenol A - Scientific Opinion of the Panel on Food additives, Flavourings, Processing aids and Materials in Contact with Food (AFC). European Food Safety Authority July, 2008. http://www.efsa.europa.eu/EFSA/efsa_locale-1178620753812_1211902017492.htm. Accessed September 2, 2014.

30. FDA (2008a) Food and Drug Administration draft assessment of bisphenol A for use in Food contact applications, August 14, 2008. http://www.fda.gov/food/ingredientspackaginglabeling/foodadditivesingredients/ucm166145.htm. watAccessed August 6, 2014.

31. Patterson TA, Twaddle NC, Roegge CS, Callicott RJ, Fisher JW, et al. (2013) Concurrent determination of bisphenol A pharmacokinetics in maternal and fetal rhesus monkeys. Toxicol Appl Pharmacol 267: 41–48.

32. Vandenberg LN, Hauser R, Marcus M, Olea N, Welshons WV (2007) Human exposure to bisphenol A (BPA). Reproductive Toxicology 24: 139–177.

33. Gayrard V, Lacroix MA, Collet SH, Viguié C, Bousquet-Melou A, et al. (2013) Interpreting Bbsphenol A absorption in the canine oral cavity. Environ Health Perspect 121: A323–A324.

34. Rochester JR (2013) Bisphenol A and human health: A review of the literature. Reprod Toxicol 42: 132–155.

35. Lang IA, Galloway TS, Scarlett A, Henley WE, Depledge M, et al. (2008) Association of urinary bisphenol A concentration with medical disorders and laboratory abnormalities in adults. JAMA 300: 1303–1310.

36. Melzer D, Osborne NJ, Henley WE, Cipelli R, Young A, et al. (2012) Urinary bisphenol a concentration and risk of future coronary artery disease in apparently healthy men and women. Circulation 125: 1482–1490.

37. Trasande L (2014) Further limiting bisphenol a in food uses could provide health and economic benefits. Health Aff (Millwood) 33: 316–323.

38. Ehrlich S, Calafat AM, Humblet O, Smith T, Hauser R (2014) Handling of thermal receipts as a source of exposure to bisphenol A. JAMA 311: 859–860.

39. Braun JM, Kalkbrenner AE, Calafat AM, Bernert JT, Ye X, et al. (2011) Variability and predictors of urinary bisphenol A concentrations during pregnancy. Environ Health Perspect 119: 131–137.

40. Zalko D, Jacques C, Duplan H, Bruel S, Perdu E (2011) Viable skin efficiently absorbs and metabolizes bisphenol A. Chemosphere 82: 424–430.

41. Vanderford BJ, Drewes JE, Eaton A, Guo YC, Haghani A, et al. (2014) Results of an Interlaboratory Comparison of Analytical Methods for Contaminants of Emerging Concern in Water. Anal Chem 86: 774–782.

42. Volkel W, Colnot T, Csanady GA, Filser JG, Dekant W (2002) Metabolism and kinetics of bisphenol a in humans at low doses following oral administration. Chem Res Toxicol 15: 1281–1287.

43. Churchwell MI, Camacho L, Vanlandingham MM, Twaddle NC, Sepehr E, et al. (2014) Comparison of life-stage-dependent internal dosimetry for bisphenol A, ethinyl estradiol, a reference estrogen, and endogenous estradiol to test an estrogenic mode of action in Sprague Dawley rats. Toxicol Sci 139: 4–20.

44. Ye X, Zhou X, Hennings R, Kramer J, Calafat AM (2013) Potential external contamination with bisphenol A and other ubiquitous organic environmental chemicals during biomonitoring analysis: an elusive laboratory challenge. Environ Health Perspect 121: 283–286.

45. Vinas R, Watson CS (2013) Bisphenol S disrupts estradiol-induced nongenomic signaling in a rat pituitary cell line: effects on cell functions. Environ Health Perspect 121: 352–358.

46. Welshons WV, Nagel SC, vom Saal FS (2006) Large effects from small exposures. III. Endocrine mechanisms mediating effects of bisphenol A at levels of human exposure. Endocrinol 147: S56–S69.

47. Liao C, Liu F, Alomirah H, Loi VD, Mohd MA, et al. (2012) Bisphenol S in Urine from the United States and Seven Asian Countries: Occurrence and Human Exposures. Environ Sci Technol 46: 6860–6866.

48. EnvironmentCanada (2008) Screening Assessment for The Challenge Phenol, 4,4′ -(1-methylethylidene) bis- (Bisphenol A) Chemical Abstracts Service Registry Number 80-05-7. Environment Canada and Health Canada, Ottawa, October 2008.

49. Liao C, Kannan K (2012b) High levels of bisphenol a in paper currencies from several countries, and implications for dermal exposure. Environ Sci Technol 45: 6761–6768.

8

Relationship between Air Pollutants and Economic Development of the Provincial Capital Cities in China during the Past Decade

Yunpeng Luo[1], Huai Chen[1,2]*, Qiu'an Zhu[1], Changhui Peng[1,3]*, Gang Yang[1], Yanzheng Yang[1], Yao Zhang[1]

1 State Key Laboratory of Soil Erosion and Dryland Farming on the Loess Plateau, College of Forestry, Northwest A&F University, Yangling, Shaanxi, China, 2 Chengdu Institute of Biology, Chinese Academy of Sciences, Chengdu, China, 3 Center of CEF/ESCER, Department of Biology Science, University of Quebec at Montreal, Montreal, Canada

Abstract

With the economic development of China, air pollutants are also growing rapidly in recent decades, especially in big cities of the country. To understand the relationship between economic condition and air pollutants in big cities, we analysed the socioeconomic indictorssuch as Gross Regional Product per capita (GRP per capita), the concentration of air pollutants (PM_{10}, SO_2, NO_2) and the air pollution index (API) from 2003 to 2012 in 31 provincial capitals of mainland China. The three main industries had a quadratic correlation with NO_2, but a negative relationship with PM_{10} and SO_2. The concentration of air pollutants per ten thousand yuan decreased with the multiplying of GRP in the provinical cities. The concentration of air pollutants and API in the provincial capital cities showed a declining trend or inverted-U trend with the rise of GRP per capita, which provided a strong evidence for the Environmental Kuznets Curve (EKC), that the environmental quality first declines, then improves, with the income growth. The results of this research improved our understanding of the alteration of atmospheric quality with the increase of social economy and demonstrated the feasibility of sustainable development for China.

Editor: Xiaoyan Yang, Chinese Academy of Sciences, China

Funding: This study was supported by 100 Talents Program of The Chinese Academy of Sciences, by Program for New Century Excellent Talents in University (NCET-12-0477), by the National Natural Science Foundation of China (No. 31100348), by a Natural Sciences and Engineering Research Council of Canada (NSERC) discovery grant, and by China's QianRen program. The funders had no role in study design, data collection and analysis, decision to publish, or preparation of the manuscript.

Competing Interests: The authors have declared that no competing interests exist.

* Email: chenhuai81@gmail.com (HC); cpeng86@gmail.com (CP)

Introduction

China has seen economic soaring in the past three decades, with its gross domestic product (GDP) expanding 140 times during 1978–2012 (National Bureau of Statistics, 2013). However, such economic soaring is accompanied with deterioration of the atmospheric quality. In the first three months of 2013, just like what happened in London in 1952 [1], long-time haze influenced large area of China (Fig S1), which further stimulated the strong demand for improvement of air quality.

Air pollution has significant influence on both climate and human health [2,3]. Oxidising air pollutants like ozone stimulate reactions to produce more greenhouse gases which exacerbate global warming [4]. Besides, decreasing precipitation and increasing dimness [5,6], widening of the tropics [7], weakening of summer moonsoon in South Asian [4,8], as well as large-scale ocean circulation and some extreme weather like hurricane [9,10], are all linkd to air pollution. Moreover, anthropogenic air pollutants, especially particulate matter is extremely harmful to human health. According to Silva *et al.* (2013), more than 2 million premature deaths are associated with $PM_{2.5}$-related diseases [11]. Research results from Spain and England reported that long-term exposure to air pollution mainly explained heart disease morbidity and mortality [12,13]. Similarly, the heating policy in Northern China was found to cause reduction in life expectancies of Northern residents by about 5.52 years [14]. Given the great influece of air pollution on natural environment and human life, researches are attaching ever greater importance to the causes and effects of air pollution [4,15–17].

Environmental problems result from economic expansion which increases extraction of natural resources and accumulation of waste, in the end exceeding the carrying capacity of the biosphere to the pollutants [18]. From the perspective of the history of human society economy development, the environmental quality is not fixed along a country's development path [19–21]. In the 1990s, scientists found an inverted U-shaped relationship between environmental quality and social income [22–25]. Such relationship was defined as Environmental Kuznets Curve (EKC), showing that the environmental quality would first deteriorate with the increase in revenue, and then it would improve when incomes rise to a certain level [25]. Numerous research results related to developed countries have identified EKC curve between income and air pollutants, especially in the Organization for Economic Co-operation and Development (OECD) countries

[18,23,26]. However, the relationship between income and air pollutants varies considerably among developing countries. For example, the same air pollutant sulfur dioxide (SO_2) showed an inverted U-shape relationship with income for Tunisia, but an N-shape relationship for Turkey [27,28].

China is the biggest developing country in the world, whose high-speed economic development as well as environmental changes and protection may provide experiences and lessons for other developing countries in this respect. There are also some studies regarding EKC in China [29,30]. However, most of them focused on econometrics, without relating air pollutants to specific levels of economic development. Considering the disparate economy development paces of different provinces in this big country, analysis about how particular air pollution is related to economic development of each region is needed [31]. Moreover, EKC researches in China concentrated on comprehensive indictors like the total amount of atmospheric emission [32–35]. Although some researches studied specific pollutants like SO_2 or PM [36–37], the relationship between the most important three categories of air pollutants and socioeconomic indicators is not adequately reported.

In this study, we aimed to establish regression models to fit the relationship between air pollutants and the three major industries (the primary, secondary and tertiary industries), so as to reveal the relationship between industry and air quality deterioration. We also caculated the ratio of air pollutant concentration to Gross Regional Product (GRP) per capita in order to know the contribution of economic development to air pollution over time. Finally, regression analysis was conducted to verify the existence of EKC in Chinese cities, or to define the otherwise relationship between air pollutants and revenue of Chinese citizens.

Datasets and Methods

1. Study area and data source

Data were collected for 31 provincial capitals in mainland China, which are representative of the general condition of each province (Fig. 2). In order to investigate the relationship between social economy and concentration of environmental pollutants in China, we downloaded data about these two aspects in the database of the National Bureau of Statistics of China (Table S1). The economic data included GRP, population, primary industry output, secondary industry output and tertiary industry output. The pollutant data collected included the concentration of PM_{10}, SO_2 and NO_2 (the three most important air pollutants in China [38]). The air pollution index (API) was calculated with the following formula:

$$I_j = \frac{I_{high} - I_{low}}{C_{high} - C_{low}}(C - C_{low}) + I_{low}$$

Where:

I = (Air pollution) index of one specific pollutant,
C = pollutant concentration,
C_{low} = the concentration breakpoint $\leq C$,
C_{high} = the concentration breakpoint $\geq C$,
I_{low} = the index breakpoint corresponding to C_{low},
I_{high} = the index breakpoint corresponding to C_{high},
j = Air pollutants indicators (PM_{10}, SO_2, NO_2).

$$API = Max(I_j)$$

The criteria of breakpoints for air pollutants were taken from the website of Ministry of Environment Protection of the People's Republic of China [39].All data from 2003 to 2012 used for statistical analyses were retrieved from National Bureau of Statistics of China. As the demographic data for Lhasa during 2003–2006 and 2010 was missing, we did not do analyze the city for these years.

2. Relationship between air pollutants and the three main industries

Linear, quadratic and cubic regression analysis was conducted to examine whether there existed simple positive or negative relationship between the concentration of air pollutants (dependent variables) and the output per capita of the three industries (independent variables) in the provincial capital cities. We chose the best appropriate regression model for each air pollutant and industry and plotted the regression line for those which were significantly correlated.

3. Trend analysis for the ratio of pollutant concentration to industry output

For comparing socioeconomic development level in different regions in China, we classified all the 31 provincial capital cities of mainland China into four economic regions including East Coast (East), Central China (Central), Northeastern China (Northeast), and Western China (West), according to strategies promulgated by the Central People's Government [40]. The ratio of annual air pollutant concentration to GRP per capita (c PM_{10}/GRP per capita, c SO_2/GRP per capita, and c NO_2/GRP per capita) was calculated year by year for each region. Hereafter, the line trend plots of the ratios were constructed to illustrate the variation of energy efficiency during 2003 to 2012.

4. Analysis associated with EKC

In order to investigate whether EKC exists in China, regression methods were applied to the panel data of GRP per capita and pollutants' indicators in all provincial capital cities. We also conducted regression fitting for the four economic regions for further information. The relationships between air pollutants and GRP per capita were estimated by the simplified EKC model provided below, which was also described by Agras et al. (41) and Li et al. (42)[41–42]:

$$E_{ij} = \alpha_{ij} + \beta_1 X_{ij} + \beta_2 X_{ij}^2 + \beta_3 X_{ij}^3$$

Where E is the concentration of air pollutant; X is GRP per capita; α_{ij} is a fixed effect; ε_{ij} is a stochastic error term; i is a region index (region values are "East, Central, Northeast, West and All provincial capital cities"); j is an air pollutant indicator (PM_{10}, SO_2, NO_2 or API); β_1, β_2, β_3 are the coefficient for the income variable, for the income squared variable and for the income cubic term, respectively.

5. Statistical analysis

All the regression analysis related to air pollutants and socioeconomic indicators was performed with SPSS for Windows (IBM SPSS statistics; Version 20). The effect of a certain variable was considered statistically significant for $P < 0.05$. Annual mean values of data used for trend analysis of energy effiency between 2003 to 2012 were caculated by Excel 2010.

Table 1. Regression for concentration of PM_{10}, SO_2, NO_2 and the three main industries.

Described Relationship	Model summary					Coefficients T test	
	Regression						
	Model	n	R^2	SE	Sig.	Independent variable	Constant
PM_{10} & Primary industry	Linear	305	0.147	0.026	0.000**	0.000**	0.000**
	Quadratic		0.178	0.025	0.000**	0.000**	0.000**
	Cubic		0.203	0.025	0.000**	0.000**	0.000**
SO_2 & Primary industry	Linear		0.141	0.020	0.000**	0.000**	0.000**
	Quadratic		0.163	0.020	0.000**	0.000**	0.000**
	Cubic		0.172	0.020	0.000**	0.001**	0.000**
NO_2 & Primary industry	Linear		0.004	0.013	0.280		
	Quadratic		0.007	0.013	0.348		
	Cubic		0.016	0.013	0.172		
PM_{10} & Secondary industry	Linear	305	0.025	0.027	0.005**	0.005**	0.000**
	Quadratic		0.034	0.027	0.006**		
	Cubic		0.046	0.027	0.003**	0.035*	0.000**
SO_2 & Secondary industry	Linear		0.018	0.022	0.018*	0.018*	0.000**
	Quadratic		0.022	0.022	0.041*	0.873	0.000
	Cubic		0.021	0.022	0.092		
NO_2 & Secondary industry	Linear		0.097	0.012	0.000**	0.000**	0.000**
	Quadratic		0.118	0.012	0.000**	0.000**	0.000**
	Cubic		0.118	0.012	0.000**	0.100**	0.000**
PM_{10} & Tertiary industry	Linear	305	0.029	0.027	0.003**	0.003**	0.000**
	Quadratic		0.037	0.027	0.003**	0.012*	0.000**
	Cubic		0.044	0.027	0.003**	0.015*	0.000**
SO_2 & Tertiary industry	Linear		0.039	0.021	0.001**	0.001**	0.000**
	Quadratic		0.039	0.022	0.002**	0.235	0.000**
	Cubic		0.046	0.021	0.003**	0.064	0.000**
NO_2 & Tertiary industry	Linear		0.120	0.012	0.000**	0.000**	0.000**
	Quadratic		0.135	0.012	0.000**	0.000**	0.000**
	Cubic		0.136	0.012	0.000**	0.225	0.000**

* $P<0.05$; ** $P<0.01$.

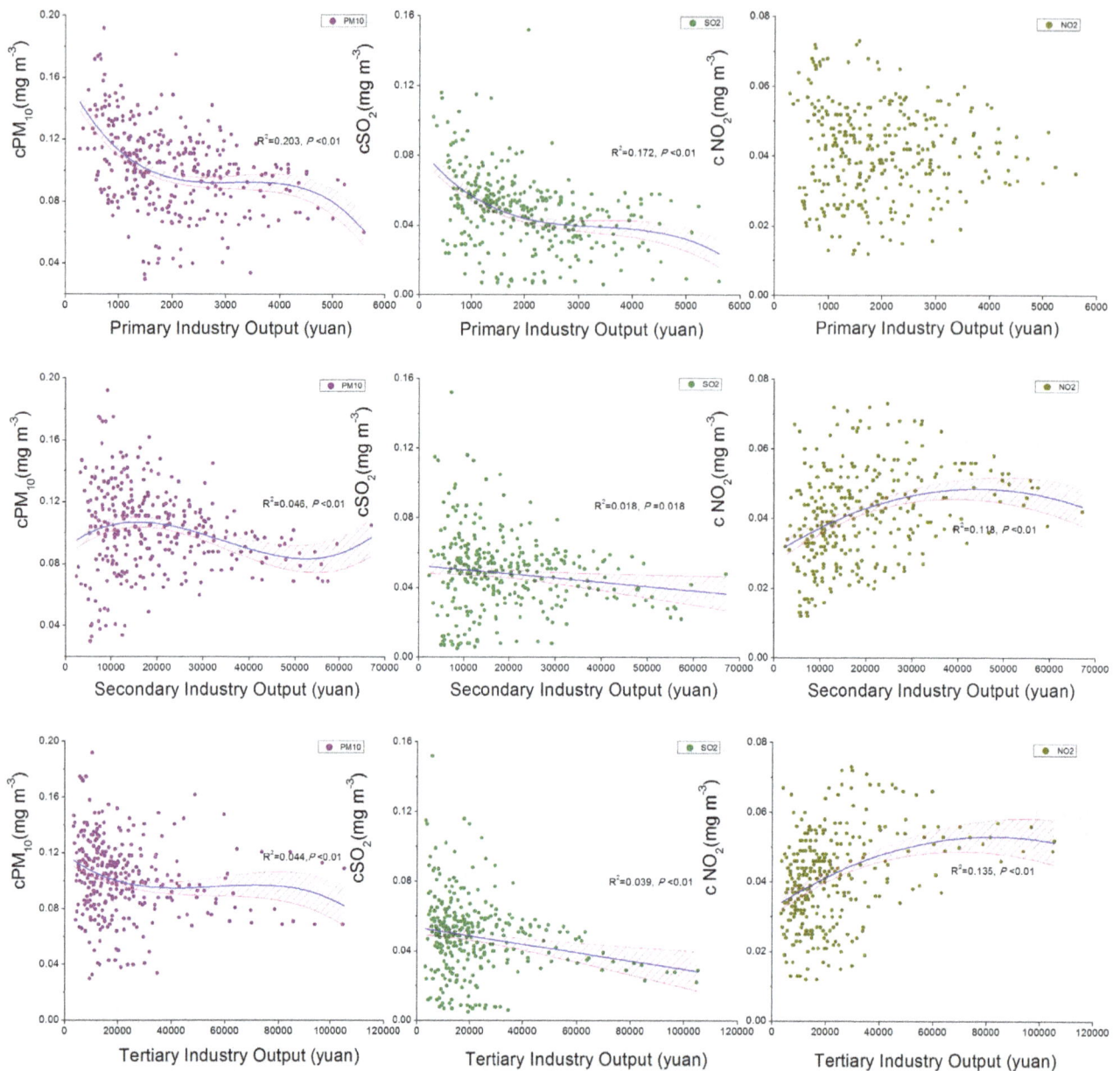

Figure 1. Air pollutant concentrations as related to the output per capita of three industries in the provincial capitals of China. (a) The output per capita of the primary industry; (b) The output per capita of the secondary industry; (c) The output per capita of the tertiary industry.

Results

1. Relationship between air pollutants and the three main industries in Chinese cities

Analysis on the provincial capital cities illustrated quadratic relationships between concentration of NO_2 and the output per capita of secondary and tertiary industries (Table 1, Fig. 1). The NO_2 concentration rose with the increase of the output per capita of the secondary and tertiary industries at the first stage, then began to decrease when the output reached around 45,000 and 70,000 yuan respectively. However, there was no remarkable relationship between that of the primary industry and the NO_2 concentration. The results also indicated that all the three industries had significantly negative relationship with the concentration of PM_{10} and SO_2.

2. Variation of efficiency ratio in recent years

The ratio of the air pollutant concentration to GRP per capita (cPM_{10}/GRP per capita, cSO_2/GRP per capita and cNO_2/GRP per capita) had a steady declining trend in the four economic regions, especially in the western mainland China, showing a notable enhancement of energy efficiency (Fig. 2). The cPM_{10}/GRP per capita ratio fell from 0.103 mg m^{-3} (ten thousand yuan)$^{-1}$ in 2003 to 0.018 in the year of 2012 by 470% in the West, from 0.064 to 0.015 in the Northeast by 320%, from 0.081 to 0.014 in the Central by 470%, and from 0.043 to 0.009 mg m^{-3} (ten thousand yuan)$^{-1}$ in the East by 370%. The ratios of cSO_2/GRP per capita and cNO_2/GRP per capita also showed analogous disparity among the four regions, with the variation range of efficiency ratio the smallest in the East (Fig. 2).

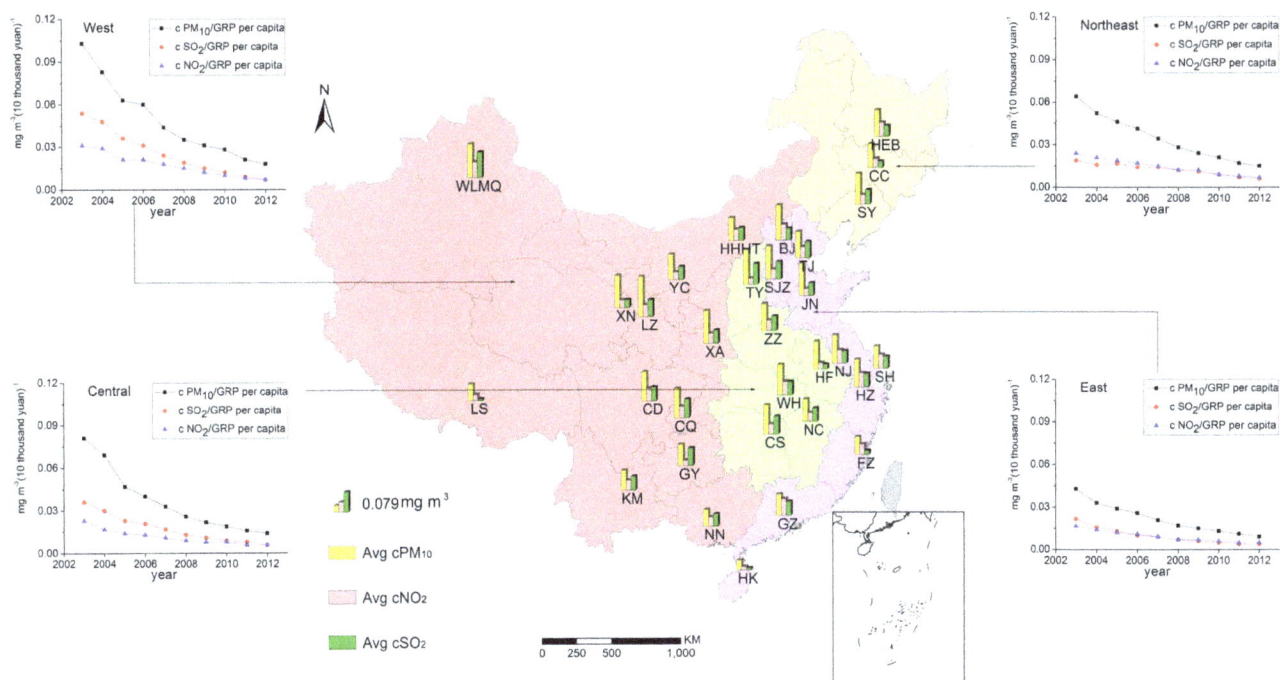

Figure 2. Annual mean concentration of PM$_{10}$, SO$_2$ and NO$_2$ from 2003 to 2012 in different province capitals of mainland China (bar charts on the Chinese map). Four line charts represent the relationships between annual mean air pollutant and GRP per capita of the East, Central, Northeast and West China respectively from 2003 to 2012.

3. EKC analysis in all the provincial capital cities and the four economic regions

The relationship between air pollutants and GRP per capita in all the provincial capitals is presented in Fig. 3. The concentration of PM$_{10}$ and SO$_2$ or API had a significantly negative linear relationship with GRP per capita; meanwhile the concentration of NO$_2$ had a quadratic relationship with GRP per capita. However, the relationship between air pollutants and GRP per capita was not the same for the four economic regions (Table 2, 3). The PM$_{10}$ concentration was significantly related to GRP per capita only in the Central. Similarly, API was also significantly related to GRP per capita only in the Central. The SO$_2$ concentration had significant negative linear relationship with GRP per capita in the Central but positive linear relationship in the Northeast. The NO$_2$ concentration was positively related to the GRP per capita in the Central, West, and quadratic for the East region, but not significantly related to that in the Northeast.

Discussion

1. The relationship between the three main industries and air pollutants

Our results (Fig. 1) showed the quadratic relationship between the secondary and tertiary industries and NO$_2$ in the provincial capital cities of China. The increase of NO$_2$ concentration was probably caused by the continuous increase of civil vehicles (The civil vehicles number increased from 1.36 million to 78.0 million according to the National Bureau of Statistics) and the widespread use of transportation in many fields such as tourism. This was in agreement with the point of view ascribing anthropogenic pollution to combustion of fossil fuel [43–44]. Fortunately, the concentration of NO$_2$ began to descend as the output of secondary and tertiary industries came to a certain level, probably due to the

increasing energy efficiency (Fig. 2) and environmental-friendly measurements such as transportation control during traffic peaks [45]. Different from NO$_2$, the concentration of PM$_{10}$ and SO$_2$ decreased with the increase of the industry output per capita, which could also be explained by the improved energy efficiency. The negative relationship between the output per capita of tertiary industry and PM$_{10}$ and SO$_2$ might, to a large part, attributable to the rapid development of low energy-consumption industries such as high-tech industry, though this explanation needed further confirmation. Besides, unadvanced managements such as straw burning were restricted in suburbs with the improved living standard in cities [46], which also helped to decrease the concentration of PM$_{10}$ and SO$_2$.

2. Variation of energy efficiency in Chinese cities

The results showed that pollutant emissions at every ten thousand yuan fell with sustainable growth of GRP in the provincial capital cities from 2003 to 2012 (Fig. 2), which was coincident with the improvement in energy and technology in Chinese industries [47–48]. Besides, there was a distinct difference between the high income regions and less developed ones: the more developed cities had lower concentration of air pollutants with smaller variation ranges than the less developed cities. This was probably because of the lower energy intensity and more advanced technology [49–50] as well as better-implemented environment-friendly policies [51] in developed cities like Beijing.

3. The EKC in China

The relationship curves of social economy and some air pollution indicators in a period of time present different shape at different stages of development level of the country or state [52]. SO$_2$ per capita, for instance, seemed to tail off in 12 selected European countries when GRP per capita reached around 10000

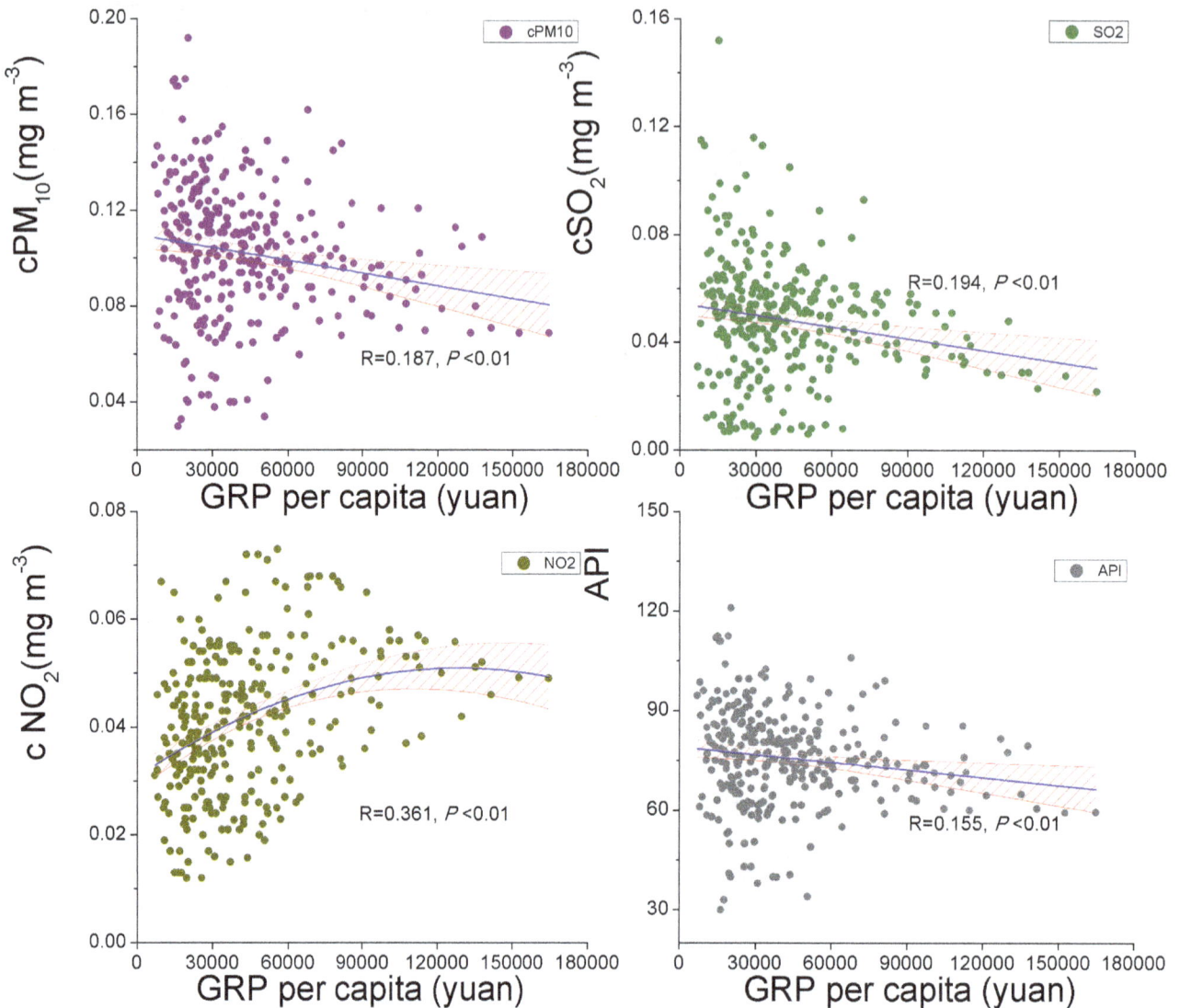

Figure 3. Regression curves between GRP per capita and air pollutant index (PM$_{10}$, SO$_2$, NO$_2$, API) in all the provincial capital cities during 2003–2012. The blue line is the regression line and the pink area the 95% confidence limits.

dollars. Consequently, the relationship curve of SO$_2$ per capita and GRP per capita appeared a declining trend [53]. Besides, the relationship curve of one air pollutant varied from another for its particular features. Carbon emissions like CO$_2$ was found to increase at ever-decreasing rates, with predicted peaks beyond reasonable income level because of its cross-border externalities which result in no sufficient incentives to urge countries to regulate emission [18,54–55]. These findings remind us to view the relationship of economy and air pollution with consideration of the time period and specific air pollutants [52,56–57].

In the national scale, our results showed a negative relationship between GRP per capita and PM$_{10}$, SO$_2$ or API while inverted-U shape relationship with NO$_2$ (Fig. 3). The EKC did exist for Chinese cities because the concentration of PM$_{10}$ and SO$_2$ stop rising from mid-1990s [58]. However, the turning point of EKC for air pollutants seemed to vary with the place, or, with the economical level. Taking SO$_2$ as an example, the turning point of EKC was approximately 20 thousand yuan in Changsha-Zhuzhou-Xiangtan Urban Agglomeration [59], but 37 thousand

in Beijing [60]. Peng & Bao [61] reported a national EKC knee point of around 36 thousand yuan, close to that of 30 thousand yuan claimed by Li et al [62]. Though we lacked data about SO$_2$ concentration before 2003, our analysis made an estimated turning point of less than 30 thousand yuan, also consistent with other results.

Such EKC pattern was probably caused by the following: (1) The structure of Chinese economy has changed from energy-intensive heavy industry to a more market-oriented service-based economy [62], which, with its lower environmental damage [25], helped China in ameliorating the environment rather than aggravating pollution. Furthermore, in order to stay competitive, firms are keen on investing new and improved technology to enhance cost effectiveness. One of the most significant consequences of this trend is an improvement in resource use efficiency within industrial sector which cut the industrial energy intensity by 50 percent during 1990s [62]. (2) Citizens' environmental awareness is improved. As Chinese people get richer and more educated, they become more concerned about the ambient

Table 2. Regression for concentration of PM_{10}, SO_2, NO_2, API and GRP per capita (panel data of all provincial cities).

| Region | Described | | Model summary | | | | | | Coefficients T test | |
| | | | Regression | | | | | | | |
	Relationship	Model	n	R	SE	Sig.			Independent variable	Constant
	PM10	Linear		0.187	0.027	0.001**			0.001**	0.000**
	&	Quadratic	305	0.187	0.027	0.005**			0.267	0.000**
	GDP per capita	Cubic		0.204	0.027	0.005**			0.076	0.000**
	SO2	Linear		0.194	0.021	0.001**			0.001**	0.000**
	&	Quadratic	305	0.194	0.022	0.003**			0.439	0.000**
All	GDP per capita	Cubic		0.213	0.021	0.003**			0.084	0.000**
Provincial										
City	NO2	Linear		0.344	0.012	0.000**			0.000**	0.000**
	&	Quadratic	305	0.361	0.012	0.000**			0.000**	0.000**
	GDP per capita	Cubic		0.363	0.012	0.000**			0.361	0.000**
	API	Linear		0.155	14.327	0.007**			0.007**	0.000**
	&	Quadratic	305	0.155	14.260	0.026*			0.346	0.000**
	GDP per capital	Cubic		0.179	14.277	0.021*			0.071	0.000**

* P<0.05; ** P<0.01.

Table 3. Regression for concentration of PM_{10}, SO_2, NO_2, API and GRP per capita.

Region	Described Relationship		Model summary Regression Model	n	R	SE	Sig.	Coefficients T test Independent variable	Constant
East	PM_{10}	Linear	0.078		0.030	0.441		
	&	Quadratic	100	0.192	0.030	0.163		
	GDP per capita	Cubic		0.192	0.030	0.305		
	SO_2	Linear		0.096	0.022	0.343		
	&	Quadratic	100	0.228	0.022	0.076		
	GDP per capita	Cubic		0.228	0.022	0.161		
	NO_2	Linear		0.383	0.014	0.000**	0.000**	0.000**
	&	Quadratic	100	0.497	0.013	0.000**	0.000**	0.000**
	GDP per capita	Cubic		0.495	0.013	0.000**	0.016*	0.164
	API	Linear		0.283	6.517	0.129		
	&	Quadratic	100	0.285	6.633	0.318		
	GDP per capita	Cubic		0.296	6.737	0.490		
Central	PM_{10}	Linear		0.449	0.018	0.000**	0.000**	0.000**
	&	Quadratic	60	0.465	0.018	0.001**	0.050*	0.000**
	GDP per capita	Cubic		0.474	0.018	0.002**	0.870	0.000**
	SO_2	Linear		0.297	0.019	0.021*	0.021*	0.000**
	&	Quadratic	60	0.332	0.019	0.036*	0.063	0.000**
	GDP per capita	Cubic		0.391	0.019	0.082		
	NO_2	Linear		0.386	0.010	0.002**	0.002**	0.000**
	&	Quadratic	60	0.386	0.010	0.010**	0.464	0.000**
	GDP per capita	Cubic		0.405	0.010	0.018*	0.244	0.116
	API	Linear		0.449	9.204	0.000**	0.000**	0.000**
	&	Quadratic	60	0.465	9.917	0.001**	0.050*	0.000**
	GDP per capita	Cubic		0.474	9.230	0.002**	0.871	0.000**
Northeast	PM_{10}	Linear		0.283	0.013	0.129		
	&	Quadratic	30	0.285	0.013	0.318		
	GDP per capita	Cubic		0.296	0.013	0.490		
	SO_2	Linear		0.463	0.012	0.010**	0.010**	0.000**
	&	Quadratic	30	0.463	0.013	0.038*	0.642	0.013*
	GDP per capita	Cubic		0.465	0.013	0.091		
	NO_2	Linear		0.333	0.009	0.072		

Table 3. Cont.

| Region | Described | | Model summary | | | | Coefficients T test | |
	Relationship	Regression Model	n	R	SE	Sig.	Independent variable	Constant
	&	Quadratic	30	0.334	0.010	0.202		
	GDP per capita	Cubic		0.348	0.010	0.332		
	API	Linear		0.283	6.517	0.129		
	&	Quadratic	30	0.285	6.633	0.318		
	GDP per capita	Cubic		0.296	6.737	0.490		
	PM$_{10}$	Linear		0.149	0.030	0.113		
	&	Quadratic	115	0.161	0.031	0.230		
	GDP per capita	Cubic		0.184	0.031	0.227		
	SO$_2$	Linear		0.169	0.023	0.071		
	&	Quadratic	115	0.222	0.022	0.059		
West	GDP per capita	Cubic		0.225	0.022	0.122		
	NO$_2$	Linear		0.211	0.012	0.024*	0.024*	0.000**
	&	Quadratic	115	0.233	0.012	0.044*	0.087	0.000**
	GDP per capita	Cubic		0.251	0.012	0.065		
	API	Linear		0.152	15.321	0.104		
	&	Quadratic	115	0.166	15.354	0.208		
	GDP per capita	Cubic		0.190	15.356	0.250		

Only significant P-values of T test are listed.
* $P < 0.05$; ** $P < 0.01$.

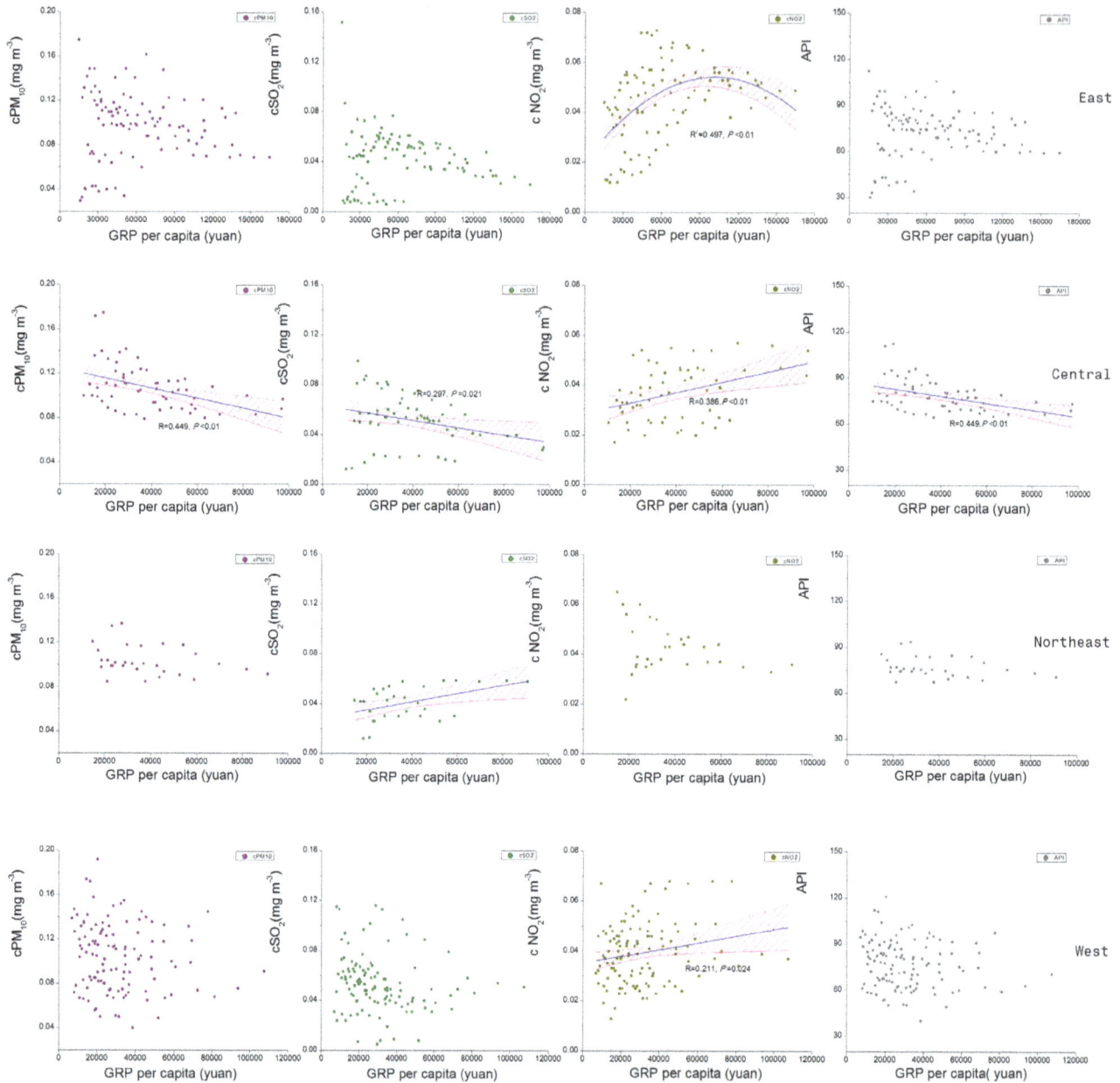

Figure 4. Regression curves between GRP per capita and air pollutant index (PM$_{10}$, SO$_2$, NO$_2$, API) in four economic regions during 2003–2012. The blue line is the regression line and the pink area the 95% confidence limits.

environment they dwell [63–65]]. At this time, their behaviors to protect environment and striving for more governmental support to do so contribute to the emergence of EKC [64,66]. For example, a gigantic demonstration against production of p-xylene in Dalian on August 14th, 2011 [67] reflected the strong demand for better living environment. (3) Regulatory policy for environment protection has been established and effectively implemented, which is another important factor to spur EKC [68-69]. In China, the first law against air pollution took into effect in 1987 and was amended in 1995 and 2000. The environment-friendly measures conducted by the government also provide significant support to air quality. The central government, for example, adopted drastic new pollution control measures for town and village industrial enterprises (TVIEs) and closed 65,000 high-pollution TVIEs in a

national campaign in 1996. Therefore, with strict and effective regulatory measures, air pollutants such as SO$_2$, soot and industrial fugitive dust began to decrease since mid-1990s [57]. It's believed that the environment will continue to improve with Chinese central government policies making efforts to promote ecological progress [70]. However, the rise of the SO$_2$ concentration in the Northeastern China simultaneous with the enhancement of civil revenues in this area (Fig. 4) might be a result of the policy of "revitalizing the old Northeast industry" by the Central Government [71].

The emergence of the knee point of EKC in the developed cities of Eastern China may be a result of well-implemented environmental policies and high investment in pollution control. But in the meantime, the Central and Northeast regions did not show a

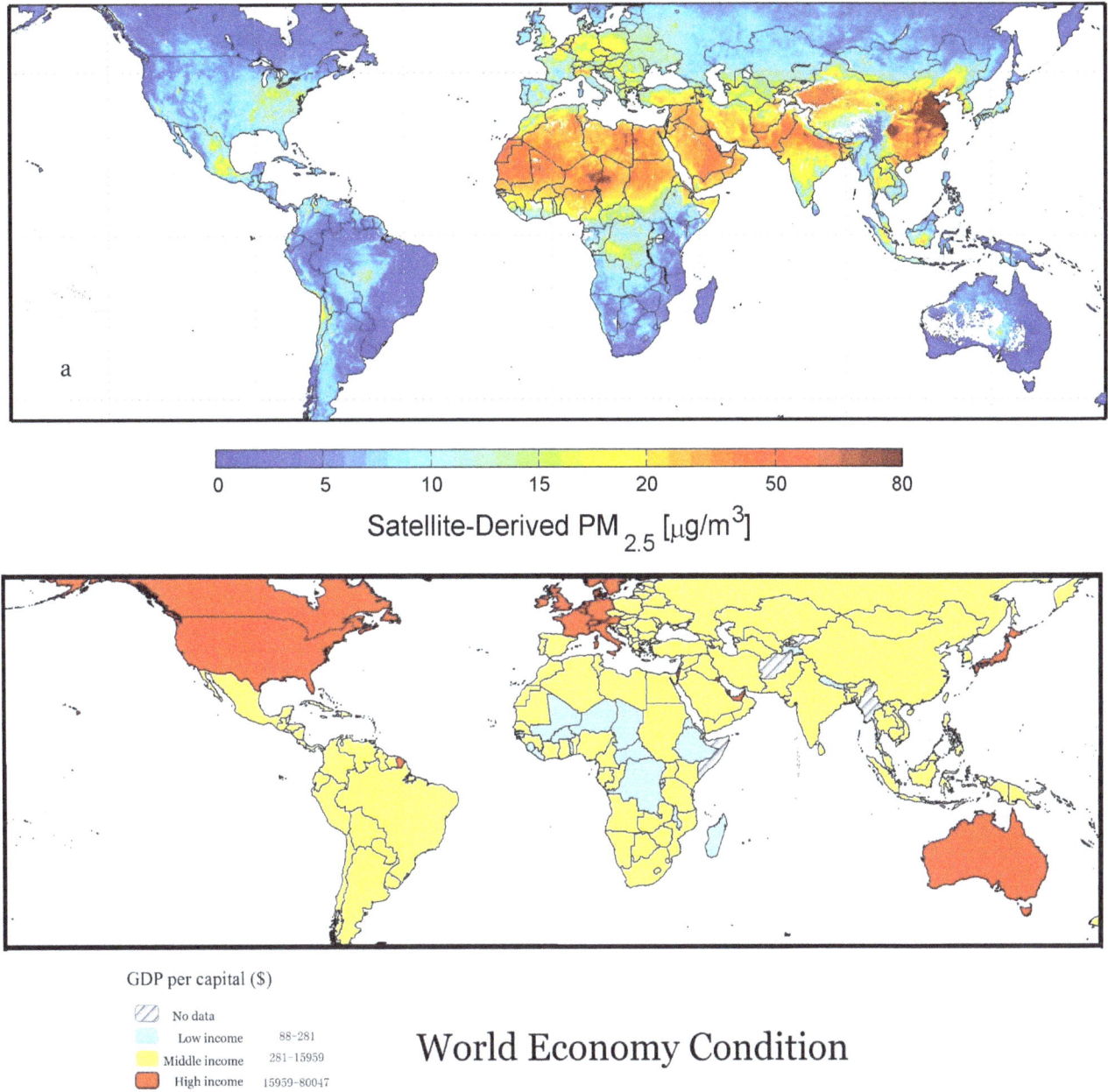

Figure 5. Maps of world PM$_{2.5}$ (μg m^{-3}) and GRP per capita ($) during 2001 to 2006. (a) PM$_{2.5}$, downloaded from NASA website and reproduced with permission from its authors and publisher (van Donkelaar et al., 2010); (b) GRP per capital, derived from the World Development Indicators of the World Bank (http://data.worldbank.org/country).

downward trend of NO$_2$ concentration (Fig. 4), which was probably attributable to the growing impact of vehicular emissions [51,72]. NO$_2$ is one of the dominant components in vehicle exhaust [73]. The ever-increasing civil vehicles, particularly the surge of vehicles in the cities after 2000, probably emit enough NO$_2$ to compensate the decrease of the pollutant from technical advancement of the industries [74]. It would be difficult to decrease NO$_2$ concentration in most cities if civil vehicles continue to increase in the near future, despite the controlling measures already taken [75].

Since API is a simple and generalized indicator, its variation can reflect the general trend of air pollutants. The negative linear trend of API and GRP per capita in the Central region (Fig. 4) was probably attributable to the overall decreasing trend of the three categories of air pollutants (Fig. 3) [76].

It is worth noticing air pollutants were not significantly related to GRP per capita in any of the four economic regions. Since the classification criterion of the four economic regions was not only the economic development level but also including geographical location, variance of economic levels within a region might have obscured the relationship between air pollutants and GRP per capita. Some detailed classification is needed to improve the accuracy of analysis.

Comparative qualitative analysis of the world also illustrated the existence of EKC and pointed out the developmental status of China in the world scale (Fig. 5 and Table 4). Two comparative

Table 4. Concentration of PM_{10} in cities of different continents.

Continent	Country(time period)	Mean PM_{10} concentration (mg m^{-3})	Scale	Reference
Asia	China(2003–2010)	0.1056 ± 0.0259	National	This study
	Japan (2007–2008)	0.0151 ± 0.0078	Yokohama	[81]
	India (1998–1999,2002)	0.2317 ± 0.0815	New Delhi	[82]
Africa	South Africa (winter of 1997)	0.0933 ± 0.0188	National	[83]
	Tanzania (2005)	0.0510 ± 0.0210	National	[84]
	Guinea (2004)	0.1453 ± 0.1092	Conakry	[85]
South America	Brazil (2008)	0.064 ± 0.0190	São Paulo	[86]
	Argentina (2008)	0.0470 ± 0.0120	Buenos Aires	
	Columbia (2008)	0.0640 ± 0.0490	Bogotá	
Europe	(1992–2009)	0.0306 ± 0.0084	Continental	[87]
	Netherlands (1985–2008)	0.0180	Rotterdam	[88]
	Greece (1999–2000)	0.0755 ± 0.0275	Athens,	[89]
	German (2002–2005)	0.0663 ± 0.0105	National	[90]
North America	US (1992–2009)	0.0276 ± 0.0081	National	[87]
	Canada (1993–2009)	0.0155 ± 0.0052	National	
Oceania	Australia (1998–2001)	0.0175 ± 0.0018	National	[91]
	New Zealand (1999–2007)	0.0299 ± 0.0132	National	[92]

"World Map" in Fig. 5 identified the existence of EKC at the global level. Emerging economics like China and India have the highest concentration of $PM_{2.5}$ (Fig. 5) and PM_{10} (Table 4), mainly resulting from fossil fuel and biomass burning in order to meet the energy demand for rapid economic growth [7]. Developed countries in the Europe, Oceania and North America, however, have lower concentration of particulate matter, probably because of the following two reasons: (1) The developed countries have already accomplished the transition of industrialization which is currently taking place in developing countries like China [77]. (2) The developed countries possess more environment-friendly technology to enhance energy efficiency and reduce pollution [78,79]. As for the undeveloped countries, most of them in Africa, the high concentration of pollutants are mainly caused by the Sahara Desert, which brings them seasonal dry, dust-laden wind known as Harmattan [80]. Of course there are other undeveloped countries with low level of pollutants, simply because they are still at the very early stage of economic development.

Conclusions

The quadratic relationship between the concentration of NO_2 and the output per capita of the secondary or tertiary industry, as well as the negative correlation between the concentration of PM_{10}, SO_2 and industry output per capita, indicate the declining trend of the pollutant concentrations with the improvement of energy efficiency and implementation of environment protection

policies. With technology innovation and modulation in industries together with policy implementation from 1990s, ratios of pollutants to GRP in the 31 provincial capitals in mainland China shows a downward trend. Such negative or invert-U quadratic relationship curve between air pollutants and GRP per capita verifies the existence of EKC in China.

Supporting Information

Figure S1   A hazy day (a: January 29, 2013) and a fine day (b: February 1, 2013) in downtown Beijing. (Pictures from http://ndphotos.oeeee.com/album/201302/01/2140.html?id = 1).

Table S1   Emission inventories of provincial cities in mainland China.

Acknowledgments

We thank Mingxu Li for doing a favor with transcribing data in order. The authors give special thanks to Ms. Wan Xiong for her editing and valuable comments on the manuscript.

Author Contributions

Conceived and designed the experiments: HC YPL QAZ CHP. Analyzed the data: YPL. Contributed reagents/materials/analysis tools: YPL YZY GY YZ. Wrote the paper: YPL GY.

References

1. Davis DL (2002) A look back at the London smog of 1952 and the half century since. Environmental health perspectives 110: A734.
2. Akimoto H (2003) Global air quality and pollution. Science 302: 1716–1719.
3. Kopp RE, Mauzerall DL (2010) Assessing the climatic benefits of black carbon mitigation. Proc Natl Acad Sci U S A 107: 11703–11708.
4. Ramanathan V, Feng Y (2009) Air pollution, greenhouse gases and climate change: Global and regional perspectives. Atmospheric Environment 43: 37–50.
5. Clarke A, Kapustin V (2010) Hemispheric Aerosol Vertical Profiles: Anthropogenic Impacts on Optical Depth and Cloud Nuclei. Science 329: 1488–1492.

6. Ramanathan V, Chung C, Kim D, Bettge T, Buja L, et al. (2005) Atmospheric brown clouds: Impacts on South Asian climate and hydrological cycle. Proc Natl Acad Sci U S A 102: 5326–5333.
7. Allen RJ, Sherwood SC, Norris JR, Zender CS (2012) Recent Northern Hemisphere tropical expansion primarily driven by black carbon and tropospheric ozone. Nature 485: 350–U393.
8. Bollasina MA, Ming Y, Ramaswamy V (2011) Anthropogenic aerosols and the weakening of the South Asian summer monsoon. Science 334: 502–505.

9. Booth BB, Dunstone NJ, Halloran PR, Andrews T, Bellouin N (2012) Aerosols implicated as a prime driver of twentieth-century North Atlantic climate variability. Nature 484: 228–232.

10. Menon S, Hansen J, Nazarenko L, Luo YF (2002) Climate effects of black carbon aerosols in China and India. Science 297: 2250–2253.

11. Silva RA, West JJ, Zhang Y, Anenberg SC, Lamarque J-F, et al. (2013) Global premature mortality due to anthropogenic outdoor air pollution and the contribution of past climate change. Environmental Research Letters 8: 034005.

12. Dadvand P, de Nazelle A, Triguero-Mas M, Schembari A, Cirach M, et al. (2012) Surrounding greenness and exposure to air pollution during pregnancy: an analysis of personal monitoring data. Environmental health perspectives 120: 1286.

13. Tonne C, Wilkinson P (2013) Long-term exposure to air pollution is associated with survival following acute coronary syndrome. European heart journal 34: 1306–1311.

14. Chen Y, Ebenstein A, Greenstone M, Li H (2013) Evidence on the impact of sustained exposure to air pollution on life expectancy from China's Huai River policy. Proceedings of the National Academy of Sciences.

15. Crandall RW (1983) Controlling industrial pollution: The economics and politics of clean air.

16. Seaton A, Godden D, MacNee W, Donaldson K (1995) Particulate air pollution and acute health effects. The Lancet 345: 176–178.

17. Stieb DM, Judek S, Burnett RT (2002) Meta-analysis of time-series studies of air pollution and mortality: effects of gases and particles and the influence of cause of death, age, and season. Journal of the Air & Waste Management Association 52: 470–484.

18. Galeotti M (2003) Economic Development and Environmental Protection. Fondazione Eni Enrico Mattei 2003:89.

19. Carson RT (2010) The environmental Kuznets curve: seeking empirical regularity and theoretical structure. Review of Environmental Economics and Policy 4: 3–23.

20. Selden TM, Song D (1994) Environmental quality and development: is there a Kuznets curve for air pollution emissions? Journal of Environmental Economics and management 27: 147–162.

21. Shafik N (1994) Economic development and environmental quality: an econometric analysis. Oxford Economic Papers: 757–773.

22. Cole MA, Rayner AJ, Bates JM (1997) The environmental Kuznets curve: an empirical analysis. Environment and development economics 2: 401–416.

23. De Bruyn SM, van den Bergh JC, Opschoor JB (1998) Economic growth and emissions: reconsidering the empirical basis of environmental Kuznets curves. Ecological Economics 25: 161–175.

24. Grossman GM, Krueger AB (1991) Environmental impacts of a North American free trade agreement. National Bureau of Economic Research.

25. Panayotou T (1993) Empirical tests and policy analysis of environmental degradation at different stages of economic development. International Labour Organization.

26. Wang Y-C (2011) Short-and Long-run Environmental Kuznets Curve: Case Studies of Sulfur Emissions in OECD Countries International Journal of Economic Research 9: 1–18.

27. Akbostancı E, Türüt-Aşık S, Tunç Gİ (2009) The relationship between income and environment in Turkey: Is there an environmental Kuznets curve? Energy Policy 37: 861–867.

28. Fodha M, Zaghdoud O (2010) Economic growth and pollutant emissions in Tunisia: An empirical analysis of the environmental Kuznets curve. Energy Policy 38: 1150–1156.

29. De Groot HL, Withagen CA, Minliang Z (2004) Dynamics of China's regional development and pollution: an investigation into the Environmental Kuznets Curve. Environment and development economics 9: 507–537.

30. Jalil A, Mahmud SF (2009) Environment Kuznets curve for CO$_2$ emissions: A cointegration analysis for China. Energy Policy 37: 5167–5172.

31. Vincent JR (1997) Testing for environmental Kuznets curves within a developing country. Environment and development economics 2: 417–431.

32. Song T, Zheng T, Tong L (2008) An empirical test of the environmental Kuznets curve in China: A panel cointegration approach. China Economic Review 19: 381–392.

33. Ren W-X, Xue B, Zhang L, Ma Z-X, Geng Y (2013) Spatiotemporal variations of air pollution index in China's megacities. Chinese Journal of Ecology 32 (10):2788–2796, in Chinese with english abstract.

34. GAO B-L (2009) Relationship Between Environmental Pollution and Economic Growth in Jiangsu Province. POLLUTION CONTROL TECHNOLOGY 22(6):36–39, in Chinese with english abstract.

35. Li H-Y, Li J, Zhang D-H, Chen S-S (2013) The Environmental Kuznets Curve in Xi'an. ARID ZONE RESEARCH 30(3): 556–562, in Chinese with english abstract.

36. Shen J (2006) A simultaneous estimation of Environmental Kuznets Curve: Evidence from China. China Economic Review 17: 383–394.

37. Hu M-X, Hu H, Liu Z (2005) Air Environmental Kuznets Curve (EKC) Characteristics in Wuhan City. Journal of Wuhan Polytechnic University 24(1):72–75, in Chinese with english abstract.

38. He K, Huo H, Zhang Q (2002) Urban air pollution in China: current status, characteristics, and progress. Annual Review of Energy and the Environment 27: 397–431.

39. Ministry of Environmental Protection of the People's Republic of China (2012) http://kjs.mep.gov.cn/hjbhbz/bzwb/dqhjbh/jcgffbz/201203/t20120302_224166.htm.

40. Wikipedia. (2013). http://en.wikipedia.org/wiki/List_of_regions_of_China.

41. Agras J, Chapman D (1999) A dynamic approach to the Environmental Kuznets Curve hypothesis. Ecological Economics 28: 267–277.

42. LI R-e, ZHANG H-j (2008) An empirical analysis on the regional discrepancy and tendency of EKC in China (1981–2004). Journal of Xi'an Jiaotong University (Social Sciences) 4: 007.

43. Kampa M, Castanas E (2008) Human health effects of air pollution. Environmental Pollution 151: 362–367.

44. Gustafsson Ö, Kruså M, Zencak Z, Sheesley RJ, Granat L, et al. (2009) Brown clouds over South Asia: biomass or fossil fuel combustion? Science 323: 495–498.

45. Zhou Y, Wu Y, Yang L, Fu L, He K, et al. (2010) The impact of transportation control measures on emission reductions during the 2008 Olympic Games in Beijing, China. Atmospheric Environment 44: 285–293.

46. Kim Oanh NT, Ly BT, Tipayarom D, Manandhar BR, Prapat P, et al. (2011) Characterization of particulate matter emission from open burning of rice straw. Atmospheric Environment 45: 493–502.

47. Sinton JE, Fridley DG (2000) What goes up: recent trends in China's energy consumption. Energy Policy 28: 671–687.

48. Hu J-L, Wang S-C (2006) Total-factor energy efficiency of regions in China. Energy Policy 34: 3206–3217.

49. Dhakal S (2009) Urban energy use and carbon emissions from cities in China and policy implications. Energy Policy 37: 4208–4219.

50. Fan J (2006) Industrial Agglomeration and Difference of Regional Labor Productivity: Chinese Evidence with International Comparison [J]. Economic Research Journal 11: 72–81.

51. Li L, Chen C, Xie S, Huang C, Cheng Z, et al. (2010) Energy demand and carbon emissions under different development scenarios for Shanghai, China. Energy Policy 38: 4797–4807.

52. Levinson A (2002) The ups and downs of the environmental Kuznets curve. Recent Advances in Environmental Economics Edward Elgar, Cheltenham: 119–139.

53. Markandya A, Golub A, Pedroso-Galinato S (2006) Empirical analysis of national income and SO$_2$ emissions in selected European countries. Environmental and resource economics 35: 221–257.

54. Lieb CM (2004) The environmental Kuznets curve and flow versus stock pollution: the neglect of future damages. Environmental and resource economics 29: 483–506.

55. Miah D, Masum FH (2010) Global observation of EKC hypothesis for CO$_2$, SO$_2$ and NO$_2$ emission: A policy understanding for climate change mitigation in Bangladesh. Energy Policy 38: 4643–4651.

56. Figueroa B, Pastén C (2009) Country specific environmental Kuznets curves: A random coefficient approach applied to high-income countries. Estudios de Economía 36: 5–32.

57. Harris PG (2006) Environmental Perspectives and Behavior in China Synopsis and Bibliography. Environment and Behavior 38: 5–21.

58. Zeng H-X, Qin D-L, Luo F, Zhang H, Luo Y-P (2010) Study on the Relationship between Enironmetal Pollution and Economic Growth in the Changsha - Zhuzhou - Xiangtan Urban Agglomera tion. Ecological Economy 1:347–350, in Chinese with english abstract.

59. Zhou Y-M, Huang S-P (2010) Research on Relationship between Economic Growth and Environmental Pollution: an Empirical Analysis Based on Panel Data of Beijing. in Chinese with english abstract.

60. Peng S-J, Bao Q (2006) Economic Growth and Environmental Pollution: An Empirical Test for the Environmental Kuznets Curve Hypothesis in China. Research on Financial and Economic Issues 8: 3–17.

61. Li R-P, Wang G-S, Wang A-J, Luo J-H, Geng N (2010). Factor Analysis of SO2 Emission Trend in Typical Industrialized Countries and Its Revelation to China. Acta Geoscientica Sinica 31(5):749–758, in Chinese with english abstract.

62. World Bank (2001) China: air, land, and water: environmental priorities for a new millennium. Washington, DC: World Bank.

63. Liu J, Diamond J (2005) China's environment in a globalizing world. Nature 435: 1179–1186.

64. Guo X, Marinova D (2011) Environmental awareness in China: Facilitating the greening of the economy.

65. Martens 1 S (2006) Public participation with Chinese characteristics: citizen consumers in China's environmental management. Environmental politics 15: 211–230.

66. Xie L (2011) China's environmental activism in the age of globalization. Asian Politics & Policy 3: 207–224.

67. BBC News. (2011) China protest closes toxic chemical plant in Dalian. http://www.bbc.co.uk/news/world-asia-pacific-14520438

68. He J, Wang H (2012) Economic structure, development policy and environmental quality: An empirical analysis of environmental Kuznets curves with Chinese municipal data. Ecological Economics 76: 49–59.

69. Kijima M, Nishide K, Ohyama A (2011) EKC-type transitions and environmental policy under pollutant uncertainty and cost irreversibility. Journal of Economic Dynamics and Control 35: 746–763.

70. Xinhua News Agency. (2012) Full text of Hu Jintao's report at 18th Party Congress. http://ncws.xinhuanet.com/english/special/18cpcnc/2012-11/17/c_131981259.htm

71. Zhang P, Ma Y, Liu W, Chen Q (2004) New Urbanization Strategy for Revitalizing the Tradianal Industrial Base of Northeast China. Acta Geographica Sinica: S1.

72. World Bank. (1997). Clear Water, blue skies: China's environmental pollution in the new century. Washington, DC: World Bank.

73. Brugge D, Durant JL, Rioux C (2007) Near-highway pollutants in motor vehicle exhaust: a review of epidemiologic evidence of cardiac and pulmonary health risks. Environmental Health 6: 23.

74. National Bureau of Statistics. (2011) http://www.stats.gov.cn/tjsj/ndsj/2011/indexch.htm.

75. Li X-F, Zhang M-J, Wang S-J, Zhao A-F, Ma Q (2012) Variation Characteristics and Influencing Factors of Air Pollution Index in China. Environmental Science 33(6):1936-1943, in Chinese with english abstract.

76. Wang L, Jang C, Zhang Y, Wang K, Zhang Q, et al. (2010) Assessment of air quality benefits from national air pollution control policies in China. Part II: Evaluation of air quality predictions and air quality benefits assessment. Atmospheric Environment 44: 3449-3457.

77. Great Lakes Invitational Conference Association (GLICA). (2008) Promoting Sustainable Industry in the Developing World. http://www.glica.org/topics/show/53

78. Dechezleprêtre A, Glachant M, Haščič I, Johnstone N, Ménière Y (2011) Invention and transfer of climate change–mitigation technologies: a global analysis. Review of Environmental Economics and Policy 5: 109–130.

79. Popp D (2011) International technology transfer, climate change, and the clean development mechanism. Review of Environmental Economics and Policy 5: 131–152.

80. Afeti G, Resch F (2000) Physical characteristics of Saharan dust near the Gulf of Guinea. Atmospheric Environment 34: 1273–1279.

81. Khan MF, Shirasuna Y, Hirano K, Masunaga S (2010) Characterization of PM2.5, PM2.5–10 and PM> 10 in ambient air, Yokohama, Japan. Atmospheric Research 96: 159–172.

82. Monkkonen P, Uma R, Srinivasan D, Koponen IK, Lehtinen KEJ, et al. (2004) Relationship and variations of aerosol number and PM10 mass concentrations in a highly polluted urban environment - New Delhi, India. Atmospheric Environment 38: 425–433.

83. Engelbrecht JP, Swanepoel L, Chow JC, Watson JG, Egami RT (2001) PM2.5 and PM10 concentrations from the Qalabotjha low-smoke fuels macro-scale experiment in South Africa. Environmental Monitoring and Assessment 69: 1–15.

84. Mkoma SL, Maenhaut W, Chi XG, Wang W, Raes N (2009) Characterisation of PM10 atmospheric aerosols for the wet season 2005 at two sites in East Africa. Atmospheric Environment 43: 631–639.

85. Weinstein JP, Hedges SR, Kimbrough S (2010) Characterization and aerosol mass balance of PM2.5 and PM10 collected in Conakry, Guinea during the 2004 Harmattan period. Chemosphere 78: 980–988.

86. Vasconcellos PC, Souza DZ, Avila SG, Araujo MP, Naoto E, et al. (2011) Comparative study of the atmospheric chemical composition of three South American cities. Atmospheric Environment 45: 5770–5777.

87. Wang KC, Dickinson RE, Su L, Trenberth KE (2012) Contrasting trends of mass and optical properties of aerosols over the Northern Hemisphere from 1992 to 2011. Atmospheric Chemistry and Physics 12: 9387–9398.

88. Keuken M, Zandveld P, van den Elshout S, Janssen NAH, Hoek G (2011) Air quality and health impact of PM10 and EC in the city of Rotterdam, the Netherlands in 1985-2008. Atmospheric Environment 45: 5294–5301.

89. Chaloulakou A, Kassomenos P, Spyrellis N, Demokritou P, Koutrakis P (2003) Measurements of PM10 and PM2.5 particle concentrations in Athens, Greece. Atmospheric Environment 37: 649–660.

90. Buns C, Klemm O, Wurzler S, Hebbinghaus H, Steckelbach I, et al. (2012) Comparison of four years of air pollution data with a mesoscale model. Atmospheric Research 118: 404–417.

91. Barnett AG, Williams GM, Schwartz J, Neller AH, Best TL, et al. (2005) Air pollution and child respiratory health - A case-crossover study in Australia and new Zealand. American Journal of Respiratory and Critical Care Medicine 171: 1272–1278.

92. Trompetter WJ, Davy PK, Markwitz A (2010) Influence of environmental conditions on carbonaceous particle concentrations within New Zealand. Journal of Aerosol Science 41: 134–142.

9

Elucidating Turnover Pathways of Bioactive Small Molecules by Isotopomer Analysis: The Persistent Organic Pollutant DDT

Ina Ehlers[1], Tatiana R. Betson[1], Walter Vetter[2], Jürgen Schleucher[1]*

1 Department of Medical Biochemistry & Biophysics, Umeå University, Umeå, Sweden, **2** Department of Food Chemistry, University of Hohenheim, Stuttgart, Germany

Abstract

The persistent organic pollutant DDT (1,1,1-trichloro-2,2-bis(4-chlorophenyl)ethane) is still indispensable in the fight against malaria, although DDT and related compounds pose toxicological hazards. Technical DDT contains the dichloro congener DDD (1-chloro-4-[2,2-dichloro-1-(4-chlorophenyl)ethyl]benzene) as by-product, but DDD is also formed by reductive degradation of DDT in the environment. To differentiate between DDD formation pathways, we applied deuterium NMR spectroscopy to measure intramolecular deuterium distributions (^{2}H isotopomer abundances) of DDT and DDD. DDD formed in the technical DDT synthesis was strongly deuterium-enriched at one intramolecular position, which we traced back to ^{2}H/^{1}H fractionation of a chlorination step in the technical synthesis. In contrast, DDD formed by reductive degradation was strongly depleted at the same position, which was due to the incorporation of ^{2}H-depleted hydride equivalents during reductive degradation. Thus, intramolecular isotope distributions give mechanistic information on reaction pathways, and explain a puzzling difference in the whole-molecule ^{2}H/^{1}H ratio between DDT and DDD. In general, our results highlight that intramolecular isotope distributions are essential to interpret whole-molecule isotope ratios. Intramolecular isotope information allows distinguishing pathways of DDD formation, which is important to identify polluters or to assess DDT turnover in the environment. Because intramolecular isotope data directly reflect isotope fractionation of individual chemical reactions, they are broadly applicable to elucidate transformation pathways of small bioactive molecules in chemistry, physiology and environmental science.

Editor: Dariush Hinderberger, Martin-Luther-Universität Halle-Wittenberg, Germany

Funding: This work was funded by The Swedish Research Council for Environment, Agricultural Sciences and Spatial Planning; JS; grant 216-2008-764 (http://www.formas.se/en/). The funders had no role in study design, data collection and analysis, decision to publish, or preparation of the manuscript.

Competing Interests: The authors have declared that no competing interests exist.

* Email: jurgen.schleucher@chem.umu.se

Introduction

The chloropesticide DDT (1,1,1-trichloro-2,2-bis(4-chlorophenyl)ethane) is one of the most controversial chemicals developed in the twentieth century. Its application as the first effective pesticide to combat malaria has saved the lives of millions of people. This success triggered its general use as a pesticide on a million-ton scale in agriculture and by households worldwide. Due to its detrimental properties (lipophilicity and persistency), the use of DDT has led to its ubiquitous occurrence in the environment even in remote areas where it has never been used [1,2]. In addition, serious toxicological effects such as thinning of egg shells of birds in the Baltic region have been linked to DDT [3]. As a consequence, DDT has strongly influenced the public perception and regulation of man-made chemicals. Although the use of DDT has been drastically reduced for decades, it is still abundant in the environment and is still having detrimental effects. For this reason, DDT has been classified as a persistent organic pollutant (POP) by the Stockholm Convention, with the purpose of phasing out its use globally. However, it is still used against vector-borne diseases [4], primarily malaria transmitted by *Anopheles* mosquitos.

Since the discovery of the persistence of DDT, numerous studies have addressed environmental and human health hazards of DDT and its congeners [5,6]. Recent studies have detected p, p'-DDE (1,1-bis-(4-chlorophenyl)-2,2-dichloroethene) in human blood serum [6], even in countries that stopped DDT usage decades ago, and link high p, p'-DDE blood serum levels with prevalent hypertension [7] and an increased risk of Alzheimer disease [8]. Technical DDT is a mixture of compounds, the major components being the p, p' and o, p' isomers of DDT and some of the minor components being the p, p' and o, p' isomers of DDD (1-chloro-4-[2,2-dichloro-1-(4-chlorophenyl)ethyl]benzene) [9] (Fig. 1). DDD not only occurs as a by-product of technical DDT production but has itself been used as a pesticide, and is formed in the environment by hydrodechlorination of DDT [9–11]. DDD is of particular ecotoxicological relevance because it accumulates in the environment and has been implicated as a possible endocrine disruptor [5,12]. As for POPs in general [13], tracing DDT in the environment is critical to understand its transformation processes, to judge environmental and human health risks, and to develop remediation strategies.

A

B

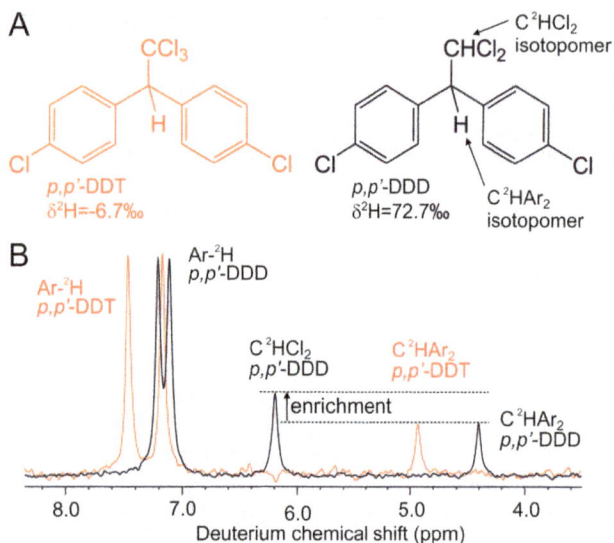

Figure 1. Comparison of p, p'-DDT with its congener p, p'-DDD.
(A) Chemical structures and whole-molecule δ^2H values. (B) Overlay of deuterium NMR spectra of reference samples of p, p'-DDD (black, reference 1) and p, p'-DDT (red). Integrals of the signals are proportional to the abundance of the respective ^2H isotopomers. Ar-^2H denotes isotopomers carrying ^2H in the aromatic moieties of the respective compound. Positions of signals ("chemical shifts") differ between DDT and DDD because they reflect the stereoelectronic properties of each molecule; these chemical shift differences do not influence the integration of signals.

Increasingly, stable heavy isotopes, which occur naturally in the environment with varying abundance, are used to trace compounds [14]. In most applications, the abundance ratio between a heavy isotope and a light isotope of a chemical element is measured by isotope-ratio mass spectrometry (IRMS). Variation observed in isotope ratios is usually expressed as δ value (in ‰), the abundance deviation of the respective heavy isotope relative to a standard with known isotope ratio [15]. However, IRMS only allows measurement of the average isotope ratio of a whole molecule, because any molecule to be studied must be converted into defined gases for IRMS analysis. It is known, though, that the abundance of heavy isotopes varies among intramolecular groups of non-symmetric molecules [16], but the consequences of this intramolecular variation cannot be assessed by whole-molecule δ measurements. A molecule carrying a specific isotope at a specified intramolecular position is called an isotopomer. Intramolecular isotope variation therefore means that the isotopomers of non-symmetric molecules differ in their abundances. This variation in abundance is mainly caused by kinetic isotope effects during chemical reactions. Thus, isotopomer abundances are directly related to reaction mechanisms and turnover along chemical pathways [17]. As has been observed for several types of compounds, kinetic protium/deuterium ($^1H/^2H$) isotope effects can easily double the abundance of particular ^2H isotopomers [18–20]. Relative abundances of ^2H isotopomers can be measured by deuterium NMR, because each ^2H isotopomer creates one signal in the deuterium NMR spectrum. If the signals can be resolved and using suitable experimental conditions, the integrals of the deuterium NMR spectrum directly reflect relative isotopomer abundances [18,21,22]. Furthermore, when isotopomer abundance can be linked to defined chemical steps, isotopomer variation yields mechanistic information on reaction

pathways. Here we use deuterium NMR to identify ^2H fractionations due to individual reactions during DDD formation, as basis to distinguish DDD formed in technical DDT production from DDD formed by hydrodechlorination of DDT in the environment.

Results & Discussion

Reference samples of p, p'-DDT and p, p'-DDD have been found to differ by as much as 79‰ in δ^2H [23] (Fig. 1A). Variation in this range would normally rule out a common source of the compounds. We measured deuterium NMR spectra of the same p, p'-DDT and p, p'-DDD reference samples, which show indistinguishable abundance patterns for the C^2HAr_2 and Ar-^2H isotopomers (Fig. 1B). In contrast, the C^2HCl_2 isotopomer in DDD was nearly twice as abundant as the C^2HAr_2 isotopomer (Table 1, p, p'-DDD reference sample 1). By combining relative isotopomer abundances from NMR spectra with the δ^2H values of the compounds [23], we calculated how much the high abundance of the C^2HCl_2 isotopomer influenced the δ^2H of DDD. This isotope balance calculation (see Materials and Methods) shows that the high abundance of the C^2HCl_2 isotopomer ($\delta^2H = 739‰$) completely explains the 79‰ difference in whole-molecule δ^2H between p, p'-DDT and p, p'-DDD. Thus, the molecular part that is common to DDD and DDT (CHAr$_2$ fragment) showed indistinguishable δ^2H values (Table 1). Consequently, the isotopomer analysis demonstrates that both compounds can have a common source.

Measurements on two further independent samples of p, p'-DDD (reference 2 and 3 in Table 1) gave indistinguishable results. Furthermore, analogous experiments on reference material of the o, p' isomer of DDD showed that the C^2HCl_2 isotopomer was again almost twice as abundant as the C^2HAr_2 isotopomer (Table 1). These results suggest that the C^2HCl_2 isotopomer of synthetic DDD generally has a very high abundance, which significantly shifts the whole-molecule δ^2H value.

To apply isotopomer analysis to DDT as used in practice, we recorded deuterium NMR spectra of technical DDT (Fig. 2B) to analyze the by-product p, p'-DDD. Technical DDT is synthesized by reacting CCl_3CHO with chlorobenzene (Fig. 2A). The technical product contains only 4% DDD, and the amount of DDD in the technical DDT sample was only approximately 15 µmol. Deconvolution allowed quantifying the signals originating from p, p'-DDD, although the signal of the C^2HCl_2 isotopomer partly overlapped with the signal of an unknown impurity. Measurements on two replicate NMR samples of the same technical DDT gave similar results (Table 1), which showed that its C^2HCl_2 isotopomer was about twice as abundant as the C^2HAr_2 isotopomer. This indicates that acceptable signal-to-noise ratios can be obtained using sample amounts of 15 µmol, using a cryogenically cooled probe [24]. The synthesis of CCl_3CHO proceeds via chlorination of CH_3CHO. Incomplete chlorination gives rise to residual $CHCl_2CHO$ from which the corresponding DDD isomers are formed. Chlorination of carbonyl compounds is associated with a kinetic $^1H/^2H$ isotope effect of up to seven [25], which is close to the maximum classical kinetic $^1H/^2H$ isotope effect for C–H bond breakage [26]. According to the theory of isotope effects [26], chlorination of CH_3CHO will - with increasing turnover - progressively enrich deuterium in residual $CHCl_2$ groups. Supporting this explanation, a commercial $CHCl_2CHO$ product showed a 1.27-fold enrichment of the C^2HCl_2 isotopomer compared to the C^2HO isotopomer (Table 1). Thus, we conclude that the high abundance of the C^2HCl_2 isotopomer of DDD is the result of the $^1H/^2H$ isotope effect

Table 1. Deuterium abundance data of DDT and related compounds.

Sample	C^2HCl_2 isotopomer abundance[a]	δ^2H, ‰	
		Whole molecule[b]	CHAr$_2$ fragment[c]
p, p'-DDT reference	–	−6.7±3.4	−6.7±3.4
p, p'-DDD reference 1	1.77±0.10	72.7±7.0	−1.3±11.4
p, p'-DDD reference 2	1.80±0.04	–	–
p, p'-DDD reference 3	1.76±0.11	–	–
o, p'-DDD reference	1.89±0.09	74.8±7.6	−6.8±10.2
Commercial CHCl$_2$CHO	1.27±0.02	–	–
p, p'-DDD in technical DDT		–	–
replicate 1	2.17±0.44	–	–
replicate 2	1.78±0.26	–	–
p, p'-DDD from DDT hydrodechlorination:			
Experiment 1	0.69±0.05	–	–
Experiment 2	0.62±0.07	–	–
Experiment 3	0.58±0.09	–	–

[a] Relative to $C^2HR_2 = 1$.
[b] from reference [23].
[c] See Materials and Methods for calculations.

during the chlorination of CH_3CHO. The original enigma of how p, p'-DDD and p, p'-DDT can have a difference in δ^2H of 79‰ can be attributed to the isotope effect during synthesis of CCl_3CHO.

DDT is known to be degraded into DDD in the environment through hydrodechlorination under anaerobic conditions [10]. In this reaction, a chlorine atom in the CCl_3 moiety of DDT is replaced by a hydride equivalent to form the $CHCl_2$ group of DDD. The hydride equivalents are thought to be derived from the biological cofactors $FADH_2$ or reduced Fe porphyrins [10]. An analogous hydrodechlorination reaction using Fe(0) has been proposed for remediation of DDT contamination [11]. Thus, we experimentally produced DDD by hydrodechlorination of p, p'-DDT with Fe(0). The deuterium NMR spectrum of the resulting reaction mixture (Fig. 2C) shows that the p, p'-DDD formed by hydrodechlorination was approximately 30% depleted in the C^2HCl_2 isotopomer. To test the reproducibility of the method, we carried out triplicate experiments (see Material S1), which gave statistically indistinguishable results with an average abundance of the C^2HCl_2 isotopomer of 0.63±0.05, relative to the C^2HPh_2 isotopomer. The strong ^2H depletion of the $CHCl_2$ group formed by hydrodechlorination is for a fundamental biophysical reason. Because $^1H^2HO$ is a weaker acid than H_2O, the hydronium ions in water are ^2H-depleted by approximately −700 ‰ [27,28]. When these ions enter reduction reactions leading to abiogenic or biogenic hydride equivalents, the resulting hydride equivalents become accordingly ^2H-depleted. The resulting depletion, of the order of −300‰, has been observed in hydride-derived isotopomers of biochemical metabolites [17]. Thus, a strong ^2H depletion of the $CHCl_2$ group, similar to our Fe(0) reduction experiment, can be expected for DDD formed by hydrodechlorination of DDT in the environment. This depletion can serve to identify DDD formed in situ, either as result of natural degradation, or of remediation efforts, and can therefore be used to monitor breakdown of DDT.

In summary, DDD from technical synthesis is characterized by an increased abundance of the C^2HCl_2 isotopomer, while DDD

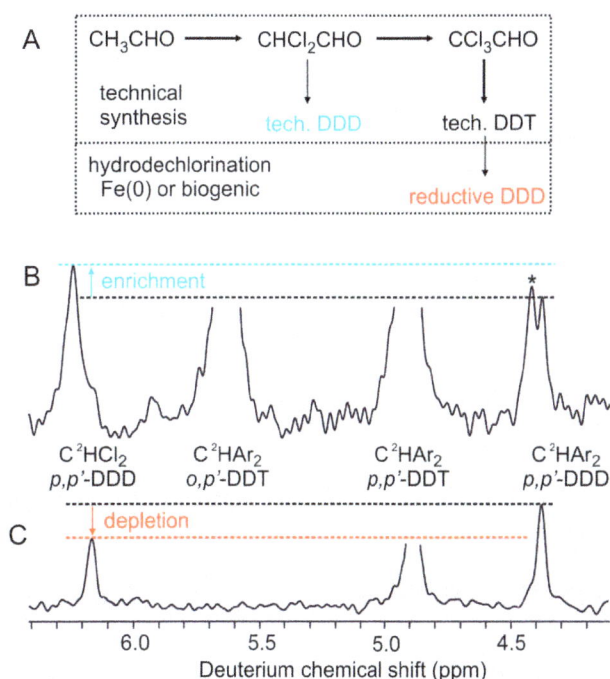

Figure 2. Pathways of DDD formation and their ^2H fractionation. (A) Technical synthesis of DDT, and pathways of formation of DDD. (B) Deuterium NMR spectrum of technical DDT (replicate 1). (C) Deuterium NMR spectrum of a reaction mixture containing DDD formed by hydrodechlorination (experiment 1). Blue and red lines qualitatively illustrate enrichment or depletion of the C^2HCl_2 isotopomer in technical and reductive DDD, respectively. To take differing linewidths and overlap into account, numerical values for isotopomer abundances were obtained by deconvolution of the signals. Note that signal positions differ slightly between samples, because of differing analyte concentrations. The asterisk denotes an unknown impurity in technical DDT.

formed by reductive dechlorination is depleted in the C^2HCl_2 isotopomer, and both isotopomer fractionations are inherent properties of the respective pathway of DDD generation. In light of the opposite deuterium fractionations, technical DDD and DDD formed by hydrodechlorination should generally differ by approximately 1,000‰ in the abundance of the C^2HCl_2 isotopomer (Table 1). This isotopomer difference can be used to estimate fractional contributions of both pathways to environmental samples of DDD. Unfortunately, the sample amount currently required for ^2H isotopomer measurements by NMR (approximately 10 mg per congener, see above) restricts the application of this approach to highly contaminated sites where large quantities of DDT congeners can be isolated. However, the C^2HCl_2 isotopomer signatures of technical and environmental DDD formation respectively induce more positive or negative whole-molecule δ^2H values of DDD, as compared to DDT. Compound-specific IRMS methods for δ^2H measurements for polyhalogenated compounds [29,30] have recently been introduced; therefore comparison of δ^2H values of DDD and DDT may allow tracing sources of DDD to technical synthesis or breakdown of DDT in the environment.

Conclusion

Our D isotopomer measurements represent a striking demonstration that individual ^2H isotopomer abundances can have a decisive effect on whole-molecule δ^2H values. Hence isotopomer abundances give essential insights into causes of variation in δ^2H, and must be taken into account in interpretations of whole-molecule isotope ratios, even when structural differences between related compounds appear to be minor. For DDT, sources of DDD may be traced based on isotopomer results. In general, isotopomer analysis of POPs can be applied: (1) to identify the responsible polluter; (2) to understand POP transformation processes in the environment; and (3) to gauge the efficacy of remediation approaches.

Isotope fractionation of chemical reactions directly affects isotopomer abundances, which in turn affects whole-molecule isotope ratios. However, without prior knowledge a change in a whole-molecule isotope ratio cannot be related to fractionation by a chemical reaction. Therefore isotopomer measurements are particularly valuable for deducing mechanistic detail of reaction pathways. The availability of cryogenically cooled probes has reduced the sample amount needed for isotopomer analysis to only micromoles, which enables applications in fields ranging from organic and pharmaceutical chemistry, to metabolic studies, and environmental science.

Materials and Methods

Sources of compounds
See Material S1.

Deuterium NMR
Several conditions must be fulfilled for isotopomer measurements by NMR. First, line widths must be minimized to resolve the signals and to optimize sensitivity. To achieve this goal, we used low-viscosity solvents C_6F_6 or hexane/C_6F_6 and elevated measurement temperatures around 50°C [21]. Second, the NMR signals must have pure Lorentzian lineshapes, so that their integrals can be determined by lineshape fitting. This goal was achieved by careful homogenizing the magnetic field of the NMR spectrometer ("shimming"). Third, because many individual spectra ("scans") are added to obtain each NMR spectrum, the

influence of the previous scan on the deuterium nuclei must have decayed when the next scan is acquired. To achieve this goal, we determined deuterium relaxation times (T_1) of all compounds using an inversion-recovery pulse sequence, and acquired spectra with appropriate recycle times between scans (6 T_1) to ensure complete relaxation. A DRX600 spectrometer (Bruker, Karlsruhe, Germany) equipped with a 5-mm broadband probe and a ^{19}F lock device was used to record ^2H NMR spectra of DDT and DDD reference materials; C_6F_6 was used for field-frequency locking. Spectra of dichloroacetaldehyde diethyl acetal were acquired on a DRX500 spectrometer in unlocked mode using a 10-mm broadband probe. Spectra of technical DDT and of the hydrodechlorination reaction mixture were obtained on an Avance III 850 spectrometer equipped with a cryogenic probe optimized for deuterium detection and equipped with ^{19}F lock. Parameter settings for all samples are given in Material S1. All spectra were recorded with proton decoupling. Spectra were processed with exponential line-broadening, and signal integrals were obtained by deconvolution in the program topspin (version 3.0; Bruker). Typically, five spectra were recorded. The standard deviation of the C^2HCl_2/C^2HR_2 abundance ratio among replicate spectra was used to estimate the overall precision of our measurements. For three independent hydrodechlorination experiments, the standard deviations of replicate spectra were 0.05, 0.07 and 0.09, respectively. The experiments led to results which did not significantly differ from each other (ANOVA), and the standard deviation among the experiments was 0.05, indicating that the standard deviation among spectra is a realistic estimate of the overall precision (Table 1).

DDT hydrodechlorination
p, p'-DDT was reduced with Fe(0) under anaerobic conditions following Sayles et al. [11]. A 1-L flask was filled with 1 L 20 mM MOPS buffer (pH 7, deoxygenated by purging with N_2); 15 g Fe(0) powder and 60 mg p, p'-DDT (dissolved in 20 mL acetone) were added. The flask was sealed and mixed by rotation. After 18 h at room temperature, the reaction mixture was extracted consecutively with n-hexane and chloroform. The organic phases were dried with Na_2SO_4, the solvent was removed under vacuum, and the resulting product was used for NMR measurements. Analysis of the product using ^1H NMR showed that the main components of the product were un-reacted p, p'-DDT and approximately 30% p, p'-DDD.

Isotope balance calculations
Deuterium NMR spectra yield relative abundances of ^2H isotopomers, but position-specific $^2H/^1H$ ratios cannot be derived from NMR alone. When the whole-molecule $^2H/^1H$ ratio is available and if all the signals in the ^2H NMR spectrum can be integrated, position-specific $^2H/^1H$ ratios can be calculated according to the equation 1,

$$\left(\frac{^2H}{^1H}\right)_i = \frac{f_i}{F_i}\left(\frac{^2H}{^1H}\right)_G \qquad (1)$$

where $(^2H/^1H)_i$ and $(^2H/^1H)_G$ are the isotope ratios of isotopomer i and of the whole molecule, respectively. f_i is the experimentally determined molar fraction of isotopomer i, that is, the integral of its NMR signal compared to the integral of all isotopomers of the molecule. F_i is the statistical molar fraction [31], that is, the stoichiometric fraction hydrogen in position i according to the molecular formula, and the term f_i/F_i represents the abundance of the ^2H isotopomer i, relative to the $^2H/^1H$ ratio of the whole molecule. Note that the isotope ratios $(^2H/^1H)_i$ and $(^2H/^1H)_G$ must be expressed as true ratios, i.e. $\delta^2H = 72.7‰$ for p, p'-DDD (Table 1) must be converted to $(^2H/^1H)_G = 1.0727$, i.e. the DDD sample has a 1.0727-fold higher $^2H/^1H$ ratio than VSMOW. For example, the C^2HCl_2 isotopomer of p, p'-DDD contributed fraction $f_{C2HCl2} = 0.1621$ to the integral of the ^2H NMR spectrum, while F_{C2HCl2} is 0.1 (1 hydrogen out of 10 in the molecule). Therefore the $^2H/^1H$ ratio of the CHCl2 group is $(^2H/^1H)_{C2HCl2} = 1.621 \times 1.0727 = 1.739$ and means that this isotopomer has an abundance corresponding to a position-specific δ^2H of 739‰. An analogous calculation for the rest of the molecule-- the CHAr$_2$ fragment ($f_{rest} = 1 - f_{C2HCl2} = 0.8379$, $F_{rest} = 0.9$) which is common to p, p'-DDT and p, p'-DDD-- yields a $^2H/^1H$ ratio of 0.9987, corresponding to a δ^2H of $-1.3‰$ for this fragment. The 74‰ difference between this number and the whole-molecule δ^2H shows how much the whole-molecule δ^2H is influenced by the high abundance of the C^2HCl_2 isotopomer. In the isotope balance results in Table 1, error estimates reflect the combined uncertainties of the whole-molecule δ^2H measurements and of NMR-derived ^2H isotopomer distributions. For the DDD component of technical DDT, this calculation is not possible because the aromatic signals of all congeners overlap.

Acknowledgments

We thank several colleagues for valuable advice on the manuscript.

Author Contributions

Conceived and designed the experiments: IE TRB WV JS. Performed the experiments: IE TRB JS. Analyzed the data: IE TRB JS. Contributed reagents/materials/analysis tools: TRB WV. Contributed to the writing of the manuscript: IE WV JS.

References

1. Sladen WJL, Menzie CM, Reichel WL (1966) DDT residues in Adelie penguins and a crabeater seal from Antarctica. Nature 210: 670–673.
2. Sheng JS, Wang X, Gong P, Joswiak DR, Tian L, et al. (2013) Monsoon-driven transport of organochlorine pesticides and polychlorinated biphenyls to the Tibetan Plateau: Three year atmospheric monitoring study. Environ. Sci. Technol. 47: 3199–3208.
3. Vos JG, Dybing E, Greim HA, Ladefoged O, Lambré C, et al. (2000) Health effects of endocrine-disrupting chemicals on wildlife, with special reference to the European situation. Crit. Rev. Toxicol. 30: 71–133.
4. World Health Organization (2013) The use of DDT in malaria vector control.
5. Turusov V, Rakitsky V, Tomatis L (2002) Dichlorodiphenyltrichloroethane (DDT): Ubiquity, persistence, and risks. Environ. Health Persp. 110: 125–128.
6. Eskenazi B, Chevrier J, Goldman Rosas L, Anderson HA, Bornman MS, et al. (2009) The Pine River Statement: Human health consequences of DDT Use. Environ. Health Persp. 117: 1359–1367.
7. Lind PM, Penell J, Salihovic S, van Bavel B, Lind L (2014) Circuling levels of p, p'-DDE are related to prevalent hypertension in the elderly. Environ. Res. 129: 27–31.
8. Richardson JR, Roy A, Shalat SL, von Stein RT, Hossain MM, et al. (2014) Elevated Serum Pesticide Levels and Risk for Alzheimer Disease. JAMA Neurol. do i:10.1001/jamaneurol.2013.6030.
9. Fishbein L (1974) Chromatographic and biological aspects of DDT and its metabolites. J. Chromatogr. 98: 177–251.
10. Aislabie JM, Richards NK, Boul HL (1997) Microbial degradation of DDT and its residues – a review. New Zeal. J. Agr. Res. 40: 269–282.
11. Sayles GD, You G, Wang M, Kupferle MJ (1997) DDT, DDD, and DDE dechlorination by zero-valent iron. Environ. Sci. Technol. 31: 3448–3454.
12. Guillette LJ Jr, Pickford DB, Crain DA, Rooney AA, Percival HF (1996) Reduction in penis size and plasma testosterone concentrations in juvenile alligators living in a contaminated environment. Gen. Comp. Endocr. 101: 32–42.
13. Fenner K, Canonica S, Wackett LP, Elsner M (2013) Evaluating Pesticide Degradation in the Environment: Blind Spots and Emerging Opportunities. Science 341: 752–758.
14. Elsner M (2010) Stable isotope fractionation to investigate natural transformation mechanisms of organic contaminants: principles, prospects and limitations. J. Environ. Monitor. 12: 2005–2031.
15. Wolfsberg M, Van Hook WA, Paneth P, Rebelo LPN (2010) Isotope effects in the chemical, geological, and bio sciences. Dordrecht, London: Springer. 466 p.
16. Schmidt HL (2003) Fundamentals and systematics of the non-statistical distributions of isotopes in natural compounds. Naturwissenschaften 90: 537–552.
17. Schmidt HL, Werner RA, Eisenreich W (2003) Systematics of ^2H patterns in natural compounds and its importance for the elucidation of biosynthetic pathways. Phytochem. Rev. 2: 61–85.
18. Remaud GS, Martin YL, Martin GG, Martin GJ (1997) Detection of sophisticated adulterations of natural vanilla flavors and extracts: Application of the SNIF-NMR method to vanillin and p-hydroxybenzaldehyde. J. Agric. Food Chem. 45: 859–866.
19. Schleucher J, Vanderveer P, Markley JL, Sharkey TD (1999) Intramolecular deuterium distributions reveal disequilibrium of chloroplast phosphoglucose isomerase. Plant Cell Environ. 22: 525–533.
20. Markai S, Marchand PA, Mabon F, Baguet E, Billault I, et al. (2002) Natural deuterium distribution in branched-chain medium-length fatty acids is nonstatistical: A site-specific study by quantitative ^2H NMR spectroscopy of the fatty acids of capsaicinoids. ChemBioChem 3: 212–218.
21. Betson TR, Augusti A, Schleucher J (2006) Quantification of deuterium isotopomers of tree-ring cellulose using nuclear magnetic resonance. Anal. Chem. 78: 8406–8411.
22. Lesot P, Courtieu L (2009) Natural abundance deuterium NMR spectroscopy: Developments and analytical applications in liquids, liquid crystals and solid phases. Prog. Nucl. Mag. Res. Sp. 55: 128–159.
23. Armbruster W, Lehnert K, Vetter W (2006) Establishing a chromium-reactor design for measuring δ^2H values of solid polyhalogenated compounds using direct elemental analysis and stable isotope ratio mass spectrometry. Anal. Bioanal. Chem. 384: 237–243.
24. Kovacs H, Moskau D, Spraul M (2005) Cryogenically cooled probes – a leap in NMR technology. Prog. Nucl. Magn. Reson. Spectrosc. 46: 131–155.
25. Wiberg KB (1955) The deuterium isotope effect. Chem. Rev. 55: 713–743.
26. Melander L, Saunders WH Jr (1980) Reaction Rates of Isotopic Molecules. Wiley, New York.
27. Luo Y-H, Sternberg L, Suda S, Kumazawa S, Mitsui A (1991) Extremely low D/H ratios of photoproduced hydrogen by cyanobacteria. Plant Cell Physiol. 32: 897–900.
28. Walter S, Laukenmann S, Stams AJM, Vollmer MK, Gleixner G, et al. (2012) The stable isotopic signature of biologically produced molecular hydrogen (H$_2$). Biogeosciences 9: 4115–4123.
29. Kuder T, Philp P (2013) Demonstration of compound-specific isotope analysis of hydrogen isotope ratios in chlorinated ethenes. Environ. Sci. Technol. 47: 1461–1467.
30. Shouakar-Stash O, Drimmie RJ (2013) Online methodology for determining compound-specific hydrogen stable isotope ratios of trichloroethene and 1,2-cis-dichloroethene by continuous-flow isotope ratio mass spectrometry. Rapid Commun. Mass Spectrom. 27: 1335–1344.
31. Martin GJ, Martin ML, Mabon F, Michon MJ (1982) Identification of the origin of natural alcohols by natural abundance hydrogen-2 nuclear magnetic resonance. Anal. Chem. 54: 2380–2382.

10

The Comparative Photodegradation Activities of Pentachlorophenol (PCP) and Polychlorinated Biphenyls (PCBs) Using UV Alone and TiO$_2$-Derived Photocatalysts in Methanol Soil Washing Solution

Zeyu Zhou[1], Yaxin Zhang[2], Hongtao Wang[1]*, Tan Chen[1], Wenjing Lu[1]

1 Department of Environmental Science and Engineering, Tsinghua University, Beijing, P.R. China, **2** College of Environmental Science and Engineering, Hunan University, Hunan, P.R. China

Abstract

Photochemical treatment is increasingly being applied to remedy environmental problems. TiO$_2$-derived catalysts are efficiently and widely used in photodegradation applications. The efficiency of various photochemical treatments, namely, the use of UV irradiation without catalyst or with TiO$_2$/graphene-TiO$_2$ photodegradation methods was determined by comparing the photodegadation of two main types of hydrophobic chlorinated aromatic pollutants, namely, pentachlorophenol (PCP) and polychlorinated biphenyls (PCBs). Results show that photodegradation in methanol solution under pure UV irradiation was more efficient than that with either one of the catalysts tested, contrary to previous results in which photodegradation rates were enhanced using TiO$_2$-derived catalysts. The effects of various factors, such as UV light illumination, addition of methanol to the solution, catalyst dosage, and the pH of the reaction mixture, were examined. The degradation pathway was deduced. The photochemical treatment in methanol soil washing solution did not benefit from the use of the catalysts tested. Pure UV irradiation was sufficient for the dechlorination and degradation of the PCP and PCBs.

Editor: Hans-Joachim Lehmler, The University of Iowa, United States of America

Funding: The authors appreciate the generous financial support for this work from the National Natural Science Foundation of China (No. 41371472) and the Major Science and Technology Program for Water Pollution Control and Treatment of the Ministry of Environmental Protection of China (No. 2011ZX07317-001). The funders had no role in study design, data collection and analysis, decision to publish, or preparation of the manuscript.

Competing Interests: The authors have declared that no competing interests exist.

* Email: htwang@tsinghua.edu.cn

Introduction

Hydrophobic chlorinated aromatic pollutants, such as pentachlorophenol (PCP) and polychlorinated biphenyls (PCBs), are among the most important environmental pollutants in the twentieth century. For several decades after their commercial production, these compounds had been widely used for numerous applications, such as in wood protection, pesticides, and dielectric fluids in capacitors and transformers, before their global biota accumulation and genotoxic activity were gradually noticed [1]. Given their potential health hazard for humans and wildlife, the Stockholm Convention in 2001 classified PCP and PCBs as Persistent Organic Pollutants. Despite a comprehensive production ban since that time, millions of tons of these compounds continue to circulate in the environment [2,3]. Thus, the dechlorination and degradation of PCP and PCBs have emerged as major issues.

The cleanup of hydrophobic chlorinated aromatic pollutants is a challenging task. Various remediation technologies have been developed because of the extremely slow natural degradation of these compounds [4]. Some of these techniques are based on a "dig and dump approach," such as landfilling or capping. This method immobilizes the contaminant to prevent it from entering the aqueous phase [5]. Other methods are based on a "dig and incinerate approach," such as thermal treatment in which heat is used to remove or destroy the contaminants [6]. Among all the remediation technologies, soil washing combined with photochemical treatments has become increasingly advantageous because this method does not release toxic by-products into the environment and the cost is reasonable [7]. Soil washing is a well-developed technology that removes contaminants from polluted solid phase. The target contaminant can be extracted from the soil by a soil washing solvent, and the following photocatalytic degradation can be altered with different soil washing solvents [8,9]. Among the various solvents used to extract the target contaminant from the soil matrix, alcohol, such as methanol and ethanol, has been successfully used to remove PCP and other contaminants [10,11]. The solvent contains target contaminants after soil washing, which are usually treated under ultraviolet light photolysis, and the organic contaminants inside can be decomposed by photodegradation [12]. During UV irradiation, •OH can be formed from water molecules or other highly reactive solvents, initiating the decomposition of the pollutants [13].

To enhance the degradation rate, different types of photocatalysts such as TiO_2 and ZnO have been investigated for photodegradation [14–17]. In these studies, the photocatalytic degradations of PCP have been enhanced by various catalysts compared with TiO_2 or pure UV irradiation. Although remediation systems using a photocatalyst combined with UV irradiation have been successfully demonstrated to treat heterogeneous polluted soil or water [18,19], whether the addition of photocatalyst provided more efficient photodegradation in any condition than using pure UV irradiation has not been proved. Considering that TiO_2 has been used extensively for water treatment and control of organic contaminants, the heterogeneous modifications of TiO_2 have been evaluated [20–23]. To demonstrate the improvements of modified TiO_2 catalysts, numerous studies have conducted photodegradation competitions between new catalysts and the original TiO_2 catalyst. Degradations using pure UV irradiation under analogous conditions were rarely investigated [21,23,24].

The different methods for treating PCP and PCBs in methanol soil washing solvent under UV irradiation with or without a TiO_2-derived catalyst were compared in this study. TiO_2 coupled with graphene was used as modified TiO_2 catalyst in this study. Graphene has high electrical conductivity and efficient electron storage and shuttling capabilities. In this study, the degradation rates under UV irradiation with and without a TiO_2-derived catalyst are presented. Although the degradation rates using modified TiO_2 catalysts were all remarkably higher compared with the original TiO_2 catalyst, they were not as high as that when only UV irradiation was used. The main factors influencing the degradation that were taken into consideration included UV light illumination, addition of methanol to the solution, catalyst dosage, and the pH value of the solution. The beneficial effect of adding a catalyst for the contaminant photodegradation in methanol soil washing solvent was not confirmed.

Materials and Methods

Reagents and materials

Commercial P25 TiO_2 (80% anatase, 20% rutile) was supplied by the Degussa (Germany). Hexane and methanol (HPLC grade) were purchased from Fisher Scientific (USA). Graphene oxide was purchased from XFNano (China). Water, which was used as a solvent, was obtained using a Milli-Q Water Purification System. Pentachlorophenol (PCP, purity >98.0%) and pentachlorophenol sodium salt (PCP-Na) were purchased from the Sigma-Aldrich (USA). PCBs were obtained directly from a waste transformer factory in China. Methyl Orange (MO, analytical grade) was purchased from Sino Chemical Reagent (China).

Graphene-TiO_2 was prepared via a hydrothermal method. 25 mg of graphene oxide were dispersed in 50 mL water via sonication for 1 h to form a stable solution. Then, 1 g of TiO_2 was added into the graphene oxide suspension, which was stirred to mix the solution thoroughly. The mixture was moved into a Teflon-lined autoclave and heated at 180°C for 6 h. The resulting grey slurry was filtered and dried prior to use [24,25].

Photoirradiation procedure

All UV irradiation parts of the experiments were conducted in a 500 mL glass reactor (length 310 mm, diameter 70 mm), which contained a UV lighting system, as shown in Fig. 1. The lighting system included a high-pressure mercury lamp (GGZ300, Phillips, maximum wavelengths at 254, 292, 313, 334, 365, 436 and 546 nm) for 100 W and 300 W UV irradiation; a low-pressure mercury lamp (Hagende, maximum wavelength at 254 nm) for

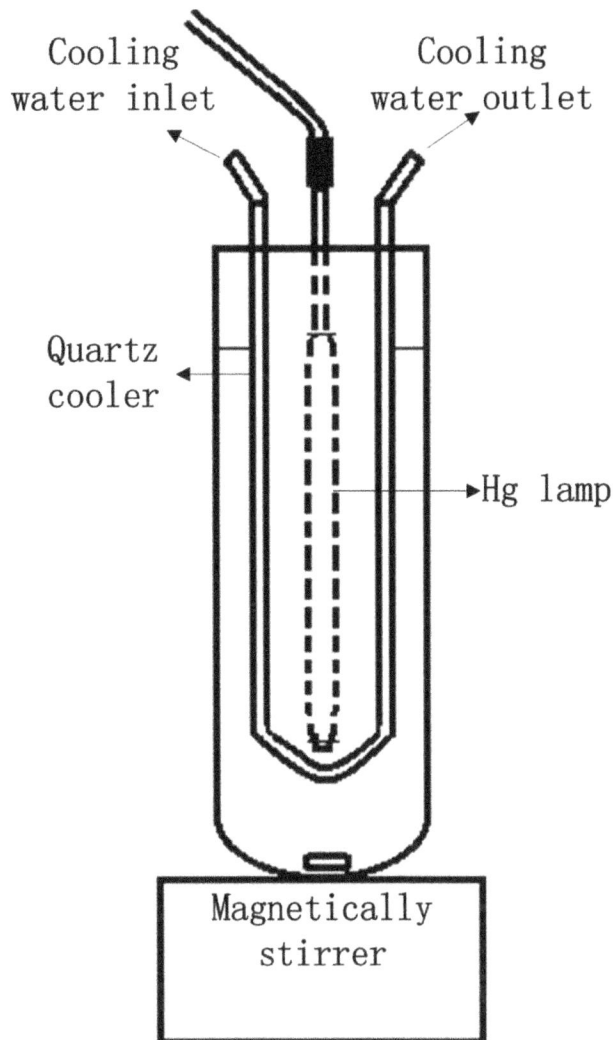

Figure 1. Schematic experimental system for photo degradation experiments.

9 W UV irradiation; a quartz well (length 300 mm, diameter 55 mm) equipped with a circulating water unit to maintain the system at 20°C; and a magnetic stirrer to promote uniform mixing of the catalyst in the solution. The UV light intensities at the reaction point were $2.0 * 10^3$ Lx for the 9 W mercury lamp, $7.2 * 10^4$ Lx for the 100 W mercury lamp, and approximately $3 * 10^5$ Lx for the 300 W mercury lamp, as measured by a Lux tester (Hioki, Japan). The radiation powers of UV light at the reaction point were $5.2 * 10^2$ μW cm^{-2} for the 9 W mercury lamp, $14.7 * 10^3$ μW cm^{-2} for the 100 W mercury lamp, and $72.3 * 10^3$ μW cm^{-2} for the 300 W mercury lamp, as measured by a radiometer (Handy, China).

Typically the reactor contained 500 mL of the soil washing solution with a concentration of 10 mg L^{-1} of target contaminant (e.g., 0.0375 mM PCP solution) with or without 200 mg catalyst prior to photodegradation [26]. After 2 h of equilibrating the solutions in the dark, UV light irradiation was initiated to start the reactions.

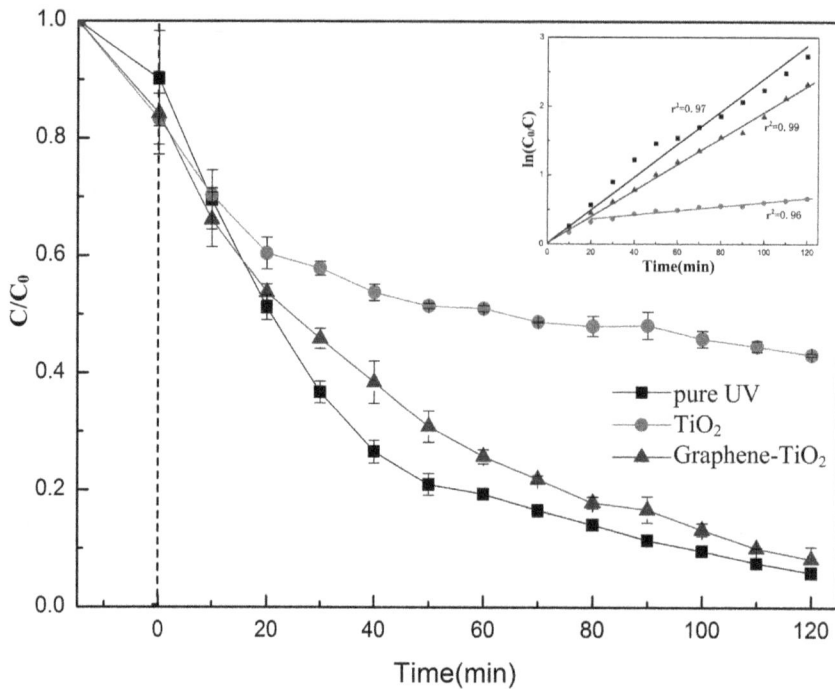

Figure 2. Photocatalytic degradation of PCP with TiO₂, graphene-TiO₂ and without catalyst under 100 W UV light. The inset represents the logarithmic transform for each curve.

Analytical methods

To follow the degradation process, 1 mL of mixture was extracted from the soil washing solution at the beginning, after equilibration, and after UV irradiation at predetermined time-points (5, 10, and 20 min for the 300, 100, and 9 W mercury lamps, respectively). Every mixture was adequately shaken with the same volume of methanol and then filtered through a 0.45 μm

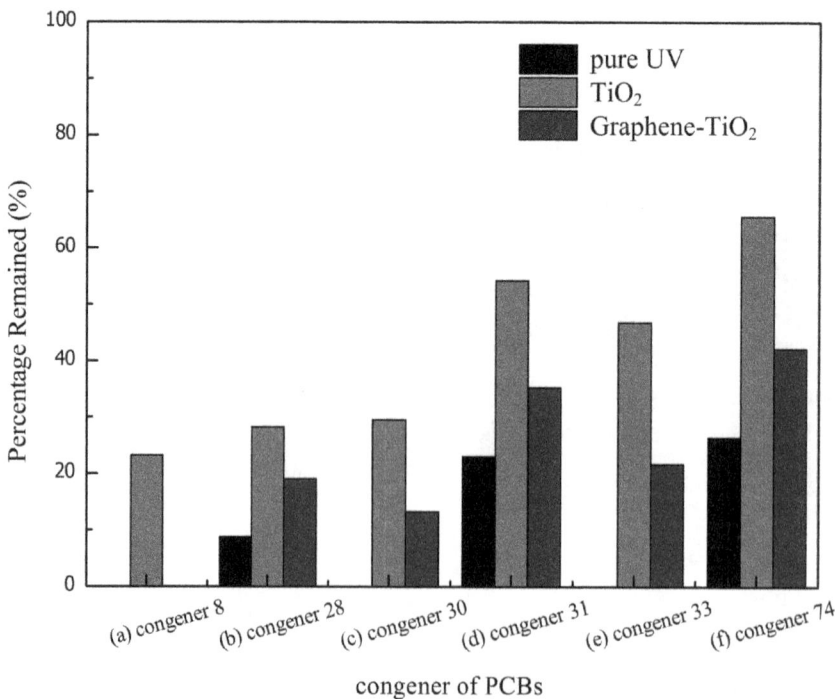

Figure 3. Percentage remained of PCBs after 2 hours photodegradation under UV light with TiO₂, graphene-TiO₂ and without catalyst: (a) congener 8; (b) congener 28; (c) congener 30; (d) congener 31; (e) congener 33; (f) congener 74.

Table 1. Comparison of photocatalytic rate constants (k) of PCP with different photocatalyst conditions under 3 kinds of UV irradiation.

Photocatalyst	k (min^{-1}, 100 W UV)	k (min^{-1}, 300 W UV)	k (min^{-1}, 9 W UV)
Pure UV	0.0236	0.1317	0.0208
TiO$_2$	0.0031	0.0211	0.0032
Graphene-TiO$_2$	0.0191	0.0493	0.0116

pore size membrane filter (for PCBs, every mixture was shaken with 2 mL of hexane).

PCP and phenol were analyzed using high-performance liquid chromatography (HPLC, Agilent 1260) equipped with an Agilent TC-C18 reverse phase column. UV detection was performed at 249, 270, and 254 nm for PCP, phenol, and other by-products, respectively [26]. A mixture of methanol and water was used as the mobile phase, with a gradient mixture and a flow rate of 1.0 mL min^{-1}. The quantification of the target analytes was based on the calibration curve in the HPLC analysis, and the linear range was 0 mg L^{-1} to 10 mg L^{-1}. Analytical results were verified by standard PCP and phenol solution with certain concentration.

PCBs were determined using a gas chromatography – mass spectrometry instrument (GC/MS-QP2010, Shimadzu, Japan) equipped with an HP-5MS capillary column. The temperature of the GC oven was held at 150°C for 1 min, increased to 185°C at a rate of 20°C min^{-1}, followed by an increase to 245°C at a rate of 2°C min^{-1}, and then held at 245°C for 3 min prior to further increase to 290°C at a rate of 6°C min^{-1}. The injector and detector temperatures were 250 and 290°C, respectively. The carrier gas was helium, which was utilized at a flow rate of 1.0 mL min^{-1}. The MS ion source and interface temperatures were 200 and 220°C, respectively.

The identification results were confirmed by GC/MS with the same HP-5MS capillary column and temperature control procedure as the GC analysis. Quantification of different PCB congeners was caculated based on the peak areas of their respective response factors of an authentic standard [27].

Results and Discussion

Photodegradation of PCP under UV irradiation

To explore the photodegradation rate under UV illumination, variations in the concentrations of PCP were investigated. The PCP photoactivities under 100 W UV light with TiO$_2$, with graphene-TiO$_2$, and without catalyst are compared in Fig. 2. The initial PCP concentration in the solution was 10 mg L^{-1} (0.0375 mM) and the solvent was a 1:100 methanol–water soil washing solution. The photocatalytic activity using TiO$_2$ was considerably lower than those in the other two treatments, and pure UV (without a catalyst) showed the highest activity. After 120 min of UV irradiation, the removal rates of PCP by pure UV (without a catalyst), UV with graphene-TiO$_2$, and UV with TiO$_2$ were 94%, 92%, and 57%, respectively.

The logarithm of the ratio between the initial concentration (C$_0$) of PCP and its concentration (C) at a specific given time is shown in the inset of Fig. 2. The slope of these straight lines provided the apparent rate constant. All correlation coefficients (r^2) obtained were higher than 0.95. As shown in this figure, pure UV irradiation of the PCP solution provided the highest photocatalytic rate constant, namely, 0.0236 min^{-1}. The photodegradation rate constant using graphene-TiO$_2$ was slightly lower at 0.0191 min^{-1}. The data for the photodegradation using TiO$_2$ was not linear at the beginning of the irradiation procedure. Thus, the slope was measured using data corresponding to 20 min or more of photoirradiation, which provided a photocatalytic rate constant of 0.0031 min^{-1}. This value was almost 1/8 of that of the highest measured rate constant. The result showed that the PCP can react with the •OH in a solution formed by UV irradiation and undergo

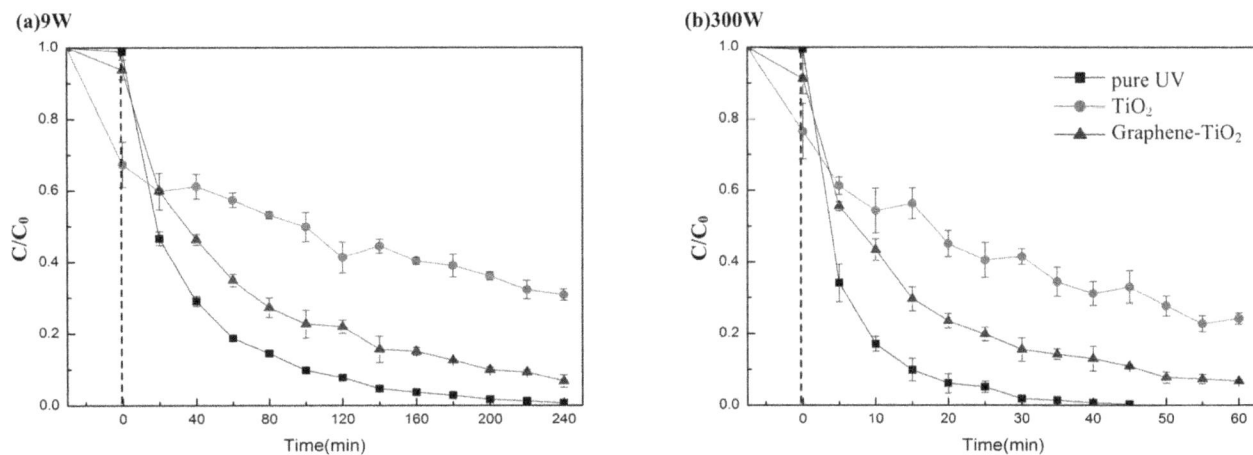

Figure 4. Photocatalytic degradation of PCP with TiO$_2$, graphene-TiO$_2$ and without catalyst under: (a) 9 W UV light; (b) 300 W UV light.

Table 2. Comparison of phenol production under different photocatalyst conditions after 2 hours UV irradiation and its percentage of initial PCP.

Photocatalyst	Concentration (mM)	Percentage phenol production of initial PCP
No catalyst	0.0057	15%
TiO_2	0.0067	18%
Graphene-TiO_2	0.0050	13%

degradation. The higher photodegradation rate using pure UV light than those with catalyst could be attributed to two main reasons. First, the C–O bond energy of methanol compound was lower than the C–Cl bond energy of the PCP compound, and the methanol compound is much easier attached to the surface of TiO_2 than hydrophobic chlorinated aromatic pollutant. When photo irradiation began, the methanol attached to the TiO_2 surface was degraded first. Second, the UV light, which could act on PCP was blocked and absorbed by TiO_2 catalyst and led to an energy loss. The high photocatalytic rate constant at the beginning of the photodegradation procedure was also caused by two reasons. First, soil washing solutions were equilibrated with the photocatalyst prior to irradiation. The high concentration of the target contaminant resulted in their adsorption on the catalyst surface, which reacted easily with the photoexcited electrons when UV irradiation began. Thus, the energy wasted was minimal. After a period of reaction, the pollution on the catalyst surface was degraded, and the photodegradation rate decreased. Second, when mercury lamp was turned on briefly, the energy in UV region was slightly higher, which also led a higher constant rate at the beginning.

Photodegradation of PCBs under UV irradiation

The experiments with the mixture of PCBs were complex because various PCB congeners were present in the sample, and excessive by-products were created during photodegradation. The experiments were simplified by monitoring the concentrations of six major components to explore the photodegradation rate of PCBs in methanol soil washing solution under 100 W UV irradiation. The initial concentration of PCBs in the solution was 10 mg L^{-1}, and the solvent was a 1:10 methanol–water soil washing solution. The concentrations of the six major components in the solution were 0.67 mg L^{-1} 2,4'-dichlorobiphenyl (congener 8), 1.48 mg L^{-1} 2,4,4'-trichlorobiphenyl (congener 28), 0.61 mg L^{-1} 2,4,6-trichlorobiphenyl (congener 30), 0.83 mg L^{-1} 2,4',5-trichlorobiphenyl (congener 31), 0.68 mg L^{-1} 2',3,4-trichlorobiphenyl (congener 33), and 0.58 mg L^{-1} 2,4,4',5-tetrachlorobiphenyl (congener 74). Figure 3 compares the percentage of photodegradation for each PCB congener under UV irradiation either with TiO_2, with graphene-TiO_2 or without catalyst.

The results in Fig. 3 indicated that all congeners obtained the highest photodegradation rates using pure UV irradiation. The half-life of congener 8 was the shortest among the six congeners.

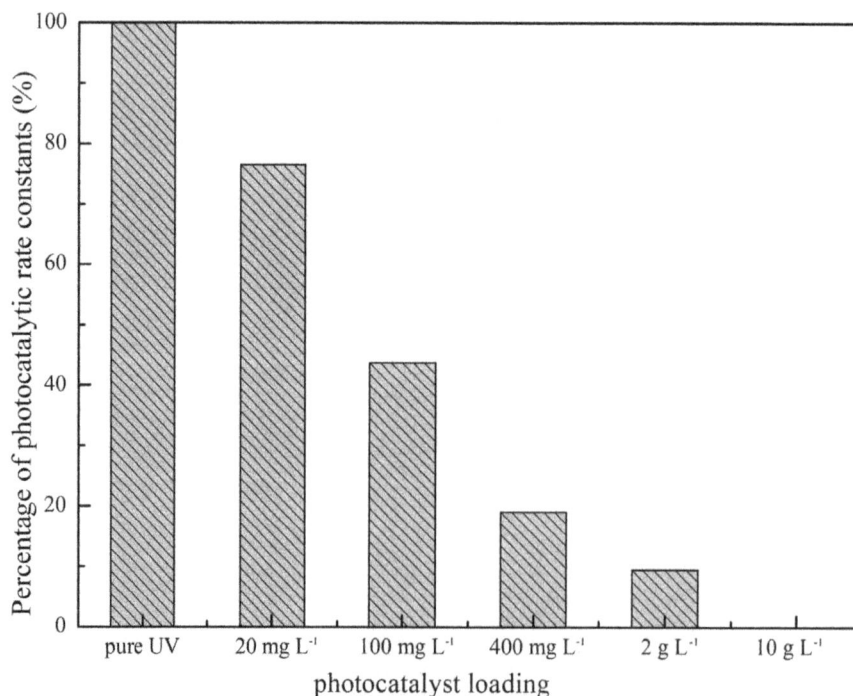

Figure 5. Photocatalytic rate constants of PCP degradation with different TiO₂ loading, compared with the one of pure UV irradiation.

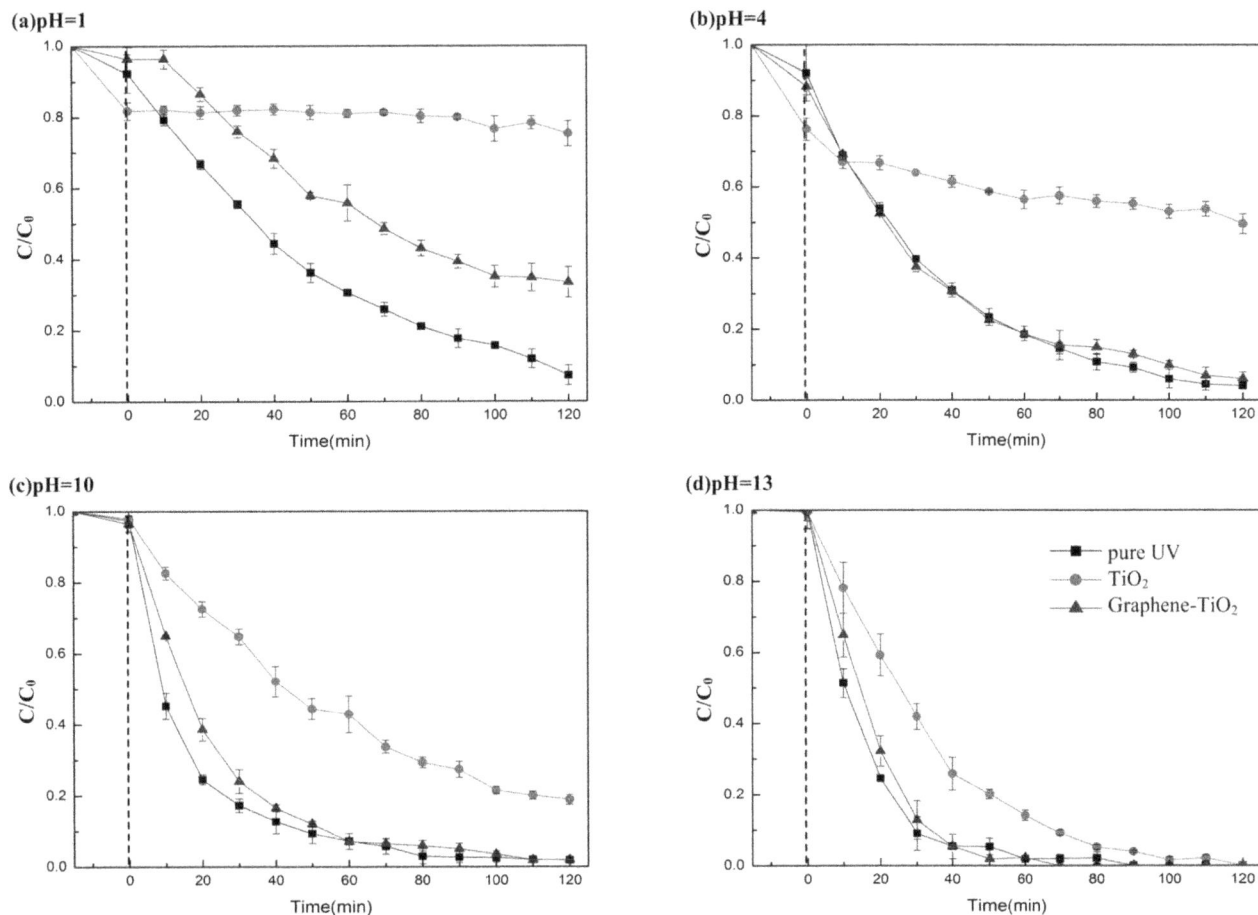

Figure 6. Photocatalytic degradation of PCP with TiO₂, graphene-TiO₂ and without catalyst under different pH values: (a) pH = 1; (b) pH = 4; (c) pH = 10; (d) pH = 13.

Congener 8 in the mixture of PCBs was fully photodegraded in 20 min using pure UV irradiation but required 100 min if graphene-TiO_2 was present in the mixture. For reactions that contained TiO_2, congener 8 was degraded by 77% after 2 h of reaction time. Congener 30 was completely removed after 110 min using pure UV irradiation. The degradation rates were 71% and 87% after 2 h of reaction for systems with extra TiO_2 and graphene-TiO_2, respectively. Congener 33 was eliminated after 100 min, which was slightly faster than congener 30. However, the degradation rates of congener 30 using TiO_2 and graphene-TiO_2 as the catalysts were 53% and 78% after the reaction, respectively. These values were lower than those observed for congener 30. Congeners 28, 31, and 74 remained in each of the mixtures after the photodegradation, and their degradation rates under these experimental conditions were similar. The degradation rates were 91%, 72%, and 81% for congener 28, 77%, 46%, and 65% for congener 31, and 74%, 34%, and 58% for congener 74. The photodegradation of PCBs did not quite follow the pseudo-first order kinetics as described in previous articles [9,12]. This finding may be caused by the fact that some PCB congeners may be parts of the pathway of another reagent, such as congeners 28 and congener 31, which can be formed from congener 74 via the dechlorination of one chlorine atom. The concentration of biphenyl could not be tested for the entire duration of all three procedures because its degradation rate was higher than those of the PCBs [9]. Chlorine atoms in the PCB

molecules break apart from biphenyl during UV irradiation. When the number of the chlorine atoms in the PCB congeners decreases, they separate easily. PCB molecules will be dechlorinated gradually to biphenyl and then be decomposed to small molecule. Similar to PCP degradation, the photocatalytic activity of TiO_2-derived catalysts was based on creating electron-hole pairs when exposed to UV radiation. The photoexcited electrons and holes were mainly reacting with the methanol in the solution, and the decomposition of target pollutants was delayed. The electron-hole pairs created on the photocatalyst surface were barely acting on the PCBs, and the block of UV light caused an energy loss, especially for the original TiO_2 catalyst.

Photodegradation of Methyl orange under UV irradiation

To certify that TiO_2 could increase the photodegradation rate in other conditions under UV irradiation, Methyl orange (MO), a commonly used dye, was photodegraded. This dye could be easily trapped by the holes on the catalyst surface [28]. The reactor contained 500 mL of 50 mg/L MO (0.15 mM) with/without 1 g L^{-1} of TiO_2 before photo degradation. This system was illuminated with 300 W UV lamp after adsorption-desorption equilibrium. To evaluate the discoloration rate of MO, the mixture withdrawn from the MO solution at every predetermined time-point was filtered through 0.45 μm pore size membrane filter and analyzed by UV-Vis spectroscopy at 463 nm (UV-2401PC, Shimadzu, Japan).

The degradation of MO solution fitted the pseudo first-order kinetic as reported. The significant degradation rates between pure UV light and UV with TiO_2 were consistent with previously reported results [28]. Photo degradation using TiO_2 as catalyst obtained a photocatalytic rate constant of 0.0759 min^{-1}, which was significantly higher than that of pure UV irradiation which was 0.0010 min^{-1}. After 60 min UV irradiation, the removal efficiency of MO using TiO_2 was 99%, whereas that of MO using pure UV was only 6%. This result proved that TiO_2 is actually useful in another condition.

Effect of different UV illumination sources

To investigate the relationship between the photodegradation rate and the UV illumination type, two groups of experiments were conducted to degrade PCP using mercury lamps of different intensities, as shown in Fig. 4. The same 300 W high-pressure mercury lamp was used. The power was increased from 100 W to 300 W. The results were compared with those obtained using a 9 W low-pressure mercury lamp. The apparent photocatalytic rate constant for each UV irradiation procedure is listed in Table 1. Pure UV photodegradation resulted in the highest photocatalytic rate constant at each of the different mercury lamp power intensities. Systems with TiO_2 as a catalyst resulted in the lowest rate constants. The photocatalytic rate constants measured using the 9 W UV lamp were almost similar to those using the 100 W UV lamp. This result was ascribed to the fact that the low-pressure mercury lamp offered a maximum wavelength at 254 nm, which concentrated all the energy in the UV region. By contrast, the energy from the high-pressure mercury lamp was separated both in the UV and the visible areas. In other words, the energy from a low-pressure mercury lamp (in this case, the 9 W UV lamp) was fully utilized in the photodegradation procedure.

Effect of adding methanol and the mineralization

Methanol – water mixture was used as soil washing solvent to extract PCP or PCBs from contaminated soil for former photodegradation experiments. To estimate the influence of methanol and evaluate the mineralization of the pollution, PCP-Na was used for UV irradiation because it is soluble in water. A solution that contained 10 mg L^{-1} PCP-Na (0.0347 mM) was degraded without additional methanol using a 100 W UV light source. Following the aforementioned method for irradiation and data analysis, a photocatalytic rate constant of 0.0429 min^{-1} was measured, which was higher than that (0.0236 min^{-1}) when 10 g L^{-1} methanol was included in the solution. The photodegradation rate of 10 mg L^{-1} PCP in pure methanol soil washing solution was also tested. After the same photo irradiation method, the photocatalytic rate constant was measured at 0.0143 min^{-1}. This result indicated that methanol does not participate in the degradation pathway of PCP. The reaction of methanol consumes energy from the UV irradiation, which reduces the photocatalytic rate constant for target contaminant. The total organic carbon (TOC) of the solution was tested. TOC was 2.0 mg L^{-1} at the beginning while there is only PCP-Na in the solution. TOC decreased to approximately 80% when the PCP-Na in the solution was almost fully degraded, which indicated that dechlorination, degradation and mineralization were coexistent. After 2 h of UV irradiation, PCP-Na was fully removed and TOC decreased a half to 1.0 mg L^{-1}, which showed that mineralization go along after PCP degraded.

Phenol production after UV irradiation of PCP

The main contaminant in the solution was PCP during the UV irradiation procedure, because the photodegradation rate of PCP was lower than those of other types of chlorophenols on its degradation pathway [29,30]. Chlorophenols produced in the degradation pathway of PCP were dechlorinated or decomposed because of their comparatively high photodegradation rate constant, and their concentration was undetectable by HPLC in the overall mixture. During UV irradiation, PCP lost chlorine atoms one by one until all Cl atoms were separated from the benzene ring at the beginning of the process. The dechlorination product of chlorophenols was phenol before further conversion, which resulted in the splitting of the benzene ring. The final concentration of phenol was investigated after each UV irradiation, as shown in Table 2. The results indicated that the final concentration of phenol in solution were similar for the three treatments. Given that the concentrations of other chlorophenols were negligible in the solution, PCP degradation did not end after dechlorination, whereas benzene ring continued to split. The photodegradation with pure UV irradiation was similar to the treatments using a catalyst, which did not simply terminate after dechlorination.

Effect of catalyst dosage

The effect of photocatalyst loading was investigated to determine the relationship between the absorption of photons and UV light energy blocked by the excess amount of catalyst. Therefore, a series of experiments with different amounts of TiO_2 was conducted. The same 1:100 methanol – water soil washing solution that contained 10 mg L^{-1} PCP (0.0375 mM) was degraded using a 100 W UV light source. All experiments followed pseudo-first order kinetics. The photocatalytic rate constant of each experiment was compared with that of pure UV irradiation, as shown in Fig. 5. Photodegradation rate decreased with increasing TiO_2 loading. TiO_2 catalyst almost blocked the whole photon energy when loading was up to 10 g L^{-1}. The result indicated that the reaction with •OH in methanol soil washing solution is the main source of PCP degradation under UV light, and TiO_2 added reduced the degradation.

Effect of pH

The pH values of the soil washing solution remained at approximately 4. To investigate the effect of pH on the photodegradation rate, PCP solutions with four different pH levels ranging from pH 1 to 13 were tested in the photoirradiation experiments using a 100 W UV light source, as shown in Fig. 6. The pH values were adjusted using either HCl or NaOH. The photodegradation rate increased as the pH value increased in each of the solutions. At high initial pH, the elevated concentration of the hydroxide ions (OH^-) is assumed to result in increased •OH production that could accelerate the photodegradation rate. The pure UV irradiation and graphene-TiO_2 catalyzed systems provided significantly higher degradation rates compared with those measured using TiO_2 at each of the different pH values. The degradation rate of PCP under pure UV irradiation was slightly higher than that of graphene-TiO_2 catalyzed systems at not so high pH level, whereas their rates became almost similar under alkaline conditions. When pH was too low, •OH was difficult to form, especially when considerable photon energy was blocked by TiO_2. Thus, the photocatalytic activity of TiO_2 was lost when the pH of the solution was adjusted to 1 using HCl.

Conclusions

TiO_2-derived catalysts are increasingly being used to treat samples that contain environmental pollutants, However, they are not well suited for the photodegradation of PCP or PCBs in methanol soil washing solution. Using graphene-TiO_2 as a photocatalyst evidently enhanced the photodegradation rate under UV light irradiation compared with using TiO_2 as the photocatalyst. However, pure UV irradiation showed the highest photocatalytic rate among the three treatment conditions. PCP and PCBs exhibited photoreactivities via •OH in the solution and were decomposed directly under UV irradiation. Addition of TiO_2-derived catalysts led to a loss of energy in this condition, thereby decelerating photodegradation. The photodegradation abilities were similar under UV irradiation regardless of whether a photocatalyst was added because all treatments completed the dechlorination and degradation, which were similar to published results [9,12]. The fact that biphenyl in the PCB solutions was undetectable during the entire reaction progress and the TOC test of PCP also illustrates that photodegradation under UV irradiation without a catalyst allows both dechlorination and degradation for the contaminants. Prior degradation of methanol and UV light blocked by the photocatalyst suspension caused photoenergy loss and a decrease in the photodegradation rate.

Author Contributions

Conceived and designed the experiments: ZZ YZ. Performed the experiments: ZZ. Analyzed the data: ZZ TC. Contributed reagents/materials/analysis tools: HW WL. Contributed to the writing of the manuscript: ZZ.

References

1. Lu GN, Tao XQ, Huang W, Dang Z, Li Z, et al. (2010) Dechlorination pathways of diverse chlorinated aromatic pollutants conducted by Dehalococcoides sp. strain CBDB1. Sci. Total Environ. 408: 2549–2554.

2. Quan X, Zhao X, Chen S, Zhao H, Chen J, et al. (2005) Enhancement of p,p'-DDT photodegradation on soil surfaces using TiO_2 induced by UV-light. Chemosphere 60: 266–273.

3. Clarke BO, Porter NA, Marriott PJ, Blackbeard JR. (2010) Investigating the levels and trends of organochlorine pesticides and polychlorinated biphenyl in sewage sludge. Environ. Int. 36: 23–329.

4. Gomes H, Dias-Ferreira C, Ribeiro A. (2013) Overview of in situ and ex situ remediation technologies for PCB-contaminated soils and sediments and obstacles for full-scale application. Sci. Total Environ. 445–446: 237–260.

5. Eek E, Cornelissen G, Kibsgaard A, Breedveld GD. (2008) Diffusion of PAH and PCB from contaminated sediments with and without mineral capping; measurement and modeling. Chemosphere 71: 1629–1638.

6. Sato T, Todoroki T, Shimoda K, Terada A, Hosomi M. (2010) Behavior of PCDDs/PCDFs in remediation of PCBs-contaminated sediments by thermal desorption. Chemosphere 80: 184–189.

7. Zhao S, Ma H, Wang M, Cao C, Xiong J, et al. (2010) Study on the mechanism of photo-degradation of p-nitrophenol exposed to 254 nm UV light. Hazard. Mater. 180: 86–90.

8. Zhang YX, Wang HT. (2014) Study of PCP photodegradation by TiO_2 catalyst based on different properties of soil washing effluents. Adv. Mater. Res. 878: 791–796.

9. Zhu X, Zhou D, Wang Y, Cang L, Fang G, et al. (2012) Remediation of polychlorinated biphenyl-contaminated soil by soil washing and subsequent TiO_2 photocatalytic degradation. Soils Sediments. 12: 1371–1379.

10. Jonsson S, Lind H, Lundstedt S, Haglund P, Tysklind M. (2010) Dioxin removal from contaminated soils by ethanol washing. J. Hazard. Mater. 179: 393–399.

11. Khodadoust AP, Sudan MT, Acheson CM, Brenner RC. (1999) Solvent extraction of pentachlorophenol from contaminated soils using water-ethanol mixtures. Chemosphere. 38: 2681–2693.

12. Miao X, Chu S, Xu X. (1999) Degradation pathways of PCBs upon UV irradiation in hexane. Chemosphere 39: 1639–1650.

13. Arany E, Szabo RK, Apati L, Alapi T, Ilisz I, et al. (2013) Degradation of naproxen by UV, VUV photolysis and theircombination. J. Hazard. Mater. 262: 151–157.

14. Xie J, Hao YJ, Zhou Z, Meng XC, Yao L, et al. (2014) Fabrication and characterization of ZnO particles with different morphologies for photocatalytic degradation of pentachlorophenol. Res. Chem. Intermed. 40: 1937–1946.

15. ThanhThuy TT, Feng H, Cai Q. (2013) Photocatalytic degradation of pentachlorophenol on ZnSe/ TiO_2 supported by photo-Fenton system, Chem. Eng. J. 223: 379–387.

16. Govindan K, Murugesan S, Maruthamuthu P. (2013) Photocatalytic degradation of pentachlorophenol in aqueous solution by visible light sensitive NF-codoped TiO_2 photocatalyst. Mater. Res. Bull. 48: 1913–1919.

17. Xie J, Bian L, Yao L, Hao Y, Wei Y. (2013) Simple fabrication of mesoporous TiO_2 microspheres for photocatalytic degradation of pentachlorophenol. Mater. Lett. 91: 213–216.

18. Chand R, Shiraishi F, (2013) Reaction mechanism of photocatalytic decomposition of 2,4-dinitrophenol in aqueous suspension of TiO_2 fine particles. Chem. Eng. J. 233: 369–376.

19. Dong D, Li P, Li X, Xu C, Gong D, et al. (2010) Photocatalytic degradation of phenanthrene and pyrene on soil surfaces in the presence of nanometer rutile TiO_2 under UV-irradiation. Chem. Eng. J. 158: 378–383.

20. Tada H, Kokubu A, Iwasaki M, Ito S. (2004) Deactivation of the TiO_2 photocatalyst by coupling with WO_3 and electrochemically assisted high photocatalytic activity of WO_3. Langmuir 20: 4665–4670.

21. Zyoud AH, Zaatar N, Saadeddin I, Ali C, Park D, et al. (2010) CdS-sensitized TiO_2 in phenazopyridine photo-degradation: Catalyst efficiency, stability and feasibility assessment. J. Hazard. Mater. 173: 318–325.

22. Lu S, Wu D, Wang Q, Yan J, Buekens A, et al. (2011) Photocatalytic decomposition on nano-TiO_2: Destruction of chloroaromatic compounds. Chemosphere 82: 1215–1224.

23. Shaban Y, Sayed M, Maradny A, Farawati R, Zobidi M. (2013) Photocatalytic degradation of phenol in natural seawater using visible light active carbon modified (CM)-n-TiO_2 nanoparticles under UV light and natural sunlight illuminations. Chemosphere 91: 307–313.

24. Li J, Zhou SL, Hong GB, Chang CT. (2013) Hydrothermal preparation of P25–graphene composite with enhanced adsorption and photocatalytic degradation of dyes. Chem. Eng. J. 219: 486–491.

25. Cheng P, Yang Z, Wang H, Cheng W, Chen M, et al. (2012) TiO_2-graphene nanocomposites for photocatalytic hydrogen production from splitting water. Int. J. Hydrogen Energy 37: 2224–2230.

26. Liu JW, Han R, Wang HT, Zhao Y, Chu Z, et al. (2011) Photoassisted degradation of pentachlorophenol in a simulated soil washing system containing nonionic surfactant Triton X-100 with La–B codoped TiO_2 under visible and solar light irradiation. Appl. Catal. B: Environ. 103: 470–478.

27. Sun Y, Takaoka M, Takeda N, Wang W, Zeng X, et al. (2012) Decomposition of 2,2',4,4',5,5'-hexachlorobiphenyl with iron supported on an activated carbon from an ion-exchange resin. Chemosphere 88: 895–902.

28. Li D, Zheng H, Wang Q, Wang X, Jiang W, et al. (2014) A novel double-cylindrical-shell photoreactor immobilized with monolayer TiO_2-coated silica gel beads for photocatalytic degradation of Rhodamine B and Methyl Orange in aqueous solution. Sep. Purif. Technol. 123: 130–138.

29. Nirupam K, Krishna GB. (2013) Cu(II)-kaolinite and Cu(II)-montmorillonite as catalysts for wet oxidative degradation of 2-chlorophenol, 4-chlorophenol and 2,4-dichlorophenol. Chem, Eng. J. 233: 88–97.

30. Zhang Y, Wu H, Zhang J, Wang H, Lu W. (2012) Enhanced photodegradation of pentachlorophenol by single and mixed cationic and nonionic surfactants. J. Hazard. Mater. 221–222: 92–99.

11

DNA Damage in Buccal Mucosa Cells of Pre-School Children Exposed to High Levels of Urban Air Pollutants

Elisabetta Ceretti[1], Donatella Feretti[1]*, Gaia C. V. Viola[1], Ilaria Zerbini[1], Rosa M. Limina[1], Claudia Zani[1], Michela Capelli[2], Rossella Lamera[2], Francesco Donato[1], Umberto Gelatti[1]

1 Unit of Hygiene, Epidemiology and Public Health, Department of Medical and Surgical Specialities, Radiological Sciences and Public Health, University of Brescia, Brescia, Italy, 2 Post-Graduate School of Public Health, University of Brescia, Brescia, Italy

Abstract

Air pollution has been recognized as a human carcinogen. Children living in urban areas are a high-risk group, because genetic damage occurring early in life is considered able to increase the risk of carcinogenesis in adulthood. This study aimed to investigate micronuclei (MN) frequency, as a biomarker of DNA damage, in exfoliated buccal cells of pre-school children living in a town with high levels of air pollution. A sample of healthy 3-6-year-old children living in Brescia, Northern Italy, was investigated. A sample of the children's buccal mucosa cells was collected during the winter months in 2012 and 2013. DNA damage was investigated using the MN test. Children's exposure to urban air pollution was evaluated by means of a questionnaire filled in by their parents that included items on various possible sources of indoor and outdoor pollution, and the concentration of fine particulate matter (PM10, PM2.5) and NO_2 in the 1–3 weeks preceding biological sample collection. 181 children (mean age$\pm$SD: 4.3$\pm$0.9 years) were investigated. The mean$\pm$SD MN frequency was 0.29$\pm$0.13%. A weak, though statistically significant, association of MN with concentration of air pollutants (PM10, PM2.5 and NO_2) was found, whereas no association was apparent between MN frequency and the indoor and outdoor exposure variables investigated via the questionnaire. This study showed a high MN frequency in children living in a town with heavy air pollution in winter, higher than usually found among children living in areas with low or medium-high levels of air pollution.

Editor: Susanna Esposito, Fondazione IRCCS Ca' Granda Ospedale Maggiore Policlinico, Università degli Studi di Milano, Italy

Funding: The study was funded by the Q-TECH Research and Study Centre (Quality and Technology Assessment, Governance and Communication Strategies in Health Systems), University of Brescia. The Lombardy Regional Authority provided a three-year research grant under agreements with the university for the promotion of research in the Lombardy region. The funders had no role in study design, data collection and analysis, decision to publish, or preparation of the manuscript.

Competing Interests: The authors have declared that no competing interests exist.

* E-mail: feretti@med.unibs.it

Introduction

Air pollution is a global health problem, particularly in urban areas. Close to 90% of European citizens residing in urban areas are exposed to air pollution exceeding EU limit values (Air Quality Directive 2008/50/EC) and WHO guideline levels [1–3].

Several epidemiologic studies have demonstrated the association between air pollution exposure, especially to particulate matter, and mortality and morbidity in humans [4–8].

The finest fractions of particulate matter (PM2.5 and less) play a major role in causing chronic diseases because they are retained in the alveolar regions of the lungs and diffuse into the blood stream, inducing inflammation, oxidative stress, and blood coagulation [9,10]. Extracts of urban air particles can induce cancer in animals [11,12] and are mutagenic in bacteria, plant and mammalian cells in *in-vitro* tests [13–17]. *In-vivo* mutagenicity tests have been performed on humans as well. The micronuclei (MN) formation has been used as an indicator of chromosome damage, induced by substances that cause chromosome breakage (clastogens) as well as by agents that affect the spindle apparatus (aneugens) [18]. Other nuclear anomalies were investigated as biomarkers of DNA damage and cell death, which may be useful for a more comprehensive assessment of genotoxic damage [19–22].

An increase in cancer risk has been observed in the presence of a high level of chromosomal aberrations and micronuclei in several studies [23–26]. This biomarker can be investigated in various organs, tissues and body fluids, such as leukocytes or lymphocytes in peripheral blood, though cells derived from target tissues are considered more appropriate. In particular, exfoliated buccal and nasal cells have been used in the biological monitoring of people exposed to airborne pollutants as they are representative of epithelial respiratory tract cells and are easier to collect than those of other respiratory organs [21,27–29].

An increase of micronuclei in leukocytes in peripheral blood has been observed in people exposed to urban air pollutants [30–36] and a strong correlation of MN frequency in buccal exfoliated cells and peripheral lymphocytes has been found [37,38].

Children are a high-risk group in terms of the health effects of air pollution [2,5,39–42]. Some studies suggest that early exposure during childhood can play an important role in the development of chronic diseases in adulthood: the earlier the exposure, the greater the risk of chronic disease, including cancer [43].

Few studies have considered genetic damage in mucosa buccal cells as MN frequency in cells of children exposed to air pollution, and they only involved a small number of subjects showing cytogenetic damage in children or young adults living in polluted

Table 1. Micronuclei (MN) frequencies observed in buccal mucosa cells of children according to socio-demographic features and exposure variables (N = 181).

Subjects and demographic features	N (%)	% MN (Mean±SD)
Sex		
M	103 (56.9)	0.30±0.13
F	78 (43.1)	0.29±0.13
Children's age		
3 years	34 (18.8)	0.31±0.16
4 years	59 (32.6)	0.28±0.13
5–6 years	88 (48.6)	0.29±0.12
Parents' education (at least one parent)		
Primary school or less	19 (10.5)	0.28±0.12
Secondary school	53 (29.3)	0.32±0.16
College or university	109 (60.2)	0.28±0.12
Home characteristics		
Traffic in the area		
Heavy	97 (53.6)	0.31±0.14
Moderate	59 (32.6)	0.28±0.13
Very light	25 (13.8)	0.25±0.08
Truck traffic in the area		
Heavy	31 (17.1)	0.30±0.12
Moderate	67 (37.0)	0.31±0.17
Very light	82 (45.3)	0.27±0.09
Indoor exposure		
Gas stove in home	61 (33.7)	0.28±0.11
Fireplace in home	37 (20.4)	0.28±0.11
Presence of smokers in home	28 (15.5)	0.30±0.18
School characteristics		
Traffic in the area		
Heavy	102 (56.4)	0.31±0.15
Moderate	68 (37.6)	0.27±0.09
Very light	10 (5.5)	0.29±0.15
Truck traffic in the area		
Heavy	27 (14.9)	0.33±0.19
Moderate	80 (44.2)	0.29±0.12
Very light	72 (39.8)	0.28±0.12
Child's habits		
Plays outdoors		
Less than 1 hour	78 (43.1)	0.31±0.16
More than 1 hour but less than 3 hours	67 (37.0)	0.28±0.10
3 hours or more	35 (19.3)	0.28±0.10
Remains in the kitchen while meals are cooked		
Never	24 (13.3)	0.29±0.13
Sometimes	119 (65.8)	0.30±0.14
Often/always	37 (20.4)	0.27±0.09
Consumes fried/grilled/smoked food		
Never	5 (2.8)	0.36±0.17
More than once per month	123 (68.0)	0.28±0.14
Parents' smoking habits		
Neither parent smoke	118 (65.2)	0.29±0.12
Mother smoked during pregnancy	40 (22.1)	0.29±0.14
Mother smokes	30 (16.6)	0.29±0.15

Table 1. Cont.

Subjects and demographic features	N (%)	% MN (Mean±SD)
Father smoke	49 (27.1)	0.30±0.17
Both parents smokers	42 (23.2)	0.29±0.15

areas with a high concentration of PM or oxidant pollutants [33,44–47].

The aim of this study was to investigate MN frequency, as a biomarker of DNA damage, in exfoliated buccal cells of pre-school children living in a town with high levels of air pollution during the winter season, when the highest levels of particulate matter and other pollutants are usually found.

Materials and Methods

Study design

This study is part of the RESPIRA study (Italian acronym for Rischio ESPosizione Inquinamento aRia Atmosferica), a molecular epidemiology cross-sectional study aiming to assess the presence of MN frequency in pre-school children living in Brescia, a highly polluted town in Northern Italy, located in the Po Valley, one of the most highly polluted areas of Europe. The children were recruited in 6 schools located in different areas of the town. The study enrolled children aged 3–6 years, born in Italy to European parents, without malignant tumours, who had not undergone radiotherapy or chemotherapy in the previous 12 months or X-rays in the previous 3 months.

The presence of MN and other nuclear anomalies were investigated in buccal mucosa cells taken from the children.

The biological samples were collected during two consecutive winter seasons (2012 and 2013) since the highest values of PM10 and PM2.5 in Brescia are usually observed in the winter months. The project was approved by the Ethics Committee of Local Unit Health of Brescia (Comitato Etico dell'ASL -Azienda Sanitaria Locale- di Brescia). The children's parents provided their written informed consent to participate in this study. All the data collected were treated confidentially in accordance with current Italian legislation (privacy law).

Questionnaire

The children's parents were interviewed using an *ad hoc* questionnaire designed to gather information on exposure to air pollutants from both indoor and outdoor sources, including some characteristics of the area of residence (e.g. traffic, factories), parents' smoking habits, and children's respiratory diseases and drug consumption.

Collection of air pollution data

Chemical data regarding daily concentration of the most commonly measured air pollutants (CO, NO_2, SO_2, benzene, O_3, PM10 and PM2.5) were retrieved from the freely available ARPA (Regional Agency for Environmental Protection) database to characterize urban air quality.

Collection of biological samples

All the biological samples were taken during or after a series of days with high levels of PM10, PM2.5 and NO_2.

For the collection of buccal mucosa cells, the children rinsed their mouths twice with mineral water. Interdental brushes were used to collect epithelial buccal cells for the micronucleus test, by gently scraping the inside of both cheeks and dipping them into tubes containing 15 ml of PBS (phosphate buffered saline solution) [21]. This method is simple and non-invasive and therefore easily acceptable by both children and parents.

Micronucleus test

After shaking and removing the brush, the epithelial buccal cells in PBS were centrifuged for 10 minutes at 1100 g at 4°C and re-suspended in warm PBS (37°C). In order to determine whether enough cells had been collected to perform the test, 10 µl of cell suspension was applied in a Burker chamber on which the number of cells was scored. The PBS cell suspension was then centrifuged for 4 minutes at 8700 g and the pellet was re-suspended in 700 µl of warm hypotonic solution (KCl 0.56%, 37°C). After 1 minute, 700 µl of cold methanol/acetic acid (14:1, −20°C) was added. This fixed suspension was centrifuged for 4 minutes at 8700 g, and the pellet was re-suspended in 40 µl of warm PBS and dropped onto two frosted slides, which were dried and stained with Giemsa dye (5 minutes at room temperature). The slides were then washed with distilled water, dried and mounted with Eukitt.

The slides were examined at 1000X magnification for microscope analysis. Before MN frequency was assessed, cells were divided into two categories: "normal" cells and ones that are considered "abnormal" based on their cytobiological and nuclear features, which are indicative of DNA damage (MN and nuclear buds), cytokinetic failure (binucleated cells), proliferative potential (basal cells) or cell death (condensed chromatin, karyorrhexis, pyknotic, karyolitic and without nucleus cells), according to the Buccal Micronucleus Cytome (BMCyt) assay [22]. To assess MN, nuclear buds, binucleated and basal cells frequency, at least 2000 "normal" cells per slide (two slides per subject, 4000 cells per subject) were scored by two expert operators with duplicate reading. The results are given as the percentage of cells with MN and buds, and binucleated and basal cells [19,22]. Moreover, condensed chromatin, karyorrhexis, pyknotic, karyolitic, and without nucleus cells were evaluated scoring 2000 total cells per subject, and expressed as percentages.

Statistical analysis

All the data were processed to investigate the associations between air pollution parameters and MN frequency. As buccal cells have a short life, no more than 3 weeks [22], the associations between frequency of MN and other biomarkers frequency and air pollutants concentration (CO, NO_2, SO_2, benzene, O_3, PM10 and PM2.5) were analysed using air pollutants mean levels at 0 (sampling day), 1, 2 and 3 weeks before sampling. Both univariate analysis and multivariate analysis (multiple regression, logistic regression) were performed to assess the association investigated, adjusting for confounding factors. Particularly, linear regression with biological parameters as dependent variables and the concentration of air pollutants as independent variables were fitted. In order to improve the interpretation of these findings the coefficients of the linear regression were computed on a 10 µg/m³

Table 2. Frequency of micronuclei (MN) and other nuclear anomalies in all children and according to sex and age.

	N° children	% MN (Mean±SD)	% Nuclear buds (Mean±SD)	% Binucleated cells (Mean±SD)	% Basal cells (Mean±SD)	% Condensed chromatin cells (Mean±SD)	% Karyorrhetic cells (Mean±SD)	% Pyknotic cells (Mean±SD)	% Karyolytic cells (Mean±SD)	% Without nucleus cells (Mean±SD)
All children	181	0.29±0.13	0.02±0.04	0.11±0.08	0.72±1.15	17.20±6.13	7.44±4.29	0.57±0.48	3.72±3.37	1.42±1.28
Sex										
Males	103	0.30±0.13	0.02±0.04	0.11±0.09	0.84±1.38	17.10±6.26	7.88±4.72	0.61±0.50	3.96±3.97	1.42±1.18
Females	78	0.29±0.13	0.02±0.03	0.10±0.08	0.56±0.73	17.33±6.00	6.86±3.60	0.52±0.46	3.42±2.34	1.43±1.40
Children's age										
3 years	34	0.31±0.16	0.02±0.06	0.10±0.09	0.56±0.80	16.70±4.52	7.48±4.84	0.65±0.53	4.07±2.91	1.65±1.27
4 years	59	0.28±0.13	0.01±0.02	0.13±0.10	0.64±0.69	16.58±6.18	7.96±4.41	0.52±0.49	3.21±2.31	1.23±1.15
5–6 years	88	0.29±0.12	0.02±0.03	0.09±0.06	0.83±1.46	17.80±6.62	7.07±3.99	0.57±0.46	3.94±4.05	1.47±1.35

scale of measures of air pollutants. Two-tailed statistical tests were performed, with 0.05 p-value as the threshold for rejecting the null hypothesis. All the analyses were performed using the Stata TM 12.0 statistical package (Stata Statistical Software Release 12.0, 2012; Stata Corporation, College Station, Texas, USA).

Results

A total of 222 children were recruited, of whom 181 (mean age ±SD: 4.35±0.84 years; 56.9% males) were examined because 41 samples were not eligible for MN analysis due to an insufficient number of cells collected.

The mean±SD MN frequency expressed as a percentage was 0.29±0.13 (median 0.28 and range 0.085–0.990).

Table 1 shows the results of the MN analysis, according to children's socio-demographic characteristics and habits, and indoor and outdoor exposure data. No variable was associated with MN frequency in exfoliated buccal cells.

Table 2 sets out all the biomarkers evaluated in the buccal mucosa cells of children according to children's sex and age. No statistically significant difference was observed according to sex and age for each parameter.

The daily levels of PM10, PM2.5 and NO_2 from January to March 2012 and 2013, and biological sampling days are shown in Figure 1. During these months, the concentrations of PM10 and PM2.5 were almost always over the EU limit values for daily means (50 and 25 $\mu g/m^3$, respectively). Likewise, the annual EU limit for NO_2 (40$\mu g/m3$) was exceeded on all the days. On the contrary, the concentration of CO, ozone, SO_2 and benzene remained low throughout the period considered and were always below the EU limit values (data not shown).

The mean MN frequency according to concentrations of PM10, PM2.5 and NO_2 on the day of sampling and one, two and three weeks before sampling are shown in Table 3. A fair variation (from 0.21% to 0.62%) was found in MN frequency from one day to another, without a clear relationship with concentration of air pollutants.

The coefficients of linear regressions of the biological markers on the concentration of PM10, PM2.5 and NO_2 on the same day and at 1, 2 and 3 weeks preceding biological sampling computed for 10 $\mu g/m^3$ units of increase are shown in Table 4. A modest, though statistically significant, increase of the frequency of nuclear anomalies (MN, buds and binucleated cells) and of basal cells for an increase of PM10, PM2.5 and NO_2 was found. No clear pattern was evident for the other parameters.

Discussion

The main finding of this study was a surprisingly high level of MN frequency in exfoliated buccal cells of pre-school children living in Brescia.

The MN frequency (mean: 0.29%) observed in our study was higher than that observed in healthy children "without important exposure" (0.108%) as a result of a pooled analysis of 321 children aged up to 9 years [48]. Furthermore, the value observed in children living in Brescia was higher than that shown in adolescents or young adults working in an engine repair workshop (mean age: 15.5 years, MN frequency: 0.07%) [49] or exposed to ozone air pollution (university students, MN frequency: 0.12%) [46]. However, these studies are not comparable to ours due to the different age of the people investigated. Age is in fact one of the most important factors affecting MN data in both lymphocytes and oral cells, with a progressively increase of MN with age [33,36,48,50]. However, the MN frequency observed in our study was higher than that found in children (aged 6–17) living in the

A

B

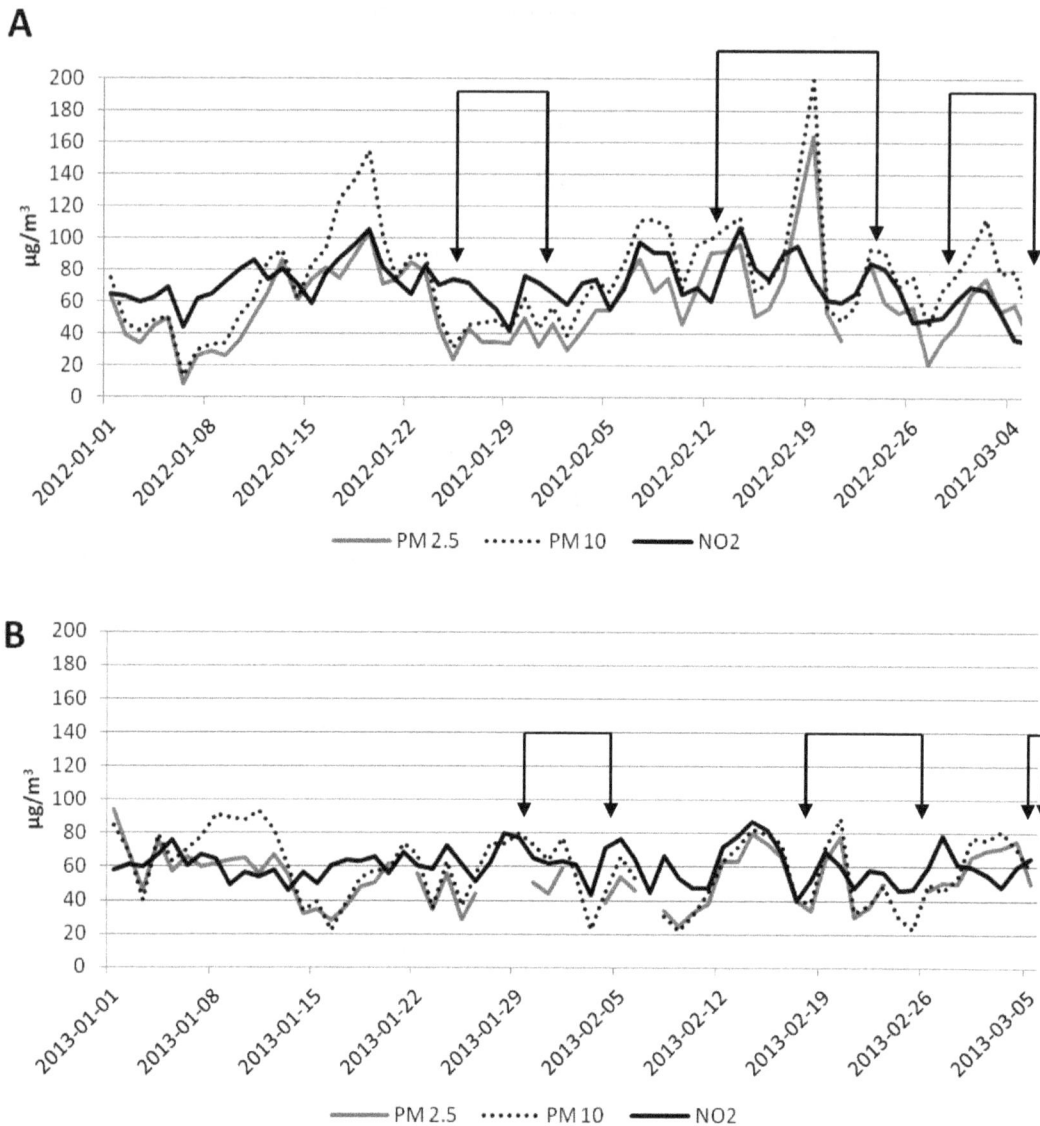

Figure 1. PM10, PM2.5 and NO$_2$ concentration in 2012 (A) and 2013 (B). The arrows indicate the days on which biological sampling was performed. The dotted lines refer to the days without measurement of air pollutants.

urban area of Calcutta [44], which was higher than that found in those living in rural area (0.22% vs 0.17%). Children of a similar age to those in our study were included in a study on genotoxicity of air pollutants generated by biomass burning [47], which found a significant difference in MN frequency in oral cells between children exposed to high PM2.5 concentration and those living in a control area with a lower PM2.5 concentration (0.12% vs 0.02% in children under 7 years of age). A MN frequency similar to, or higher than, that observed in our study was reported in some studies carried out in Eastern Europe, as revised by Holland et al. [36].

According to a wide dataset which includes adults and children, a MN frequency interval of between 0.03% and 0.17% can be considered as a range of spontaneous MN frequency in buccal exfoliated cells of "unexposed" children [48]. Other authors have reported a range of 0.05–0.08% as the baseline MN frequency in exfoliated cells in healthy people [46]. Therefore, the mean MN frequency of 0.29% found in our study is about two-three-fold

higher than that considered as a "reference" value for children of this age.

The MN frequency did not vary according to sex, age, parental education and all the variables investigated through the questionnaire. A weak, though statistically significant, association of MN frequency with concentration of air pollutants in the week preceding the buccal mucosa cell collection, but not on the same day and in the second or third week before sampling, was found. These results are partially in agreement with findings from previous studies comparing people at different exposure levels to air pollutants, which showed that industrial or urban air pollutants had a genotoxic effect on mucosa cells [44,47,49]. It should be pointed out, however, that the concentration of PM10, PM2.5 and NO$_2$ was high throughout the study period, and always above the EU proposed limits for daily levels. The range of variation of pollutant concentration may therefore have been too narrow to determine a substantial change in MN frequency in the children. Alternatively, a threshold instead of a dose-response mechanism

Table 3. Micronuclei (MN) frequency and PM10, PM2.5 and NO$_2$ concentrations on the day of sampling and in the 1–3 weeks preceding biological sampling.

| Day of sampling | N° children | % MN (Mean±SD) | Same day | | | Weeks preceding biological sampling | | | | | | | | |
| | | | | | | 1 week | | | 2 weeks | | | 3 weeks | | |
			PM10 (µg/m³)	PM2.5 (µg/m³)	NO$_2$ (µg/m³)	PM10 (µg/m³)	PM2.5 (µg/m³)	NO$_2$ (µg/m³)	PM10 (µg/m³)	PM2.5 (µg/m³)	NO$_2$ (µg/m³)	PM10 (µg/m³)	PM2.5 (µg/m³)	NO$_2$ (µg/m³)
2012-01-24	8	0.62±0.23	52.0	44	70.3	110.2	82.6	84.5	92.9	73.8	80.0	73.9	59.6	74.1
2012-01-25	4	0.27±0.10	31	24	73.9	99.8	78.1	82.0	93.0	74.4	79.3	74.4	60.0	74.6
2012-01-26	4	0.30±0.14	45.5	43	71.8	84.7	68.7	78.9	90.8	72.4	78.4	73.5	59.0	75.1
2012-01-30	4	0.18±0.03	62.0	50	76.1	50.9	42.0	65.4	80.7	62.4	74.6	76.2	60.7	74.7
2012-01-31	7	0.27±0.10	43.0	32	71.9	46.8	37.9	64.5	78.5	60.2	74.5	77.5	61.8	74.9
2012-02-13	12	0.26±0.15	106.5	92	87.2	97.1	72.0	77.4	77.1	58.1	72.5	68.4	52.8	70.2
2012-02-14	12	0.34±0.08	113.0	96	107.1	100.1	75.0	80.1	80.3	61.1	73.3	69.1	53.4	70.4
2012-02-20	12	0.25±0.09	57.0	53	61.3	112.3	93.1	87.0	104.7	82.6	82.2	88.9	69.8	77.4
2012-02-22	11	0.29±0.04	57.5	nd	65.4	95.9	79.0	76.6	98.2	67.3	79.0	88.9	70.1	76.1
2012-02-23	10	0.23±0.05	93.5	83	84.4	94.6	71.7	74.4	94.3	72.9	77.2	88.9	68.0	76.1
2012-02-28	11	0.24±0.09	67.5	36	50.5	68.7	44.3	65.0	87.0	65.9	74.2	91.4	69.0	76.1
2012-02-29	12	0.30±0.13	77.5	46	60.8	71.5	44.3	63.7	83.7	61.6	70.1	89.3	66.5	73.9
2012-03-01	12	0.24±0.07	92.0	65	69.7	74.4	50.9	63.3	84.5	61.3	68.7	87.7	65.6	72.5
2012-03-05	6	0.25±0.04	51.0	38	34.5	78.8	50.7	55.6	74.6	49.8	61.2	87.2	64.2	69.8
2013-01-29	6	0.40±0.19	80	nd	77.4	57.9	36.7	63.8	53.6	40.1	62.5	61.3	46.4	60.0
2013-01-30	3	0.21±0.03	73.5	51	65.5	59.8	32.8	66.2	56.5	40.5	64.5	60.8	45.6	60.6
2013-02-04	1	0.32±0.00	45.0	39	71.4	63.7	38.5	64.7	60.8	41.6	63.4	55.0	41.7	62.1
2013-02-18	9	0.31±0.05	39.0	34	52.2	65.2	60.6	67.9	50.3	46.8	64.3	54.8	44.9	64.4
2013-02-19	5	0.33±0.06	71.0	64	69.5	64.4	60.0	68.6	49.9	46.4	62.9	53.1	46.8	63.1
2013-02-25	7	0.41±0.15	23	nd	46.9	58.5	48.7	55.8	55.7	55.1	61.8	50.1	47.8	61.4
2013-03-04	16	0.25±0.07	70.5	75	60.1	58.0	59.0	58.6	53.9	53.8	57.2	57.7	56.3	60.7
2013-03-05	9	0.21±0.04	58.0	50	65.0	64.8	61.3	60.5	56.1	57.3	57.8	58.9	58.3	61.4

Table 4. Coefficients of linear regression of the biological markers for 10 μg/m³ units of increase of PM10, PM2.5 and NO$_2$ concentration.

| | Same day | | | 1 week | | | 2 weeks | | | 3 weeks | | |
	PM 10	PM 2.5	NO$_2$	PM 10	PM 2.5	NO$_2$	PM 10	PM 2.5	NO$_2$	PM 10	PM 2.5	NO$_2$
MN	-0.01	-0.00	0.00	0.01*	0.01	0.03**	0.00	0.00	0.02	-0.01	-0.03**	-0.01
Nuclear buds	-0.00	0.00	0.08*	0.01***	0.01**	0.01***	0.00	0.00	0.01**	-0.00	-0.00	0.00
Binucleated cells	-0.00	-0.00	0.01	0.01**	0.01*	0.02**	0.01**	0.02**	0.03***	0.00	0.00	0.03**
Basal cells	0.02	0.00	0.10	0.20***	0.15**	0.40***	0.15**	0.26**	0.42***	0.06	0.03	0.40**
Condensed chromatin cells	0.20	0.06	1.07***	1.13***	1.06***	2.93***	1.26***	1.94***	3.61***	0.74**	0.71	3.79***
Karyorrhetic cells	0.04	-0.36**	-0.26	1.73	-0.29	0.20	0.66***	0.60	1.22**	1.05***	1.32**	2.22***
Pyknotic cells	0.01	0.04**	0.02	0.02	0.03	0.03	-0.03	-0.01	-0.05	-0.07**	-0.11**	-0.12*
Karyolytic cells	-0.04	0.02	-0.25	-0.48***	-0.36**	-0.93***	-0.61***	-0.93***	-1.40***	-0.61**	-0.71*	-2.05***
Without nucleus cells	0.09*	0.07	0.12	-0.01	-0.08	0.05	-0.01	-0.04	0.03	-0.01	-0.12	-0.04

Weeks preceding biological sampling

statistically significant: *p<0.05;
**p<0.01;
***p<0.001.

may be proposed, so similar genotoxic effects on children's buccal mucosa cells may have been produced by air pollutant values above a defined level. Lastly, the role of confounding factors, such as children's diet, physical activity and others, which was not investigated in our study, cannot be excluded. Indeed, some studies have observed an association between MN frequency in buccal mucosa cells and diet (e.g. fruit consumption, supplementation with B vitamins or antioxidants) and some specific behaviours, such as smoking habit, intake of alcohol, and betel quid chewing [21,37,48,51,52].

Other nuclear anomalies apart from MN were associated with the PM10, PM2.5 and NO$_2$ concentration, particularly during the week preceding exfoliated cells sampling confirming findings of MN analysis on the possible effects of air pollutant exposure. However, the biological significance of these biomarkers in exfoliated oral cells is still unclear. Some nuclear anomalies, as buds, could be the result of genotoxic damage but these events are also associated with natural degenerative processes in these short-lived cells [53,54]. The frequency of binucleated cells was related to air pollutants concentration too. This parameter is primarily an indicator of cytotoxicity because of failures in cell division. Although these findings should be considered with caution because of paucity of data on the relationship between these biomarkers and air pollutants levels, overall they suggest that air pollution exposure may induce both genotoxic and cytotoxic damage in buccal mucosa cells of children. This observation strengthens the need to investigate these additional biomarkers of DNA damage together with MN for a more comprehensive evaluation of these issues.

This study has various strengths. First, the number of subjects recruited (181 children) is relatively high compared to previous studies of MN frequency in oral cells, particularly in this age group.

Second, it is unlikely that other important factors biased our results: there is no major industrial exposure in the area and few children could have been exposed to indoor pollution sources, including passive smoking at home, according to data collected via the questionnaire. Furthermore, the inclusion and exclusion criteria, e.g. residence in town, born to European parents, no malignant tumours, no radiotherapy or chemotherapy in the previous 12 months, no X-ray exposure in the three months before buccal cell collection, allowed us to rule out a possible role of other mutagenic factors.

Third, the collection of exfoliated buccal cells during two consecutive winter seasons (2012 and 2013) provided information on two periods characterized by similar environmental conditions and it showed consistent results, with no substantial difference in MN frequency in the two years.

Lastly, 4000 cells per child were scored for assessing MN frequency, representing a very high number of observations, higher than normally performed in current practice, in order to reduce the variability of the estimate, as suggested by Ceppi et al. [37].

This study has some limitations, however, mainly the lack of a control group of subjects living in a less polluted area or another collection of biological samples from the same or similar children in a period with lower airborne pollutant levels, for evaluating the role of spatial or seasonal differences in influencing MN frequency in buccal mucosa cells. Differences between more and less polluted areas have been found in some research [47] and seasonal differences in air pollution composition could be relevant for genotoxic damage, such as the summer increase of ozone concentration [46]. Nevertheless, this study was designed to assess MN frequency in a sufficiently large sample of children regularly exposed to high levels of air genotoxic pollutants (PM and PAH), and not seasonal differences. We are, however, planning to extend the research to summer months and to children living in less polluted areas.

Only one sampling of oral cells was performed for each child, so the intra-individual variability of this early effect biomarker was not evaluated. The use of only one effect measure does not cause an important bias, however, because it reflects the mean exposure of the three preceding weeks, when the climate situation had not substantially changed. On the other hand, previous studies which found an association between this biomarker and air pollution also used a single measure of effect [31,44,47,49,50,55,56].

MN are a marker of early biological effect able to detect both clastogens and aneuploidy-inducing chemicals [18]. They are formed from acentric chromosomal fragments or whole chromosomes that are not included in the main daughter nuclei during nuclear division and their induction therefore reflects clastogenic and/or aneugenic events. They represent stable cytogenetic alterations which are the result of recent exposure of buccal mucosa cells, in a rapidly dividing tissue [18,28]. These cells are short-lived and they are the first barrier for substances introduced into the body by inhalation or ingestion and may be an excellent target tissue for detecting early genotoxic effects induced by mutagenic airborne compounds. Their use can, therefore, be proposed for assessing exposure to airborne mutagens, especially in the paediatric population as they are easy to collect, considering also the strong correlation of MN frequency in these cells and in lymphocytes, which have been shown to be related to the subsequent risk of developing cancer [37,48].

Brescia is a highly industrialized area with a high level of motor vehicle traffic. It is located in the Po Valley, one of the most highly polluted areas of Europe, where the concentrations of PM10, PM2.5 and NOx are usually above the EU reference values for many days of the year, as in the two years of this study (2012 and 2013), similar to those found in other towns and cities in the Po Valley [3].

In conclusion, this study shows that children living in a town with high levels of air pollutants in a Western country have a high level of MN in buccal mucosa cells, confirming previous findings of a mutagenic effect of urban air pollution on human beings.

Acknowledgments

We wish to thank the Brescia Municipality for its logistic support and the school personnel and all children and parents involved in the research.

Author Contributions

Conceived and designed the experiments: EC DF FD UG. Performed the experiments: EC DF GCVV IZ CZ. Analyzed the data: RML MC RL FD. Wrote the paper: EC DF FD UG.

References

1. WHO (2006) Air Quality Guidelines. Global Update 2005. World Health Organization Office for Europe, Copenhagen, Denmark.
2. WHO (2007) Children's health and the environment in Europe: a baseline assessment. World Health Organization Regional Office for Europe, Copenhagen, Denmark.
3. EAA (2012) Air Quality in Europe 2011. European Environment Agency, Technical report No 4/2012.
4. Pope CA 3rd, Burnett RT, Thun MJ, Calle EE, Krewski D, et al. (2002) Lung cancer, cardiopulmonary mortality, and long-term exposure to fine particulate air pollution. JAMA 287(9):1132–1141.
5. ERS (2010) Air quality and health. European Respiratory Society. Lausanne, Switzerland.
6. Janssen NA, Fischer P, Marra M, Ameling C, Cassee FR (2013) Short-term effects of PM2.5, PM10 and PM2.5-10 on daily mortality in the Netherlands. Sci Total Environ 463–464: 20–26.
7. Raaschou-Nielsen O, Andersen ZJ, Beelen R, Samoli E, Stafoggia M, et al. (2013) Air pollution and lung cancer incidence in 17 European cohorts: prospective analyses from the European Study of Cohorts for Air Pollution Effects (ESCAPE).. Lancet Oncol 14: 813–822.
8. Shah ASV, Langrish JP, Nair H, McAllister DA, Hunter AL, et al. (2013) Global association of air pollution and heart failure: a systematic review and meta-analysis. Lancet 382: 1039–1048.
9. Sørensen M, Autrup H, Møller P, Hertel O, Jensen SS, et al. (2003) Linking exposure to environmental pollutants with biological effects. Mutat Res 544: 255–271.
10. Lewtas J (2007) Air pollution combustion emissions: characterization of causative agents and mechanisms associated with cancer, reproductive, and cardiovascular effects. Mutat Res 636: 95–133.
11. Claxton LD, Woodall GM Jr (2007) A review of the mutagenicity and rodent carcinogenicity of ambient air. Mutat Res 636: 36–94.
12. Møller P, Folkmann JK, Forchhammer L, Bräuner EV, Danielsen PH, et al. (2008) Air pollution, oxidative damage to DNA, and carcinogenesis. Cancer Lett 266: 84–97.
13. Monarca S, Crebelli R, Feretti D, Zanardini A, Fuselli S, et al. (1997) Mutagens and carcinogens in size-classified air particulates of a northern Italian town. Sci Total Environ 205: 137–144.
14. Monarca S, Feretti D, Zanardini A, Falistocco E, Nardi G (1999) Monitoring of mutagens in urban air samples. Mutat Res 426: 189–192.
15. Claxton LD, Matthews PP, Warren SH (2004) The genotoxicity of ambient outdoor air, a review: Salmonella mutagenicity. Mutat Res 567: 347–399.
16. Traversi D, Degan R, De Marco R, Gilli G, Pignata C, et al. (2009) Mutagenic properties of PM2.5 urban pollution in the northern Italy: the nitro-compounds contribution. Environ Int 35: 905–910.
17. de Brito KC, de Lemos CT, Rocha JA, Mielli AC, Matzenbacher C, et al. (2013) Comparative genotoxicity of airborne particulate matter (PM2.5) using Salmonella, plants and mammalian cells. Ecotoxicol Environ Saf 94: 14–20.
18. Kirsch-Volders M, Plas G, Elhajouji A, Lukamowicz M, Gonzalez L, et al. (2011) The in vitro MN assay in 2011: origin and fate, biological significance,

protocols, high throughput methodologies and toxicological relevance. Arch Toxicol 85: 873–899. doi:10.1007/s00204-011-0691-4
19. Tolbert PE, Shy CM, Allen JW (1992) Micronuclei and other nuclear anomalies in buccal smears: methods development. Mutat Res 271: 69–77.
20. Fenech M, Crott JW (2002) Micronuclei, nucleoplasmic bridges and nuclear buds induced in folic acid deficient human lymphocytes – evidence for breakage – fusion-bridge cycles in the cytokinesis-block micronucleus assay. Mutat Res 504: 131–136.
21. Holland N, Bolognesi C, Kirsch-Volders M, Bonassi S, Zeiger E, et al. (2008) The micronucleus assay in human buccal cells as a tool for biomonitoring DNA damage: the HUMN project perspective on current status and knowledge gaps. Mutat Res 659: 93–108.
22. Thomas P, Holland N, Bolognesi C, Kirsch-Volders M, Bonassi S, et al. (2009) Buccal micronucleus cytome assay. Nat Protoc 4(6):825–837.
23. Hagmar L, Brøgger A, Hansteen IL, Heim S, Högstedt B, et al. (1994) Cancer risk in humans predicted by increased levels of chromosomal aberrations in lymphocytes: Nordic study group on the health risk of chromosome damage. Cancer Res 4(11):2919–2922.
24. Bonassi S, Abbondandolo A, Camurri L, Dal Prá L, De Ferrari M, et al. (1995) Are chromosome aberrations in circulating lymphocytes predictive of future cancer onset in humans? Preliminary results of an Italian cohort study. Cancer Genet Cytogenet 79(2):133–135.
25. Bonassi S, Znaor A, Ceppi M, Lando C, Chang WP, et al. (2007) An increased micronucleus frequency in peripheral blood lymphocytes predicts the risk of cancer in humans. Carcinogenesis 28(3):625–631.
26. Bonassi S, El-Zein R, Bolognesi C, Fenech M (2011) Micronuclei frequency in peripheral blood lymphocytes and cancer risk: evidence from human studies. Mutagenesis 26(1):93–100.
27. Coronas MV, Pereira TS, Rocha JA, Lemos AT, Fachel JM, et al. (2009) Genetic biomonitoring of an urban population exposed to mutagenic airborne pollutants. Environ Int 35(7):1023–1029.
28. Kashyap B, Reddy PS (2012) Micronuclei assay of exfoliated oral buccal cells: means to assess the nuclear abnormalities in different diseases. J Cancer Res Ther 8(2):184–191.
29. Samanta S, Dey P (2012) Micronucleus and its applications. Diagn Cytopathol 40: 84–90.
30. Bolognesi C, Merlo F, Rabboni R, Valerio F, Abbondandolo A (1997) Cytogenetic biomonitoring in traffic police workers: micronucleus test in peripheral blood lymphocytes. Environ Mol Mutagen 30: 396–402.
31. Maffei F, Hrelia P, Angelini S, Carbone F, Cantelli Forti G, et al. (2005) Effects of environmental benzene: Micronucleus frequencies and haematological values in traffic police working in an urban area. Mutat Res 583: 1–11.
32. Cavallo D, Ursini CL, Carelli G, Iavicoli I, Ciervo A, et al. (2006) Occupational exposure in airport personnel: characterization and evaluation of genotoxic and oxidative effects. Toxicology 223: 26–35.
33. Huen K, Gunn L, Duramad P, Jeng M, Scalf R, et al. (2006) Application of a geographic information system to explore associations between air pollution and

micronucleus frequencies in African American children and adults. Environ Mol Mutagen 47(4):236–246.

34. Pedersen M, Vinzents P, Petersen JH, Kleinjans JC, Plas G, et al. (2006) Cytogenetic effects in children and mothers exposed to air pollution assessed by the frequency of micronuclei and fluorescence in situ hybridization (FISH): a family pilot study in the Czech Republic. Mutat Res 608: 112–120.

35. Tovalin H, Valverde M, Morandi MT, Blanco S, Whitehead L, et al. (2006) DNA damage in outdoor workers occupationally exposed to environmental air pollutants. Occup Environ Med 63: 230–236.

36. Holland N, Fucic A, Merlo DF, Sram R, Kirsch-Volders M (2011) Micronuclei in neonates and children: effects of environmental, genetic, demographic and disease variables. Mutagenesis 26: 51–56.

37. Ceppi M, Biasotti B, Fenech M, Bonassi S (2010) Human population studies with the exfoliated buccal micronucleus assay: statistical and epidemiological issues. Mutat Res 705(1):11–19.

38. Desai SS, Ghaisas SD, Jakhi SD, Bhide SV (1996) Cytogenetic damage in exfoliated oral mucosal cells and circulating lymphocytes of patients suffering from precancerous oral lesions. Cancer Lett 109: 9–14.

39. Landrigan PJ, Kimmel CA, Correa A, Eskenazi B (2004) Children's health and the environment: public health issues and challenges for risk assessment. Environ Health Perspect 112(2):257–265.

40. WHO (2005) Effects of air pollution on children's health and development. World Health Organization Regional Office for Europe, Copenhagen, Denmark.

41. Bateson TF, Schwartz J (2008) Children's response to air pollutants. J Toxicol Environ Health A 71(3):238–243.

42. Grigg J (2009) Particulate matter exposure in children: relevance to chronic obstructive pulmonary disease. Proc Am Thorac Soc 6(7):564–569.

43. Wild CP, Kleinjans J (2003) Children and increased susceptibility to environmental carcinogens: evidence or empathy? Cancer Epidemiol Biomarkers Prev 12(12):1389–1394.

44. Lahiri T, Roy S, Basu C, Ganguly S, Ray MR, et al. (2000) Air pollution in Calcutta elicits adverse pulmonary reaction in children. Indian J Med Res 112: 21–26.

45. Montero R, Serrano L, Dávila V, Segura Y, Arrieta A, et al. (2003) Metabolic polymorphisms and the micronucleus frequency in buccal epithelium of adolescents living in an urban environment. Environ Mol Mutagen 42: 216–222.

46. Chen C, Arjomandi M, Qin H, Balmes J, Tager I, et al. (2006) Cytogenetic damage in buccal epithelia and peripheral lymphocytes of young healthy individuals exposed to ozone. Mutagenesis 21(2):131–137.

47. Sisenando HA, Batistuzzo de Medeiros SR, Artaxo P, Saldiva PH, Hacon Sde S (2012) Micronucleus frequency in children exposed to biomass burning in the Brazilian Legal Amazon region: a control case study. BMC Oral Health 12: 6. doi:10.1186/1472-6831-12-6

48. Bonassi S, Coskun E, Ceppi M, Lando C, Bolognesi C, et al. (2011) The HUman MicroNucleus project on eXfoLiated buccal cells (HUMN(XL)): the role of life-style, host factors, occupational exposures, health status, and assay protocol. Mutat Res 728(3):88–97.

49. Karahalil B, Karakaya AE, Burgaz S (1999) The micronucleus assay in exfoliated buccal cells: application to occupational exposure to polycyclic aromatic hydrocarbons. Mutat Res 442(1):29–35.

50. Rossnerova A, Spatova M, Pastorkova A, Tabashidze N, Veleminsky M Jr, et al. (2011) Micronuclei levels in mothers and their newborns from regions with different types of air pollution. Mutat Res 715: 72–78.

51. Mondal NK, Ghosh S, Ray MR (2011) Micronucleus formation and DNA damage in buccal epithelial cells of Indian street boys addicted to gasp 'Golden glue'. Mutat Res 721: 178–183.

52. Thomas P, Wu J, Dhillon V, Fenech M (2011) Effect of dietary intervention on human micronucleus frequency in lymphocytes and buccal cells. Mutagenesis 26: 69–76.

53. Nersesyan AK (2005) Nuclear buds in exfoliated human cells. Letter to the Editor. Mutat Res 588: 64–68.

54. Cerqueira EMM, Gomes-Filho IS, Trindade S, Lopes MA, Passos JS, et al. (2004) Genetic damage in exfoliated cells from oral mucosa of individuals exposed to X-rays during panoramic dental radiographies. Mutat Res 562: 111–117.

55. Neri M, Ugolini D, Bonassi S, Fucic A, Holland N, et al. (2006) Children's exposure to environmental pollutants and biomarkers of genetic damage. II. Results of a comprehensive literature search and meta-analysis. Mutat Res 612: 14–39.

56. Pedersen M, Wichmann J, Autrup H, Dang DA, Decordier I, et al. (2009) Increased micronuclei and bulky DNA adducts in cord blood after maternal exposures to traffic-related air pollution. Environ Res 109: 1012–1020.

12

Bisphenol A Exposure and Asthma Development in School-Age Children: A Longitudinal Study

Kyoung-Nam Kim[1], Jin Hee Kim[2], Ho-Jang Kwon[3], Soo-Jong Hong[4], Byoung-Ju Kim[5], So-Yeon Lee[6], Yun-Chul Hong[1,2], Sanghyuk Bae[1]*

1 Department of Preventive Medicine, Seoul National University College of Medicine, Seoul, Korea, **2** Institute of Environmental Medicine, Medical Research Center, Seoul, Korea, **3** Department of Preventive Medicine and Public Health, Dankook University College of Medicine, Cheonan, Korea, **4** Department of Pediatrics, Childhood Asthma Atopy Center, Research Center for Standardization of Allergic Diseases, Asan Medical Center, University of Ulsan College of Medicine, Seoul, Korea, **5** Department of Pediatrics, Haeundae Paik Hospital, Inje University College of Medicine, Busan, Korea, **6** Department of Pediatrics, Hallym University Sacred Heart Hospital, Hallym University College of Medicine, Anyang, Korea

Abstract

Background: Although the effect of bisphenol A on various health outcomes has been extensively examined, few studies have investigated its effect on asthma.

Objective: We hypothesized that exposure to bisphenol A in school-age children was associated with wheezing and asthma.

Methods: Participants included 127 children aged 7–8 years without a previous asthma diagnosis in an elementary school in Seoul, Korea. Three surveys were conducted, each 2 years apart. Bisphenol A concentration was measured at the baseline survey, and PC_{20}, which is defined as the methacholine concentration that induces a decrease in FEV_1 of 20% from baseline, was measured at every survey. Associations between bisphenol A concentration at 7–8 years of age and wheezing, asthma, and PC_{20} at ages up to 11–12 years were examined using generalized estimating equations, a marginal Cox regression model, and a linear mixed model.

Results: The log-transformed creatinine-adjusted urinary bisphenol A concentration at 7–8 years was positively associated with wheezing (odds ratio, 2.48; 95% confidence interval, 1.15–5.31; $P = .02$) and asthma (hazard ratio, 2.13; 95% confidence interval, 1.51–3.00; $P < .001$) at ages up to 11–12 years. Bisphenol A was also negatively associated with PC_{20} ($ß = -2.33$; $P = .02$). When stratified by sex, the association between bisphenol A and asthma remained significant only in girls (hazard ratio, 2.45; 95% confidence interval, 2.18–2.76; $P < .001$).

Conclusion: Increased urinary bisphenol A concentrations at 7–8 years old were positively associated with wheezing and asthma and negatively associated with PC_{20} at ages up to 11–12 years.

Editor: David O. Carpenter, Institute for Health & the Environment, United States of America

Funding: This research was funded by the Korean Ministry of Environment. The funders had no role in study design, data collection and analysis, decision to publish, or preparation of the manuscript.

Competing Interests: The authors have declared that no competing interests exist.

* Email: sanghyukbae@snu.ac.kr

Introduction

Asthma is one of the most common childhood diseases with prevalence of 9.1% in the United States and 7.6% in Korea [1–3]. The global prevalence of asthma in children has risen significantly over the recent decades [4,5]. The exact cause of this increase is not known, but associations with increasing urbanization have been reported [6,7]. Furthermore, the increase in global asthma prevalence has occurred within approximately the same timeframe as the widespread use of industrial chemicals like bisphenol A (BPA) [8]. BPA is one of the chemicals produced in the highest volumes worldwide [9] and is used in the production of polycarbonate plastics and epoxy resins. Polycarbonate plastics are used to make products such as water bottles, toys, dental sealants, and compact discs, whereas epoxy resins are used to coat the insides of cans for food and beverages [8,10]. Human exposure to BPA is extensive, and 95% of the United States population has detectable urinary BPA concentrations [11].

Previous studies have shown that BPA could have various health effects, including diabetes [12,13], coronary artery stenosis [14,15], heart rate variability and blood pressure [16], abnormal liver function [17], childhood neurobehavioral problems [18], oxidative stress and inflammation [10], male sexual dysfunction [19,20], decreased semen quality [21,22], and adverse birth outcomes [23]. Especially in women, BPA has been associated with abnormal pubertal development [24,25], externalizing behaviors [26], recurrent miscarriages [27], and premature delivery [28].

Studies have also reported a potential association between BPA exposure and asthma. Animal studies suggest that BPA might affect the development of asthma-related conditions by promoting allergic immune responses [29–33]. BPA has also been shown to promote eosinophilic bronchial inflammation and airway responsiveness in mice [9,34] and production of airway secretary proteins, which is one of the hallmarks of asthma, in rhesus monkeys [35]. In humans, a relationship between prenatal urinary BPA and wheezing has been reported among children under the age of 3 years [36]. These findings are supported by a reported association between pre- and postnatal BPA exposure and wheezing and asthma development in preschool children [37].

However, to the best of our knowledge, there is little evidence regarding the effects of BPA exposure on asthma in school-age children. Furthermore, examining the association in Korea, where relatively low BPA concentration of school-age children (1.2 µg/L) compared to that of USA (2.7 µg/L) had been reported, gives extra information [38,39]. In the present study, we hypothesized that exposure to BPA in school-age children is associated with asthma-related outcomes such as wheezing, asthma, and PC_{20}, which is defined as the methacholine concentration that causes a decrease in FEV_1 of 20% from baseline.

Materials and Methods

Study sample and data collection

In 2005, all the 1st grade (n = 92) and 2nd grade children (n = 96) in an elementary school in Seoul, Korea, were invited to the study. Of a total of 188 children aged 7–8 years, parents of 153 schoolchildren agreed to enroll for the baseline survey, which consisted of a methacholine challenge test, urinary BPA measurement, and the International Study of Asthma and Allergies in Childhood (ISAAC) questionnaire, answered by the parents or guardian. Of the original sample, the participants without a BPA measurement (n = 16) or with a previous asthma diagnosis (n = 10) at baseline were subsequently excluded from the analysis, resulting in 127 children. In 2007, 125 of the 127 children who were then aged 9–10 years participated in the first follow-up survey, and, in 2009, all of the original 127 children who were then aged 11–12 years participated in the second follow-up survey (Figure 1). The follow-up surveys consisted of the ISAAC questionnaire and the methacholine challenge test.

Written informed consent was obtained from parents or guardians of all participating children, and the study protocols were reviewed and approved by the Institutional Review Board at the Dankook University Medical Center.

Exposure assessment

All participating children were asked to fast for more than 8 h before the survey. Spot urine samples (50 mL) were collected between 0900 and 1200 h from participants at the baseline survey (n = 127) in conical tubes (SPL Lifesciences, Pocheon, Gyunggi-do, Korea). Urine samples were stored at −20°C in the freezer and sent to the laboratory (NeoDin Medical Institute, Seoul, Korea) within 90 min. Urine samples were buffered using 30 µL of 2M sodium acetate (pH 5.0), and a mixture of 10 µL ß-glucuronidase/sulfatase (Sigma, St. Louis, MO, USA) and 25 µL BPA (RING-13C12, 99%, Cambridge Isotope Lab, Inc., Tewksbury, MA, USA) was added. The samples were incubated at 37°C for 3 h to deconjugate the glucuronidated BPA, and 100 µL of 2N HCl was added after incubation. The extract was dehydrated with nitrogen gas and then reconstituted with 1 mL of HPLC-grade H_2O in a

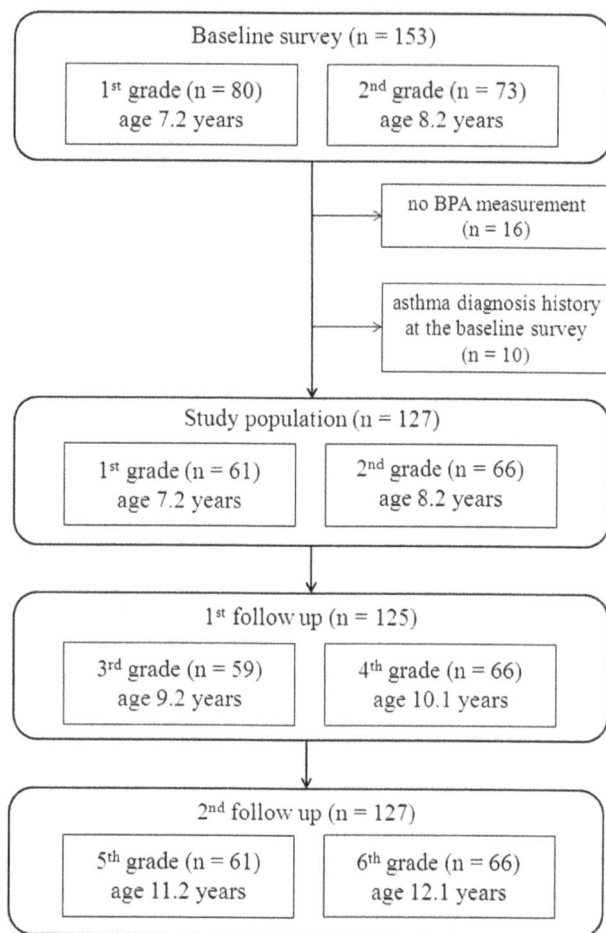

Figure 1. Overview of study population sampling and follow-up.

2 mL glass vial. Each batch of samples included a quality control sample and a blank. We added the quality control sample in pooled urine with a mixture of BPA standard. Liquid-liquid extraction was performed with Agilent Eclipse plus C18, 3.5 µm, 2.1×100 mm. The mobile phase was 60:40 (v/v) acetonitrile:-water, and the flow rate was 0.4 mL/min. Total BPA, including the free and conjugated forms, was measured using a high performance liquid chromatography-mass selective detector (HPLC-MS/MS, Agilent Triple Quad 6410, Santa Clara, CA, USA). The BPA concentrations in all the urinary samples were above the limit of detection (0.005 µg/L). The concentration of urinary BPA was adjusted for creatinine measured from the same urine sample to eliminate the influence caused by the different urinary excretion rates between participants [11].

Outcome measures

Wheezing was defined using the ISAAC question "Has your child had wheezing or whistling in the chest in the last 12 months?". Children were considered to have asthma (current asthma) when either of the following criteria was met: 1) wheezing or the use of asthma medication in the previous 12 months combined with a $PC_{20} \leq 8$ mg/mL or 2) wheezing or the use of asthma medication in the previous 12 months combined with a history of an asthma diagnosis or a history of wheezing. Incident asthma was defined as the first detection of current asthma without having satisfied the criteria of current asthma at any previous survey.

The response to the methacholine challenge is expressed as PC_{20}. To evaluate PC_{20}, which is an indicator of airway hyperresponsiveness, participants inhaled increasing concentrations of methacholine (0.625, 1.25, 2.5, 5, 12.5, and 25 mg/dL) using a nebulizer until the FEV_1 measured by a portable spirometer (Microspiro HI-298, Chest Corporation, Tokyo, Japan) decreased by at least 20% from the baseline value. The PC_{20} was calculated using a log-dose response curve [40].

Covariates

The parental asthma history was reported on the baseline questionnaire. Fetal tobacco smoke exposure was defined as active maternal smoking or the report of a smoker in the home during pregnancy. Environmental tobacco smoke exposure was defined as active maternal smoking during the first year after delivery, maternal smoking at the time of the survey, or the report of a smoker in the home from delivery until the time of the survey. Pet ownership was defined as having ever kept dogs or cats as pets from delivery until the time of the surveys.

Statistical analyses

Urinary BPA concentration was adjusted for creatinine measured from the same urine sample (i.e., urinary BPA concentration divided by urinary creatinine concentration) and log-transformed to approximate a normal distribution. The associations between urinary BPA concentration at 7–8 years of age and the dichotomous outcome variable, e.g., wheezing and current asthma, were analyzed using generalized estimating equations (GEE) with a logit link. A marginal Cox model with a robust sandwich estimator of variance was applied to analyze the association between BPA exposure and risk of incident asthma considering the grade-at-enrollment dependence due to the possibility of clustering within the grade. The proportional hazards assumption was tested using a time-dependent explanatory variable. For those with incident asthma, the time at risk was considered as the number of years from the baseline survey to the mid-point between the previous survey and the survey when the incident asthma was observed and, for those without incident asthma, as the number of years between the baseline survey and the last follow-up survey. Due to the difficulty of assigning the exact time at risk and the relatively long period between the surveys, the association between BPA concentration and incident asthma was also analyzed using a logistic regression model.

The associations between urinary BPA concentration at 7–8 years of age and PC_{20} at the 3 time points (7–8, 9–10, and 11–12 years of age) were analyzed in two steps. First, a generalized additive mixed model was constructed to graphically examine the linearity of the association. Then, a linear mixed model using repeated-measures analysis and a random effect of grade-at-enrollment was constructed to analyze the relationship between BPA concentration at 7–8 years of age and PC_{20} at 3 time points. Logistic and linear regression models were used to analyze the association between BPA concentration and wheezing, current asthma, and PC_{20} at each time point.

The child's sex, parental asthma history, fetal tobacco smoke exposure, and pet ownership were selected as covariates based on earlier literature reviews [36,37]. Potential confounders, such as environmental tobacco smoke exposure, history of breastfeeding, cockroach sensitization, and maternal education level were evaluated using bivariate analyses. Covariates that predicted wheezing at $P \le .20$ in the bivariate analyses were added to the initial multivariable GEE model and retained if the estimate of the

association between BPA at 7–8 years of age and wheezing changed >10%, which resulted in the child's sex, parental asthma history, fetal tobacco smoke exposure, environmental tobacco smoke exposure, and pet ownership as the covariates in the analysis. The GEE with a logit link, logistic regression, and linear regression models were also adjusted for grade at enrollment.

In the secondary analyses, potential interaction was tested by adding cross product term between BPA concentration and each covariate in the main analysis. The participants were stratified into subgroups based on sex and analyzed using the marginal Cox model for clustered data and a linear mixed model.

SAS version 9.3 (SAS Institute Inc., Cary, NC, USA) was used for statistical analyses, and R version 2.14.2 (The Comprehensive R Archive Network: http://cran.r-project.org) was used for visualization. Two-sided P values <.05 were used to indicate statistical significance.

Results

Of the 127 participants, 54.3% were male, 4.7% had a parental asthma history, 17.3% experienced fetal tobacco smoke exposure, and 26.8% experienced environmental tobacco smoke exposure. Difference between baseline characteristics of children who had current asthma during the study period and those of children who did not was not observed (Table 1). When comparing the children that were included in the analysis with those excluded, there were no differences except for a slightly higher level of cockroach sensitization in the children that were included in the analysis (Table S1 in File S1).

The geometric mean of urinary BPA concentration was 1.02 µg/L, 1st quartile 0.63 µg/L, median 0.97 µg/L, 3rd quartile 1.67 µg/L, and maximum 21.37 µg/L. The distribution of urinary BPA concentrations was positively skewed. In the present study, 9 children were assessed to have current asthma only at 9–10 years, 7 children only at 11–12 years, while 2 children both at 9–10 years and 11–12 years of age. The one-unit increase in log-transformed, creatinine-adjusted urinary BPA concentration measured at 7–8 years of age was associated with wheezing (odds ratio [OR], 2.48; 95% confidence interval [CI], 1.15–5.31; $P = .02$) and current asthma (OR, 2.35; 95% CI, 1.03–5.32; $P = .04$) at ages up to 11–12 years. A relationship between urinary BPA concentration and the risk of incident asthma was also observed (hazard ratio [HR], 2.13; 95% CI, 1.51–3.00; $P<.001$; Table 2). A statistically significant association was also observed between BPA concentration and incident asthma in the logistic regression model (OR, 2.44; 95% CI, 1.11–5.36; $P = .03$).

The penalized regression spline showed an almost linear association between BPA at 7–8 years of age and PC_{20} at ages up to 11–12 years (Figure 2). In the linear mixed model, the association between BPA at 7–8 years of age and PC_{20} at ages up to 11–12 years was significant ($\beta = -2.33$; $P = .02$).

The analysis of the associations between BPA at 7–8 years of age and wheezing, PC_{20}, and current asthma at each time point resulted in significant relationships with wheezing, PC_{20}, and current asthma at 9–10 years of age only (Table 3). When the interaction term between BPA and each covariate was added to the multivariable model, significant interactions were not found. When stratified by sex, a significant association between BPA and incident asthma (HR, 2.45; 95% CI, 2.18–2.76; $P<.001$) and a marginally significant association between BPA and PC_{20} ($\beta = -2.59$, $P = .09$; Table S2 in File S1) were observed only in girls.

For the sensitivity analysis, we conducted the analyses after including the children who had been diagnosed with asthma before the baseline survey (n = 10). This did not change the result

Table 1. Baseline characteristics of the study participants stratified by children who had current asthma during the study period and children who did not [n (%)].

Characteristic	Children with asthma (n=18)	Children without asthma (n=109)	P value*
Sex			.53
Boy	11 (61.1)	58 (53.2)	
Girl	7 (38.9)	51 (46.8)	
Parental asthma history			.09
No	14 (77.8)	99 (90.8)	
Yes	1 (5.6)	5 (4.6)	
Missing	3 (16.7)	5 (4.6)	
Fetal tobacco smoke exposure†			.94
No	15 (83.3)	90 (82.6)	
Yes	22 (16.7)	19 (17.4)	
Environmental tobacco smoke exposure§			.50
No	12 (66.7)	81 (74.3)	
Yes	6 (33.3)	28 (25.7)	
Pet ownership‡			.45
No	15 (83.3)	82 (75.2)	
Yes	3 (16.7)	27 (24.8)	
Breast-fed			.11
No	5 (27.8)	35 (32.1)	
<3 months	4 (22.2)	18 (16.5)	
3–5 months	0 (0)	16 (14.7)	
≥6months	4 (22.2)	30 (27.5)	
Did not answer	5 (27.8)	10 (9.2)	
Cockroach sensitization			.67
No	17 (94.4)	103 (94.5)	
Yes	0 (0)	3 (2.8)	
Did not answer	1 (5.6)	3 (2.8)	
Maternal education			.67
< High school	3 (11.9)	13 (11.9)	
High school	7 (38.9)	51 (46.8)	
>High school	5 (27.8)	34 (31.2)	
Did not answer	3 (16.7)	11 (10.1)	
Paternal education			.19
< High school	3 (16.7)	6 (5.5)	
High school	6 (33.3)	50 (45.9)	
>High school	6 (33.3)	43 (33.3)	
Did not answer	3 (16.7)	3 (9.2)	

* P value was estimated based on Chi-square test or Fisher's exact test.
†Active maternal smoking during pregnancy or presence of a smoker in the home during pregnancy.
‡Active maternal smoking during the first year after delivery, current maternal smoking, or presence of a smoker in the home after delivery until the present time.
§Having had a pet dog or cat after delivery until the present time.
Size of the wheal produced by the cockroach antigen ≥3 mm and larger than size of the wheal produced by histamine.

substantially (Table S3 in File S1). When excluding the children who satisfied the criteria of current asthma at the baseline survey, BPA concentration at 7–8 years of age was significantly associated with increased risk of incident asthma (HR, 1.64; 95% CI, 1.10–2.45; $P = .02$). Significant association of BPA with PC_{20} at 9–10 years and marginally significant association with wheezing at 9–10 years of age was also observed. The trend of association between BPA and current asthma was similar although attenuated, partly due to small sample size (Table S4 in File S1).

Discussion

In the present study, we found associations between the urinary BPA concentration in the earlier years of children in elementary school and wheezing, asthma, and PC_{20} in the later years. A potential modifying effect of sex on these associations was also observed.

Previous studies have reported that prenatal or postnatal exposure to BPA is associated with an increased risk of wheezing

Table 2. Association of urinary BPA concentrations (log transformed, µg/g creatinine) at 7–8 years with wheezing and asthma over 11–12 years of age, by longitudinal analyses.

Outcome	No.*	OR† or HR‡ (95% CI)	P value
Wheeze	28/335	2.48 (1.15–5.31) †	.02
Current Asthma	20/252	2.35 (1.03–5.32) †	.04
Incident Asthma	18/127	2.13 (1.51–3.00) ‡	<.001

HR, hazard ratio.
* Number with outcome/total number for analysis.
†Generalized estimating equation with a logit link model adjusted for gender, parental asthma history, fetal and environmental tobacco smoke exposure, pet ownership, and grade at enrollment.
‡Marginal Cox model considering grade-at-enrollment clustering adjusted for gender, parental asthma history, fetal and environmental tobacco smoke exposure, and pet ownership.

and asthma. One birth cohort study of 398 mother-child pairs demonstrated that prenatal urinary BPA concentrations above, versus below, the median are associated with the child's wheezing at 6 months (OR, 2.27; 95% CI, 1.28–4.06) but not at older ages until 3 years of age [36]. Another birth cohort study of 568 mother-child pairs showed that urinary BPA concentrations at 3 years are associated with wheezing at 5 and 6 years of age, and urinary BPA concentrations at 7 years are associated with wheezing at 7 years of age. A one-unit increase in log-transformed, creatinine-adjusted urinary BPA concentrations at 3, 5, and 7 years is associated with asthma at a single assessment by a physician between 5 and 12 years of age (OR, 1.5; 95% CI, 1.1–2.0 for BPA at 3 years; OR, 1.4; 95% CI, 1.0–1.9 for BPA at 5 years; OR, 1.5; 95% CI, 1.0–2.1 for BPA at 7 years of age) [37]. Our results are similar to the previous findings, although the effect size is relatively large in the current study, which may be explained

by the differences in outcome definition, study population, and adjusting covariates.

In the current study, urinary BPA at 7–8 years was associated with asthma-related outcomes such as wheezing, asthma, and PC_{20} at 9–10 years, but not at 7–8 years or 11–12 years of age. Due to the lack of mechanistic studies investigating the time lag between BPA exposure and occurrence of asthma-related outcomes, we have no definite explanation for this finding. Pubertal stage may play a role; it has been reported that pubertal stage is related to the development and progression of asthma [41–44]. Alternatively, observed null association with current asthma at 11–12 years of age might reflect the fact that only a small number of incident asthma cases occur in later years in elementary school, and the incidence of wheezing and asthma continues to decline from childhood to adolescence or greater non-differential misclassification due to longer time interval [45]. Further research considering the pubertal stage is warranted.

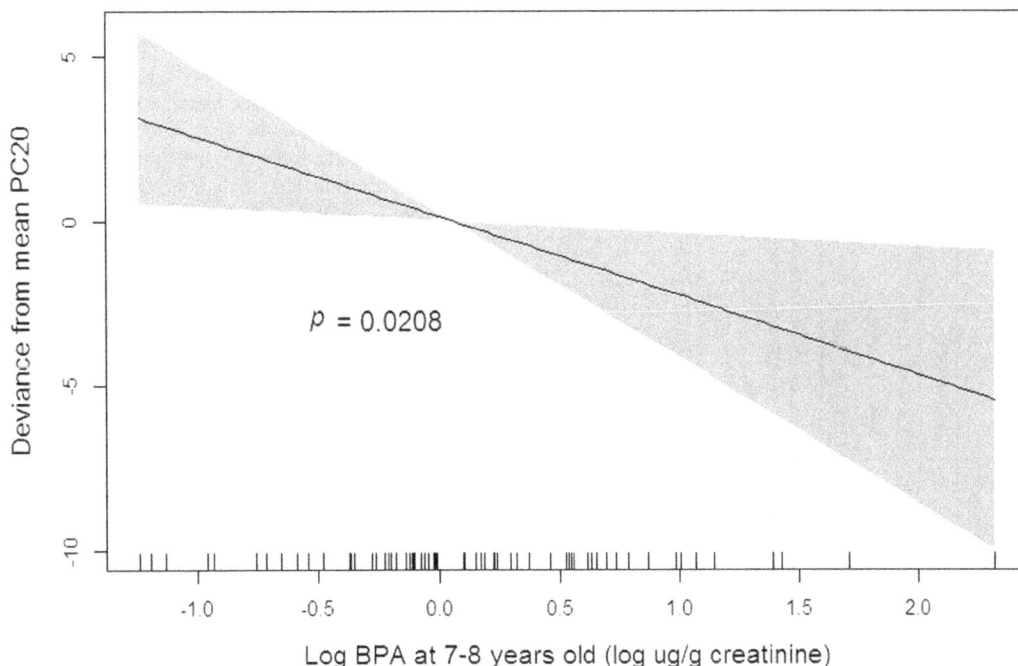

Figure 2. Relationship between urinary BPA concentration and PC_{20}. Penalized regression spline of log-transformed urinary BPA concentrations at 7–8 years on PC_{20} at ages up to 11–12 years. Solid lines, spline curve; shaded area, 95% confidence intervals. The model is adjusted for gender, parental asthma history, fetal and environmental tobacco smoke exposure, and pet ownership.

Table 3. Association of urinary BPA concentrations (log transformed, μg BPA/g creatinine) at 7–8 years with wheezing, PC_{20}, and current asthma at 7–8, 9–10, and 11–12 years of age.

| | Log BPA (μg/g creatinine) at 7–8 y | | | | | | | | |
| | Wheeze* | | | PC_{20}† | | | Current asthma* | | |
Age (years)	No.‡	OR (95%CI)	P value	No.§	ß (SE)	P value	No.‡	OR (95%CI)	P value
7–8	8/120	3.08 (0.86–11.09)	.09	64	−2.01 (1.22)	.11	NA		
9–10	11/121	4.24 (1.35–13.28)	.01	19	−6.37 (2.26)	.02	11/125	3.64 (1.23–10.76)	.02
11–12	9/122	1.80 (0.69–4.71)	.23	14	−3.64 (2.50)	.20	9/127	1.87 (0.72–4.86)	.20

*Logistic regression model adjusted for gender, parental asthma history, fetal and environmental tobacco smoke exposure, pet ownership, and grade at enrollment.
†Linear regression model adjusted for gender, parental asthma history, fetal and environmental tobacco smoke exposure, pet ownership, and grade at enrollment.
‡Number with outcome/total number for analysis.
§Total number for analysis.

The mechanism behind the present findings is still unclear; however, the oxidative stress pathway with BPA exposure could be suggested as a possible explanation. BPA is known to cause oxidative stress in rats [46–49] and humans [10], and growing evidence indicates that oxidative stress and subsequent mitochondrial dysfunction are associated with BPA-induced damage [50,51]. Increased production of reactive oxygen species and decreased anti-inflammatory capacity could enhance the susceptibility to the insults, such as air pollution, which results in chronic airway inflammation and asthma [52,53].

The results of the present study also suggest that the effects of BPA on asthma-related outcomes might be mediated, at least in part, by an endocrine-disrupting mechanism. It was reported that the prevalence of asthma is higher in boys before puberty and in girls and women after puberty [54]. Changes in hormonal status, such as with estrogen, have been suggested as one possible factor causing this phenomenon [45,55]. Estrogen has been demonstrated to encourage T-helper 2 (T_H2) polarization, class switching of B cells to the production of immunoglobulin E, and the degranulation of mast cells [56–59]. In epidemiologic studies, early menarche is associated with a higher prevalence of asthma in adult women [60], and the diagnosis of asthma increases in postmenopausal women who receive hormone replacement therapy [61]. Meanwhile, BPA acts imperfectly as estrogen in numerous organs [62], and female mice that were exposed to BPA prenatally demonstrate increased airway and lung inflammation, whereas the male mice exposed to BPA prenatally did not [63]. The results of these studies and ours suggest that BPA affects asthma-related outcomes by disrupting the endocrine system. However, the mechanisms underlying the sex-specific effect of and susceptibility to BPA are not yet fully understood [29,55,64].

BPA concentration in the present study was markedly lower than the previously reported BPA concentration in the United States [11,38]. Geographic variance of the BPA concentration has been reported not only in the children [38,39] but also in different age group population [11,17,65,66]. The observed lower BPA concentration in the present study might also be attributable, in part, to the study design using overnight fasting spot urine samples [67,68]. However, it has been reported that urinary BPA concentration did not decrease rapidly with fasting time [69], and another study following five fasting individual's spot urine BPA concentration has also demonstrated the decline of BPA concentration with gentle slopes during the first 24 h and fluctuated at lower levels during the next 24 h [70]. This might be due to accumulation of BPA in body tissue, such as fat or exposure to nonfood source and could lead to reduction in the variability of BPA concentration and potential misclassification [71]. Due to these traits, fasting urinary BPA concentration has been used as an indicator of exposure in the previous studies [16,72]. Further, the random variation in BPA concentrations may have shifted the association toward the null.

Exposure to BPA is thought to be mainly from dietary route [67] and the observed associations could be confounded by the chemicals that were taken with BPA. For instance, mercury and BPA share common exposure source such as canned tuna fish [73,74]. Although exposure to mercury has been associated with immunotoxic effects [75–78], we could not find previous literature supporting the association between exposure to mercury and asthma. Further study exploring the co-exposure of BPA and other environmental risk factors including phthalate, other phenolic compounds, heavy metals, and persistent organic compounds is warranted to assess potential effect modification or confounding [79].

The present study has some strengths. First, the longitudinal study design addressed the potential of temporal ambiguity and recall bias. Second, the use of the objective methacholine challenge test provided a more accurate diagnosis and reduced the potential of misclassification. Furthermore, the reliability of the results was demonstrated by the consistency of results between the parent-reported presence of wheezing and asthma and the results from the objective methacholine challenge test.

The major limitation in the current study is the small sample size. In addition, the selection of participants from one elementary school may have resulted in drawbacks in generalizability. Future studies should include a more representative sample of sufficient size to confirm the findings of the present study. Although PC_{20} is a commonly used indicator for airway hyperresponsiveness, its use is limited in participants whose decrease in FEV_1 is less than the cut-off value before the maximal concentration is reached [80], which happens with the majority of the participants in epidemiologic studies [81]. However, despite the weakness regarding censoring, previous studies have reported that PC_{20} correlates well with other indicators of airway hyperresponsiveness [82–84] and can be used as a reliable indicator of airway hyperresponsiveness [85,86]. Lastly, due to the lack of a BPA measurement before 7–8 years of age, the possibility that the observed association is due to earlier exposure and that the BPA exposure at 7–8 years of age is reflective of this could not be assessed. Further studies are required to confirm the temporal specifics regarding vulnerability.

We found that the urinary BPA concentration at 7–8 years was associated with wheezing, asthma, and PC_{20} at ages up to 11–12 years. These findings provide information about the health effects of BPA exposure in school-age children and support public health initiatives to protect the health of a susceptible population such as children.

References

1. Hong S, Son DK, Lim WR, Kim SH, Kim H, et al. (2012) The prevalence of atopic dermatitis, asthma, and allergic rhinitis and the comorbidity of allergic diseases in children. Environ Health Toxicol 27: e2012006. doi:10.5620/eht.2012.27.e2012006.
2. Lee SI (2010) Prevalence of Childhood Asthma in Korea: International Study of Asthma and Allergies in Childhood. Allergy Asthma Immunol Res 2: 61. doi:10.4168/aair.2010.2.2.61.
3. Akinbami LJ, Moorman JE, Garbe PL, Sondik EJ (2009) Status of childhood asthma in the United States, 1980–2007. Pediatrics 123 Suppl 3: S131–S145. doi:10.1542/peds.2008-2233C.
4. Hansen TE, Evjenth B, Holt J (2013) Increasing prevalence of asthma, allergic rhinoconjunctivitis and eczema among schoolchildren: three surveys during the period 1985–2008. Acta Paediatr Oslo Nor 1992 102: 47–52. doi:10.1111/apa.12030.
5. Van Schayck OCP (2013) Global strategies for reducing the burden from asthma. Prim Care Respir J J Gen Pract Airw Group 22: 239–243. doi:10.4104/pcrj.2013.00052.
6. Masoli M, Fabian D, Holt S, Beasley R, Global Initiative for Asthma (GINA) Program (2004) The global burden of asthma: executive summary of the GINA Dissemination Committee report. Allergy 59: 469–478. doi:10.1111/j.1398-9995.2004.00526.x.
7. Wong GWK, Chow CM (2008) Childhood asthma epidemiology: insights from comparative studies of rural and urban populations. Pediatr Pulmonol 43: 107–116. doi:10.1002/ppul.20755.
8. Kwak ES, Just A, Whyatt R, Miller RL (2009) Phthalates, Pesticides, and Bisphenol-A Exposure and the Development of Nonoccupational Asthma and Allergies: How Valid Are the Links? Open Allergy J 2: 45–50. doi:10.2174/1874838400902010045.
9. Nakajima Y, Goldblum RM, Midoro-Horiuti T (2012) Fetal exposure to bisphenol A as a risk factor for the development of childhood asthma: an animal model study. Environ Health Glob Access Sci Source 11: 8. doi:10.1186/1476-069X-11-8.
10. Yang YJ, Hong YC, Oh SY, Park MS, Kim H, et al. (2009) Bisphenol A exposure is associated with oxidative stress and inflammation in postmenopausal women. Environ Res 109: 797–801. doi:10.1016/j.envres.2009.04.014.
11. Calafat AM, Kuklenyik Z, Reidy JA, Caudill SP, Ekong J, et al. (2005) Urinary concentrations of bisphenol A and 4-nonylphenol in a human reference population. Environ Health Perspect 113: 391–395.
12. Shankar A, Teppala S (2011) Relationship between urinary bisphenol A levels and diabetes mellitus. J Clin Endocrinol Metab 96: 3822–3826. doi:10.1210/jc.2011-1682.
13. Silver MK, O'Neill MS, Sowers MR, Park SK (2011) Urinary bisphenol A and type-2 diabetes in U.S.adults: data from NHANES 2003–2008. PloS One 6: e26868. doi:10.1371/journal.pone.0026868.
14. Melzer D, Gates P, Osborne NJ, Osborn NJ, Henley WE, et al. (2012) Urinary bisphenol a concentration and angiography-defined coronary artery stenosis. PloS One 7: e43378. doi:10.1371/journal.pone.0043378.
15. Melzer D, Osborne NJ, Henley WE, Cipelli R, Young A, et al. (2012) Urinary bisphenol A concentration and risk of future coronary artery disease in apparently healthy men and women. Circulation 125: 1482–1490. doi:10.1161/CIRCULATIONAHA.111.069153.
16. Bae S, Kim JH, Lim YH, Park HY, Hong YC (2012) Associations of bisphenol A exposure with heart rate variability and blood pressure. Hypertension 60: 786–793. doi:10.1161/HYPERTENSIONAHA.112.197715.
17. Lee MR, Park H, Bae S, Lim YH, Kim JH, et al. (2014) Urinary bisphenol A concentrations are associated with abnormal liver function in the elderly: a repeated panel study. J Epidemiol Community Health 68: 312–317. doi:10.1136/jech-2013-202548.
18. Braun JM, Kalkbrenner AE, Calafat AM, Yolton K, Ye X, et al. (2011) Impact of early-life bisphenol A exposure on behavior and executive function in children. Pediatrics 128: 873–882. doi:10.1542/peds.2011-1335.
19. Li D, Zhou Z, Qing D, He Y, Wu T, et al. (2010) Occupational exposure to bisphenol-A (BPA) and the risk of self-reported male sexual dysfunction. Hum Reprod Oxf Engl 25: 519–527. doi:10.1093/humrep/dep381.
20. Li DK, Zhou Z, Miao M, He Y, Qing D, et al. (2010) Relationship between urine bisphenol-A level and declining male sexual function. J Androl 31: 500–506. doi:10.2164/jandrol.110.010413.
21. Meeker JD, Ehrlich S, Toth TL, Wright DL, Calafat AM, et al. (2010) Semen quality and sperm DNA damage in relation to urinary bisphenol A among men from an infertility clinic. Reprod Toxicol Elmsford N 30: 532–539. doi:10.1016/j.reprotox.2010.07.005.
22. Li DK, Zhou Z, Miao M, He Y, Wang J, et al. (2011) Urine bisphenol-A (BPA) level in relation to semen quality. Fertil Steril 95: 625–630.e1–e4. doi:10.1016/j.fertnstert.2010.09.026.
23. Chou WC, Chen JL, Lin CF, Chen YC, Shih FC, et al. (2011) Biomonitoring of bisphenol A concentrations in maternal and umbilical cord blood in regard to

Supporting Information

File S1 This file contains Table S1, Table S2, Table S3, and Table S4. Table S1, Baseline characteristics of the study population included and excluded in the current study. **Table S2**, Association of urinary BPA concentrations (log transformed, µg BPA/g creatinine) at 7–8 years with incident asthma and PC_{20} over 11–12 years of age stratified by gender. **Table S3**, Association of urinary BPA concentrations (log transformed, µg BPA/g creatinine) at 7–8 years with wheezing, PC_{20}, and asthma at 7–8, 9–10, and 11–12 years of age, including the children who had been diagnosed with asthma before 7–8 years of age. **Table S4**, Association of urinary BPA concentrations (log transformed, µg BPA/g creatinine) at 7–8 years with wheezing, PC_{20}, and current asthma at 7–8, 9–10, and 11–12 years of age, excluding the children who satisfied the criteria of current asthma at the baseline survey.

Acknowledgments

The authors thank the research workers, technicians, and participating children and their families. Without them, this work would not have been possible.

Author Contributions

Conceived and designed the experiments: K-NK JHK H-JK S-JH B-JK S-YL Y-CH SB. Performed the experiments: K-NK H-JK S-JH B-JK S-YL Y-CH SB. Analyzed the data: K-NK Y-CH SB. Contributed reagents/materials/analysis tools: JHK H-JK S-JH B-JK S-YL Y-CH SB. Contributed to the writing of the manuscript: K-NK Y-CH SB.

birth outcomes and adipokine expression: a birth cohort study in Taiwan. Environ Health Glob Access Sci Source 10: 94. doi:10.1186/1476-069X-10-94.

24. Wolff MS, Teitelbaum SL, Pinney SM, Windham G, Liao L, et al. (2010) Investigation of relationships between urinary biomarkers of phytoestrogens, phthalates, and phenols and pubertal stages in girls. Environ Health Perspect 118: 1039–1046. doi:10.1289/ehp.0901690.

25. Wolff MS, Britton JA, Boguski L, Hochman S, Maloney N, et al. (2008) Environmental exposures and puberty in inner-city girls. Environ Res 107: 393–400. doi:10.1016/j.envres.2008.03.006.

26. Braun JM, Yolton K, Dietrich KN, Hornung R, Ye X, et al. (2009) Prenatal bisphenol A exposure and early childhood behavior. Environ Health Perspect 117: 1945–1952. doi:10.1289/ehp.0900979.

27. Sugiura-Ogasawara M, Ozaki Y, Sonta S, Makino T, Suzumori K (2005) Exposure to bisphenol A is associated with recurrent miscarriage. Hum Reprod Oxf Engl 20: 2325–2329. doi:10.1093/humrep/deh888.

28. Cantonwine D, Meeker JD, Hu H, Sánchez BN, Lamadrid-Figueroa H, et al. (2010) Bisphenol a exposure in Mexico City and risk of prematurity: a pilot nested case control study. Environ Health Glob Access Sci Source 9: 62. doi:10.1186/1476-069X-9-62.

29. Bonds RS, Midoro-Horiuti T (2013) Estrogen effects in allergy and asthma. Curr Opin Allergy Clin Immunol 13: 92–99. doi:10.1097/ACI.0b013e32835a6dd6.

30. Lee MH, Chung SW, Kang BY, Park J, Lee CH, et al. (2003) Enhanced interleukin-4 production in CD4+ T cells and elevated immunoglobulin E levels in antigen-primed mice by bisphenol A and nonylphenol, endocrine disruptors: involvement of nuclear factor-AT and Ca2+. Immunology 109: 76–86.

31. Sawai C, Anderson K, Walser-Kuntz D (2003) Effect of bisphenol A on murine immune function: modulation of interferon-gamma, IgG2a, and disease symptoms in NZB X NZW F1 mice. Environ Health Perspect 111: 1883–1887.

32. Tian X, Takamoto M, Sugane K (2003) Bisphenol A promotes IL-4 production by Th2 cells. Int Arch Allergy Immunol 132: 240–247. doi:74305.

33. Yan H, Takamoto M, Sugane K (2008) Exposure to Bisphenol A prenatally or in adulthood promotes T(H)2 cytokine production associated with reduction of CD4CD25 regulatory T cells. Environ Health Perspect 116: 514–519. doi:10.1289/ehp.10829.

34. Midoro-Horiuti T, Tiwari R, Watson CS, Goldblum RM (2010) Maternal bisphenol a exposure promotes the development of experimental asthma in mouse pups. Environ Health Perspect 118: 273–277. doi:10.1289/ehp.0901259.

35. Van Winkle LS, Murphy SR, Boetticher MV, VandeVoort CA (2013) Fetal exposure of rhesus macaques to bisphenol a alters cellular development of the conducting airway by changing epithelial secretory product expression. Environ Health Perspect 121: 912–918. doi:10.1289/ehp.1206064.

36. Spanier AJ, Kahn RS, Kunselman AR, Hornung R, Xu Y, et al. (2012) Prenatal exposure to bisphenol A and child wheeze from birth to 3 years of age. Environ Health Perspect 120: 916–920. doi:10.1289/ehp.1104175.

37. Donohue KM, Miller RL, Perzanowski MS, Just AC, Hoepner LA, et al. (2013) Prenatal and postnatal bisphenol A exposure and asthma development among inner-city children. J Allergy Clin Immunol 131: 736–742. doi:10.1016/j.jaci.2012.12.1573.

38. Lakind JS, Naiman DQ (2011) Daily intake of bisphenol A and potential sources of exposure: 2005–2006 National Health and Nutrition Examination Survey. J Expo Sci Environ Epidemiol 21: 272–279. doi:10.1038/jes.2010.9.

39. Hong SB, Hong YC, Kim JW, Park EJ, Shin MS, et al. (2013) Bisphenol A in relation to behavior and learning of school-age children. J Child Psychol Psychiatry 54: 890–899. doi:10.1111/jcpp.12050.

40. Sumino K, Sugar EA, Irvin CG, Kaminsky DA, Shade D, et al. (2012) Methacholine challenge test: diagnostic characteristics in asthmatic patients receiving controller medications. J Allergy Clin Immunol 130: 69–75.e6. doi:10.1016/j.jaci.2012.02.025.

41. Fu L, Freishtat RJ, Gordish-Dressman H, Teach SJ, Resca L, et al. (2014) Natural progression of childhood asthma symptoms and strong influence of gender and puberty. Ann Am Thorac Soc. doi:10.1513/AnnalsATS.201402-084OC.

42. Clark NM, Dodge JA, Thomas LJ, Andridge RR, Awad D, et al. (2010) Asthma in 10- to 13-year-olds: challenges at a time of transition. Clin Pediatr (Phila) 49: 931–937. doi:10.1177/0009922809357339.

43. Protudjer JLP, Lundholm C, Bergström A, Kull I, Almqvist C (2014) Puberty and asthma in a cohort of Swedish children. Ann Allergy Asthma Immunol Off Publ Am Coll Allergy Asthma Immunol 112: 78–79. doi:10.1016/j.anai.2013.10.015.

44. Postma DS (2007) Gender differences in asthma development and progression. Gend Med 4 Suppl B: S133–S146.

45. Almqvist C, Worm M, Leynaert B, working group of GA2LEN WP 2.5 Gender (2008) Impact of gender on asthma in childhood and adolescence: a GA2LEN review. Allergy 63: 47–57. doi:10.1111/j.1398-9995.2007.01524.x.

46. Song S, Zhang L, Zhang H, Wei W, Jia L (2014) Perinatal BPA exposure induces hyperglycemia, oxidative stress and decreased adiponectin production in later life of male rat offspring. Int J Environ Res Public Health 11: 3728–3742. doi:10.3390/ijerph110403728.

47. Aboul Ezz HS, Khadrawy YA, Mourad IM (2013) The effect of bisphenol A on some oxidative stress parameters and acetylcholinesterase activity in the heart of male albino rats. Cytotechnology. doi:10.1007/s10616-013-9672-1.

48. Hassan ZK, Elobeid MA, Virk P, Omer SA, ElAmin M, et al. (2012) Bisphenol A induces hepatotoxicity through oxidative stress in rat model. Oxid Med Cell Longev 2012: 194829. doi:10.1155/2012/194829.

49. D'Cruz SC, Jubendradass R, Mathur PP (2012) Bisphenol A induces oxidative stress and decreases levels of insulin receptor substrate 2 and glucose transporter 8 in rat testis. Reprod Sci Thousand Oaks Calif 19: 163–172. doi:10.1177/1933719111415547.

50. Tiwari D, Kamble J, Chilgunde S, Patil P, Maru G, et al. (2012) Clastogenic and mutagenic effects of bisphenol A: an endocrine disruptor. Mutat Res 743: 83–90. doi:10.1016/j.mrgentox.2011.12.023.

51. Anjum S, Rahman S, Kaur M, Ahmad F, Rashid H, et al. (2011) Melatonin ameliorates bisphenol A-induced biochemical toxicity in testicular mitochondria of mouse. Food Chem Toxicol Int J Publ Br Ind Biol Res Assoc 49: 2849–2854. doi:10.1016/j.fct.2011.07.062.

52. Esposito S, Tenconi R, Lelii M, Preti V, Nazzari E, et al. (2014) Possible molecular mechanisms linking air pollution and asthma in children. BMC Pulm Med 14: 31. doi:10.1186/1471-2466-14-31.

53. Levy BD, Bonnans C, Silverman ES, Palmer LJ, Marigowda G, et al. (2005) Diminished lipoxin biosynthesis in severe asthma. Am J Respir Crit Care Med 172: 824–830. doi:10.1164/rccm.200410-1413OC.

54. Vollmer WM, Osborne ML, Buist AS (1998) 20-year trends in the prevalence of asthma and chronic airflow obstruction in an HMO. Am J Respir Crit Care Med 157: 1079–1084. doi:10.1164/ajrccm.157.4.9704140.

55. Vink NM, Postma DS, Schouten JP, Rosmalen JGM, Boezen HM (2010) Gender differences in asthma development and remission during transition through puberty: the TRacking Adolescents' Individual Lives Survey (TRAILS) study. J Allergy Clin Immunol 126: 498–504.e1–e6. doi:10.1016/j.jaci.2010.06.018.

56. Cai Y, Zhou J, Webb DC (2012) Estrogen stimulates Th2 cytokine production and regulates the compartmentalisation of eosinophils during allergen challenge in a mouse model of asthma. Int Arch Allergy Immunol 158: 252–260. doi:10.1159/000331437.

57. Jing H, Wang Z, Chen Y (2012) Effect of oestradiol on mast cell number and histamine level in the mammary glands of rat. Anat Histol Embryol 41: 170–176. doi:10.1111/j.1439-0264.2011.01120.x.

58. Sakai T, Furoku S, Nakamoto M, Shuto E, Hosaka T, et al. (2010) The soy isoflavone equol enhances antigen-specific IgE production in ovalbumin-immunized BALB/c mice. J Nutr Sci Vitaminol (Tokyo) 56: 72–76.

59. Zaitsu M, Narita SI, Lambert KC, Grady JJ, Estes DM, et al. (2007) Estradiol activates mast cells via a non-genomic estrogen receptor-alpha and calcium influx. Mol Immunol 44: 1977–1985. doi:10.1016/j.molimm.2006.09.030.

60. Macsali F, Real FG, Plana E, Sunyer J, Anto J, et al. (2011) Early age at menarche, lung function, and adult asthma. Am J Respir Crit Care Med 183: 8–14. doi:10.1164/rccm.200912-1886OC.

61. Barr RG, Wentowski CC, Grodstein F, Somers SC, Stampfer MJ, et al. (2004) Prospective study of postmenopausal hormone use and newly diagnosed asthma and chronic obstructive pulmonary disease. Arch Intern Med 164: 379–386. doi:10.1001/archinte.164.4.379.

62. Braun JM, Hauser R (2011) Bisphenol A and children's health. Curr Opin Pediatr 23: 233–239. doi:10.1097/MOP.0b013e3283445675.

63. Bauer SM, Roy A, Emo J, Chapman TJ, Georas SN, et al. (2012) The effects of maternal exposure to bisphenol A on allergic lung inflammation into adulthood. Toxicol Sci Off J Soc Toxicol 130: 82–93. doi:10.1093/toxsci/kfs227.

64. Tantisira KG, Colvin R, Tonascia J, Strunk RC, Weiss ST, et al. (2008) Airway responsiveness in mild to moderate childhood asthma: sex influences on the natural history. Am J Respir Crit Care Med 178: 325–331. doi:10.1164/rccm.200708-1174OC.

65. Bushnik T, Haines D, Levallois P, Levesque J, Van Oostdam J, et al. (2010) Lead and bisphenol A concentrations in the Canadian population. Health Rep 21: 7–18.

66. Becker K, Göen T, Seiwert M, Conrad A, Pick-Fuss H, et al. (2009) GerES IV: phthalate metabolites and bisphenol A in urine of German children. Int J Hyg Environ Health 212: 685–692. doi:10.1016/j.ijheh.2009.08.002.

67. Wilson NK, Chuang JC, Morgan MK, Lordo RA, Sheldon LS (2007) An observational study of the potential exposures of preschool children to pentachlorophenol, bisphenol-A, and nonylphenol at home and daycare. Environ Res 103: 9–20. doi:10.1016/j.envres.2006.04.006.

68. Völkel W, Colnot T, Csanády GA, Filser JG, Dekant W (2002) Metabolism and kinetics of bisphenol a in humans at low doses following oral administration. Chem Res Toxicol 15: 1281–1287.

69. Stahlhut RW, Welshons WV, Swan SH (2009) Bisphenol A data in NHANES suggest longer than expected half-life, substantial nonfood exposure, or both. Environ Health Perspect 117: 784–789. doi:10.1289/ehp.0800376.

70. Christensen KLY, Lorber M, Koslitz S, Brüning T, Koch HM (2012) The contribution of diet to total bisphenol A body burden in humans: results of a 48 hour fasting study. Environ Int 50: 7–14. doi:10.1016/j.envint.2012.09.002.

71. Ye X, Wong LY, Bishop AM, Calafat AM (2011) Variability of urinary concentrations of bisphenol A in spot samples, first morning voids, and 24-hour collections. Environ Health Perspect 119: 983–988. doi:10.1289/ehp.1002701.

72. Ning G, Bi Y, Wang T, Xu M, Xu Y, et al. (2011) Relationship of urinary bisphenol A concentration to risk for prevalent type 2 diabetes in Chinese adults: a cross-sectional analysis. Ann Intern Med 155: 368–374. doi:10.7326/0003-4819-155-6-201109200-00005.

73. Cao XL, Corriveau J, Popovic S (2010) Bisphenol a in canned food products from canadian markets. J Food Prot 73: 1085–1089.

74. Gerstenberger SL, Martinson A, Kramer JL (2010) An evaluation of mercury concentrations in three brands of canned tuna. Environ Toxicol Chem SETAC 29: 237–242. doi:10.1002/etc.32.

75. Vas J, Monestier M (2008) Immunology of mercury. Ann N Y Acad Sci 1143: 240–267. doi:10.1196/annals.1443.022.

76. Gardner RM, Nyland JF, Silbergeld EK (2010) Differential immunotoxic effects of inorganic and organic mercury species in vitro. Toxicol Lett 198: 182–190. doi:10.1016/j.toxlet.2010.06.015.

77. Alves MFA, Fraiji NA, Barbosa AC, De Lima DSN, Souza JR, et al. (2006) Fish consumption, mercury exposure and serum antinuclear antibody in Amazonians. Int J Environ Health Res 16: 255–262. doi:10.1080/09603120600734147.

78. Nyland JF, Fillion M, Barbosa F, Shirley DL, Chine C, et al. (2011) Biomarkers of methylmercury exposure immunotoxicity among fish consumers in Amazonian Brazil. Environ Health Perspect 119: 1733–1738. doi:10.1289/ehp.1103741.

79. Vrijheid M, Slama R, Robinson O, Chatzi L, Coen M, et al. (2014) The human early-life exposome (HELIX): project rationale and design. Environ Health Perspect 122: 535–544. doi:10.1289/ehp.1307204.

80. Marcon A, Cerveri I, Wjst M, Antó J, Heinrich J, et al. (2014) Can an airway challenge test predict respiratory diseases? A population-based international study. J Allergy Clin Immunol 133: 104–110.e1–e4. doi:10.1016/j.jaci.2013.03.040.

81. Jayet PY, Schindler C, Künzli N, Zellweger JP, Brändli O, et al. (2005) Reference values for methacholine reactivity (SAPALDIA study). Respir Res 6: 131. doi:10.1186/1465-9921-6-131.

82. Aerts JG, Bogaard JM, Overbeek SE, Verbraak AF, Thio P (1994) Extrapolation of methacholine log-dose response curves with a Cumulative Gaussian Distribution function. Eur Respir J 7: 895–900.

83. Cockcroft DW, Berscheid BA (1983) Slope of the dose-response curve: usefulness in assessing bronchial responses to inhaled histamine. Thorax 38: 55–61.

84. Koh YY, Kang EK, Min YG, Kim CK (2002) The importance of maximal airway response to methacholine in the prediction of asthma development in patients with allergic rhinitis. Clin Exp Allergy J Br Soc Allergy Clin Immunol 32: 921–927.

85. Sutherland ER, King TS, Icitovic N, Ameredes BT, Bleecker E, et al. (2010) A trial of clarithromycin for the treatment of suboptimally controlled asthma. J Allergy Clin Immunol 126: 747–753. doi:10.1016/j.jaci.2010.07.024.

86. Weiss ST, Van Natta ML, Zeiger RS (2000) Relationship between increased airway responsiveness and asthma severity in the childhood asthma management program. Am J Respir Crit Care Med 162: 50–56. doi:10.1164/ajrccm.162.1.9811005.

13

Experimental Comparison of the Reproductive Outcomes and Early Development of the Offspring of Rats Given Five Common Types of Drinking Water

Hui Zeng[1], Wei-qun Shu[1]*, Ji-an Chen[1], Lin Liu[2], Da-hua Wang[1], Wen-juan Fu[1], Ling-qiao Wang[1], Jiao-hua Luo[1], Liang Zhang[1], Yao Tan[1], Zhi-qun Qiu[1], Yu-jing Huang[1]

1 Department of Environmental Hygiene, College of Preventive Medicine, Third Military Medical University, Chongqing, P. R. China, 2 The Lundberg-Kienlen Lung Biology and Toxicology Laboratory, Department of Physiological Sciences, Oklahoma State University, Stillwater, Oklahoma, United States of America

Abstract

Tap water (unfiltered), filtered tap water and processed bottled water (purified water, artificial mineralized water, or natural water) are now the five most widely consumed types of drinking water in China. However, the constituents (organic chemicals and inorganic ingredients) of the five waters differ, which may cause them to have different long-term health effects on those who drink them, especially sensitive children. In order to determine which type of water among the five waters is the most beneficial regarding reproductive outcomes and the developmental behaviors of offspring, two generations of Sprague–Dawley rats were given these five waters separately, and their reproductive outcomes and the developmental behaviors of their offspring were observed and compared. The results showed that the unfiltered tap water group had the lowest values for the maternal gestation index (MGI) and offspring's learning and memory abilities (OLMA); the lowest offspring survival rate was found in the purified water group; and the highest OLMA were found in the filtered tap water group. Thus, the best reproductive and offspring early developmental outcomes were found in the group that drank filtered tap water, which had the lowest levels of pollutants and the richest minerals. Therefore, thoroughly removing toxic contaminants and retaining the beneficial minerals in drinking water may be important for both pregnant women and children, and the best way to treat water may be with granular activated carbon and ion exchange by copper zinc alloy.

Editor: Nick Ashton, The University of Manchester, United Kingdom

Funding: This project was supported by National Natural Science Foundation of China (No. 81001234) and the Project of Chongqing Municipal Health Bureau (Grant No. 2012-2-447). The funders had no role in study design, data collection and analysis, decision to publish, or preparation of the manuscript.

Competing Interests: The authors have declared that no competing interests exist.

* Email: xm0630@sina.com

Introduction

Global environmental and economic changes have led to the diversification of human drinking water. Traditional tap water is the most popular drinking water in the world. The addition of chlorine to tap water is one of the most common treatments to ensure its bacteriological quality. However, tap water remains susceptible to biological or chemical contamination [1]: if the water contains organic matter, this may produce disinfection by-products (DBPs), especially trichloromethane (THMs), in the water [2–4]. In addition, heavy metals such as lead and copper can be leached from pipes into the potable water stream [5–8]. Therefore, unpleasant tastes such as a chlorine flavor, DBPs and lead exposure in tap water may be the most common reasons driving people to choose alternative drinking water options such as bottled water or filtered tap water.

Bottled water's consumption has been steadily growing for the past 30 years. In 2011, the consumption was approximately 40,000 million liters in China (ranked number 1), 32,500 million liters in the United State of America (ranked number 2) and 262 billion liters in total around the world (90 countries) [9]. Three major types of bottled water are sold in Chinese groceries and supermarkets: bottled purified water, bottled mineralized water,

and bottled natural water [10]. Bottled purified water, including distilled water, demineralized water, deionized water and reverse osmosis water, is usually tap water that has been treated by a series of filtration processes to remove nearly all minerals and electrolytes, disinfected by ozone or chlorine and finally packaged in a bottle (table 1) [11]. Thus, the purified water in theory is only H_2O. However, the purified water tastes bad and may not quench thirst [12]. In order to improve the taste, small quantities of mineral salts such as potassium chloride and magnesium sulfate are added to the purified water, resulting in mineralized (or low-mineral drinking) water (table 1) [11]. Bottled natural water comes from high-quality underground or surface water sources. This water is also treated by serial filtration, usually disinfected by ozone and then packaged in bottles *in situ* (table 1) [11]. As such, bottled natural water generally contains certain amounts of minerals. Thus, it is clear that different bottled waters contain different minerals, and the mineral levels in these bottled waters are lower than those in the tap water.

In a previous study that reported that drinking water is an important source of essential elements such as Ca and Mg [13], Sabatier suggested that magnesium and calcium in water is more bioavailable to a higher content (from 40% to 60%) than the magnesium and calcium obtained through diet because calcium

Table 1. The drinking water treatment process for the five drinking waters in China.

	water treatment process
tap water	surface water → preliminary sedimentation → coagulation (aluminium polychlorid) → sedimentation → filtration → disinfected by chlorine → clean water tank → water pipes → user
bottled purified water	municipal tap water → quarts sand filtration → activated carbon → reverse osismis or nano filtration (0.0001 μm) → disinfected by ozone → packaged by plastic bottle(disinfect by chlorine) → user
bottled mineralized water	municipal tap water → quarts sand filtration → activated carbon → reverse osismis or nano filtration (0.0001 μm) → minerals added → disinfected by ozone → packaged by plastic bottle(disinfected by chlorine) → user
bottled natural water	surface/underground water → quarts sand filtration → activated carbon → ultrafiltration (0.001–0.1 μm)→ disinfect by ozone → packaged by plastic bottle(disinfected by chlorine) → user
filtered tap water	municipal tap water → activated carbon → KDF filtration → user

and magnesium are mainly present as the simple ions Ca^{2+} and Mg^{2+} in water [14]. Furthermore, Gillies reported that tap water supplies 10% of the average individual's zinc intake [15]. Additionally, consumers want to have a drinking water option that has sufficient quantities of beneficial minerals but no pollutants, and filtered tap water may meet these requirements. Water filtration via a terminal water processor can not only remove chlorine and other impurities [16] but also significantly improve the taste and odor of public tap water. Therefore, it is suitable for home or anywhere where the water quality is poor. Currently, Chinese water pollution is widespread, and water filters are used in more and more residential buildings and private kitchens to improve public tap water quality. At present, more than 15% of the families in Beijing, Guangzhou and Shanghai have a household water purifier. Many materials can be used for water filtration: food-grade cocoanut active charcoal (CAC) and kinetic degradation fluxion (KDF) are the most popular in China. CAC can remove residual chlorine; KDF is a high-purity copper-zinc formulation that uses redox (oxidation/reduction) to remove chlorine, lead, mercury, iron, and hydrogen sulfide from water. The process also has mild anti-bacterial, algicidal, and fungicidal effects [17–18].

Therefore, the constituents in all five of these popular types of water are different and may have different biological effects in humans. Several studies have reported that many factors in drinking water have negative effects on human reproduction and development. For example, epidemiologic studies have reported that low calcium and magnesium intake from drinking water significantly increase the risk of delivering a very low birth weight baby [19–20]. Rats given water containing high levels of zinc have deficits in spatial and working memory [21]. Drinking purified water cannot obviously affect the rats' reproductive outcomes, but it can induce the occurrence of development retardation in the offspring [22]. Eliminating harmful effects on humans over their lifetime and in future generations is the ultimate goal of improving drinking water. However, up to now, there have been no comparative studies regarding the effects of drinking water on the reproductive system and offspring development. In this study, the reproductive and developmental outcomes of rats given the five types of water consumed most widely in China were compared to address which drinking water is the best for pregnant women and infants.

Materials and Methods

Ethics Statement

The Sprague–Dawley rats (150 females, 70~90 g; 75 males, 110~130 g) used in our experiments were obtained from the Laboratory Animal Center, Third Military Medical University (Chongqing, China) and were treated humanely according to the criteria outlined in the "Guide for the Care and Use of Laboratory Animals" prepared by the National Academy of Sciences. The

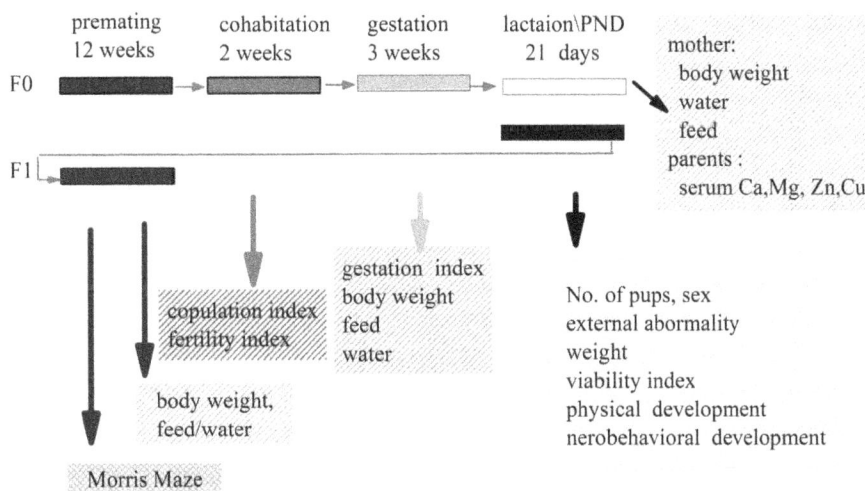

Figure 1. Schematic illustration of the study design (PND, postnatal day).

Table 2. The water quality indices of the five drinking waters.

	bPW	bMW	bNW	FTW	TW	WHO Guideline (2011)	GB5749-2006 in China	LDL	Unit
pH	6.8	6.8	7.55	7.72	7.57	6.5–8.5	6.5–8.5	0	
TDS	1.2	10.9	87.2	291	229	<1000	<1000	0.1	mg/L
TH_{CaCO3}	0.8	2.3	69.6	202.4	200.3	——	<450	0.05	mg/L
COD_{Mn}	0.5	0.6	0.6	0.6	1.0	——	<3	0.05	mg/L
Potassium	<0.5	3.4	<0.5	2.1	2.5	——	——	0.5	mg/L
Sodium	0.1	0.1	0.1	17.0	12.4	——	200	0.1	mg/L
Calcium	0.04	0.02	0.04	40.4	52.9	——	——	0.01	mg/L
Magnesium	0.02	0.4	0.02	11.0	12.7	——	——	0.01	mg/L
Zinc	0.01	0.01	0.01	0.03	0.07	<1.0	<1.0	0.01	mg/L
Copper	0.05	0.05	0.06	0.05	0.05	<1.0	<1.0	0.01	mg/L
Iron	<0.01	<0.01	<0.01	<0.01	0.13	<0.3	<0.3	0.01	mg/L
Mercury	<0.0001	0.0002	<0.0001	0.0001	0.0003	<0.006	<0.001	0.0001	mg/L
Arsenic	<0.01	<0.01	<0.01	<0.01	0.01	<0.01	<0.01	0.01	mg/L
Lead	<0.005	<0.005	<0.005	<0.005	<0.005	<0.01	<0.01	0.005	mg/L
Nitrite (nitrogen)	0.002	<0.001	0.001	0.026	<0.001	<0.9	<0.005	0.001	mg/L
Nitrate (nitrogen)	<0.5	<0.5	0.5	0.8	1.2	<11	<10	0.5	mg/L
Fluoride	<0.1	<0.1	<0.1	0.2	0.2	<1.5	<1.0	0.1	mg/L
Perchlormethane	0.0013	0.0009	0.0011	<0.0001	<0.0001	——	<0.002	0.0001	mg/L
Trichlormethane	0.021	0.033	0.015	<0.001	0.029	<0.3	<0.06	0.001	mg/L

Abbreviation: bMW, bottled mineralized water; bNW, bottled natural water; bPW, bottled purified water; COD, Chemical Oxygen Demand; FTW, filtered tap water; TDS, Total Dissolved Solids; TH, Total hardness; TW, tap water; WHO, world health organization; LDL, lowest detectable limit; —— not establishing guideline values.

Table 3. The reproductive outcomes of the rat parents given the five drinking waters.

	bPW	bMW	bNW	FTW	TW	P
No. of rats (male/female)	30/15	30/15	30/15	30/15	30/15	
Copulation index (%)$_a$, male	100 (15/15)	100 (15/15)	100 (15/15)	100 (15/15)	100 (15/15)	1.00
Copulation index (%)$_a$, female	100 (30/30)	100 (30/30)	97 (29/30)	100 (30/30)	100 (30/30)	1.00
Maternal Fertility index (%)$_b$	67 (20/30)*	100 (30/30)	83 (24/29)	73 (22/30)	100 (30/30)	0.00
Maternal Gestation index (%)$_c$	100 (20/20)*	100 (30/30)*	100 24/24)*	100 (22/22)*	87 (26/30)	0.01
Gestation length (days)$_d$	22 (0)	22 (0)	22 (0)	22 (0)	22 (1)	0.13

$_a$Copulation index (%) = (no. of animals with successful copulation/no. of animals paired) × 100.
$_b$Fertility index (%) = (no. of animals with pregnant/no. of animals with successful copulation) ×100.
$_c$Gestation index (%) = (no. of females that delivered live pups/no. of pregnant females) ×100.
$_d$Values are given as the median (interquartile range).
*$p<0.05$, statistically significant difference from TW group.
Abbreviation: bMW, bottled mineralized water; bNW, bottled natural water; bPW, bottled purified water; FTW, filtered tap water; TW, tap water.

protocol was approved by the Committee on the Ethics of Animal Experiments of Chongqing Experimental Animal Management Center. All serum sample collections were performed under sodium pentobarbital anesthesia, and all efforts were made to minimize suffering.

This study is part of a non-profit project supported by the National Natural Science Foundation of China. All necessary permits were obtained for the described field studies and approved by Chongqing Municipal Health Bureau. The study location is not privately owned, and the field studies did not involve endangered or protected species.

Data Availability Statement

All data were uploaded as Data S1, accompanying the manuscript.

Drinking water and diet

All the animals were acclimated to the laboratory environment for 1 week before the beginning of the study and had free access to food and water.

The tap water used was the municipal water of Chongqing city, which originated from the Yangzi River and was disinfected by liquid chlorine in the water treatment plant. The filtered tap water was municipal tap water filtered by a KDF-CAC-purifier. The tap water and filtered tap water were collected every day for the animals to drink. Three types of bottled water were purchased from supermarkets at a single time point, and each of the same type of bottled water was produced at the same time and was of a single origin. Furthermore, the quality guarantee period for the bottled water is 12 months, which ensured that the samples did not expire before the completion of the animal experiment. One box of bottled water was randomly selected from each type of bottled water to analyze the water quality parameters. The tap water and filtered tap water were collected once at the same time to analyze the water quality parameters. The water quality parameters were determined according to Chinese GB/5750-2006 [23].

The rat feed was prepared bi-monthly in the Laboratory Animal Center of the Third Military Medical University (license number: SCX 2007-018) and strictly followed the GB 14924-2001 in China for experimental animal feed nutrition (calcium 1.11%, magnesium 0.22%).

Parent's reproductive procedure

The animals were randomly divided into 5 groups, and each group was given one type of water from 28 days of age of the parents to 8 weeks of age of the pups (Fig. 1). The reproductive procedures were practiced according to the procedure modified from OECD 415 and GB15193.15-2003 in China [24–25]. At 119 days of age, two females were paired with one male (2:1) from the same water group for 14 days until mating was confirmed by a copulatory plug or sperm in a vaginal rinse. The mating day was recorded as gestation day (GD) 0. The mated females were weighed on GD 0, 6, 12 and 18 and on lactation day (LD) 0, 4, 7, 14 and 21. Water and diet consumption were also recorded on the same days. The day of offspring birth was identified as postnatal day (PND) 0.

Parents' serum calcium, magnesium, phosphorus, copper and zinc levels

Following lactation, parental rats were anesthetized by i.p. pentobarbital, and blood was collected by heart puncture into heparinized syringes. The total serum concentrations of calcium, magnesium phosphorus, and copper were determined in an Olympus AU 600 auto-analyzer(Japan) using the commercial kits of the same brand. Serum zinc was determined in an air–acetylene flame atomic absorption spectrometer (Puxi TAS-986, Peking).

Parents' serum estradiol and testosterone levels

Serum estradiol and testosterone levels were determined in a radioimmunoassay counter (GC-911γ, Zhongjia photoelectric company) using commercial kits (Northern Beijing Institute of Biotechnology).

Offspring studies

All the pregnant rats were allowed to give birth and nurture their offspring normally. On PND 0, the pups were examined for gross malformations, and the numbers of live and stillborn pups were recorded. Each litter was examined twice daily for survival. On PND 4, 8 surviving pups (four males and four females, if possible) were retained randomly, and the remaining pups were culled from each nest. Each pup was weighed on PND 0, 4, 7, 14 and 21.

All behavioral development parameters of the pups were assessed between 9:00 and 10:00 a.m. Pups were separated from the mothers for the time of observation and then immediately

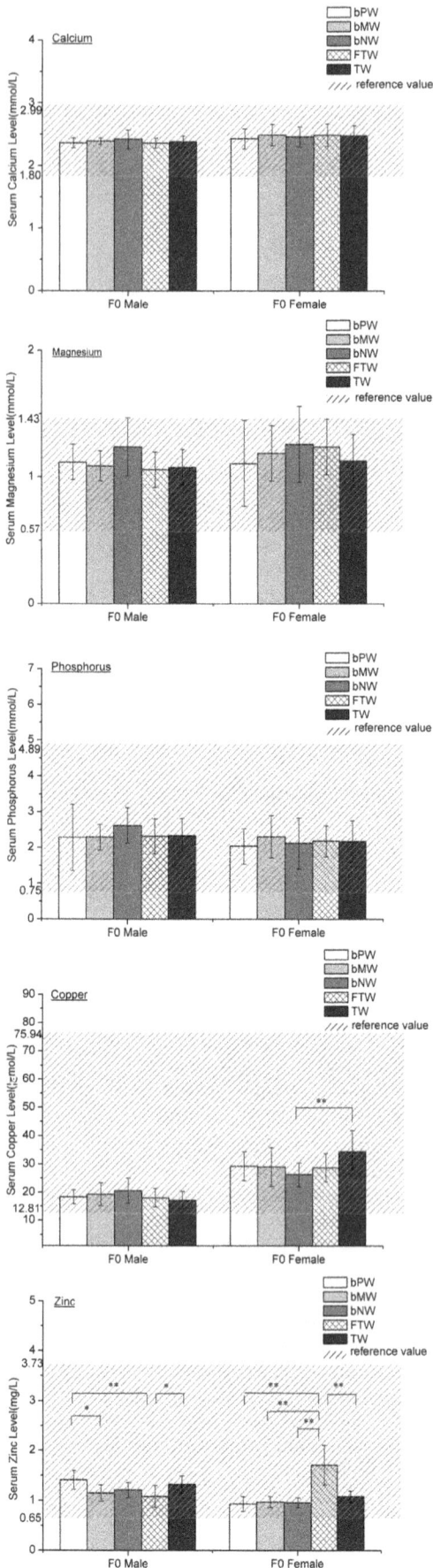

Figure 2. The serum mineral levels of the maternal rats after consuming the five waters ($\bar{x} \pm s$). Statistically significant differences between groups TW, bPW, bMW, bNW, and FTW are marked with asterisks: *p<0.05 and **p<0.01.

returned to their home cages. Pinna detachment, incisor eruption, eye opening, cliff avoidance and surface righting [26–27] were recorded according to the method previously described by our group [22].

The acquisition of spatial learning and memory was assessed via three components in the Morris water maze (MWM) (including hidden platform acquisition, probe trial and subsequent visible platform test) according to a modified version of the procedure of Morris [28]. Ten pups were chosen randomly from ten different litters per group for the MWM. On PND 28, all of them were tested in the MWM [22]. In the hidden platform acquisition test, the rat was given two trials a day for 5 days with an inter-trial period of 15 min, and the time to reach the escape platform was measured. Thus, the learning test refers to the time to reach a hidden platform on successive trials over 5 days. The probe trial was conducted on the sixth day. The platform was removed from the pool, and each rat was allowed to swim for 90 s in the water maze. The number of platform area crossings was recorded, and the memory test refers to the number of goal crossings (traversing the actual location of the escape platform). In the visible platform test, each rat was placed in the pool to find the visible platform in 90 s, and the time to find the platform and the swimming velocity were observed and measured.

Statistical analysis

Following an assessment for homogeneity of variance, the data for quantitative and continuous variables (e.g., body weights) collected from all the rats were analyzed by one-way ANOVA. Data were transformed to achieve approximate normality if the data were not normally distributed. Data are presented as medians (interquartile range). For the hidden platform test, a repeated-measure-ANOVA was used to determine the significance of the difference among the groups. The frequency data were analyzed by a nonparametric χ^2-test. The Pearson bivariate correlation analysis test was used to analyze the relationship between the levels of minerals (in water and serum) and reproductive or neurobehavioral parameters. Unless otherwise noted, the presented data are the mean values $\pm$ standard errors of the means (SEM). Statistical analyses were performed using SPSS 20.0 Statistical Software. In all the experiments, p<0.05 was considered to be statistically significant.

Results

The constituents of the five types of drinking waters

Table 2 shows that the levels of the constituents in the 5 waters were all within the levels established by the GB5749-2006 in China[29] and the WHO guidelines (2011). The three bottled waters all contained relatively low levels of total dissolved solids (TDS) and total hardness (TH), indicating that the levels of inorganic components were lower in the bottled waters. Macro elements such as calcium, magnesium, and sodium were higher in the tap water and the filtered tap water than in the three bottled waters. Furthermore, the calcium and magnesium ratio was 1:20 in mineralized water but 2:1 in natural water and 4:1 in tap water or filtered tap water. Zinc, another important component, was highest in tap water. Fluoride and nitrate concentrations were also higher in tap water and filtered tap water than in the three bottled waters. Arsenic is a toxic element in water, and its level in tap

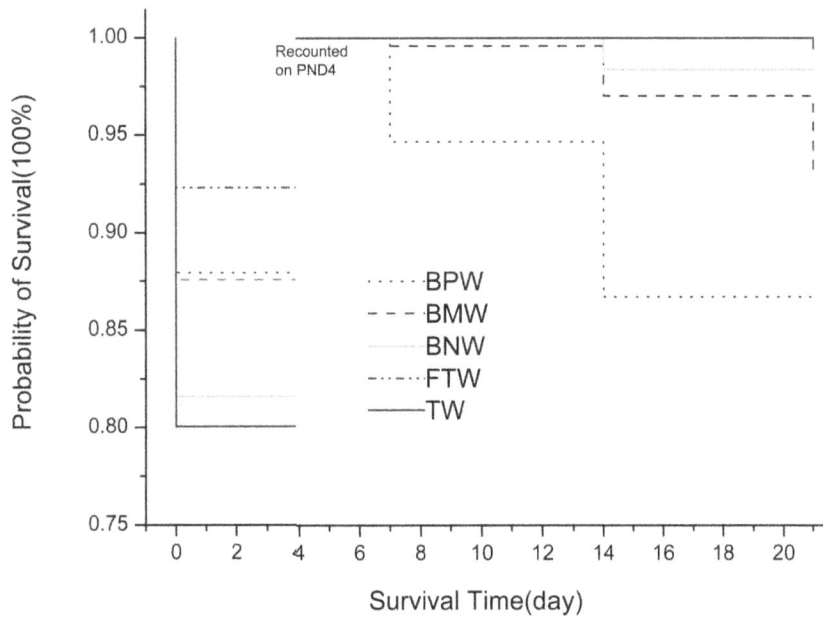

Figure 3. The survival plot of the F1 pups on lactation day 0, 4, 7, 14 and 21. Statistically significant differences between groups TW, bPW, bMW, bNW, and FTW on LD 0, LD 4 and LD 21 (p<0.05). The survival rats were recounted because 8 surviving pups (4 males and 4 females, if possible) were retained randomly, and the remaining pups were culled from each nest on PND 4.

water reached the limit of GB5749-2006 in China. Chemical oxygen demand (COD), an indicator of the level of organic components, was the highest in tap water among five waters (Table 2). Trichloromethane, a conventional water quality index of the DBPs, was highest in bottled mineralized water and lowest in the filtered tap water. Perchloromethane, another conventional water quality index in Chinese GB5749-2006 that may originate from raw water, was found at high levels in three types of water.

Parents' body weight, feed consumption and water consumption

No significant differences in body weight, water consumption or feed consumption were observed in maternal rats in the premating, mating, gestation and lactation periods (data are included in Data S1).

Parents' reproductive parameters and serum levels of estradiol and testosterone

Table 3 shows that the maternal fertility index in the bPW group was significantly lower than that in the TW group, but the maternal gestation index in the TW group was statistically lower than in the other groups. There were no significant differences in the copulation index (male and female), fertility index (females) or the gestation length (females). Furthermore, there were also no significant differences in the levels of testosterone (male) and estradiol (females) among all groups (data are included in Data S1).

Parental serum concentrations of calcium, magnesium, phosphorus, copper and zinc

There were no significant differences in parental serum levels of calcium, magnesium or phosphorus among all five groups (Fig. 2). The maternal serum copper levels in the bNW group were statistically lower than in the TW group (Fig. 2). The paternal serum zinc levels in the FTW group were lower than in the TW and bPW groups, but the levels in the bPW group were statistically

significantly higher than in the bMW group. Furthermore, the maternal serum zinc levels in the FTW group were the highest among the five groups (Fig. 2).

Offspring's developmental parameters

On LD 0, there were no significant differences in external malformation, litter size, pup weight, sex ratio (data are included in Data S1) and survival rate (Fig. 3). On LD 4, survival rates in the TW and bNW groups were significantly lower than that in the FTW group (Fig. 3). On LD 21, the survival rate in the bPW group was the lowest among the 5 groups (Fig. 3). From LD 0 to LD 21, there were also no significant differences in pups' body weights, physiological development (pinna detachment, incisor eruption, and eye opening), and reflex development (surface righting and cliff avoidance reflex) (data are included in Data S1).

Offspring's learning and memory ability after lactation

During the learning period (days 1–5), the time to reach the platform in the place navigation test was shorter day by day in all groups (Fig. 4A), indicating that all the pups could generate space allocation memory regarding the platform. Based on the repeated measure ANOVA, the TW group showed a longer time to reach the platform than the bPW, bMW, bNW, and FTW groups (Fig. 4A). The pups' memory can be tested by the spatial probe test, and the number of goal crossings was recorded to denote the memory ability. Based on one-way ANOVA, the FTW group exhibited statistically significantly higher memory ability than the TW group (Fig. 4B). There were no significant differences in the visible platform trial (negative data not shown).

The relationship of water constituents, serum mineral levels and reproductive and developmental parameters

Table 4 shows that the maternal gestation index was negatively associated with water COD (p = 0.005), zinc (p = 0.016), and arsenic levels (p = 0.000). It also showed that the pups' memory

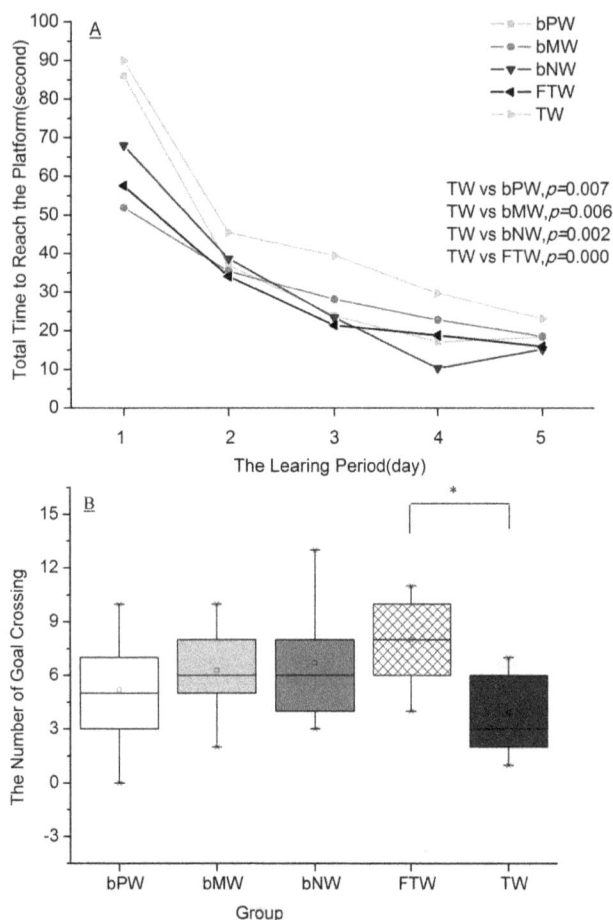

Figure 4. The pups' learning (A) and memory ability (B) in the place navigation test after the mothers drank each of the five waters (median). The data showed that the time to reach the platform of rat pups in the TW group was statistically longer than that of rat pups in the bPW, bMW, bNW and FTW groups, and the total number of rats to reach the platform in the FTW group was statistically higher than in the TW group. Statistically significant differences in the TW and FTW groups are marked with asterisks: *p<0.05.

ability was also negatively associated with water COD (p = 0.012), arsenic (p = 0.009), and perchloromethane levels (p = 0.012) but was positively associated with maternal serum zinc levels (p = 0.027). Furthermore, maternal estradiol was negatively associated with perchloromethane (p = 0.042).

Maternal serum copper and zinc levels were significantly associated with water constituents, including calcium and magnesium. However, maternal serum calcium and magnesium levels had no correlation with water constituents. No correlation was observed between other water constituents and other maternal reproductive or pup developmental parameters.

Discussion

Water constituents

Different water treatment procedures may result in different water compositions. In this study, the data showed the different TDS levels in five types of water. There were no minerals or very low mineral levels in the bottled purified water and bottled mineralized water, which suggested that the "mineralized water" on the market appears not to actually be mineralized water. The

calcium and magnesium ratio also indicated that the "mineralized" water did not meet the natural rule. The TDS and TH were also lower in bottled natural water, which indicated that the filtration treatment removed the majority of the minerals. In contrast, house water filters removed only minimal amounts of beneficial minerals but effectively removed toxic metals by ion exchange. It should be noted that the organic constituent was detected in all five waters. Trichloromethane and perchloromethane were the most common DBPs detected in all three bottled waters, and the levels were very similar, indicating that the main source of organic pollutants may be the plastic bottle and cap, which were always disinfected by chlorine. To our knowledge, COD is commonly used to indirectly measure the amount of organic compounds in water [29] because it is difficult or impossible to identify each contaminant in water, especially the trace organic contaminants. The present study showed that the COD of the tap water was the highest, although it did not exceed the GB19298-2003 and GB5749-2006 in China [30–31]. The high COD in tap water indicated that trichloromethane was not the only organic pollutant. Previous studies in our laboratory reported that 30 types of non-volatile organic pollutants were detected in the tap water that came from the same water plant, while 46 types of non-volatile organic pollutants were detected in the source waters, which were the Yangzi River, in 2000–2001 [32]; 50 types of non-volatile organic pollutants were detected in the same source water in the year 2005 [33]. The water treatment technology in that water plant has remained the same to the present day. However, because the pollution of the water source has not stopped, the number and variety of pollutants may increase. Fortunately, it is obvious that trichloromethane and other pollutants were effectively removed from the tap water through the use of a house water filter. Thus, the three bottled waters were just soft water with low TDS, the tap water was full of not only minerals but also pollutants, and the filtered tap water was full of minerals but with fewer pollutants. Furthermore, the bottled mineralized water did not meet the natural rule with regard to the ratio of calcium to magnesium.

Water constituents exert effect on reproduction

Previous studies have shown that macro elements such as calcium and magnesium and trace elements such as zinc and copper have great impact on reproduction [34–36]. In the present study, the data showed varied levels of such mineral elements in the five drinking waters (Table 2) and the lowest female gestation index but a high fertility index in the TW group (Table 3). These results suggest that embryo implantation in early pregnancy may be affected. Macro elements in the body such as calcium, magnesium and phosphorus are fully supplied in the diet, but when the micro elements are marginally deficient in the diet, water becomes the main source [37]. In this study, the calcium level was 1.11% and the magnesium level was 0.22% in the rat diet, and those levels can fulfill the rats' requirements for growth and development. Therefore, the serum levels of calcium and magnesium were not significantly different between the 5 groups, and the correlation analysis also showed no significant relationship between the levels of water macro elements (calcium, magnesium and phosphorus) and reproductive or developmental parameters. Micro elements such as zinc may be derived from the galvanized plumbing materials. In the present study, zinc levels were highest in the tap water. Unfortunately, we did not determine the zinc content in the feed, as nutritional requirements may be fully met by the rats' feed. However, the data also showed that the serum zinc level was statically different in all 5 groups, and there were no correlations between the serum zinc level and the water zinc level.

Table 4. Correlation analysis of the relationship among water constituents, serum mineral levels, maternal reproductive and pups' developmental parameters.

	Maternal Serum Calcium	Maternal Serum Magnesium	Maternal Serum Phosphorus	Maternal Serum Copper	Maternal Serum Zinc	Maternal Serum estradiol	Paternal serum Testosterone	Maternal Fertility Index	Maternal Gestation Index	Pup's Memory Ability
Water TDS	0.075	0.053	−0.002	0.163	0.675**	0.196	−0.047	0.020	−0.450	0.09
Water COD$_{Mn}$	0.061	−0.070	0.020	0.356**	−0.009	−0.049	−0.172	0.669	−0.975**	−0.305**
Water Calcium	0.072	−0.035	0.003	0.308**	0.500**	0.141	−0.092	0.220	−0.738	−0.096
Water Magnesium	0.077	−0.025	0.007	0.291**	0.551**	0.157	−0.085	0.189	−0.683	−0.061
Water Zinc	0.055	−0.082	−0.001	0.372**	0.201	0.038	−0.133	0.437	−0.943**	−0.261
Water Copper	−0.042	0.148	−0.042	−0.260**	−0.243	−0.099	0.028	−0.059	0.250	0.118
Water Arsenic	0.034	−0.112	−0.004	0.381**	−0.079	−0.054	−0.150	0.567	−1.00**	−0.368**
Water Trichlormethane	−0.105	−0.011	−0.041	−0.256*	−0.597*	−0.157	0.104	0.715	−0.406	−0.010
Water Perchlormethane	0.009	−0.127	0.060	0.181	−0.652**	−0.237*	−0.130	−0.295	−0.001	−0.353*
Maternal Serum Calcium	1.00	0.718**	0.671**	0.229*	−0.047	0.038	—	0.144	0.049	−0.166
Maternal Serum Magnesium	0.718**	1.00	0.734**	0.088	−0.001	0.067	—	−0.101	0.442	−0.130
Maternal Serum Phosphorus	0.671**	0.734**	1.00	0.005	−0.086	0.075	—	−0001	0.378	−0.132
Maternal Serum Copper	0.229*	0.088	0.005	1.00	0.165	−0.154	—	0.879**	−0.826	−0.079
Maternal Serum Zinc	−0.047	−0.001	−0.086	0.165	1.00	−0.227	—	0.176	−0.225	0.316**

*, $p < 0.05$,
**, $p < 0.01$ indicate significant difference level.
No statistically correlation between other water constituents and pups' development parameters, between maternal serum elements and pups' development parameters (data not shown).

The correlation analysis also showed that the zinc level in water was significantly and negatively correlated to the maternal gestation index, although maternal serum zinc levels had no significant relation to the maternal gestation index. Serum zinc values vary diurnally, decrease after meals, and appear to be related to gender and age [37]. More than 90% of zinc is stored in the muscle, and only 10% is present in the serum. Because the zinc content of the muscle was not determined in the current study, we cannot rule out the possibility that the differences in the serum zinc levels are partly attributed to zinc mobilization from muscle tissue. It appears that there was a complicated link between zinc levels in the water and serum. The relationship between water zinc levels and reproductive index parameters requires further study. Thus, the micro elements such zinc, but the not macro elements in the water, may affect the rats' reproduction.

Clearly, toxic chemical elements such as arsenic exert negative reproductive effects [38]. In the present study, we found that only the arsenic levels in tap water reached the limit of the GB19298-2003 and GB5749-2006 in China. A correlation analysis showed that arsenic level in the water was negatively associated with maternal gestation index. However, the arsenic values were very close to each other, suggesting that arsenic levels may have not been sufficient to decrease maternal reproductive capacity in the TW group but may have increased the effects of other factors; these findings require further study.

Organic contaminants in water also have clear reproductive toxicity [39–40]. There was a negative correlation between water COD and maternal gestation index parameters. It may be possible that the lowest MGI in the TW group may have been the result of organic pollutants in the tap water other than trichloromethane, which was present in similar levels in each of the 4 waters. The organic pollutants including DBPs in the tap water may also exert an effect on reproductive function, as 40 types of trace organic pollutants have been detected in tap water, and the organic extract can increase endometrial thickness in rats [41–42]. How the mixture exerts these effects require further study.

Water constituents exert effects on development

Previous studies have suggested that zinc can enhance learning and memory ability [43] and that organic pollutants [44] and arsenic [45–46] in drinking water can decrease learning and memory abilities. The results from the present study are consistent with these findings: organic pollutants and arsenic were the highest in the TW group, and the learning ability of the pups in the TW group was statistically lower than in the other groups; organic pollutants were the lowest and the maternal serum zinc level was the highest in the FTW, and the memory ability of the pups in the FTW group was the highest among the 5 groups. A correlation analysis also showed that the pups' memory ability was positively correlated to maternal serum zinc levels and negatively correlated to COD_{mn}, arsenic and perchloromethane levels in water. Thus,

the better memory ability in the FTW group may be due to the near absence of organic compounds or toxic metals and the rich mineral content in the filtered tap water. However, the reason for the lower learning ability in the TW group may be complicated because COD_{mn}, arsenic and perchloromethane levels were all below the Chinese water standards GB5749-2006 and WHO Guidelines (2001) for tap water; in particular, the arsenic levels were very close in all five waters, which suggests that the effect on learning ability is most likely due to other pollutants we did not detect or a complex combination of all pollutants. These results require further study.

Survival rates on PND 4 and PND 21 can be used as indicators of the maternal nutritional status and maternal instinct. On PND 0, survival rates were not significantly different between the 5 groups. However, on PND 4, survival rates in all 5 groups decreased, and the survival rates in the TW and bNW groups were statistically lower than in the FTW group, which indicated that the maternal nutrition may have been deficient in the TW and bNW groups because the mothers nursed all pups. After PND 4, the pups were culled to 6–8 pups per litter in order to assure each mother had enough nutrition to nurse all pups. Therefore, the survival rate increased on PND 21. However, the survival rate in the bPW group was the lowest among the 5 groups, suggesting that maternal nutrition in the bPW group may have been insufficient.

Overall, among the five drinking waters, filtered tap water had the lowest levels of pollutants, had the highest hardness, was the richest in minerals, and showed the best benefit for maternal reproductive parameters and pups' development parameters. Removing toxic contaminants and maintaining minerals are both important for drinking water, especially when consumed by pregnant women and children. Granular activated carbon and ion exchange by copper zinc alloy may be the best way to treat water, but how water produced with these methods exerts beneficial effects needs further study. This result may have important implications for the selection of healthy drinking water and for water plants to optimize their treatment processes, as organic pollutants and toxic metals (arsenic) in drinking water may decrease maternal reproductive parameters.

Supporting Information

Data S1 All data underlying the findings described in this manuscript.

Author Contributions

Conceived and designed the experiments: WS LL. Performed the experiments: HZ JC DW WF LW JL LZ YT ZQ YH. Analyzed the data: HZ LZ. Contributed reagents/materials/analysis tools: HZ JC DW WF LW JL LZ YT ZQ YH. Wrote the paper: HZ WS.

References

1. Rosa C, Franco F, Paolo T. (2011) Drinking water quality: Comparing inorganic components in bottled water and Italian tap water. J Food Compost Anal 24: 184–193.
2. Richardson SD. (2003) Disinfection by-products and other emerging contaminants in drinking water. Trends Analyt Chem 22: 666–684.
3. Wolf A, Bergmann A, Wilken RD, Gao X, Bi Y, et al. (2013) Occurrence and distribution of organic trace substances in waters from the Three Gorges Reservoir, China. Environ Sci Pollut Res Int 20: 7124–39.
4. Richardson SD, Plewa MJ, Wagner ED, Schoeny R, DeMarini DM. (2007) Occurrence, genotoxicity, and carcinogenicity of regulated and emerging disinfection byproducts in drinking water: a review and roadmap for research. Mutat Res 636: 178–242.

5. Clement M, Seux R, Rabarot S. (2000) A practical model for estimating total lead intake from drinking water. Water Res 34: 1533–1542.
6. Cartier C, Nour S, Richer B, Deshommes E, Prévost M. (2012) Impact of water treatment on the contribution of faucets to dissolved and particulate lead release at the tap. Water Res 46: 5205–5216.
7. Kim EJ, Herrera JE, Huggins D, Braam J, Koshowski S. (2011) Effect of pH on the concentrations of lead and trace contaminants in drinking water: A combined batch, pipe loop and sentinel home study. Water Res 45: 2763–2774.
8. Fertmann R, Hentschel S, Dengler D, Janßen U, Lommel A. (2004) Lead exposure by drinking water: an epidemiological study in Hamburg, Germany. Int J Hyg Environ Health 207: 235–244.

9. Richard H. (2012) global bottled water congress and market trends. Available at: http://zenithinternational.com/pdf/events/00101slides.pdf. Accessed 8 April 2013.

10. Deng XQ, Liang JH, Cheng CY. (2001) A superficial view on the development of bottled water market at home and abroad. Food and Fermentat ion Industries 27(4): 70–74.

11. No author. (2008) bottled water: the question of the source. Popular Standardization 9: 6–9. Available at: http://www.cqvip.com/QK/96243X/200809/28415476.html. Accessed 8 April 2013.

12. Frantisek K. Health risks from drinking demineralised water. Available at: http://www.who.int/water_sanitation_health/dwq/nutrientschap12.pdf. Accessed 8 April 2013.

13. Widdowson EM. (1944) Health significance of drinking water calcium and magnesium Minerals in human nutrition. British Medical Bulletin 10: 221–222.

14. Sabatier M, Arnaud MJ, Kastenmayer P, Rytz A, Barclay DV. (2002) Meal effect on magnesium bioavailability from mineral water in healthy women. American Journal of Clinical Nutrition 75: 65–71.

15. Gillies ME, Paulin HV. (1982) Estimations of daily mineral intakes from drinking water. Human nutrition:applied nutrition 36: 287–292.

16. Xu W, Wang DX, Zhang JW, Shen JY. (2011) Function Comparison of Four Types of Household Water Purifier. Occup and HeMth 27: 2004–2005.

17. Teakle JT. (1997) Pretreatment of reverse osmosis systems with KDF55 process medium. Michigan:KDF Fluid Treatment, Inc pp: 1–5.

18. Xiong R, Liu W, Xi X, Xiao S. (2004) Application and amelioration prospect of copper-zinc alloy in water treatment. Industrial Safety and Dust Control 30: 5–8.

19. Yang CY, Chiu HF, Tsai SS, Chang CC, Sung FC. (2002) Magnesium in drinking water and the risk of delivering a child of very low birth Weight. Magnes Res 15: 207–213.

20. Yang CY, Chiu HF, Chang CC, Wu TN, Sung FC. (2002) Association of very low birth weight with calcium levels in drinking water. Environ Res 89: 189–194.

21. Linkous DH, Smith LN, Conko KM, Jones BF, Flinn JM. (2004) Differential effects of zinc association in drinking waters on spatial memory. Met Ions Biol Med 8: 390–395.

22. Zeng H, Shu WQ, Zhao Q, Chen Q. (2008) Reproductive and neurobehavioral outcome of drinking purified water under magnesium deficiency in the rat's diet. Food Chem Toxicol 46: 1495–1502.

23. Ministry of Health of China, Standardization Administration of China. National Standard of the People's Republic of China—standard examination methods for drinking water (GB/T5750-2006). Issued on December 29, 2006. Implemented on July 1, 2007.

24. OECD Publishing, 1983. Test No. 415: One-Generation Reproduction Toxicity Study. In: OECD Guidelines for the Testing of Chemicals/Section 4: Health Effects. Available: http://www.oecdbookshop.org/oecd/display.asp?lang=EN&sf1=identifiers&st1=5lmqcr2k7nnq. Accessed 24 August 2013.

25. Ministry of Health of China, Standardization Administration of China. National Standard of the People's Republic of China—reproductive study (GB15193.15-2003). Issued on September 24, 2003. Implemented on May 1, 2004.

26. Zbinden G. (1981) Experimental methods in behavioral teratology. Arch Toxicol 48: 69–88.

27. Pantaleoni GC, Fanini D, Sponta AM, Palumbo G, Giorgi R, et al. (1988) Effects of maternal exposure to polychlorobiphenyls(PCBs) on F1 generation behavior in the rat. Fundam Appl Toxicol 11: 440–449.

28. Morris R. (1984) Developments of water-maze procedure for studying spatial learning in the rat. J Neurosci Methods 11: 47–60.

29. Devi R, Dahiya RP. (2006) Chemical oxygen demand (COD) reduction in domestic wastewater by fly ash and brick kiln ash. Water Air Soil Pollut 174: 33–46.

30. Ministry of Health of China, Standardization Administration of China. National Standard of the People's Republic of China—hygienic standard of bottled water for drinking (GB19298-2003). Issued on September 24, 2003. Implemented on May 1, 2004.

31. Ministry of Health of China, Standardization Administration of China. National Standard of the People's Republic of China—standards for drinking water quality (GB 5749–2006). Issued on December 29, 2006. Implemented on July 1, 2007(replacing GB 5749–1985).

32. Tian HJ, Shu WQ, Zhang XK, Wang YM, Cao J. (2002) Organic pollutants in source water in Jialing and Yangze River (Chongqing Section). Resources and Environment in the Yangtze basin 24(4): 226–239.

33. Guo ZS, Luo CH, Zhang WD, Lu Y, Sun J, et al. (2006) The analysis of the persistent organic pollution in the Three Gorges Region in Chongqing. Environmental Monitoring in China 22(4): 45–48.

34. Mao YX. (2003) Analysis magnesium ions content in normal pregnancy and abnormal pregnancy. Chinese Journal of Birth Health and Heredity 11: 73–75.

35. Cheng F, Wang Y. (2012) The correlation analysis on pregnant women's and neonate's serum levels of calcium,ferrum,zinc,copper and weight. Journal of Chongqing Medical University 37: 89–91.

36. Dutt B, Mills CF. (1960) Reproductive Failure in Rats Due to Copper Deficiency. J Comp Pathol 70: 120–125.

37. Maret W, Sandstead HH. (2006) Zinc requirements and the risks and benefits of zinc supplementation. J Trace Elem Med Biol 20: 3–18.

38. Bloom MS, Fitzgerald EF, Kim K, Neamtiu I, Gurzau ES. (2010) Spontaneous pregnancy loss in humans and exposure to arsenic in drinking water. Int J Hyg Environ Health 213: 401–413.

39. Yang CY, Cheng BH, Tsai SS, Wu TN, Lin MC, et al. (2000) Association between chlorination of drinking water and adverse pregnancy outcome in Taiwan. Environ Health Perspect 108: 765–768.

40. Wright JM, Schwartz J, Dockery DW. (2003) Effect of trihalomethane exposure on fetal development. Occup Environ Med 60: 173–180.

41. Cao B, Ren Q, Qiu ZQ, Zhao Q, Shu WQ. (2009) Evaluation of reproductive toxicity in rats caused by organic extracts of Jialing River water of Chongqing, China. Environ Toxicol Pharmacol 27: 357–365.

42. Ren Q, Cao B, Qiu ZQ, Zhao Q, Shu WQ. (2005) A study of toxicity of organic extract of tap water on male reproductive system in mice. Acta Academiae Medicinae Militaris Tertiae 27: 1012–1015.

43. Nakashima AS, Dyck RH. (2009) Zinc and cortical plasticity. Brain Res Rev 59: 347–373.

44. Janulewicz PA, White RF, Winter MR, Weinberg JM, Gallagher LE, et al. (2008) Risk of learning and behavioral disorders following prenatal and early postnatal exposure to tetrachloroethylene (PCE)-contaminated drinking water. Neurotoxicol Teratol 30: 175–185.

45. Jing J, Zheng G, Liu M, Shen X, Zhao F, et al. (2012) Changes in the synaptic structure of hippocampal neurons and impairment of spatial memory in a rat model caused by chronic arsenite exposure. Neurotoxicology 33: 1230–1238.

46. O'Bryant SE, Edwards M, Menon CV, Gong G, Barber R. (2011) Long-term low-level arsenic exposure is associated with poorer neuropsychological functioning: a Project Frontier study. Int J Environ Res Public Health 8: 861–874.

14

Effectiveness of Low Emission Zones: Large Scale Analysis of Changes in Environmental NO$_2$, NO and NO$_x$ Concentrations in 17 German Cities

Peter Morfeld[1,2]*, **David A. Groneberg**[3], **Michael F. Spallek**[3,4]

1 Institute for Occupational Epidemiology and Risk Assessment (IERA) of Evonik Industries, Essen, Germany, 2 Institute and Policlinic for Occupational Medicine, Environmental Medicine and Preventive Research, University of Cologne, Cologne, Germany, 3 Institute of Occupational Medicine, Social Medicine and Environmental Medicine, Goethe-University, Frankfurt am Main, Germany, 4 European Research Group on Environment and Health in the Transport Sector (EUGT), Berlin, Germany

Abstract

Background: Low Emission Zones (LEZs) are areas where the most polluting vehicles are restricted from entering. The effectiveness of LEZs to lower ambient exposures is under debate. This study focused on LEZs that restricted cars of Euro 1 standard without appropriate retrofitting systems from entering and estimated LEZ effects on NO$_2$, NO, and NO$_x$ (= NO$_2$+ NO).

Methods: Continuous half-hour and diffuse sampler 4-week average NO$_2$, NO, and NO$_x$ concentrations measured inside and outside LEZs in 17 German cities of 6 federal states (2005–2009) were analysed as matched quadruplets (two pairs of simultaneously measured index values inside LEZ and reference values outside LEZ, one pair measured before and one after introducing LEZs with time differences that equal multiples of 364 days) by multiple linear and log-linear fixed-effects regression modelling (covariables: e.g., wind velocity, amount of precipitation, height of inversion base, school holidays, truck-free periods). Additionally, the continuous half-hour data was collapsed into 4-week averages and pooled with the diffuse sampler data to perform joint analysis.

Results: More than 3,000,000 quadruplets of continuous measurements (half-hour averages) were identified at 38 index and 45 reference stations. Pooling with diffuse sampler data from 15 index and 10 reference stations lead to more than 4,000 quadruplets for joint analyses of 4-week averages. Mean LEZ effects on NO$_2$, NO, and NO$_x$ concentrations (reductions) were estimated to be at most -2 µg/m^3 (or -4%). The 4-week averages of NO$_2$ concentrations at index stations after LEZ introduction were 55 µg/m^3 (median and mean values) or 82 µg/m^3 (95th percentile).

Conclusions: This is the first study investigating comprehensively the effectiveness of LEZs to reduce NO$_2$, NO, and NO$_x$ concentrations controlling for most relevant potential confounders. Our analyses indicate that there is a statistically significant, but rather small reduction of NO$_2$, NO, and NO$_x$ concentrations associated with LEZs.

Editor: Qinghua Sun, The Ohio State University, United States of America

Funding: The study was funded by the European Research Group on Environment and Health in the Transport Sector; www.eugt.org. The funder supported data collection and preparation of the manuscript.

Competing Interests: PD Dr. Peter Morfeld and Prof Dr. David Groneberg are members of the scientific advisory group of the European Research Group on Environment and Health in the Transport Sector (EUGT, www.eugt.org). PD Dr. Spallek is managing director of EUGT e.V. EUGT performs and supports research on emissions, immissions and health effects related to the transport sector.

* Email: Peter.Morfeld@evonik.com

Introduction

Low Emission Zones (LEZs) are areas or roads where the most polluting vehicles are restricted from entering. They are currently introduced in 13 European countries [1]. In Europe, vehicle emissions are classified by the so-called "Euro Standards" with a current range from Euro 1 to Euro 6 regarding the technical features of the vehicles which are fixed in several EU-Directives for passenger cars and heavy-duty trucks [e.g., 2]. Basically, this means that vehicles are restricted in relation to their Euro emission level. The configuration of LEZs is extremely different and heterogeneous in Europe, for example in Italy, where the entry standards, the subsistent regulations and the daily duration of LEZ conditions differ substantially from town to town. However, most LEZs in Europe operate 24 hours a day, 365 days a year [see 1].

One of the most developed applications is found in Germany. Low emission zones have been introduced in Germany since 2008 in different stages, resulting in meanwhile 48 LEZs with restrictions for pollutant groups 2 or 3 in 11 Federal states by the end of January 2014 [3]. In this study we analysed the effect of introducing the "LEZ of pollutant group 1" which restricts from entering Diesel cars of an European emission standard below Euro 2 without particulate reduction system and gasoline cars of an European emission standard below Euro 1 without appropriate exhaust gas catalytic converters [3].

Traffic emissions are considered to be a relevant source of air pollution [4] and LEZs are believed to be the most effective measure that cities can take to reduce vehicle-induced air pollution problems in their area [5–7]. The emissions that are aimed to be reduced by LEZs are mainly fine particles like PM10 or smaller [8–12]. The effectiveness of LEZs to reduce traffic-related exposures is still under debate [13] and there is an open discussion in the public about the "outcome" and cost-benefit ratio of LEZs [14–16]. Most of the published information refers to particulate matter.

Additionally, nitrogen dioxide is discussed to be a major traffic-related pollutant as well as an epidemiologic marker of air quality and related adverse health effects [17–21]. On the other hand, a systematic literature review showed only moderate evidence for adverse health effects at a long-term exposure below an annual mean of 40 $\mu g/m^3$ NO2 [22].

According to EU rules [23,24] limits were additionally imposed for NO2 and are enforced in Germany since 2010: 200 $\mu g/m^3$ as an 1 hour average (acceptable: 18 excursions/year) and 40 $\mu g/m^3$ as an annual average. Values were and are in excess: about 69% of all stations near to traffic showed annual averages higher than 40 $\mu g/m^3$ in Germany [7,25]. This non-compliance is not restricted to Germany but the European limit value for NO2 is exceeded in many European cities [26–28]. The LEZ concept was extended and it was assumed that LEZs are an effective measure not only to lower PM10 dust levels but also to reduce NO2 concentrations [6,29]. There are indications that LEZs may indeed reduce NOx concentrations effectively [30–32], but ozone has to be considered a confounder in NO2 measurements [e. g. 33], and the gases NO and NO2 rapidly interconvert, too [34]. Furthermore, national emission ceilings were defined for NOx, i.e., the sum of NO2 and NO [35]. Thus, there is interest in the impact of LEZs on concentrations of NO and NOx also [36].

However, a scientific proof of the LEZ concept targeting at NO_2, NO, and NO_x is still missing. In order to test the views of legislators and researchers that LEZs are effective measures to reduce nitrogen oxide concentrations [29,37], this study focused on the potential effects of LEZs on ambient concentrations of NO_2, NO, and NO_x in LEZ areas of 17 German cities.

We reported on the effect of LEZs on PM_{10} concentrations elsewhere [32].

Methods

Target parameters

The aim of the study was to analyse the effectiveness of German LEZs (as many as eligible) to lower NO_2, NO, and NO_x ($= NO_2 + NO$) concentrations. The first analysis series of NO_2, NO, and NO_x were based on continuous half-hour measurement data of NO_2 and NO. Second, measurement data for NO_2 and NO concentrations collected by diffuse samplers and determined over longer sampling periods were available. These data were allocated

to 4-week periods. Third, we collapsed the half-hour measurement data to four-week averages and pooled these collapsed continuous data and the diffuse sampler data to perform joint analyses over 4-week periods. The original NO_2 and NO measurements were performed by the Environmental State Institutions in Germany (Landesumweltämter). A federal data base [38] reports on the applied measurement procedures.

Measuring procedure

Two measuring procedures were applied: continuous measurement devices (chemiluminescence), data stored as half-hour averages and diffuse samplers (Palmes tubes, chromatography), data stored as long-term averages over weeks. The chemiluminescence method relies on the reaction of NO with O3: NO+ O3→NO2*+O2. Chemiluminescence is generated in the range of 600 nm to 3,000 nm when the excited molecules return to the ground state. The light intensity is proportional to the concentration of NO molecules. A deoxidation converter is used to reduce NO2 to NO. Thus, the NO2 concentration is determined as the difference between the NOx concentration measured when the sample gas is directed through a deoxidation converter and the NO concentration measured when the gas is not run through the converter. The diffuse samplers were Palmes type tubes modified with a glass frit as turbulence barrier. In these passive samplers molecules diffuse because of a concentration gradient through an intake opening with a defined cross-section along a fixed diffusion path to a sampling medium by which they are adsorbed. This process is described by Fick's first diffusion law. The chemical analysis is done by chromatography. [More details on both methods may be found in 39,40–43].

Period of investigation

The period of investigation was from 2005 until the end of 2009 (31 December 2009), starting at least from the introduction of the individual LEZ minus the length of the respective LEZ phase (or earlier if restrictions of truck traffic were enforced before the introduction of the LEZ).

Low Emission Zones

There were 34 German active LEZs until the end of 2009 and 774 monitoring stations in use. With introduction of these LEZs, as a main effect, only those diesel vehicles with an exhaust emission standard better than Euro 1 (with sticker) were allowed to enter the zone. In principle, the German "LEZ of pollutant group 1" restricts from entering

– Diesel passenger cars, trucks and buses of an European emission standard below Euro 2 without particulate reduction system, and

– Gasoline passenger cars, trucks and buses of an European emission standard below Euro 1 without appropriate exhaust gas catalytic converters.

Local authorities can set up exception permits especially for light duty vehicles, trucks and buses due to local necessities [3].

According to protocol LEZs were included into the study if and only if

– monitoring stations existed, that operated before and after the LEZ introduction and measured inside the LEZ area (*index stations*) and

– monitoring stations existed, that operated before and after the LEZ introduction and measured outside the LEZ area – in a circle around the centre with a radius of about 25 km – and if outside the city area, than in no other LEZ (*reference stations*) and

– these monitoring stations measured NO_2 or NO (continuous measurements or diffuse samplers).

(For the terminology and the use of index and reference values in comparisons if exposures levels see Rothman et al. [44])

Seventeen cities with LEZs in 6 German Federal states could be included into the study (Baden-Württemberg: Herrenberg, Ludwigsburg, Mannheim, Reutlingen, Stuttgart, Tübingen; Bavaria: Augsburg, Munich; Berlin: Berlin; Hesse: Frankfurt; Lower Saxony: Hannover; North Rhine-Westphalia: Dortmund, Duisburg, Düsseldorf, Essen, Cologne, Wuppertal). Figure 1 shows all active 34 German active LEZs in December 2009 and the 17 LEZs included for study. File S1 entails maps of all LEZs eligible for study with all index and reference stations marked (Figure S1 in S1 to Figure S19 in S1).

In total, these 17 LEZs, eligible for study, contained 108 eligible monitoring stations with 53 index stations and 55 reference stations. The data base constructed from transferred data encompassed a total of 9,517,911 data lines which were used as input to analysis. An overview is given in Table 1.

Data analysis

The data set structured for analysis consisted of matched quadruplets. A matched quadruplet comprises four pairwise corresponding measurement values consisting of two index- and two reference values. One index value and the simultaneously measured reference value were obtained during the active LEZ period, the other pair of values was obtained before introducing the LEZ. The pairs of values had a 364 days difference in time of or a multiple of 364 days, hence keeping the season, day of the week and time of day constant within the quadruplets. The allocation of reference stations to index stations was done pairwise, i.e., quadruplets were constructed by the data of one index station and allocating to it all appropriate reference stations with their data without a prior collapsing ("collapsing" is a technical term widely used in statistics describing the summary of a table in marginal, http://www.stata.com/manuals13/dcollapse.pdf). The method has been described in detail before [45] and is a refined approach in comparison to other analytical strategies [46]. The analysis plan was critically reviewed by a chair of statistics.

The quadruplets were analysed by the "difference score method in the two period case" [47]: Differences in index values were regressed on differences in reference values while other data were taken into account as covariates in fixed-effect regression analyses. Two types of models were fitted: a linear (additive) model and a log-linear (multiplicative) model. The difference of the index concentration data was used as the response variable in the linear model. The log of this response variable was entered into the log-linear regression model after applying an appropriate positive offset calculated from the data [48]. The two model types differ in the assumption on how covariables may influence the index station concentration data: on an additive scale or on a multiplicative scale [49,50].

The following covariables were taken into account in the basic fixed effects regression analyses: differences at reference stations in $\mu g/m^3$ (to control e.g. for large-scale meteorological changes and seasonal effects), baseline data at reference stations in $\mu g/m^3$ (to control for time-dependent effects of reference data, Allison [47], and baseline data at index stations in $\mu g/m^3$ (to control for "regression to the mean" [51]. This structure defines the basic regression approach. The covariables were entered into the log-

linear (multiplicative) models after adding an appropriate offset if indicated [48] and then taking logs.

The following equation describes the analysis of matched quadruples in the basic fixed-effect linear ("additive") regression model [47]

$$\Delta x_{mdh} = E + \sum_{k=1}^{Z} E_k \cdot z_k + b_x \cdot x_{0mdh,cent} + b_{\Delta r} \cdot \Delta r_{zdh} + b_r \cdot r_{z0dh,cent} + \varepsilon.$$

Δx_{mdh} describes the difference of the index station data at monitoring station m between days d and d-364 (= day d+1 in the year before), always at time (hour) h, i.e., x_{1mdh} - x_{0mdh} (compare Figure 2). $x_{0mdh,cent}$ denotes the baseline value at station m on day d at time h, centred at the mean of all baseline values at station m. The terms Δr_{dh} and $r_{z0dh,cent}$ are the corresponding reference value data. The coefficient of major interest is the intercept of the regression model because it estimates the LEZ effect: E measures the mean effect across all LEZs, $E+E_k$ the mean effect in zone k, $1 \leq k \leq Z$. The coefficient b_x accounts for "regression to the mean", $b_{\Delta r}$ for the bias in annual levels (e.g., changed meteorological conditions), b_r for a time-dependent effect of reference values and ε is the residual error of the concentration difference at the index stations. The second model type had the same structure but used logs of the terms ("log-linear", "multiplicative"). An appropriate small offset was added to avoid undefined logarithms [48].

The equation of the basic fixed-effect linear ("additive") regression model can be justified as follows (to keep the notation simple we suppress the time index: we write, eg, x_{z0} instead of x_{z0dh}).

Let us start with assuming an ideal hypothetical situation: measurements are without any distortions and random errors, no covariates are operating. In that case we will measure the index concentration at time point 0 before the LEZ was introduced as a constant value c_0 at all index stations of the LEZ. After the introduction of the LEZ we will measure at time point 1 at the same index station the constant value c_1. The effectiveness of the zone is simply $E = c_1 - c_0$.

But even if there are no biases, random errors and no covariates we do not expect to see the same E for all LEZs. The effectiveness may depend on characteristics of the zone k like the area of the LEZ, A_k (e.g., we may expect a larger effect if the LEZ area is larger). The concentration at time point 1 may be written more appropriately as $c_1 + f \times A_k$ (with a multiplicative coefficient f mapping the effect of the area into the concentration scale). The effect of zone k can be described as $E = c_1 + f \times A_k - c_0$. Note that A_k operates as an effect modifier.

We can take account of differences between the zones without referring to a specific characteristic of LEZ k, like the area: We may describe the effect of zone k in more abstract terms as $E+E_k$ [$E+E_k$ means $E+E_k \times z_k$ with a multiplicative indicator z_k, that takes the value 1 for zone k and zero otherwise, $1 \leq k \leq Z$]. E_k ist the specific effect offset of LEZ k in comparison to the overall mean E of the LEZ effects. It is simple to extend the notation to cover different baseline concentrations for the different LEZs. Thus, $E+E_k = c_1+(c_{z1}-c_1) - [c_0+(c_{z0}-c_0)] = c_{z1}-c_{z0}$, i.e., the effect of zone k is the difference between the zone specific measurement values after (c_{z1}) and before (c_{z0}) the introduction of the LEZ k (c_1 and c_0 now denote the averages of the concentrations across all LEZs under study).

Still, the approach is not very realistic. We should take into account background variations of the intensities, resulting from e.g. large area changes of the concentrations. These large area variations are reflected in the values r_{z0} and r_{z1} at the reference

Figure 1. Active and investigated LEZs in Germany, December 2009. The 17 LEZs included into the study are marked with open red circles. The 17 LEZs that were active but were excluded according to protocol are indicated by full red circles. Capital cities of the Federal states are shown by black squares. This is a modification of the map as published at URL www.umweltbundesamt.de/umweltzonen. Date of access: 6th November 2009.

stations. Despite all efforts to measure the concentrations as precisely as possible we always will have random errors ε_1 and ε_0. In this extended approach the measurement values for zone k before introducing the LEZ are $x_{z0} = c_0 + (c_{z0} - c_0) + g \times r_{z0} + \varepsilon_0$ and $x_{z1} = c_1 + (c_{z1} - c_1) + g \times r_{z1} + \varepsilon_1$ after the introduction. The factor g measures how strong the reference values do influence the index values. It follows that $x_{z1} - x_{z0} = E + E_k + g^* (r_{z1} - r_{z0}) + (\varepsilon_1 - \varepsilon_0)$. With $\Delta x_z = x_{z1} - x_{z0}$, $\Delta r_z = r_{z1} - r_{z0}$, $\varepsilon = \varepsilon_1 - \varepsilon_0$ and $b_{\Delta r} = g$ we yield the major part of the equation of the basic fixed-effect linear ("additive") regression model. Note that we substituted z by m which means that we apply the approach in a refined way to every

index monitoring station m. We have demonstrated above that a potential confounder/adjuster, like the concentration at a reference station, enters the equation in terms of the difference of the values across time (e.g., $\Delta r_z = r_{z1} - r_{z0}$). And we have seen that the model can be extended by potential modifiers of the LEZ effect ("interaction terms"), like the area, by adding terms like $f \times A_k$ (in contrast to adjusters not as a difference in time). We will now explain in more detail why we included the effect modifying variables x_{z0} and r_{z0} additionally.

Altman and Bland [52] and Bland and Altman [53] suggested including the mean value x_{zm} of the concentrations at index station

Table 1. Overview of low emission zones (LEZs) in German cities included for analysis: 17 LEZs in 6 federal states.

Federal State	City	Data Since	Introduction of LEZ	NO$_2$ Index	NO$_2$ Reference	NO Index	NO Reference	NO$_x$ Index	NO$_x$ Reference	Area LEZ/km²	Area City/km²	Population LEZ 2007	Population City 2011
Baden-Württemberg[a]	Herrenberg	2006	01Jan2009	1	1[d]	1	1[d]	1	1[d]	17	64.71	15000	29935
	Ludwigsburg	2005	01Mar2008	2	1[d]	2	1[d]	2	1[d]	30	43.33	55000	86939
	Mannheim	2005	01Mar2008	2	2	2	2	2	2	7.5	144.96	93900	291458
	Reutlingen	2004	01Mar2008	1	2[d]	1	2[d]	1	2[d]	4.5	87.06	24500	110084
	Stuttgart	2004	01Mar2008	5	2[d]	5	2[d]	5	2[d]	207	207	590000	591015
	Tübingen	2005	01Mar2008	1	1[d]	1	1[d]	1	1[d]	11.9	108.12	64000	83248
Total Baden-W[a]				12	9	12	9	12	9	277.9	655.18	842400	1192679
Bavaria[a]	Augsburg	2007	01Jul2009	3	1	3	1	3	1	5.2	146.93	40000	272699
	Munich	2006	01Oct2008	5	3	5	3	5	3	44	310.71	431000	1388308
Total Bavaria[a]				8	4	8	4	8	4	49.2	457.64	471000	1661007
Berlin "BLUME"[a]	Berlin	2004	01Jan2008	5	9	5	9	5	9	88	891.85	1100000	3375222
Hesse[a]	Frankfurt a.M.	2003	01Oct2008	1	3	1	3	1	3	110	248.31	550000	687775
Lower Saxony[a]	Hannover	2005	01Jan2008	1	4	1	4	1	4	50	204.14	218000	514137
North Rhine-Westphalia[a]	Dortmund	2006	01Jan2008	2	4[d]	2	4[d]	2	4[d]	19.1	280.71	ca. 158000	572087
	Duisburg	2006	01Oct2008	2	1	2	1	2	1	ca. 50	232.83	ca. 376500	486816
	Düsseldorf	2006	15Feb2009	1	3	1	3	1	3	13.8	217.41	36500	593682
	Essen	2006	01Oct2008	3	2[d]	3	2[d]	3	2[d]	84	210.34	ca. 291000	566862
	Cologne	2006	01Jan2008	2	4	2	4	2	4	16	405.17	130000	1023373
	Wuppertal	2006	01Feb2009	1	2[d]	1	2[d]	1	2[d]	35.3	168.39	194000	342885
Total North Rhine-W[a]				11	16	11	16	11	16	218.2	1514.85	1186000	3585705
Total (contin)[a]				38	45	38	45	38	45	793.3	3971.97	4367400	11016525
Berlin "RUBIS"[a,b]	Berlin	2004	01Jan2008	8	9	8	9	8	9	88	891.85	1100000	3375222
North Rhine-Westphalia[b]	Essen	2006	01Oct 2008	7	1	–	–	–	–	84	210.34	ca.291000	566862
Total (diffuse)[b]				15	10	8	9	8	9	172	1102.19	1391000	3942084
Total (overall)[c]				53	55	46	54	46	54	965.3	5074.16	5758400	14958609

Number of index and reference stations per gaseous component (NO$_2$, NO, NO$_x$) suitable for analysis of the first tier of LEZs until Dec 31, 2009.
[a] continuous measurements, half-hour concentration averages.
[b] diffuse samplers, longer sampling periods (over weeks).
[c] all data combined (continuous, diffuse), adjusted to concentration averages over 4-weeks periods.
[d] reference station used more than once in comparisons to index stations.

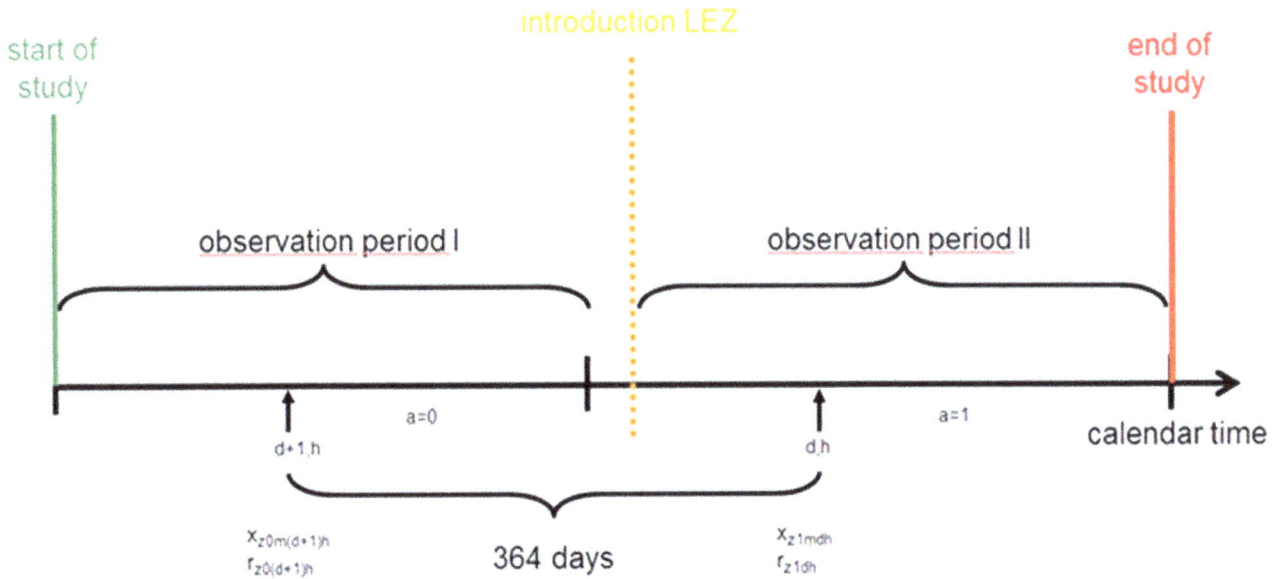

Figure 2. Index (x_{zamdh}) and reference concentration (r_{zadh}) at index measurement station m and LZ z in observation period II with active LEZ (a = 1) and in observation period I with inactive LEZ (a = 0): matched quadruplets consisting of two index measurement values and two reference measurement values. The time difference between compared measurement values at day d of the year in period 2 and d+1 of the year in period 1, always with starting time h, is not a full year but 364 days to keep the weekday constant.

m as another covariate: this allows the difference Δx_{zm} to depend on the average concentration at the station. This inclusion of x_{zm} operates again a distortion due to "regression to the mean" [54]. This phenomenon is inevitably complicating longitudinal comparisons. Baseline values that are very high due to random errors will probably not be reproduced but lower values will be measured, and this is so even if the null hypotheses of no effect is true [51,55–57]. A better strategy to correct for this potential distortion is to include $x_{z0m,cent}$, i.e., the baseline values at the index station [58,59]. Including additionally $r_{z0, cent}$ was exercised in Allison [47], p. 10. This approach allows for a flexible adjustment of the annual level bias because we get rid of the assumption of a time-invariant effect of the reference station values on the index stations values. The covariates $x_{z0m, cent}$ and $r_{z0, cent}$ are centered on the mean of the values of each measuring station so that the terms E und E_k can be interpreted without further transformations.

Since the impact of meteorological conditions is extremely relevant [e.g., 60], the following data were collected to be used in addition to the reference station data to control for distortions due to meteorological changes. We took over the height of the inversion base H in m, the wind velocity V in m/s, the amount of precipitation P in mm/h from the PAREST project for all investigated measurement stations and half hours in the follow-up period [61–65].

We extended the basic regression models to the regression model 1 approach by adjusting additionally for the change of the three meteorological variables at the index stations According to the box model of meteorology [66] the differences were calculated on the additive scale after transforming the variables into 1/H, 1/(V+0.1 m/s), and 1/(P+0.1 mm/h). The smallest unit of scale was 0.1 throughout, hence this value was used as an offset to avoid divisions by zero [48]. Differences were determined on the multiplicative scale after taking logs of these terms [48]. We adjusted for the time span (in years) between measurements

considered within a quadruplet in order to adjust for trends in concentration levels before the LEZ was introduced. In multiplicative models log of the time span was used after applying an appropriate offset [48].

In regression model 2 approach, the following time-dependent binary indicators were additionally adjusted for: period of school holidays (yes/no), period of environmental bonus paid (yes/no) and periods when trucks were not allowed to enter the area where the measurement station was located (yes/no). In Germany a bonus was paid to car owners between January 14, 2009 and November 2, 2009 if they bought a new car with a reduced exhaust emission (http://www.bafa.de/bafa/de/wirtschaftsfoerderung/umweltpraemie/index.html). These binary indicators were entered also into the extended log-linear (multiplicative) models.

This statistical approach was successfully validated in advance to the study in an analysis of simulated data from FU Berlin [67]. The simulated data was produced by the PAREST project [61,63, www.parest.de]. The major aim of this project was the identification of emission reducing strategies by simulation. Transport and distribution models were developed and applied, the so called REM-CALGRID approach [68–70]. The model was applied to the city of Munich, and simulated half-hour PM_{10} data were generated for each of the five index and three reference stations (see Figure S10 in File S1). Data of the year 2005 were simulated twice, with and without adding an LEZ effect (the value of the imprinted effect was unknown to the analyzing working group). 280,320 data lines were transferred. The simulated PM_{10} concentrations were analyzed and showed a mean value of about 21 µg/m³ at the index stations and 18 µg/m³ at the reference stations. The basic additive regression model estimated an LEZ effect of −0.130 µg/m³, the multiplicative model a relative change of −0.7%. The PAREST research report [63, www.parest.de] described the LEZ effect that was imprinted: PM_{10} mean values are reduced in Munich city by at most 0.2 µg/m³ or at most 1%.

Additive and multiplicative regression models were fitted to subsets of the data to perform sensitivity analyses: continuous measurement data, continuous measurement data collapsed to four-week averages, diffuse sampler data, pooled diffuse sampler and collapsed continuous data, always with and without excluding times with restrictions of truck traffic; quadruplets produced by index traffic stations only. The pooled continuous and diffuse sampler measurement data determined for four weeks periods was of major interest in this study because the annual average is the most critical endpoint to consider (see Introduction section) and these data cover both types of measurement data. Because annual data are generally too coarse for LEZ effect estimation, we followed-up on averages over about a month. The additive and multiplicative regression models analysing these sets of data were specified with three sets of covariables as described above. All basic models, the models evaluating continuous data collapsed to 4-week averages and the models fitted to single index stations were not used for statistical testing of the LEZ effects. Tests for effects measured by NO_2, NO, and NO_x quadruplets were not considered as independent. According to this structure, we evaluated $2*3*2*2*2 = 48$ statistical tests for each of the three endpoints. Due to this multiple testing scenario we applied an adapted significance level of $5\%/50 = 0.1\%$ ["family wise error rate", 71].

We fitted additionally explorative models that estimated the size of the LEZ effect at each index station enrolled. In addition, we estimated mean effects of the LEZs across the Federal states. The results of these exploratory analyses were mainly used for internal discussions of the project steering committee (see Acknowledgement).

All regression models used robust estimators of coefficient variances. All data analyses were performed using Stata 11 [72] on a 64-bit PC.

Results

NO_2 - continuous measurements

The basic data consisted of 6,412,864 data lines leading to 3,038,781 quadruplets of continuous NO_2 measurement (half-hour averages) from 6 Federal states and 17 LEZs with 38 index stations and 45 reference stations. Table 2 gives an overview of the distributions observed: on average, NO_2 concentrations were between $50 \mu g/m^3$ and $52 \mu g/m^3$ at the index stations and between $26 \mu g/m^3$ and $27 \mu g/m^3$ at the reference stations. The differences at the stations varied substantially in a range of hundreds of $\mu g/m^3$ upwards and downwards. A comparison of mean and median differences at index and reference stations indicated a crude LEZ effect estimate of about $-1 \mu g/m^3$. In the linear model 1 the absolute effect estimate was similar: $-1.11 \mu g/m^3$ (Table 3). The model 1 results showed a time-dependent impact of reference station data, a pronounced "regression to the mean", a clear influence of the three meteorological variables (independently from the crude adjustment by reference station data), and a downward trend of concentrations before the LEZs were introduced. The direction of impact of the meteorological variables was as expected: the smaller H, V, or P the larger the index NO2 concentrations. In linear model 2 the LEZ effect estimate was slightly more pronounced: $-1.85 \mu g/m3$. In the log-linear model 1 (multiplicative approach), the relative effect estimate was 0.979, i.e., a reduction of 2.1% was found (Table 4). The estimated impact of covariables agreed with the finding in the corresponding linear model. When applying regression model 2 the relative LEZ effect estimate was 0.961, i.e., the reduction was estimated to be 3.9%.

NO_2 - pooled continuous and diffuse sampler measurement data

6,133 data lines and 4,095 quadruplets of NO_2 pooled continuous and diffuse sampler measurement data (averaging period: four weeks) were examined from 17 LEZs with 53 index stations and 55 reference stations. A crude comparison based on the observed distributions revealed a LEZ effect of about $-0.2 \mu g/m^3$ to $-0.6 \mu g/m^3$ (Table 5). Using the linear model 1 approach the absolute effect estimate was $-0.826 \mu g/m^3$ (Table 6). The meteorological variables showed no substantial impact due to the long averaging period. Model 2 estimated the LEZ effect as $-1.73 \mu g/m^3$. The log-linear modelling led to a relative effect of 0.980 (Table 7, model 1) or 0.961 (model 2). Table S1 in File S1 provides a detailed overview of the results when fitting a series of models to analyse the NO_2 measurements. LEZ effect estimates were about $-1 \mu g/m^3$ to $-2 \mu g/m^3$ (additive models) or -2% to -4% (multiplicative models).

NO - pooled continuous and diffuse sampler measurement data

A total of 5,790 data lines from 17 LEZs with 46 index stations and 54 reference stations were available to analyse pooled continuous and diffuse sampler NO measurement data. A descriptive analysis of the 4,005 quadruplets indicated a LEZ effect of about $0 \mu g/m^3$ to $-1 \mu g/m^3$ (Table 8). Using the additive approach the absolute effect estimate was $-1.13 \mu g/m^3$ in model 1 (Table 9). When the model specification 2 was applied the LEZ effect estimate changed the sign: $+0.38 \mu g/m^3$, i.e., no reduction was indicated in this extended model type. The log-linear regression model of type 1 yielded a relative effect estimate of 0.968 (Table S2 in File S1). The direction of the estimated relative effect changed when model 2 was applied: $+1.20$.

NO_x - pooled continuous and diffuse sampler measurement data

The analysis of pooled continuous and diffuse sampler NO_x measurement data was performed using 4,005 quadruplets that originated from a set of 5,790 data lines generated by 46 index stations and 54 reference stations of 17 LEZs. According to the distributions of differences a crudely estimated LEZ effect (based on averages or medians) was present of about $-0.2 \mu g/m^3$ to $-1.3 \mu g/m^3$ (Table S3 in File S1). Adjusting for covariables in linear model 1 returned an absolute effect estimate of $-1.74 \mu g/m^3$ (Table S4 in File S1). The adjustment for further covariables (regression model 2) led to an effect estimate of $-0.89 \mu g/m^3$. When the log-linear model 1 was used (Table S5 in File S1), a relative LEZ effect of 0.976 was found. The adjustment for additional covariables (model 2) led to a change in direction: the relative effect was estimated as 1.048.

Summary of Results for NO_2, NO, and NO_x

Table 10 gives an overview of the findings for NO_2, NO, and NO_x. The mean concentration levels at the index stations were about $50 \mu g/m^3$ for NO_2 and for NO, thus, about $100 \mu g/m^3$ for NO_x. Model 1 analyses showed reductions of the concentrations after introducing the LEZs. Although small, all effect estimates were statistically significant at the 0.1% level. Model 1 estimates based on an additive structure gave compatible findings to the log-linear multiplicative approach (e.g., 2% of $50 \mu g/m^3 = 1 \mu g/m^3$). The model 1 LEZ effect estimates were similar to, but slightly more pronounced than crude LEZ effect estimates based on direct comparisons of the measurement differences at index stations and reference stations within the quadruplets while ignoring the impact

Table 2. NO$_2$: Quadruplets of continuous NO$_2$-measurements: index stations (Ind), reference stations (Ref) before (pre) and after (post) introduction of LEZ.

Statistic	Ind,pre	Ind,post	Ref,pre	Ref,post	Ind.diff.	Ref.diff
N	3038781	3038781	3038781	3038781	3038781	3038781
min	0.4	1.3	0.4	0.5	−330	−215
p5	12	11	4.0	4.0	−52	−33
p50	45.0	43.8	20.7	20.6	−1.0	0.0
mean	51.959	50.831	26.383	26.17	−1.128	−0.212
p95	115	114	68	67	49	33
max	392	436	248	434	375	317

Ind.diff and Ref.diff denote differences between index measurements and between reference measurements (negative post-pre differences indicate lower values after introduction of LEZ). Concentrations measured in μg/m^3.
N: number of quadruplets; Min: minimum, p5: 5th percentile, p50: median, mean: arithmetic average, p95: 95th percentile, max: maximum, unit:

of covariables (compare Tables 2, 5, and 8 and Table S1 in S1). All analyses point to the conclusion that on average the concentration reducing effect of LEZs was smaller than 2 μg/m^3 for each of the three components NO$_2$, NO, and NO$_x$, i.e., not higher than about 4%, when considering all investigated index stations. However, breaking down the analyses by Federal states or LEZs yielded heterogeneous estimates of effects.

The NO$_2$ analysis was based on 192 comparisons of index vs reference stations, among them were 31 index stations characterized as "background", one characterized as "industry" and 160 as "traffic" stations. We performed a sensitivity analysis by restricting the evaluation to the stations close to traffic. The additive linear type 2 model estimated an effect of −1.73 μg/m^3 at all index stations (see last line in Table S1 in S1). When the analysis only accounted for the traffic stations we got a slightly more pronounced LEZ effect estimate of −2.26 μg/m^3 (3,406 quadruplets, pooled data: four week averages). An analysis of the continuous data yielded almost the same result: −2.35 μg/m^3 (2,105,702 quadruplets, half-hour averages).

Discussion

In this study we analysed the effect of introducing the "Low Emission Zone (LEZ) of pollutant group 1" (which restricts from

entering Diesel cars of an European emission standard below Euro 2 without particulate reduction) on NO$_2$, NO, and NO$_x$ concentrations in Germany. We included as many LEZs as possible (17 out of 34 in 2009 met our inclusion criteria) into a homogeneous analysis of nitrogen oxide data measured before and after the introduction of LEZs of pollutant group 1 until the end of 2009. We used matched quadruplets of index and reference station values and analysed the changes in concentrations with fixed-effect regression models while adjusting for important covariables. We performed sensitivity analyses by applying two model structures (additive and multiplicative) with varying sets of covariables to different subsets of the data. We based our study on precisely matched quadruplets to avoid distortions and to increase validity. A potential downside of the increased validity is a loss in precision due to the reduced data set eligible for analysis. However, the loss in power was negligible in this application because P-values were small even when taking multiple testing into account [73]. The statistical approach was successfully validated in advance to the study in an analysis of simulated data from FU Berlin [67]. We checked whether the adjustment in one model that analyzed all LEZs simultaneously and assumed unknown but identical covariate coefficients was appropriate for all LEZs. To do so we evaluated each LEZ separately and performed a meta-analysis on the findings. The precision weighted mean of the effect estimates

Table 3. NO$_2$: Linear (additive) model 1 evaluating the quadruplets of continuous NO$_2$-measurements.

Ind.diff	Coef.	Std. Err.	t	p	95% Conf.	Interval
Ref.diff	0.677	0.001	623	<0.001	0.675	0.679
Ref.base	0.509	0.001	417	<0.001	0.507	0.512
Ind.base	−0.644	0.001	−845	<0.001	−0.645	−0.642
Diff 1/H	564	2.8	200	<0.001	558	569
Diff 1/V	3.12	0.025	126	<0.001	3.07	3.17
Diff 1/P	0.088	0.004	23.6	<0.001	0.081	0.095
Time.diff	−0.399	0.023	−17.7	<0.001	−0.443	−0.354
E	−1.112	0.013	−87.3	<0.001	−1.137	−1.087

Regression coefficient, robust standard errors of coefficient, t-statistic, two-sided P-value, and 95%-confidence interval of coefficient. The absolute LEZ effect estimate is given by the coefficient E in μg/m^3 (<0: concentration is lowered by LEZ).
Covariables: difference in reference stations Ref.diff in μg/m^3, centered reference baseline concentration Ref.base in μg/m^3, centered index baseline concentration Ind.base in μg/m^3, Diff 1/H = difference in 1/(height of inversion layer) in 1/m, Diff 1/V = 1/(wind velocity+0.01) in (m/s)$^{-1}$, difference in 1/P = 1/(amount of precipitation+0.01) (mm/h)$^{-1}$, centered difference zones in time Time.diff in years.

Table 4. NO_2: Log-linear (multiplicative) model 1 evaluating the quadruplets of continuous NO_2-measurements.

ln Ind.diff	Coef.	Std. Err.	t	p	95% Conf.	Interval
ln Ref.diff	0.296	0.001	582	<0.001	0.295	0.297
ln Ref.base	0.265	0.002	141	<0.001	0.262	0.269
ln Ind.base	−0.981	0.002	−412	<0.001	−0.986	−0.977
Diff ln 1/H	0.071	0.001	194	<0.001	0.070	0.071
Diff ln 1/V	0.12	0.001	301	<0.001	0.119	0.121
Diff ln 1/P	0.014	0.001	29	<0.001	0.013	0.015
ln Time.diff	−0.025	0.001	−32	<0.001	−0.027	−0.024
ln E	−0.021	0.0	−77	<0.001	−0.022	−0.021

Regression coefficient, robust standard errors of coefficient, t-statistic, two-sided P-value, and 95%-confidence interval of coefficient. The relative LEZ effect estimate is given by the coefficient E (<1: concentration is lowered by LEZ).
relative effect E = 0.979, 95% Conf. Interval = 0.979, 0.980.
Covariables, before taking logs: difference in reference stations Ref.diff in $\mu g/m^3$, centered reference baseline concentration Ref.base in $\mu g/m^3$, centered index baseline concentration Ind.base in $\mu g/m^3$, 1/H = 1/(height of inversion layer) in 1/m, 1/V = 1/(wind velocity+0.01) in $(m/s)^{-1}$, 1/P = 1/(amount of precipitation+0.01) $(mm/h)^{-1}$, centered difference in time Time.diff in years.

at all index stations (n = 192, −1.71 $\mu g/m^3$) was almost identical to the overall additive linear type 2 model (−1.73 $\mu g/m^3$, see last line in Table S1 in S1). We conclude that the fitted single model that evaluated all LEZs simultaneously was appropriate and did not suffer from an insufficient adjustment.

As an overall finding the average effect of LEZ introduction on nitrogen oxide concentrations (NO_2, NO, and $NO_x = NO_2+NO$) was not higher than 2 $\mu g/m^3$ at all index stations, i.e., not higher than about 4%. The effect was only slightly larger when we restricted the analyses to stations close to traffic. In the main analyses the coefficients describing the reductions were statistically significant on the 0.1% level, i.e., after taking multiple testing into account. We note, however, that the P-values calculated are potentially too small because autocorrelations in the data were not taken into account.

We detected a substantial heterogeneity of effects across the investigated LEZs and Federal states. However, this finding is not surprising because

- the realisation of LEZs differed between states and within states (e.g., date of introduction, covered population and area of LEZs differ (compare Table 1), some operate together with an additional restriction of van traffic)

- the degree of representativeness of monitoring stations inside the LEZs differs across LEZs (index stations: distances from centre/border of LEZ differ, used as background or hot spot stations and sometimes placed in street canyons)

- the degree of representativeness of monitoring stations outside the LEZs differs across LEZs (reference stations: distances from LEZ differ, traffic conditions differ)

- the applied measuring systems differ (continuous chemiluminescense procedure vs diffuse long-term sampling with chromatography).

The large variation of LEZ effect estimates across the LEZs should be put into perspective by considering the phenomenon of "regression-to-the-mean" [51]. Due to this phenomenon we expect that single observations with high baseline values show potentially decreasing trends – and low baseline values potentially increasing trends. This is true even under the null hypothesis of no causal LEZ effects on nitrogen oxide concentrations. "Regression-to-the-mean" has been shown to be rather pronounced in this study. Thus, the interpretation of single LEZs effect estimates is clearly limited and we will not report any details with the consent of the involved state institutions who performed the measurements (see Acknowledgements).

Table 5. NO_2: Quadruplets of pooled continuous and diffuse sampler NO_2-measurements: index stations (Ind), reference stations (Ref) before (pre) and after (post) introduction of LEZ.

Statistic	Ind,pre	Ind,post	Ref,pre	Ref,post	Ind.diff.	Ref.diff
N	4095	4095	4095	4095	4095	4095
min	13	11	6	5	−32	−23
p5	26	24	11	10	−14	−13
p50	57.31	54.23	41.04	39.78	−2.34	−1.71
mean	56.298	54.246	40.007	38.18	−2.053	−1.824
p95	84	82	66	64	12	8
max	134	136	75	76	41	25

Ind.diff and Ref.diff denote differences between index measurements and between reference measurements (negative post-pre differences indicate lower values after introduction of LEZ). Concentrations measured in $\mu g/m^3$.
N: number of quadruplets, Min: minimum, p5: 5th percentile, p50: median, mean: arithmetic average, p95: 95th percentile, max: maximum.

Table 6. NO$_2$: Linear (additive) model 1 evaluating the quadruplets of pooled continuous and diffuse sampler NO$_2$-measurements.

Ind.diff	Coef.	Std. Err.	t	p	95% Conf.	Interval
Ref.diff	0.564	0.018	31.7	<0.001	0.529	0.599
Ref.base	0.394	0.025	15.9	<0.001	0.345	0.442
Ind.base	−0.695	0.019	−36.7	<0.001	−0.732	−0.657
Diff 1/H	−65.1	20.8	−3.13	0.002	−106	−24.3
Diff 1/V	0.034	0.148	0.23	0.816	−0.256	0.325
Diff 1/P	0.009	0.028	0.30	0.764	−0.047	0.064
Time.diff	0.369	0.189	1.95	0.051	−0.001	0.739
E	−0.826	0.123	−6.71	<0.001	−1.068	−0.585

Regression coefficient, robust standard errors of coefficient, t-statistic, two-sided P-value, and 95%-confidence interval of coefficient. The absolute LEZ effect estimate is given by the coefficient E in µg/m^3 (<0: concentration is lowered by LEZ).
Covariables: difference in reference stations Ref.diff in µg/m^3, centered reference baseline concentration Ref.base in µg/m^3, centered index baseline concentration Ind.base in µg/m^3, Diff 1/H = difference in 1/(height of inversion layer) in 1/m, Diff 1/V = 1/(wind velocity+0.01) in (m/s)$^{-1}$, difference in 1/P = 1/(amount of precipitation+0.01) (mm/h)$^{-1}$, centered difference in time Time.diff in years.

Models of type 2 showed more instability and returned positive effect estimates in some situations (see Table 10). Regression model 2 included as additional variables time-dependent binary indicators for period of school holidays, period of environmental bonus paid and periods when trucks were not allowed to enter the area where the measurement station was located. In some LEZs these variables were highly correlated with the active LEZ periods so that unstable findings due to collinearities can be expected. Such collinearities can introduce a bias away from the null and may generate exaggerated negative or positive model coefficients even if the true effects are near to zero [74]. Log-linear models showed to be more sensitive to these distortions. This may indicate a less appropriate modelling of the data when assuming multiplicative effects of covariates.

There are evaluations available concerning potential effects of LEZs on NO$_2$ concentrations summarized by the German Federal Environmental Agency [7]: A total NO$_2$ reduction by 5% and a local traffic-related NO$_2$ reduction by 12% may be reached given "LEZ of pollutant group 3" so that only cars with a green sticker (Diesel vehicles of Euro 6, 5, 4 or Euro 3 with particle filter,

gasoline cars with catalytic converter) are allowed to enter the LEZ [3]. This statement is based mainly on preliminary evaluations of the Berlin LEZ data by Rauterberg-Wulff and Lutz [29]. Puls and Jäger-Ambrozewicz [75] reported for the Frankfurt LEZ and an observation period until the end of 2011 effects of less than 3% which is closer to our present findings although they also cover a period of "LEZ of pollutant group 2" after Jan1, 2010. Only cars with a yellow sticker (Diesel vehicles of Euro 3 or 4 standard or Euro 2 with particle filter, gasoline cars with catalytic converter) were allowed to enter the Frankfurt LEZ after Jan 1, 2010 [3]. Bruckmann et al. [6] reported reductions of the annual average of NO$_2$ concentrations up to 2% associated with the introduction of "LEZs of pollutant group 1" in North-Rhine Westphalia, and an absolute LEZ effect of about −1.2 µg/m^3. In Hannover no NO$_2$ reduction could be shown after introducing an "LEZ of pollutant group 1" [76]. All of these statements, however, were based on crude comparisons without sufficiently adjusting for important covariates like weather conditions, and traffic restrictions etc. Only Puls and Jäger-Ambrozewicz [75] applied a more sophisticated approach. They performed a time-series analysis and fitted

Table 7. NO$_2$: Log-linear (multiplicative) model 1 evaluating the quadruplets of pooled continuous and diffuse sampler NO$_2$-measurements.

ln Ind.diff	Coef.	Std. Err.	t	p	95% Conf.	Interval
ln Ref.diff	0.359	0.015	23.9	<0.001	0.329	0.388
ln Ref.base	0.024	0.005	4.49	<0.001	0.014	0.034
ln Ind.base	−0.039	0.004	−10.1	<0.001	−0.047	−0.032
Diff ln 1/H	−0.016	0.003	−5.32	<0.001	−0.022	−0.01
Diff ln 1/V	0.013	0.003	4.21	<0.001	0.007	0.02
Diff ln 1/P	0.011	0.003	3.68	<0.001	0.005	0.017
ln Time.diff	0.005	0.006	0.95	0.34	−0.006	0.017
ln E	−0.02	0.003	−7.82	<0.001	−0.025	−0.015

Regression coefficient, robust standard errors of coefficient, t-statistic, two-sided P-value, and 95%-confidence interval of coefficient. The relative LEZ effect estimate is given by the coefficient E (<1: concentration is lowered by LEZ).
relative effect E = 0.980, 95% Conf. Interval = 0.975, 0.985.
Covariables, before taking logs: difference in reference stations Ref.diff in µg/m^3, centered reference baseline concentration Ref.base in µg/m^3, centered index baseline concentration Ind.base in µg/m^3, 1/H = 1/(height of inversion layer) in 1/m, 1/V = 1/(wind velocity+0.01) in (m/s)$^{-1}$, 1/P = 1/(amount of precipitation+0.01) (mm/h)$^{-1}$, centered difference in time Time.diff in years.

Table 8. NO: Quadruplets of pooled continuous and diffuse sampler NO-measurements: index stations (Ind), reference stations (Ref) before (pre) and after (post) introduction of LEZ.

Statistic	Ind,pre	Ind,post	Ref,pre	Ref,post	Ind.diff.	Ref.diff
N	4005	4005	4005	4005	4005	4005
min	2.7	2.5	1.0	1.0	−86	−50
p5	4.6	4.4	2.2	2.2	−22	−21
p50	48.54	43.44	31.83	26.32	−2.59	−1.26
mean	49.479	46.373	34.153	31.025	−3.105	−3.129
p95	95	95	86	81	17	13
max	230	251	139	104	77	56

Ind.diff and Ref.diff denote differences between index measurements and between reference measurements (negative post-pre differences indicate lower values after introduction of LEZ). Concentrations measured in $\mu g/m^3$.
N: number of quadruplets, Min: minimum, p5: 5th percentile, p50: median, mean: arithmetic average, p95: 95th percentile, max: maximum.

regression models for the Frankfurt LEZ. These models', however, were not correctly specified as they did not include differences of the covariables but the absolute values only, and so they could not control for potential confounding effects although this was intended by the authors. All publications cited above reported only on individual LEZs or certain Federal states in Germany and not on the LEZ effect on the national level. Generalisations from these data are problematic because of the heterogeneous configurations of LEZs. A realistic estimate should be based on a homogeneous analytical approach covering as many LEZs and Federal states simultaneously as possible, as performed in this study.

Table 11 presents an overview of other study results published in the peer-reviewed literature on forecasted or measured LEZ effects on NO_2 concentrations.

Our results are in good accordance with the prognosis study PAREST of FU Berlin [61,63]. An extensive description of the project is available [77]. The prognoses of PAREST are comparable with our estimates at all index stations because PAREST worked with an area coarseness defined by grid square of about 1 km×1 km and, thus, cannot estimate changes at single stations. Duyzer et al. [78] studied whether monitoring station data are representative for the population living in the area and concluded that the background station data are more appropriate

to describe the impact on the citizens than the hot spot traffic stations. We conclude that the findings of PAREST and our results about the effect at all index stations should be preferred in an evaluation (not the effect estimates restricted to the traffic stations). PAREST predicted LEZ effects on NO_2 levels assuming that only cars with green stickers are allowed to enter (LEZ of pollutant level 3). For the Berlin LEZ the authors calculated a reduction of about 1 $\mu g/m^3$ to 1.3 $\mu g/m^3$ in the city centre (relative: 3% to 5%), for the Munich LEZ a reduction of 1 $\mu g/m^3$ in the city centre (relative: up to 5%), for the Ruhr area a reduction of 1 $\mu g/m^3$ to 1.7 $\mu g/m^3$ (relative: 3% to 4%). Setting the whole Ruhr area to a LEZ of pollutant level 3 lead to the prognosis of a reduction in NO_2 concentrations of 1 $\mu g/m^3$ to 2 $\mu g/m^3$ (relative: 3% to 6%). It needs to be taken into account that these prognoses by PAREST are based on the pollutant level 3 LEZ scenario. We do not expect, therefore, that our findings from this study may change relevantly if the LEZs are extended to cover larger areas or if stricter traffic restrictions are applied.

A very large LEZ was introduced in London as a congestion charging zone. However, only prognoses of the potential LEZ effect on nitrogen oxide concentrations are available. NO_x reductions between 3.8% in 2008 up to 7.3% in 2012 along roadways were predicted in a modelling scenario for the London LEZ with vehicles and buses required to meet Euro 4 standards

Table 9. NO: Linear (additive) model 1 evaluating the quadruplets of pooled continuous and diffuse sampler NO-measurements.

Ind.diff	Coef.	Std. Err.	t	p	95% Conf.	Interval
Ref.diff	0.666	0.022	30.5	<0.001	0.623	0.708
Ref.base	0.496	0.026	19.1	<0.001	0.445	0.547
Ind.base	−0.473	0.022	−21.5	<0.001	−0.516	−0.430
Diff 1/H	75.8	29.5	2.57	0.010	18.0	134
Diff 1/V	−0.571	0.216	−2.65	0.008	−0.994	−0.148
Diff 1/P	0.028	0.048	0.59	0.557	−0.066	0.123
Time.diff	0.245	0.313	0.78	0.433	−0.369	0.860
E	−1.128	0.218	−5.19	<0.001	−1.555	−0.702

Regression coefficient, robust standard errors of coefficient, t-statistic, two-sided P-value, and 95%-confidence interval of coefficient. The absolute LEZ effect estimate is given by the coefficient E in $\mu g/m^3$ (<0: concentration is lowered by LEZ).
Covariables: difference in reference stations Ref.diff in $\mu g/m^3$, centered reference baseline concentration Ref.base in $\mu g/m^3$, centered index baseline concentration Ind.base in $\mu g/m^3$, Diff 1/H = difference in 1/(height of inversion layer) in 1/m, Diff 1/V = 1/(wind velocity+0.01) in $(m/s)^{-1}$, difference in 1/P = 1/(amount of precipitation+ 0.01) $(mm/h)^{-1}$, centered difference in time Time.diff in years.

Table 10. Summarized Results on Nitrogen Oxides.

	Evaluation Period	Mean Index Concentration/$\mu g/m^3$	Model 1 Estimates		Model 2 Estimates	
			Additive Effect/$\mu g/m^3$	Multiplicative Effect/%	Additive Effect/$\mu g/m^3$	Multiplicative Estimate/%
NO$_2$	0.5 h	50	−1.1	−2.1	−1.9	−3.9
	4 weeks	55	−0.8	−1.9	−1.7	−3.9
NO	4 weeks	48	−1.1	−3.2	+0.4	+1.9
NO$_x$	4 weeks	103	−1.7	−2.4	−0.9	+4.8

Model 1 covariables: difference in reference stations Ref.diff in $\mu g/m^3$, centered reference baseline concentration Ref.base in $\mu g/m^3$, centered index baseline concentration Ind.base in $\mu g/m^3$, Diff 1/H = difference in 1/(height of inversion layer) in 1/m, Diff 1/V = 1/(wind velocity+0.01) in $(m/s)^{-1}$, difference in 1/P = 1/(amount of precipitation+0.01) $(mm/h)^{-1}$, centered difference in time Time.diff in years.

compared to current LEZ restrictions for Euro 3 vehicles [46]. The authors stated that despite of the large area of the London LEZ, the predicted changes in NO$_2$ (and PM$_{10}$) were generally small. Their modelled results stay partly in contrast to the prognoses published by Tonne et al. [79], who estimated for the London Congesting Charge Scheme a small decrease for NO$_2$ of −0.64 $\mu g/m^3$ only, corresponding to −1.1%.

Table 11. Overview of studies estimating the effect of LEZs on NO$_2$ concentrations.

Study	Country	City/Area	Intervention	NO$_2$ effect/$\mu g/m^3$	Comments
Builtjes et al. 2012, Stern 2013	Germany	Berlin	LEZ of level 3: forbidden<Euro 4	−1.3	
		Munich		−1.0	
		Ruhr Area		−1.7	
Tonne et al. 2008,	UK	London	Congestion Charging Zone: forbidden<Euro 4	−0.64 (−1.1%)	simulation study 2008
Kelly et al 2011				reduction up to −7.3%	simulation study 2011
Briggs 2008	Italy	Rome	2 LEZs:	−2.3/−3.0	simulation study: main effect due to the exclusion of Euro 0 vehicles
			Euro 0 forbidden 2 LEZs:		
Cesaroni et al. 2012	Italy	Rome	forbidden<Euro 4	−3.0/−4.1	
Boogaard et al. 2012	The Netherlands	Amsterdam	LEZ:	−4.5	analysis of measurements: crude comparisons, no covariates taken into account
		Den Bosch	trucks forbidden	(not statistically significant)	
		The Hague	<Euro 2 and Euro 3 trucks only allowed if retrofitted		
		Tilburg			
		Utrecht			
Johansson et al. 2009	Sweden	Stockholm	Road Pricing System: Vehicles travelling into and out of the charge cordon were charged for every passage during weekdays	NO$_x$:	analysis of measurements: crude comparisons, no covariates taken into account
				−0.23 (Greater Stockholm)	
				−0.81 (inner city)	
This study	Germany	17 cities	LEZ of level 1: Diesel passenger cars, trucks and buses forbidden<Euro 2 without particulate reduction system	NO$_2$, NO, NO$_x$:	analysis of measurements: construction of matched quadruplets, regression analyses with covariates, all estimated effects statistically significant at the 0.1% level
				reduction less than −2.0 (−4%)	
				[PM$_{10}$: reduction less than − 0.2 $\mu g/m^3$ (−1%)]	

The INTARESE project [80] modeled NO_2 concentration changes for both LEZs in Rome and confirmed this finding of only small additional gains by stricter traffic restrictions. The main reductions were expected to be achieved already by excluding Euro 0 cars: -2.3 $\mu g/m^3$ or -3.0 $\mu g/m^3$. If only Euro 4 cars were allowed to enter the LEZs the reductions were expected to increase only slightly to -3.0 $\mu g/m^3$ or -4.1 $\mu g/m^3$ [81].

The "Stockholm Trial" involved a road pricing system to improve the air quality and reduce traffic congestion. The test period of the trial was January 3, 2006 to July 31, 2006. Vehicles travelling into and out of the charge cordon were charged for every passage during weekdays. Annual mean contributions to total levels of nitrogen oxides from emissions from road traffic with and without charges according to the Stockholm Trial were estimated. NO_x concentrations were lowered in periods with charges, but the study showed a small decrease only: -0.23 $\mu g/m^3$ (Greater Stockholm) and -0.81 $\mu g/m^3$ (inner city) [82]. No multivariable modeling was tried.

Boogaard et al. [83] analyzed measurements of NO_2 and NO_x conducted simultaneously at eight streets, six urban background locations and four suburban background locations before (2008) and two years after implementation of an LEZ (2010) in five cities of The Netherlands (8 index stations, 4 reference stations). Index concentrations were lower in 2010 than in 2008 (NO_2: -4.5 $\mu g/m^3$, NO_x: -6.1 $\mu g/m^3$) but the differences were not statistically different. The study performed only crude comparisons and did not apply regression techniques to adjust for covariables.

The present study can be regarded as one of the most comprehensive approaches so far, analysing measurement data of nitrogen oxides concentrations in order to assess LEZ effects. The LEZ pollutant group 1 reduction effect on nitrogen oxides (NO_2, NO, and NO_x) was estimated as being no higher than 2 $\mu g/m^3$ at all index stations and index traffic stations, i.e., no higher than about 4%. This estimate based on measurement data can be rated as the most profound currently available. This result also needs to be interpreted in the light of the existing EU limit values because LEZs are often supposed to be the most effective measure that cities can take to reduce air pollution problems in their area [84]. The respective NO_2 concentration limit [24] enforced in Germany since 2010 is 40 $\mu g/m^3$ (1 year average). Values are in excess and about 69% of all German traffic stations showed annual averages higher than 40 $\mu g/m^3$ [25]. The four week averages of NO_2 concentrations at index stations after LEZ introduction were found to be 55 $\mu g/m^3$ (median and mean) or 82 $\mu g/m^3$ (95th percentile). It follows that the estimated reduction of NO_2 concentrations in the range of 2 $\mu g/m^3$ appears to be of negligible impact when the current concentration levels should be lowered to the EU limit. The same judgement seems to apply on the EU level where the NO_2 concentrations were reported to show a pronounced excess in many cities [26].

Regarding the information from the HBEFA [85] for real driving conditions in Germany, Austria and Switzerland with respect to vehicles that meet Euro 5 and 6 emission standards, no noteworthy reductions of NO_2 and NO_x immissions are to be expected until a remarkable share of vehicles with NO_x after treatment systems (Euro 5 for HD trucks and Euro 6 for passenger cars) will be on the street [85].

The Handbook of Emission Factors for Road Transport (HBEFA) was originally developed on behalf of the Environmental Protection Agencies of Germany, Switzerland and Austria. In the meantime, further countries (Sweden, Norway, France) as well as the JRC (European Research Center of the European Commission) are supporting HBEFA. HBEFA provides emission factors, i.e. the specific emission in g/km for all current vehicle categories

(PC, LDV, HDV, buses and motor cycles), each divided into different categories, for a wide variety of traffic situations (http://www.hbefa.net/e/index.html).

Interestingly, remarkable differences in NO_x and NO_2 emissions from passenger cars and light duty vehicles are documented when low test cycle emissions were compared with relatively higher NO_x/NO_2 concentrations measured along roadsides [86,87].

We analysed PM_{10} concentrations additionally [32] from 19 German LEZs. From about 2005 until the end of 2009 continuous half-hour measurement values as well as gravimetrically determined daily measurements of PM_{10} were collected. Two continuous procedures were used to measure mean PM_{10} concentrations per half-hour intervals [38,88]:

- Absorption of β-radiation (BA). The particulate matter is deposited on a filter tape and the change in β-ray transmission is measured.
- Tapered Element Oscillating Microbalance (TEOM). An inertial balance directly measures the mass collected on an exchangeable filter cartridge by monitoring the corresponding frequency changes of a tapered element.

In addition, gravimetric samplers were used to measure daily averages of PM_{10} concentrations [49,88,89]. 2,110,803 quadruplets of continuous PM_{10} and 15,735 gravimetric quadruples were identified leading to 61,169 quadruplets based on daily PM_{10} averages. The analyses showed that best LEZ effect estimates were ≤ 0.2 $\mu g/m^3$ at all index stations, i.e., the relative PM_{10} reduction $\leq 1\%$. Best estimates at all index stations near traffic (excluding urban background and industry index stations) were below 1 $\mu g/m^3$ (less than 5%, resp). Effects were smaller than predicted prior to the introduction of LEZs. Limited data (1750 quadruplets of monthly averages) were also available to estimate the effects on soot parameters (elemental carbon, organic carbon and total carbon). The average of total carbon concentrations was estimated as 13 $\mu g/m^3$ and LEZ effect estimates were about -0.55 $\mu g/m^3$ or -4.2%. For $PM_{2.5}$ only 650 quadruplets based on half-hour data and 99 quadruplets of daily concentration averages could be analyzed. The $PM_{2.5}$ concentration mean was found at 17 $\mu g/m^3$. All LEZ effect estimates on $PM_{2.5}$ were positive, i.e., no indication of reduced concentrations after the introduction of the LEZs was found.

Due to the proven marginal reduction of nitrogen oxide concentrations (NO, NO_2, NO_x), LEZ as a regulatory action cannot be seen as an efficient measure to substantially reduce ambient nitrogen oxide exposures in the cities. Beyond that, this result is in good accordance to the effectiveness of LEZs on the reduction of PM_{10}, too [32]. As predicted [33], long-term compliance problems with ambient air NO_2 concentrations should be expected even if LEZs were introduced or enlarged for the purpose of NO_2 reductions in cities.

The approach can be extended to account for other variables that are considered relevant [45]. Such data can only be used if these data are homogeneously available at all index and reference stations and are also available before and after the introduction of LEZs. Traffic density and car fleet properties are such variables of interest that do not meet the inclusion criteria: there are almost no data available in Germany to describe differences in flow of traffic and car fleet properties between index and reference stations and across time. To put this into perspective, we like to note first that changes in traffic density and car fleet properties are potentially affected by LEZs. It follows that traffic density and fleet properties should be considered as potential outcomes of LEZ introduction and not only as confounders of LEZ effects. This means that these data must not be accounted for by covariables in regression modelling even if the data were available in such a way that the inclusion criteria were met. Anyhow, authors who described changes in traffic-flow in Berlin argued against the interpretation that LEZs caused such displacements of traffic-flow from inside the LEZ to the reference stations [31]. Second, we note that the missing information on traffic density and fleet properties can be used to argue for biases in both directions. On the ones side, traffic could be displaced from the LEZ area to the reference stations outside so that the concentrations are underestimated inside but overestimated outside the LEZ, causing a potential overestimate of the LEZ effect. On the other side, if the car fleet is renewed not only inside the LEZ but also outside at the reference stations this may lead to a potential underestimate of the LEZ effect. We cannot conclude, therefore, on the direction of the potential bias.

The data analyzed in this study are the only available longitudinal measuring data to investigate the development of nitrogen oxide concentrations before and after the introduction of LEZs in Germany. We conclude that the material used can be considered as "data best available". Interpretations are limited, however, because spatial representativeness of the measuring sites can be disputed. It is unknown whether these data can be used to reliably estimate the exposures of citizens living in the LEZs. Since this is not only a problem of German measuring networks but an issue on the European level a research project was started to investigate the representativeness of measurement sites [78]. The authors concluded that measurements at the background stations are of greater importance than the data collected at the hot spots (traffic stations). Other limitations of hot spot data result from the fact that the citizens living in the LEZ area spend most of their time indoors and that indoor pollution data differ from hot spot outdoor concentrations [90,91].

Conclusions

This is the first comprehensive approach to assess effects of LEZs on NO_2, NO and NO_x concentrations with the help of measurement data on the Federal level in Germany. Reductions due to introducing LEZs of pollutant group 1 were estimated to be limited by 2 $\mu g/m^3$ (or 4%). The 4-week averages of NO_2 concentrations at index stations after LEZ introduction were found to be 55 $\mu g/m^3$ (median and mean) or 82 $\mu g/m^3$ (95th percentile). The NO_2 concentration limit [24] enforced in Germany since 2010 is 40 $\mu g/m^3$ (1 year average). Concerning the expenditure of regulations and controls which are required to introduce and operate LEZs in cities, the proven impact of LEZs on the reduction of NO_2 ambient air concentrations with at a maximum of 4% in the first phase is very small.

Supporting Information

File S1 Contains Tables S1–S4 and Figures S1–S19. Table S1: Detailed results on NO_2 - quadruplet analyses by linear (additive) log-linear (multiplicative) regression models. **Table S2:** NO: Log-linear (multiplicative) model 1 evaluating the quadruplets of pooled continuous and diffuse sampler NO-measurements. Regression coefficient, robust standard errors of coefficient, t-statistic, two-sided P-value, and 95%-confidence interval of coefficient. The relative LEZ effect estimate is given by the coefficient E (<1: concentration is lowered by LEZ). **Table S3:** NO_x: Quadruplets of pooled continuous and diffuse sampler NO_x-measurements: index stations (Ind), reference stations (Ref) before (pre) and after (post) introduction of LEZ. Ind.diff and Ref.diff denote differences between index measurements and between reference measurements (negative post-pre differences indicate

lower values after introduction of LEZ). **Table S4:** NO_x: Linear (additive) model 1 evaluating the quadruplets of pooled continuous and diffuse sampler NO_x-measurements. Regression coefficient, robust standard errors of coefficient, t-statistic, two-sided P-value, and 95%-confidence interval of coefficient. The absolute LEZ effect estimate is given by the coefficient E in $\mu g/m^3$ (<0: concentration is lowered by LEZ). **Table S5:** NO_x: Log-linear (multiplicative) model 1 evaluating the quadruplets of pooled continuous and diffuse sampler NO_x-measurements. Regression coefficient, robust standard errors of coefficient, t-statistic, two-sided P-value, and 95%-confidence interval of coefficient. The relative LEZ effect estimate is given by the coefficient E (<1: concentration is lowered by LEZ). **Figure S1:** Low emission zone **Herrenberg** (marked area), implemented in 2009-01-01 (modified from www.map24.de). One index station: 1)DEBW135 Hindenburger Straße, no NO, no NO_x. One reference station outside the low emission zone: 2)DEBW112 Gärtringen (not included in the figure since located approx. 5 km north of low emission zone). **Figure S2:** Low emission zone **Ilsfeld** (marked area), implemented in 2008-03-01 (modified from www.map24. de). One index station: 1)DEBW133 König-Wilhelm-Straße, no NO, no NO_x. One reference station outside the low emission zone: 2)DEBW034 Waiblingen (not included in the figure since located approx. 24 km south of low emission zone). **Figure S3:** Low emission zone **Karlsruhe** (marked area), implemented in 2009-01-01 (modified from www.map24.de). One index station: 1)DEBW126 Kriegsstraße, no NO_2, no NO_x. Two reference stations outside the low emission zone: 2)DEBW001 Karlsruhe-Mitte 3)DEBW004 Eggenstein (not included in the figure since located approx. 6 km north of low emission zone). **Figure S4:** Low emission zone **Ludwigsburg** (marked area), implemented in 2008-03-01 (modified from www.map24.de). Two index stations: 1)DEBW024 Weimar-/Schweizerstraße 2)DEBW017 Friedrichstraße. One reference station outside the low emission zone: 3)DEBW034 Waiblingen (not included in the figure since located approx. 7 km south east of low emission zone). **Figure S5:** Low emission zone **Mannheim** (marked area), implemented in 2008-03-01 (modified from www.map24.de). Two index stations: 1)DEBW006 Mannheim-Mitte 2)DEBW098 Friedrichsring U2. Two reference stations outside the low emission zone: 3)DEBW005 Mannheim Nord (not included in the figure since located approx. 4 km north of low emission zone) 4)DEBW007 Mannheim-Süd (not included in the figure since located approx. 5 km south of low emission zone). **Figure S6:** Low emission zone **Reutlingen** (marked area), implemented in 2008-03-01 (modified from www.map24.de). One index station: 1)DEBW027 Ebertstraße. Two reference stations outside the low emission zone: 2)DEBW042 Bernhausen (not included in the figure since located approx. 20 km north of low emission zone) 3)DEBW117 Gärtringen (not included in the figure since located approx. 28 km west of low emission zone). **Figure S7:** Low emission zone **Stuttgart** (marked area), implemented in 2008-03-01 (modified from www.map24.de). Six index stations: 1)DEBW011 Zuffenhausen 2)DEBW013 Seuberstraße 3)DEBW099 Arnulf-Klett-Platz 4)DEBW116 Hohenheimer Straße 5)DEBW118 Am Neckartor 6)DEBW134 Waiblinger Straße. Two reference stations outside the low emission zone: 7)DEBW034 Waiblingen 8)DEBW042 Bernhausen (not included in the figure since located approx. 2 km south of low emission zone). **Figure S8:** Low emission zone **Tübingen** (marked area), implemented in 2008-03-01 (modified from www.map24.de). One index station: 1)DEBW107 Derendingerstraße. One reference station outside the low emission zone: 2)DEBW112 Gärtringen (not included in the figure since located approx. 15 km north west of low emission zone). **Figure S9:** Low emission zone **Augsburg** (marked area), implemented in 2009-07-01 (modified from www.map24.de). Three index stations: 1)DEBY007

Bourges-Platz 2)DEBY110 Karlstraße 3)DEBY006 Königsplatz. One reference station outside the low emission zone: 4)DEBY099 LfU (not included in the figure since located approx. 3 km south of low emission zone). **Figure S10:** Low emission zone **Munich** (marked area), implemented in 2008-10-01 (modified from www.map24.de). Five index stations: 1)DEBY037 Stachus 2)DEBY039 Lothstraße 3)DEBY085 Luise-Kiesselbach-Platz 4)DEBY114 Prinzregentenstraße e 5)DEBY115 Landshuter Allee. Three reference stations outside the low emission zone: 6)DEBY043 Moosach, no PM_{10} 7)DEBY089Johanneskirchen 8)DEBY109 Andechs/Rothenfeld (not included in the figure since located approx. 27 km south west of low emission zone).

Figure S11a: Low emission zone **Berlin Blume-Messnetz** (marked area), implemented in 2008-01-01 (modified from www. map24.de).Five index stations: 1)DEBE018 B Schöneberg-Belziger Straße 2)DEBE034 B Neukölln-Nansenstraße 3)DEBE064 B Neukölln-Karl-Marx-Straße 76 4)DEBE065 B Friedrichshain-Frankfurter Allee 5)DEBE067 B Hardenbergplatz. Nine reference stations outside the low emission zone: 6)DEBE061 B Steglitz-Schildhornstraße 7)DEBE062 B Frohnau, Funkturm (not included in the figure since located approx. 13 km north of low emission zone) 8)DEBE063 B Neukölln-Silbersteinstraße) 9)DEBE066 B Karlshorst-Rheingoldstraße, no PM_{10} (not included in the figure since located approx. 5 km east of low emission zone) 10)DEBE010 B Wedding-Amrumer Straße 11)DEBE027 B Marienfelde-Schichauweg (not included in the figure since located approx. 8 km south of low emission zone) 12)DEBE032 B Grunewald (not included in the figure since located approx. 4 km south west of low emission zone) 13)DEBE051 B Buch (not included in the figure since located approx. 12 km north east of low emission zone) 14)DEBE056 B Friedrichshagen (not included in the figure since located approx. 14 km south east of low emission zone). **Figure S11b:** Low emission zone **Berlin RUBIS-Messnetz** (marked area), implemented in 2008-01-01 (modified from www.map24.de). Ten index stations: 1)DEBE530 Hauptstraße 30 2)DEBE504 Beusselstraße 66 3)DEBE537 Alt Moabit 63 4)DEBE545 Sonnenallee 68 5)DEBE547 Landsberger Allee 6–8 6)DEBE517 Neukölln-Nansenstraße 7)DEBE519 Friedrichshain-Frankfurter Allee 8)DEBE555 Herrmannplatz Laterne 21 9)DEBE562 Friedrichstraße Laterne 156 10)DEBE525 Leipziger Straße 32. Twelve reference stations outside the low emission zone:11) DEBE501 Berliner Allee 118 12)DEBE577 Buch, no NO, no NO_x (not included in the figure since located approx. 12 km north of low emission zone) 13)DEBE507 Grünauer Straße 4 (not included in the figure since located approx. 9 km south east of low emission zone) 14)DEBE539 Schloßstraße 29 15)DEBE542 Tempelhofer Damm 148 16)DEBE513 Spreestraße 2 (not included in the figure since located approx. 5 km south east of low emission zone) 17)DEBE514 Alt Friedrichsfelde 8a (not included in the figure since located approx. 3 km east of low emission zone) 18)DEBE521 Steglitz-Schildhornstraße 19)DEBE559 Buschkrugallee Laterne 3 20)DEBE522 Neukölln-Silbersteinstraße1 21)DEBE573 Badstraße 22)DEBE576 Spandau, Klosterstraße 12 (not included in the figure since located approx. 6 km west of low emission zone). **Figure S12:** Low emission zone **Frankfurt a.M.** (marked area), implemented in 2008-10-01 (modified from www.map24.de). One index station: 1)DEHE041 Frankfurt-Friedb.Ldstr. Three reference stations outside the low emission zone: 2)DEHE008 Frankfurt-Ost 3)DEHE011 Hanau (not included in the figure since located approx. 13 km east of low emission zone) 4)DEHE005 Frankfurt-Höchst. **Figure S13:** Low emission zone **Hannover** (marked area), implemented in 2008-01-01 (modified from www.map24.de). One index station: 1)DENI048 Hannover Verkehr. Four reference stations outside the low emission zone: 2)DENI054 Hannover 3)DENI011 Braunschweig, Broizemer Steinberg (not included in the figure since located approx. 49 km east of low emission zone) 4)DENI041 Weserbergland/Rinteln, Brugfeldsweide (not included in the figure

since located approx. 48 km south west of low emission zone) 5)DENI052 Allertal/Walsrode, Auf dem Kamp 8 (not included in the figure since located approx. 47 km north of low emission zone). **Figure S14:** Low emission zone **Dortmund** (marked area), implemented in 2008-10-01, but Brackelerstr. 2008-01-01 (modified from www.map24.de). Four index stations: 1)DENW101 Steinstraße 2)DENW136 Brackeler Straße 3)DENW184 Westfalendamm 190, no NO, no NO_x, no PM_{10} 4)DENW185 Rheinlanddamm 5–7, no NO, no NO_x, no PM_{10}. Four reference stations outside the low emission zone: 5)DENW002 Datteln-Hagem (not included in the figure since located approx. 15 km north west of low emission zone) 6)DENW008 Do-Eving 7)DENW029 Hattingen, An der Becke (not included in the figure since located approx. 19 km south west of low emission zone) 8)DENW179 Schwerte (not included in the figure since located approx. 8 km south of low emission zone). **Figure S15:** Low emission zone **Duisburg** (marked area), implemented in 2008-10-01 (modified from www.map24.de).Three index stations: 1)DENW034 Duisburg-Walsum 2)DENW040 Duisburg-Buchholz 3)DENW112 Kardinal-Galen-Straße. One reference station outside the low emission zone: 4)DENW038 45476 Mühlheim, Neustadtstraße (not included in the figure since located approx. 5 km east of low emission zone). **Figure S16:** Low emission zone **Düsseldorf** (marked area), implemented in 2009-02-15 (modified from www.map24.de). Two index stations: 1)DENW082 Corneliusstraße 2)DENW216 Düsseldorf-Bilk, no NO, no NO_x, no PM_{10}. Four reference stations outside the low emission zone: 3)DENW042 Krefeld-Linn (not included in the figure since located approx. 14 km north west of low emission zone) 4)DENW071 Düsseldorf-Lörick (not included in the figure since located approx. 3 km west of low emission zone) 5)DENW078 Ratingen-Tiefenbroich (not included in the figure since located approx. 6 km north east of low emission zone) 6)DENW116 Krefeld Hafen (not included in the figure since located approx. 12 km north west of low emission zone). **Figure S17:** Low emission zone **Essen** (marked area), implemented in 2008-10-01 (modified from www.map24.de). Eight index stations: 1)DENW043Ost Steeler Straße 2)DENW134 Gladbecker Straße 3)DENW135 Hombrucher Straße 4)DENW161 Alfredstraße 9/11, no NO, no NO_x, no PM_{10} 5)DENW168 Gladbecker Straße 245, no NO, no NO_x, no PM_{10} 6)DENW169 In der Baumschule 7, no NO, no NO_x, no PM_{10} 7)DENW171 Hombrucherstraße 21/23, no NO, no NO_x, no PM_{10} 8)DENW215 Hausackerstraße 11, no NO, no NO_x, no PM_{10}. Three reference stations outside the low emission zone: 9)DENW024 Essen-Vogelheim 10)DENW029 Hattingen-Blankenstein (not included in the figure since located approx. 10 km south east of low emission zone), 11) DENW162 Brückstraße 29, no NO, no NO_x, no PM_{10} (not included in the figure since located approx. 4 km south of low emission zone). **Figure S18:** Low emission zone **Cologne** (marked area), implemented in 2008-01-01 (modified from www.map24.de). Seven index stations: 1)DENW148 Justinianstraße 13–15, no NO, no NO_x, no PM_{10} 2)DENW151 Neumarkt 25, no NO, no NO_x, no PM_{10} 3)DENW153 Tunisstraße/Elstergasse, no NO, no NO_x, no PM_{10} 4)DENW164 Hohenstaufenring 57A, no NO, no NO_x, no PM_{10} 5)DENW198 Gereonsdriesch 21, no NO, no NO_x, no PM_{10} 6)DENW211 Clevischer Ring 3 7)DENW212 Turiner Straße. Four reference stations outside the low emission zone: 8)DENW053 Cologne-Chorweiler (not included in the figure since located approx. 9 km north west of low emission zone) 9)DENW058 Hürth (not included in the figure since located approx. 7 km south west of low emission zone), 10)DENW059 Cologne-Rodenkirchen (not included in the figure since located approx. 4 km south of low emission zone), 11)DENW079 Leverkusen-Manfort (not included in the figure since located approx. 7 km north of low emission zone). **Figure S19:** Low emission zone **Wuppertal** (marked area), implemented in 2009-02-15 (modified from www.map24.de). Two index stations: 1)DENW114 Wuppertal-Langerfeld, no NO, no NO_x 2)DENW189 Wuppertal

Gathe. Two reference stations outside the low emission zone: 3)DENW029 Hattingen-Blankenstein (not included in the figure since located approx. 13 km north of low emission zone) 4)DENW080 Solingen-Wald (not included in the figure since located approx. 5 km south west of low emission zone).

Acknowledgments

We thank the project steering committee and the following state authorities for the intense and helpful discussion and the supply of the raw data. In specific: Peter Bruckmann and Reinhold Beier, Landesamt für Natur-, Umwelt- und Verbraucherschutz NRW, Abteilung 4, Luftqualität, Geräusche, Erschütterungen, Strahlenschutz; Stefan Jacobi and Wieslawa Stec-Lazaj, Hessisches Landesamt für Umwelt und Geologie (HLUG), Abteilung I Immissions- und Strahlenschutz; Michael Köster and Andreas Hainsch, Staatliches Gewerbeaufsichtsamt Hildesheim, Abteilung 4 -

Zentrale Unterstützungsstelle für Luftreinhaltung und Gefahrstoffe (ZUS LG); Martin Lutz and Arnold Kettschau, Berliner Senatsverwaltung für Gesundheit, Umwelt und Verbraucherschutz, Abteilung III Umweltpolitik, Referat III D Immissionsschutz; Heinz Ott, Bayerisches Landesamt für Umwelt Abteilung 2, Referat 2, 4 Luftgütemessungen Südbayern, Luftreinhaltung beim Verkehr; Werner Scholz and Christiane Lutz-Holzhauer, Landesanstalt für Umwelt, Messungen und Naturschutz Baden-Württemberg (LUBW), Referat 33 Luftqualität; Ralf Wehrse and Jan Osmers, Bremer Senator für Umwelt, Bau, Verkehr und Europa Fachbereich Umwelt, Abteilung 2, Umweltwirtschafts-, Klima- und Ressourcenschutz, Referat 22 Immissionsschutz.

Author Contributions

Analyzed the data: PM. Contributed reagents/materials/analysis tools: PM. Contributed to the writing of the manuscript: PM DG MS. Data Organization: MS DG PM. Health Aspects: MS DG.

References

1. LEZEN (2013) Low emission zone in Europe network. Available from: http://www.lowemissionzones.eu. Accessed 2014 July 8.
2. EC (2007) Regulation (EC) No 715/2007on type approval of motor vehicles with respect to emissions from light passenger and commercial vehicles (Euro 5 and Euro 6) and on access to vehicle repair and maintenance information. 1–16. Available from: http://eur-lex.europa.eu/LexUriServ/LexUriServ.do?uri=OJ:L:2007:171:0001:0001:EN:PDF. Accessed 2014 July 8.
3. Umweltbundesamt (2014) Umweltzonen in Deutschland. Available from: http://www.umweltbundesamt.de/themen/luft/luftschadstoffe/feinstaub/umweltzonen-in-deutschland. Accessed 2014 July 8.
4. Colvile RN, Hutchinson EJ, Mindell JS, Warren RF (2001) The transport sector as a source of air pollution. Atmospheric Environment 35: 1537–1565.
5. Bruckmann P, Lutz M (2010) Verbessern Umweltzonen die Luftqualität? In: Verband der Automobilindustrie (VDA), editor. 12. Technischer Kongress, 24. und 25. März. Forum am Schlosspark, Ludwigsburg: Henrich Druck+Medien GmbH. pp. 299–311.
6. Bruckmann P, Wurzler S, Brandt A, Vogt K (2011) Erfahrungen mit Umweltzonen in Nordrhein-Westfalen. In: Bundesamt für Strahlenschutz, Bundesinstitut für Risikobewertung, Robert Koch-Institut, Umweltbundesamt, editor. UMID Umwelt und Mensch - Informationsdienst. Berlin. pp. 27–33.
7. Kacsóh L (2011) Umweltzonen in Europa und in Deutschland. In: Bundesamt für Strahlenschutz, Bundesinstitut für Risikobewertung, Robert Koch-Institut, Umweltbundesamt, editor. UMID Umwelt und Mensch - Informationsdienst. Berlin. pp. 5–10.
8. Morfeld P, Keil U, Spallek M (2013) The European "Year of the Air": fact, fake or vision? Archives of Toxicology 87: 2051–2055.
9. Turner MC, Krewski D, Pope CA, Chen Y, Gapstur SM, et al. (2011) Long-term ambient fine particulate matter air pollution and lung cancer in a large cohort of never-smokers. American Journal of Respiratory and Critical Care Medicine 184: 1374–1381.
10. Hoek G, Krishnan RM, Beelen R, Peters A, Ostro B, et al. (2013) Long-term air pollution exposure and cardio- respiratory mortality: a review. Environmental Health 12: 1–15.
11. Hystad P, Demers PA, Johnson KC, Carpiano RM, Brauer M (2013) Long-term residential exposure to air pollution and lung cancer risk. Epidemiology 24: 762–772.
12. Raaschou-Nielsen O, Andersen ZJ, Beelen R, Samoli E, Stafoggia M, et al. (2013) Air pollution and lung cancer incidence in 17 European cohorts: prospective analyses from the European Study of Cohorts for Air Pollution Effects (ESCAPE). The Lancet Oncology 14: 813–822.
13. Schuster C (2009) Umweltzonen: Nutzen weiterhin umstritten. Dtsch Arztebl 106: A228.
14. Eikmann T, Herr C (2009) Ist die Einführung von Umweltzonen tatsächlich eine sinnvolle Maßnahme zum Schutz der Gesundheit der Bevölkerung? Umweltmed Forsch Prax 14: 125–126.
15. Friedrich B (2008) Umweltzonen. Straßenverkehrstechnik 11: 673.
16. Zellner R, Kuhlbusch TAJ, Diegmann V, Herrmann H, Kasper M, et al. (2009) Feinstäube und Umweltzonen. Available from: Available from: www.processnet.org/dechema_media/Downloads/Positionspapiere/Stellungnahme__Feinstaeube.pdf. Accessed 2014 July 8.
17. Downs SH, Schindler C, Liu LJ, Keidel D, Bayer-Oglesby L, et al. (2007) Reduced exposure to PM10 and attenuated age-related decline in lung function. The New England Journal of Medicine 357: 2338–2347.
18. Gauderman WJ, Avol E, Gilliland F, Vora H, Thomas D, et al. (2004) The effect of air pollution on lung development from 10 to 18 years of age. The New England Journal of Medicine 351: 1057–1067.
19. Lim SS, Vos T, Flaxman AD, Danaei G, Shibuya K, et al. (2013) A comparative risk assessment of burden of disease and injury attributable to 67 risk factors and risk factor clusters in 21 regions, 1990–2010: A systematic analysis for the Global Burden of Disease Study 2010. The Lancet 380: 2224–2260.
20. Nawrot TS, Perez L, Kunzli N, Munters E, Nemery B (2011) Public health importance of triggers of myocardial infarction: a comparative risk assessment. The Lancet 377: 732–740.
21. Samet JM, Dominici F, Curriero FC, Coursac I, Zeger SL (2000) Fine particulate air pollution and mortality in 20 U.S. cities, 1987–1994. The New England Journal of Medicine 343: 1742–1749.
22. Latza U, Gerdes S, Baur X (2009) Effects of nitrogen dioxide on human health: systematic review of experimental and epidemiological studies conducted between 2002 and 2006. International Journal of Hygiene and Environmental Health 212: 271–287.
23. EC (2008) Directive 2008/50/EC of the European Parliament and of the councilof 21 May 2008 on ambient air quality and cleaner air for Europe. 1–44. Available from: http://eur-lex.europa.eu/LexUriServ/LexUriServ.do?uri=OJ:L:2008:152:0001:0044:EN:PDF. Accessed 2014 July 8.
24. EC (1999) Council Directive 1999/30/EC of 22 April 1999 relating to limit values for sulphur dioxide, nitrogen dioxide and oxides of nitrogen, particulate matter and lead in ambient air. 41–60. Available from: http://eur-lex.europa.eu/LexUriServ.do?uri=OJ:L:1999:163:0041:0060:EN:PDF. Accessed 2014 July 8.
25. Umweltbundesamt (2012) Luftqualität 2011 - Feinstaubepisoden prägten das Bild. Dessau-Roßlau. Available from: www.umweltbundesamt.de/uba-info-medien/4211.html. Accessed 2014 July 8.
26. Giannouli M, Kalognomou E-A, Mellios G, Moussiopoulos N, Samaras Z, et al. (2011) Impact of European emission control strategies on urban and local air quality. Atmospheric Environment 45: 4753–4762.
27. European Environment Agency (2014) Indicator: Exceedance 492 of air quality limit values in urban areas. Copenhagen, Denmark: European Environment Agency. Available from: http://www.eea.europa.eu/data-and-maps/indicators/exceedance-of-air-quality-limit-1/exceedance-of-air-quality-limit-5. Accessed 2014 July 16.
28. European Environment Agency (2014) AirBase - The European air Quality database. Copenhagen, Denmark. Available from: http://acm.eionet.europa.eu/databases/airbase. Accessed 2014 July 8 http://www.eea.europa.eu/data-and-maps/data/airbase-the-european-air-quality-database-8. Accessed 2014 July 8.
29. Rauterberg-Wulff A, Lutz M (2011) Wirkungsuntersuchungen zur Umweltzone Berlin. In: Bundesamt für Strahlenschutz, Bundesinstitut für Risikobewertung, Robert Koch-Institut, Umweltbundesamt, editor. UMID Umwelt und Mensch - Informationsdienst. Berlin. pp. 11–18.
30. Lutz M, Rauterberg-Wulff A (2010) Berlin's low emission zone - top or flop? 14th ETH Conference on Combustion Generated Particles.
31. Lutz M, Rauterberg-Wulff A (2009) Ein Jahr Umweltzone Berlin: Wirkungsuntersuchungen. Berlin. 1–31. Available from: www.berlin.de/sen/umwelt/luftqualitaet/de/luftreinhalteplan/download/umweltzone_1jahr_bericht.pdf. Accessed 2014 July 8.
32. Morfeld P, Groneberg D, Spallek M (2014) Effectiveness of low emission zones: Analysis of the changes in fine dust concentrations (PM10) in 19 German cities. Pneumologie 68: 173–186.
33. Vogt R, Kessler C, Schneider C (2010) Städtische NO2 Luftqualität: Quellenanalyse und zukünftige Entwicklung. In: Verband der Automobilindustrie (VDA), editor. 12 Technischer Kongress 2010, 24 und 25 März. Forum am Schlosspark, Ludwigsburg: Henrich Druck+Medien GmbH. pp. 287–297.
34. US EPA (2008) Integrated science assessment for oxides of nitrogen - health criteria (final report). EPA/600/R-08/071. 260. Available from: http://www.epa.gov/ord/htm/whatsnew.htm. Accessed 2014 July 8.
35. EC (2001) Directive 2001/81/EC of the European Parliament and of the council of 23 October 2001 on national emission ceilings for certain atmospheric pollutants. 22–30. Available from: http://eur-lex.europa.eu/LexUriServ/LexUriServ.do?uri=OJ:L:2001:309:0022:0030:EN:PDF. Accessed 2014 July 8.

36. Atkinson RW, Barratt B, Armstrong B, Anderson HR, Beevers SD, et al. (2009) The impact of the congestion charging scheme on ambient air pollution concentrations in London. Atmospheric Environment 43: 5493–5500.

37. Bruckmann P, Brandt A, Wurzler S, Vogt K (2011) Verbessern Umweltzonen die Luftqualität? Neue Entwicklungen bei der Messung und Beurteilung der Luftqualitaet Fachtagung der Kommission Reinhaltung der Luft im VDI und DIN-Normenausschuss KRdL, Baden-Baden, 11–12 Mai. pp. 3–24.

38. Umweltbundesamt (2014) Stationsdatenbank des Umweltbundesamtes. Available from: http://www.env-it.de/stationen/public/stationList.do. Accessed 2014 July 8.

39. Pfeffer U (2010) Messtechnik für Stickstoffdioxid (NO₂). In: Kommission Reinhaltung der Luft im VDI und DIN-Normenausschuss KRdL, editor. Stickstoffdioxid und Partikel (PM₂.₅/PM₁₀). pp. 113–122.

40. Umweltbundesamt (2014) Luft und Luftreinhaltung. Available from: http://www.umweltbundesamt.de/themen/luft/messenbeobachtenueberwachen/messgeraete-messverfahren. Accessed 2014 July 8.

41. Pfeffer U, Bier R, Zang T (2006) Measurements of nitrogen dioxide with diffusive samplers at traffic-related sites in North Rhine-Westphalia (Germany). Gefahrstoffe, Reinhaltung der Luft 1/2: 38–44.

42. Pfeffer U, Zang T, Rumpf E-M, Zang S (2010) Calibration of diffusive samplers for nitrogen dioxide using the reference method – Evaluation of measurement uncertainty. Gefahrstoffe, Reinhaltung der Luft 11/12: 500–506.

43. DIN EN 13528-3 Außenluftqualität (2004) Passivsammler zur Bestimmung der Konzentrationen von Gasen und Dämpfen - Teil 3: Anleitung zur Auswahlt, Anwendung und Handhabung. Berlin: Beuth.

44. Rothman KJ, Greenland S, Lash TL (2008) Modern Epidemiology. 3. ed. Philadelphia: Lippincott Williams & Wilkins.

45. Morfeld P, Spallek M, Groneberg D (2011) Zur Wirksamkeit von Umweltzonen: Design einer Studie zur Ermittlung der Schadstoffkonzentrationsänderung für Staubpartikel (PM₁₀) und andere Größen durch Einführung von Umweltzonen in 20 deutschen Städten. Zentralblatt für Arbeitsmedizin, Arbeitsschutz und Ergonomie 61: 148–165.

46. Kelly F, Armstrong B, Atkinson R, Anderson HR, Barratt B, et al. (2011) The London low emission zone baseline study. Health Effects Institut 163. 1–96. Available from: http://pubs.healtheffects.org/view.php?id = 366. Accessed 2014 July 8

47. Allison PD (2009) Fixed effects regression models. Los Angeles: SAGE.

48. Royston P, Sauerbrei W (2008) Multivariable model-buildung. Chichester, England: John Wiley & Sons Inc. 1–303 p.

49. Lenschow P, Abraham HJ, Kutzner K, Lutz M, Preuß JD, et al. (2001) Some ideas about the sorces of PM10. Atmospheric Environment 35: S23–S33.

50. Hoek G, Meliefste K, Cyrys J, Lewné M, Bellander T, et al. (2002) Spatial variability of fine particle concentrations in three European areas. Atmospheric Environment 36: 4077–4088.

51. Senn S (1997) Editorial: regression to the mean. Statistical Methods in Medical Research 6: 99–102.

52. Altman DG, Bland AE (1983) Measurement in medicine: the analysis of method comparison studies. The Statistician 32: 307–317.

53. Bland JM, Altman DG (1986) Statistical methods for assessing agreement between two methods of clinical measurement. The Lancet 1: 307–310.

54. Gill JS, Zezulka AV, Beevers DG, Davies P (1985) Relation between initial blood pressure and its fall with treatment. The Lancet 1: 567–569.

55. Bland JM, Altman DG (1994) Regression towards the mean. British Medical Journal 308: 1499.

56. Bland JM, Altman DG (1994) Some examples of regression towards the mean. British Medical Journal 309: 780.

57. Stigler SM (1997) Regression towards the mean, historically considered. Statistical Methods in Medical Research 6: 103–114.

58. Barnett AG, van der Pols JC, Dobson AJ (2005) Regression to the mean: what it is and how to deal with it. International Journal of Epidemiology 34: 215–220.

59. Twisk JWR (2004) Applied longitudinal data analysis for epidemiology. Cambridge: Cambridge University Press. 62–77 p.

60. Klingner M, Sähn E, Anke K, Holst T, Rost J, et al. (2006) Reduktionspotenziale verkehrsbeschränkender Maßnahmen in Bezug zu meteorologisch bedingten Schwankungen der PM₁₀- und NOₓ-Immissionen. Gefahrstoffe, Reinhaltung der Luft 66: 326–334.

61. Builtjes P, Jörß W, Theloke J, Thiruchittampalam B, van der Gon HD, et al. (2012) Strategien zur Verminderung der Feinstaubbelastung. 1–160. Available from: http://www.umweltbundesamt.de/sites/default/files/medien/461/publikationen/4268.pdf. Accessed 2014 July 8

62. Reimer E, Scherer B (1992) An Operational Meteorological Diagnostic System for Regional Air Pollution Analysis and Long Term Modeling. In: Dop H, Kallos G, editors. Air Pollution Modeling and Its Application IX: Springer US. pp. 565–572.

63. Stern R (2013) Anwendung des REM-CALGRID-Modells auf die Ballungsräume Berlin, München und Ruhrgebiet. Berlin: Freie Universität Berlin, Institut für Meteorologie. Tropophärische Umweltforschung. 67/2013. 1–95. Available from: http://www.umweltbundesamt.de/sites/default/files/medien/461/publikationen/texte_67_2013_appelhans_m14_komplett_0.pdf. Accessed 2014 July 8.

64. Kerschbaumer A, Reimer E (2003) Erstellung der Meteorologischen Eingangsdaten für das REM/Calgrid-Modell: Modellregion Berlin-Brandenburg. Abschlussbericht zum UBA-Forschungsvorhaben 29943246. Freie Universität

65. Kerschbaumer A (2010) Abhängigkeit von RCG-Simulationen von unterschiedlichen meteorologischen Treibern. Forschungs-Teilbericht an das Umweltbundesamt, im Rahmen des PAREST-Vorhabens: FKZ 206 43 200/1 "Strategien zur Verminderung der Feinstaubbelastung". Institut für Meteorologie der Freien Universität Berlin. Available from: http://www.umweltbundesamt.de/sites/default/files/medien/461/publikationen/texte_55_2013_appelhans_m02_komplett_0_0.pdf. Accessed 2014 July 17.

66. Graedel TE, Crutzen PJ (1994) Chemie der Atmosphäre. Heidelberg: Spektrum Akademischer Verlag.

67. Morfeld P, Stern R, Builtjes P, Groneberg DA, Spallek M (2013) Einrichtung einer Umweltzone und ihre Wirksamkeit auf die PM₁₀-Feinstaubkonzentration – eine Pilotanalyse am Beispiel München. Zentralblatt für Arbeitsmedizin, Arbeitsschutz und Ergonomie 63: 104–115.

68. Stern R (2003) Entwicklung und Anwendung des chemischen Transportmodells REM/CALGRID. Abschlussbericht zum Forschungs- und Entwicklungsvorhaben 298 41 252 des Umweltbundesamts "Modellierung und Prüfung von Strategien zur Verminderung der Belastung durch Ozon". Available from: http://www.umweltbundesamt.de/sites/default/files/medien/publikation/long/3604.pdf. Accessed 2014 July 17.

69. Stern R (2004) Großräumige PM₁₀-Ausbreitungsmodellierung: Abschätzung der gegenwärtigen Immissionsbelastung in Europa und Prognose bis 2010. KRdL-Experten-Forum "Staub und Staubinhaltsstoffe", 2004-11-11/10. VDI-KRdL-Schriftenreihe 33.

70. Stern R (2004) Weitere Entwicklung und Anwendung des chemischen Transportmodells REM-CALGRID für die bundeseinheitliche Umsetzung der EU-Rahmenrichtlinie Luftqualität und ihrer Tochterrichtlinien. Abschlussbericht zum FuE-Vorhaben 201 43 250 des Umweltbundesamts "Anwendung modellgestützter Beurteilungssysteme für die bundeseinheitliche Umsetzung der EU-Rahmenrichtlinie Luftqualität und ihrer Tochterrichtlinien". Available from: http://www.umweltbundesamt.de/sites/default/files/medien/publikation/long/3610.pdf. Accessed 2014 July 17.

71. Benjamini Y, Hochberg Y (1995) Controlling the false discovery rate: a practical and powerful approach to multiple testing. Journal of the Royal Statistical Society, Series B 57: 289–300.

72. StataCorp (2009) Stata: Release 11. Statistical Software. College Station, TX: StataCorp LP.

73. Vickers AJ, Altman DG (2001) Statistics notes: Analysing controlled trials with baseline and follow up measurements. BMJ 323: 1123–1124.

74. Neter J, Wasserman W, Kutner MH (1985) Applied linear statistical models. Regression, Analysis of variance and experimental designs. Homewood, Illinois: Richard Dr. Irwin.

75. Puls T, Jäger-Ambrozewicz (2012) Die Auswirkungen der Umweltzone in Frankfurt auf die NO₂-Immissionen. Köln: Institut der deutschen Wirtschaft Köln. pp. 1–18.

76. ZUS LG (2010) Bewertung der Auswirkungen der Umweltzone Hannover auf Basis von Messdaten. Hildesheim: Zentrale Unterstützungsstelle Luftreinhaltung und Gefahrstoffe - Dezernat 42. 15. Available from: http://www.umwelt.niedersachsen.de/download/48880. Accessed 2014 July 8.

77. Umweltbundesamt (2014) Mediendatenbank. Available from: http://www.umweltbundesamt.de/publikationen?keys = PAREST&topic = All&series = All&sort_bef_combine = field_date_monthly_value+DESC. Accessed 2014 July 17.

78. Duyzer J, van den Hout D, Zandveld P, van Ratingen S (2013) Representativeness of air quality monitoring stations. TNO Research Report 2013 R11055. TNO Utrecht (in print).

79. Tonne C, Beevers S, Armstrong B, Kelly F, Wilkinson P (2008) Air pollution and mortality benefits of the London Congestion Charge: spatial and socioeconomic inequalities. Occupational and Environmental Medicine 65: 620–627.

80. Briggs DJ (2008) A framework for integrated environmental health impact assessment of systemic risks. Environmental Health 7: 61–77.

81. Cesaroni G, Boogaard H, Jonkers S, Porta D, Badaloni C, et al. (2012) Health benefits of traffic-related air pollution reduction in different socioeconomic groups: the effect of low-emission zoning in Rome. Occupational and Environmental Medicine 69: 133–139.

82. Johansson C, Burman L, Forsberg B (2009) The effects of congestions tax on air quality and health. Atmospheric Environment 43: 4843–4854.

83. Boogaard H, Janssen NAH, Fischer PH, Kos GPA, Weijers EP, et al. (2012) Impact of low emission zones and local traffic policies on ambient air pollution concentrations. The Science of the Total Environment 435–436: 132–140.

84. EU (2013) Low emission zones in Europe. Available from: http://www.lowemissionzones.eu/. Accessed 2014 July 8.

85. HBEFA (2010) Handbook emission factors for road transport (HBEFA 3.1). Available from: http://www.hbefa.net/e/index.html. Accessed 2014 July 8.

86. Carslaw DC, Beevers SD, Tate JE, Westmoreland EJ, Williams ML (2011) Recent evidence concerning higher NOₓ emissions from passenger cars and light duty vehicles. Atmospheric Environment 45: 7053–7063.

87. Beevers SD, Westmoreland E, de Jong MC, Williams ML, Carslaw DC (2012) Trends in NOₓ and NO₂ emissions from road traffic in Great Britain. Atmospheric Environment 54: 107–116.

88. LANUV, Landesamt für Natur, Umwelt und Verbraucherschutz NRW (2008) Vorgehensweise des LANUV zur Korrektur kontinuierlicher PM10-Messdaten im Luftmessnetz von NRW. Available from: http://www.lanuv.nrw.de/luft/immissionen/ber_trend/erlaeuterungen.pdf. Accessed 2014 July 8.

Berlin, Institut für Meteorologie. Available from: http://opus.kobv.de/zlb/volltexte/2009/7447/pdf/3729.pdf. Accessed 2014 July 17.

89. LUBW, Landesanstalt für Umwelt, Messungen und Naturschutz Baden-Württemberg (2009) Untersuchung von massenrelevanten Inhaltsstoffen in Feinstaub PM_{10}. Karlsruhe. 1–72. Available from: www.lubw.baden-wuerttemberg.de/servlet/is/207409/untersuchung_massenrelevanten_inhaltsstoffen_feinstaub_pm10.pdf?command = downloadContent&filename = untersuchung_massenrelevanten_inhaltsstoffen_feinstaub_pm10.pdf. Accessed 2014 July 8.

90. Dons E, Int Panis L, Van Poppel M, Theunis J, Willems H, et al. (2011) Impact of time-activity patterns on personal exposure to black carbon. Atmospheric Environment 45: 3594–3602.

91. Fischer PH, Hoek G, van Reeuwijk H, Briggs DJ, Lebret E, et al. (2000) Traffic-related differences in outdoor and indoor concentrations of particles and volatile organic compounds in Amsterdam. Atmospheric Environment 34: 3713–3722.

15

Evaluation of the Ecotoxicity of Sediments from Yangtze River Estuary and Contribution of Priority PAHs to Ah Receptor-Mediated Activities

Li Liu[1], Ling Chen[1], Ying Shao[1,2], Lili Zhang[1], Tilman Floehr[2], Hongxia Xiao[2], Yan Yan[2], Kathrin Eichbaum[2], Henner Hollert[1,2,3,4], Lingling Wu[1]*

1 Key Laboratory of Yangtze Water environment, Ministry of Education, Tongji University, Shanghai, China, **2** Department of Ecosystem Analysis, Institute for Environmental Research (Biology V), Aachen Biology and Biotechnology, RWTH Aachen University, Aachen, Germany, **3** College of Resources and Environmental Science, Chongqing University, Chongqing, China, **4** School of Environment, Nanjing University, Nanjing, China

Abstract

In this study, in vitro bioassays were performed to assess the ecotoxicological potential of sediments from Yangtze River estuary. The cytotoxicity and aryl hydrocarbon receptor (AhR)-mediated toxicity of sediment extracts with rainbow trout (*Oncorhynchus mykiss*) liver cells were determined by neutral red retention and 7-ethoxyresorufin-*O*-deethylase assays. The cytotoxicity and AhR-mediated activity of sediments from the Yangtze River estuary ranged from low level to moderate level compared with the ecotoxicity of sediments from other river systems. However, Yangtze River releases approximately 14 times greater water discharge compared with Rhine, a major river in Europe. Thus, the absolute pollution mass transfer of Yangtze River may be detrimental to the environmental quality of estuary and East China Sea. Effect-directed analysis was applied to identify substances causing high dioxin-like activities. To identify unknown substances contributing to dioxin-like potencies of whole extracts, we fractionated crude extracts by open column chromatography. Non-polar paraffinic components (F1), weakly and moderately polar components (F2), and highly polar substances (F3) were separated from each crude extract of sediments. F2 showed the highest dioxin-like activities. Based on the results of mass balance calculation of chemical toxic equivalent concentrations (TEQs), our conclusion is that priority polycyclic aromatic hydrocarbons indicated a low portion of bio-TEQs ranging from 1% to 10% of crude extracts. Further studies should be conducted to identify unknown pollutants.

Editor: Aditya B. Pant, Indian Institute of Toxicology Reserach, India

Funding: This work was supported by a grant from National Natural Science Foundation of China (No. 41101499), the Fundamental Research Funds for the Central Universities of China (No. 0400219213) and Chinese 111 Program. The work was also supported by the research cluster "Pollutants/Water/Sediment-Impacts of Transformation and Transportation Processes on the Yangtze Water Quality" sponsored by the Federal Ministry of Education and Research, Germany (BMBF) and supported by a cooperation project with Chinese colleagues sponsored by the BMBF DLR. The funders had no role in study design, data collection and analysis, decision to publish, or preparation of the manuscript.

Competing Interests: The authors have declared that no competing interests exist.

* Email: wulingling@tongji.edu.cn

Introduction

Marine ecosystem contamination has been greatly affected by human activities. As a coastal transitional system, an estuary serves as a recipient of environmental contaminants derived from land and rivers as well as the atmosphere. Sediments in estuaries are the main sinks of numerous potential chemical and biological pollutants. These sediments may also be considered as a secondary source of water pollution or be absorbed by benthic organisms via bioaccumulation. Thus, sediments as contaminants in rivers and estuaries may pose a potential threat to aquatic organisms and human health.

Toxicity biotests are necessary to assess sediment quality, because they can determine the effect of chemicals in sediments on organisms directly [1]. Some studies have shown many species, such as bacteria, microalgae, yeast, and fish, are widely used to assess the acute toxicity of sediments [2,3]. To obtain compre-hensive insights into potential ecotoxicological effects, the specific effects such as mutagenic, genotoxic, and dioxin-like responses should also be assessed. In vitro cell-based bioassays are of great benefit to characterize mechanism-specific activities of environ-mental contaminants because they are sensitive and require a smaller amount of sample than whole animal experiments [3]. Fish cell lines have been successfully used as a biological alternative of animal tests to assess the toxic effects of sediment extracts [4,5]. For example, rainbow trout (*Oncorhynchus mykiss*) liver cell line (RTL-W1) exhibits high sensitivity to the cytotoxicity of pure substances [5,6]; moreover, this cell line can highly express cytochrome P4501A (CYP1A)-based EROD activity upon expo-sure to dioxin-like compounds [1]. As such, RTL-W1 is commonly used to determine the acute cytotoxicity and CYP1A-based EROD activity of sediment extracts [2,7].

Effect-directed analysis (EDA) is a powerful tool used to identify toxic substances in complex environmental samples [8]. This method is based on a combination of biotests, fractionation procedures, and chemical analysis. EDA can be performed to evaluate the toxic potencies of substances in sediments. This method was applied to analyze toxic chemicals in complex environmental samples. Tetrabromobisphenol A diallyl ether was identified as an emerging neurotoxicant in environmental samples by bioassay-directed fractionation and high-performance liquid chromatography-atmospheric pressure chemical ionization-tandem mass spectrometry [9]. Applying EDA to evaluate sediment extracts in Bitterfeld, Germany, where dinitropyrenes and 3-nitrobenzanthrone are quantitatively identified as the main mutagens [10]. Other studies have also successfully applied EDA to identify EROD-inducing compounds in sediments [11,12].

Polycyclic aromatic hydrocarbons (PAHs) have been extensively investigated because these substances are widely distributed in sediments and may function as mutagens and carcinogens [13]. As dioxin-like compounds, PAHs can induce the biotransformation of CYP1A enzyme in cells by binding its ligand to aryl hydrocarbon receptor (AhR) [14]. PAH concentrations have increased as a result of the continuous and rapid development of China's economy; these compounds have been detected in the sediments of coastal embayment, continental shelf in China, and Yangtze River estuary [15,16].

The Yangtze River estuary, one of the largest estuaries in the world, is located in the east of China and adjacent to the East China Sea. With agricultural and industrial developments in this region, numerous pollutants have been discharged into this estuary [17]. The increasing fraction of wastewater in the Yangtze River increases the levels of nutrients, heavy metals, and dissolved organic carbon; in addition to wastewater, industrial organic chemicals at concentrations of 500 kg to 3,500 kg per day [18]. Approximately 206 hazardous organic chemicals have been detected in water, and 106 chemicals have been found in sediments; among these chemicals, 17 are listed as priority controlled pollutants in America [19]. Studies have also indicated that various organic contaminants, such as PAHs [16], polychlorinated dibenzo-p-dioxins (PCDDs)/polychlorinated dibenzofurans (PCDFs) [20], and polychlorinated biphenyls (PCBs) [21], are present in sediments from the Yangtze River estuary. In addition, newly emerging contaminants, such as polybrominated diphenyl ethers and perfluorinated compounds, have been extensively investigated because of their presence in the Yangtze River [22,23]. Despite the polluted state of this river, studies have been rarely conducted regarding the ecotoxicological potential of Yangtze River sediments [17,24].

In this study, RTL-W1 cells were used to determine the acute cytotoxicity and CYP1A-based EROD activity of the sediment extracts from the Yangtze River estuary. EDA was applied to identify the potential hazardous substances causing AhR-mediated activities. PAHs were chemically analyzed to obtain comprehensive information on these sediments and estimate the extent to which these sediment extract chemicals elicit EROD-inducing effects. The main objectives of this study are listed as follows: (1) to assess the ecotoxicological potential of sediments from the Yangtze River estuary by evaluating the cytotoxic and AhR-mediated effects on RTL-W1 cells; (2) to identify the sediment fractions causing such AhR-mediated effects; and (3) to determine the contribution of priority PAHs to EROD induction of sediment samples.

Materials and Methods

2.1 Sampling

No specific permissions were required for the completion of this study as the field measurements did not involve endangered or protected species nor were conducted in a specified protected area. The surface layer of sediment samples (0 cm to 5 cm) were collected from nine locations in the Yangtze River estuary in March 2012. For each sampling locations, four samples was collected and then mixed to form one sample. The sampling locations are shown in Figure 1. Samples were collected along the salinity gradient and location information is presented in Table S1. Samples Y1 to Y3 were fresh water dominated sediments. Sites Y4 and Y5 were located in the turbidity maximum zone and the samples were brackish water dominated sediments. Sites Y6 to Y9 were located in the river plume zone and the samples were marine sediments. Sediments were collected from each location by using a stainless steel grab sampler. Samples were shock-frozen at –20°C immediately until further processing. Afterward, these samples were freeze-dried at –50°C and sieved using a 100-mesh sieve. The samples were then stored in combusted glass with Teflon lined lids at –20°C in the dark until extraction.

2.2 Sediment extraction procedure

The samples were prepared according to previously described methods [25]. The freeze-dried sediment samples (20 g) were separately extracted with acetone (Merck, Darmstadt, Germany, HPLC) for 48 h by using standard reflux (Soxhlet) extractors with six cycles per hour. Elemental sulfur was removed by treating the extracts with activated copper. The extracts were then reduced in volume to obtain a final volume of 2 mL by rotary evaporation. The solvent was divided into two aliquots and concentrated close to dryness with gentle nitrogen stream. One aliquot was replaced with 1 mL of dimethylsulfoxide (DMSO; Sigma-Aldrich Chemie GmbH, Steinheim, Germany) and used for in vitro biotests, resulting in a final concentration of 10 g of sediment dry weight per mL of DMSO (10 g/mL). The other aliquot was dissolved in n-hexane (Merck, HPLC) and used for multilayer fractionation. The extracts were then stored at –20°C until further analysis.

2.3 Multilayer fractionation

To identify unknown substances that contribute to the dioxin-like potencies of whole extracts, we selected the sediments used for fractionation on the basis of maximum AhR agonist activities in crude extracts. Fractionation was performed according to previously described methods with slight modification [26]. Fractionation was also performed using a silica gel/aluminum oxide column. The column was prepared according to the following procedures. Silica gel (60 mesh to 200 mesh, Merck, Darmstadt, Germany) and aluminum oxide (50 mesh to 200 mesh, Merck, Darmstadt, Germany) were activated at 180°C for 24 h. Sodium sulfate was baked at 450°C for 5 h before use to eliminate possible production residues, such as phthalates. Glass columns (inner diameter = 10 mm and height = 50 cm) were filled from the bottom with 10 g of silica gel, 5 g of aluminum oxide, and 2 g of sodium sulfate. The gels were then washed twice with n-hexane. Afterward, the extracts were placed in the columns and eluted with different solvents at increasing polarities: F1 containing non-polar paraffinic components were eluted with 50 mL of n-hexane; F2 characterized by weakly and moderately polar components were eluted with 70 mL of n-hexane/dichloromethane (Merck, HPLC; 7:3, v/v); and F3 containing highly polar components were eluted with 50 mL of acetone/methanol (Merck, HPLC; 1:1, v/v). Using a rotary evaporator, we decreased the volumes of the eluates to

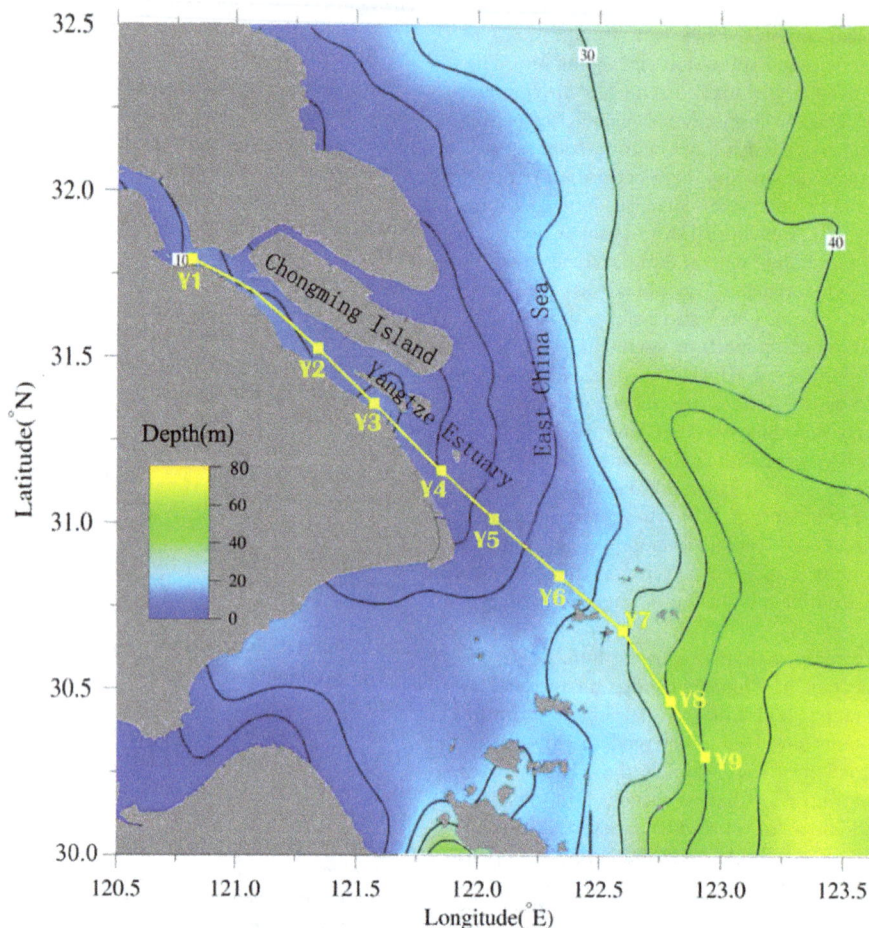

Figure 1. Map of the sampling locations in the Yangtze River estuary. Nine samples Y1–Y9 were collected along the estuary in March 2012.

2 mL and evaporated close to dryness with gentle nitrogen stream. Afterward, the eluates were replaced with 1 mL of DMSO. These fractions were then stored at –20°C in the dark until analysis.

2.4 Cell culture

Cytotoxicity was assessed using the fibroblast-like permanent cell line RTL-W1 isolated from the liver of a female *O. mykiss* [6]. The cells were kindly provided by Drs. Niels C. Bols and Lucy Lee (University of Waterloo, Canada) and cultured according to the method described by Klee et al. [27].

2.5 Neutral red retention (NR) assay

The acute cytotoxicity of sediment extracts on RTL-W1 cells were assessed by NR assay according to Borenfreund and Puerner [28] with slight modifications by Klee et al. [27]. Sediment extracts were serially diluted with L15 medium to obtain a concentration range of 1.57 mg to 100 mg of dry sediment per mL medium with six internal replicates. 3,5-Dichlorophenol (Sigma, Seelze, Germany) was used as a positive control sample at a concentration of 80 mg/L in each test plate. All of the experiments were performed in independent triplicates. The cytotoxic potentials of individual extracts that induced 50% mortality after 48 h (48 h NR_{50}) were calculated accordingly. The unit of NR_{50} was designated as mg dry weight sediment equivalent (SEQ) per mL test medium.

2.6 EROD induction assay

The dioxin-like activity of sediment extracts was determined by EROD induction assay according to previously described methods [29]. The highest test concentration of the EROD assay to determine the enzyme activity and avoid the cytotoxic effects based on NR_{80} obtained from a preliminary NR assay with RTL-W1 cells. The cells were seeded into 96-well microtiter plates (TPP, Trasadingen, Switzerland) and exposed to sediment extracts in eight dilution steps with six internal replicates. 2,3,7,8-Tetrachlorodibenzo-*p*-dioxin (TCDD, Promochem, Wesel, Germany) was serially diluted to obtain a final concentration ranging from 3.13 pM to 100 pM in two separate rows of each plate as a series of positive control sample. After the samples were incubated for 72 h, induction was terminated by removing the growth medium and freezing at –70°C to lyse the cells. The deethylation of exogenous 7-ethoxyresorufin was initiated by adding 7-ethoxyresorufin to each well; the resulting reaction mixture was then incubated in the dark at room temperature for 10 min before NADPH (Sigma) in PBS (Sigma) was added. The plates were further incubated for 10 min and the reaction was terminated by adding fluorescamine (Sigma) dissolved in acetonitrile (Merck, HPLC). EROD activity was determined fluorometrically after another 15 min at excitation and emission wavelengths of 544 and 590 nm, respectively, by using an Infinite M200 plate reader (Tecan, Crailsheim, Germany). The amount of protein was fluorometrically determined using the fluorescamine method at excitation and emission wavelengths of 360 and 465 nm, respectively [30]. The concentration-response curves of EROD induction in the RTL-W1 bioassay were designed by non-linear regression (Prism 4.0, GraphPad, San Diego, USA) using the classic sigmoid curve or Boltzmann curve as model equations.

AhR agonist activities were determined using a fixed-level approach [31]. Extract EC_{25TCDD} of each sediment extract was obtained and normalized to that of the positive control 2,3,7,8-TCDD as biological toxic equivalent concentrations (bio-TEQs) [29]. EC_{25TCDD} was used to compare the samples and calculate TCDD-equivalent concentrations in the samples. Bio-TEQ concentrations were calculated as follows:

$$[bio-TEQ](pg/g(d.w.)) = \frac{TCDD\,EC_{25}\,(pg/mL)}{extract\,EC_{25TCDD}\,(g(d.w.)/mL)} \quad (1)$$

where TCDD EC_{25} (pg/mL) is the concentration of the TCDD positive control sample causing 25% of EROD induction and extract EC_{25TCDD} (g/mL) is the concentration of the sediment extract equivalent causing 25% of EROD induction. EC_{25TCDD} was considered as a more appropriate measure than EC_{50} in this study because EC_{50} has not been well defined by dose-response curve in several cases [31].

2.7 Chemical analysis

A total of 16 priority PAHs in the crude sediment extracts described by Environmental Protection Agency (EPA) were quantified by Agilent 6890B gas chromatograph coupled to a mass selective Agilent 5977A MSD detector, which was operated in the selective ion monitoring mode. The sum parameter was calculated on the basis of 16 priority PAHs as follows: naphthalene; acenaphthylene; acenaphthene; fluorene; phenanthrene; anthracene; fluoranthene; pyrene; benzo[a]anthracene; chrysene; benzo[b,j]fluoranthene; benzo[k]fluoranthene; benzo[a]pyrene; dibenzo[a,h]anthracene; benzo[g,h,i]perylene; and indeno[1,2,3-cd]pyrene. 2-Fluorobiphenyl and terphenyl-d14 were used as surrogate standards and added before Soxhlet extraction for each sample. The detector was equipped with 30 m×0.25 μm film HP-5MS fused silica capillary column (Agilent Technologies), and 1 μL of each fraction was injected in a splitless mode. The initial temperature was set at 45°C for 2 min and then increased to 265, 285, and 320°C at a rate of 20, 6, and 10°C/min. Afterward, temperature was held constant at 320°C for 4 min. Instrumental analysis was conducted after internal standards (naphthalene-d8, acenaphthene-d10, phenanthrened10, chrysene-d12, and perylene-d12) were added.

During the analytical procedure, a procedural blank, a spiked blank, and a duplicate sample were processed. Targets were not detected in procedural blanks. The surrogate recoveries of 2-fluorobiphenyl and terphenyl-d14 were 88% (±5%) and 102% (±4%), respectively. The recoveries of targets ranged from 88% to 103% in spiked blank samples. The reported concentrations were not corrected by surrogate standard recoveries.

2.8 Chem-TEQ calculation

To determine the contribution of EPA-PAHs to the overall AhR-mediated activity, we determined chem-TEQs based on PAH potencies relative to 2,3,7,8-TCDD [32]. Chem-TEQs were calculated by multiplying the concentration of each AhR-active chemical by specific relative potency (REP) of RTL-W1 cells (Eq. 2) [29,32]. Chem-TEQ concentrations at picogram PAH per gram of SEQ were calculated as picogram per gram (pg/g).

$$Chem-TEQ[pgPAH/gSEQ] =$$
$$\sum_{i=1}^{n} concentration PAH_i(\frac{pg}{g}dwSEQ) \times REP_i \quad (2)$$

2.9 Data analysis

Data were expressed as mean ± SD. Statistical analyses were performed using SPSS 17.0 (SPSS Inc., Chicago, IL, USA). Maximum concentrations were used to assess the worst pollution scenario at each section compared with other sediment systems.

Results and Discussion

3.1 Cytotoxicity of crude extracts and multilayer fractions

The results of the NR assay on the crude extracts are shown in Figure 2. The sediment extracts from the study sites Y1, Y2, and Y3 showed very low potency and failed to elicit cytotoxic effects at a concentration of 100 mg/mL. By contrast, extracts from the six other sites revealed evident cytotoxicity to RTL-W1 cells with NR_{50} ranging from 4.1 mg/mL to 43.4 mg/mL (Table S2). The extracts from sites Y5, Y8, and Y9 revealed relatively greater cytotoxicity than those from other sites. The lowest cytotoxic potency was obtained from extracts collected from sites Y4, Y6, and Y7. The cytotoxicity of the sediment extracts from downstream regions generally showed higher activities than those from upstream regions.

In the multilayer fractionations, only fraction F1 from sites Y5 and Y9 elicited cytotoxic effects with NR_{50} of 63.2 and 60.4 mg/mL, respectively. No cytotoxic potential was detected in F2 and F3.

According to a sediment classification system of sediments found in Germany [33], the threshold values of NR biotest system are as follows: $NR_{50} > 80$ mg/mL, non-toxic; $80 \geq NR_{50} \geq 31$ mg/mL, moderately toxic; and $NR_{50} < 31$ mg/mL, strongly toxic. On the basis of this threshold values, we classified the crude sediment extracts from sites Y4 and Y7 as moderately toxic and those from sites Y5, Y6, Y8, and Y9 were strongly toxic. The cytotoxicity effects elicited by the sediment extracts in this study were compared with those observed in sediment extracts from other Asian, European, and South American locations (Table 1). The cytotoxic potential of the sediment extracts from the Yangtze River estuary was comparable to that of the suspended particulate matter (SPM) and sediment extracts from the upper Danube River and the surface sediments and sediment core extracts from the upper Rhine River [5] and the Mecklenburg Bight (Western Baltic Sea) [2]. Furthermore, the cytotoxic potential of these extracts from the Yangtze River estuary was higher than that of the sediment extracts from Tietê River [34].

The cytotoxic effect of the sediment extracts may be attributed to the organic contaminants previously polluting the Yangtze River estuary with the rapid development of industrialization. In 2006, more than 10,000 chemical enterprises, or equivalent to approximately half of the total number of enterprises in China, were situated by the river [35]. Many petroleum and chemical plants and docks were also located along the river. In this study, high PAH concentrations, pharmaceuticals, and other organic compounds have been frequently detected in the study region [36,37,38]. Organic compounds, such as PAHs with two or three rings (naphthalene, acenaphthylene, acenaphthene, fluorene and phenanthrene), can elicit cytotoxic effects on rainbow trout (gill) cells [39]. Pharmaceuticals, such as substituted phenols, are also

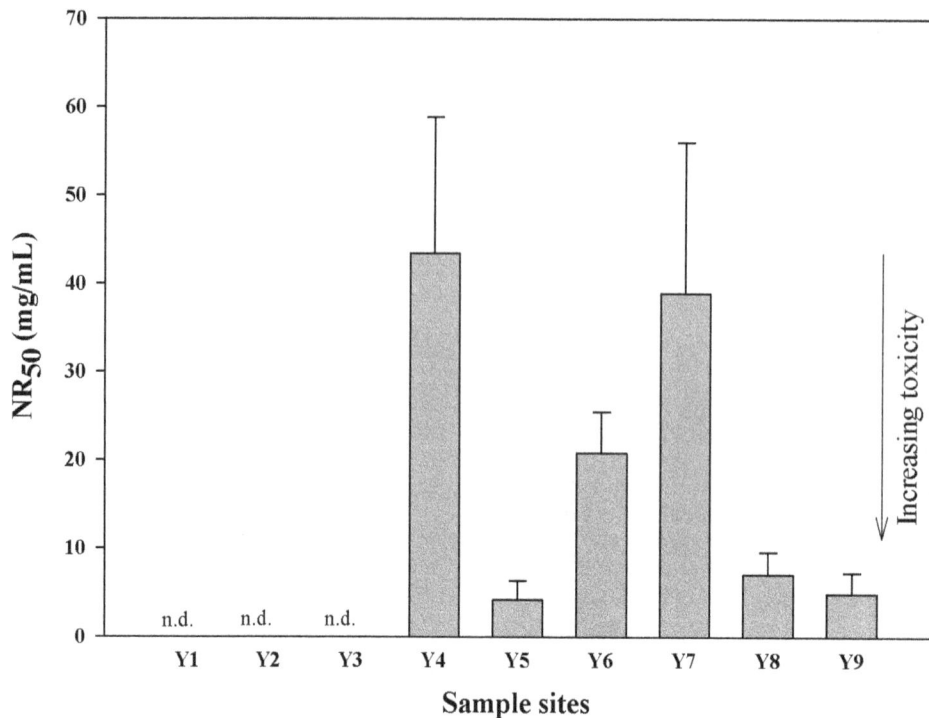

Figure 2. Cytotoxicity of sediment extracts from the Yangtze River estuary. n.d. = not detectable, indicating that no toxic effect of sediment extracts at a concentration of 100 mg/mL with RTL-W1 cells. NR_{50} values for sediment extracts are given in mg sediment equivalent per mL medium (mg/mL). All the values are expressed as means $\pm$ SD.

potential inducers of cytotoxicity [40]. In this study, toxic compounds may have accumulated in the Yangtze River estuary and may be accounted for the cytotoxic potential of sediment extracts.

3.2 AhR-mediated activities in crude sediment extracts

Figure 3 shows the results of the EROD assay of the AhR agonists in the crude sediment extracts. The dioxin-like potentials of the sediment extracts among samples varied. In particular, bio-TEQ of crude extracts ranged from 38.9 pg/g dw to 323.5 pg/g dw (Table S2). A relatively high EROD induction was observed in sediment extracts from sites Y2, Y5, and Y9. By contrast, bio-TEQs of the sediments from Y3 showed relatively low EROD induction.

We obtained low to moderate values (Table 2) and corresponding AhR-mediated activity of the investigated sediments compared with bio-TEQs from other river systems. The bio-TEQs of the sediments from different locations were compared on the basis of the reported results regardless of the difference between the extraction methods. EROD induction potential of the sediment extracts from the Yangtze River estuary was comparable to that of the sediment extracts from River Elbe estuary [41] and Saginaw Bay [42]. The sediment extracts from the Yangtze River estuary showed higher EROD induction potential than those from Masan Bay [43], Tai Lake [44], and Yellow Sea [45]. The EROD induction potential of the sediment extracts from the Yangtze River estuary was also lower than that of the sediments from Tietê River [34], Morava River [14], Dagu Rivers, and Haihe [46]. Keiter et al. [33] developed a classification system for AhR-mediated toxicity of German river systems. According to this evaluation scheme, the sediment samples of the Yangtze River estuary revealed only minor AhR-mediated effects. The concentrations of organic pollutants may be lower in Yangtze River than in Rhine River; however, a comparably higher mass transport of water and particulate matter in the Yangtze River system contributes to relatively larger amounts of organic pollutants in the receiving water body of Yangtze River [17]. The mean water discharge (30,200 m^3/s) (http://china.org.cn/english/eng-

Table 1. Cytotoxicity of sediment extracts of the Yangtze River estuary and that of other sediment extracts from previous studies.

Sampling Locations	NR_{50} (mg/ml)	References
Mecklenburg Bight (Western Baltic Sea)	≥ 13.5	[3]
Upper Rhine (Germany)	≤ 50	[6]
Upper Danube River (Germany)	<40	[6]
Tietê River (Sao Paulo, Brazil)	29~225	[35]
Yangtze River estuary (China)	4.1~43.4	In this study

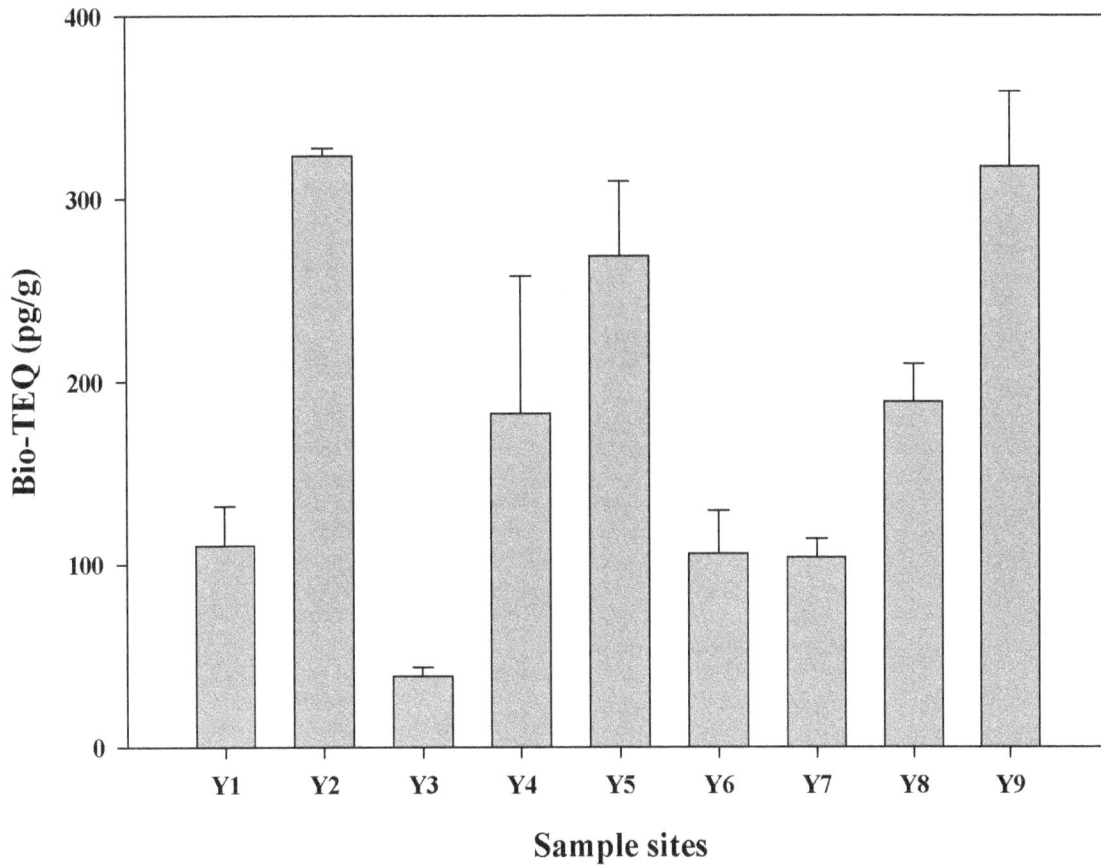

Figure 3. AhR-mediated activity of sediment extracts from the Yangtze River estuary. The values as determined by the EROD assay with RTL-W1 cells are expressed as biological toxicity equivalents (Bio-TEQ; pg/g dw). Data are given as means of 3 replicates ± SD.

shuzi2003/gq/dili5.htm) Yangtze River is approximately 14 times greater than that of Rhine River (2,200 m^3/s) [47]. As such, the total mass transfer of AhR compounds has been extensively observed in the China Sea because Yangtze River releases a higher discharge than European waters.

NR assay results showed that the Yangtze River estuary is polluted by a large variety of chemicals. A high concentration of organic pollutants, such as PAHs [48], PCBs, and PCDDs/PCDFs [49], which are potential inducers of the cytochrome P450 system, were detected in the sediments collected from Yangtze River estuary. These sediment-bound contaminants may affect the AhR activity of the samples. Discharges from the industries along the Yangtze River estuary should be dealt with, and measures should

Table 2. AhR-mediated activity of sediment extracts of the Yangtze River estuary and those of other rivers described in previous studies.

Sampling Locations	Bio-TEQ (pg/g)	References
Masan Bay (Korea)	17~275	[44]
River Elbe Estuary (Germany)	15.5~322	[42]
Tietê River (Brazil)	n.d.~24170	[35]
Saginaw Bay (USA)	11~348	[43]
Morava River (Czech Republic)	1~17000	[15]
Tai Lake (China)	n.d.~114.5	[45]
Dagu Rivers (China)	1200~13900	[47]
Haihe (China)	330~930	[47]
Yellow Sea (China)	3.4~28	[46]
Yangtze River estuary (China)	38.9~323.5	In this study

Note: n.d. = not detectable or below the detection limit.

be established to reduce the input of toxic effluents into the estuary [18].

3.3 AhR-mediated activities in multilayer fractions

Based on EROD induction assay results of the crude extracts, the sediment samples from sites Y2, Y4, Y5, Y7, Y8, and Y9, which had the highest bio-TEQ values, were selected for further identification of AhR-mediated activities using multilayer fractionation. As shown in Figure 4, AhR-mediated activities caused by multilayer fractions (F1 to F3) were compared with crude sediment extract inductions.

Fractions F1, which contained non-polar paraffinic components, showed no or very low AhR-mediated activities. Primary fractions F2, which contained weakly and moderately polar components, showed the highest AhR-mediated activities. In addition, fractions F3, which contained more polar components, also showed significantly increased activities. The results were consistent with those of other studies performed previously. Some non-priority PAHs and more polar compounds, such as heterocyclic compounds containing nitrogen, sulfur, or oxygen heteroatoms (NSO-Het), also show dioxin-like activity [50]. The highest AhR agonistic effect of the SPM collected during a flood event from the rivers Neckar and Rhine was found in fractions containing high molecular weight PAHs with more than 16 aromatic C-atoms [51]. Engwall et al. [52] indicated that the polyaromatic fractions of the bottom sediment and suspended particulate matter showed relatively higher dioxin-like potencies than the other fractions. Therefore, weakly and moderately polar components and more polar components are major inducers of AhR-mediated activities and require further study.

However, AhR-mediated activities showed site-specific differences among locations (Figure 4). Sample sites Y4, Y7, and Y8 (especially site Y4) showed significantly greater bio-TEQs for the sum of primary fractions F1 to F3 than the crude extracts, and the percentages of the exceeded parts were 62%, 40%, and 32%, respectively (Table S3). Sample site Y5 exhibited comparable bio-TEQs of crude extracts to the added primary fractions.

Interactions, such as synergistic or antagonistic effects, exist in different substances and may lead to changes in the effects of certain substances. Low EROD activities do not necessarily indicate low induction potency but may have been caused by high concentrations of inducers and non-inducers that inhibit EROD activity [53]. In crude extracts, the low bio-TEQs might be explained by EROD antagonistic and inhibiting effects because sediments contain a broad range of unspecified compounds [2]. EROD inductions with more than 100% induction in fractions compared with crude extracts may be caused by the retention of humic substances during the fractionation procedure [54]. The separation of aliphatic and polar compounds from nonpolar aromatics resulted in an activity of a fraction that exceeds that of the crude sediment extract in the Neckar River basin [11]. Thus, in this study, some antagonistically acting substances were

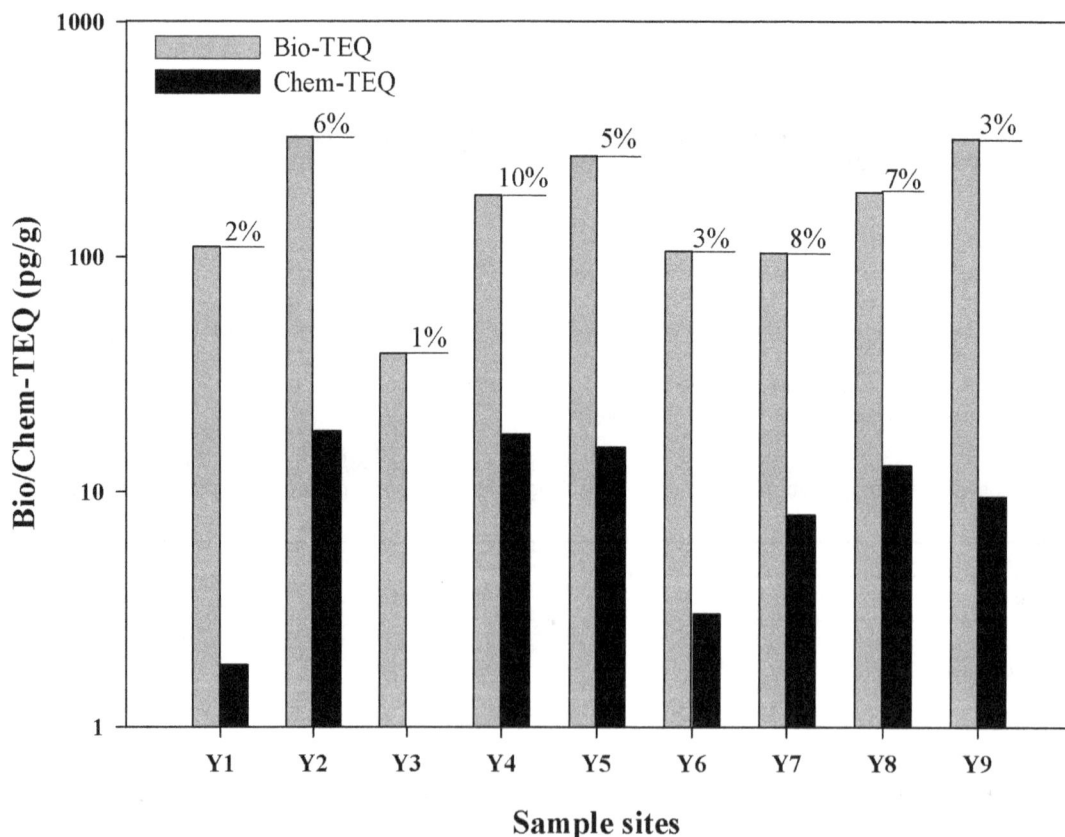

Figure 4. AhR-mediated activities of crude and multilayer fractions of sediment extracts from the Yangtze River estuary. The results are based on EROD inductions with RTL-W1 cells and presented as bio-TEQs. F1 contains non-polar paraffinic components. F2 contains weakly and moderately polar components. F3 contains highly polar components. The numbers in percentage were calculated as (F1+F2+F3–CE)/CE×100%. The difference between the combined F1 to F3 and the induced crude extracts induction is given. Bio-TEQs are given as means of $n = 3$ independent experiments. CE, crude extracts.

probably separated in the process of elution, and thus, agonists could completely display activity.

In contrast, sites Y2 and Y9 showed decreased induction when comparing bio-TEQs of F1 to F3 with the crude extract. This phenomenon can be explained as follows: fractionation of mixtures of bioactive compounds into multiple fractions may have reduced the individual bioactivity to below detection limits, thereby efficiently rendering some of the active compounds non-detectable by the bioassay [55]. Thomas et al. [56] applied the EDA method to analyze the estrogen receptor agonist of water extracts. After fractionation, the sum of activity of the individual fractions was less than 20% of the activity of the crude extract. The crude extracts are a complex mixture of unknown substances, and thus, the full recovery of all effect compounds within the samples was difficult. The AhR-mediated activities of the sediment extract fractions are no longer observed in the present study, indicating that the reduced complexity of the mixture also reduces toxicity.

3.4 Concentrations of PAHs

Chemical analysis was applied to determine the concentrations of the 16 EPA-PAHs listed in Table 3. Most of the 16 EPA-PAHs were detected in all the sediment samples from the Yangtze River estuary. The highest concentrations of total PAHs were found in site Y2, and the lowest concentrations were found in site Y1 and Y3. The concentrations of total PAHs ranged from 21.5 ng/g dw to 190.5 ng/g dw sediment. Some studies reported the levels of PAHs in the sediment from different sections of the Yangtze River. Concentrations of total PAHs at the Chongqing section was 257 ng/g dw to 723 ng/g dw, and at Jialing River, a tributary of the upper Yangtze River, was 132 ng/g dw to 349 ng/g dw [57]. Concentration of PAHs in the Wuhan section was high, ranging from 303 ng/g to 3,995 ng/g in the main stream and 4,121 ng/g to 4,262 ng/g in the tributaries [58]. Compared with other sections of the Yangtze River, the concentration of PAHs in sediment from the Yangtze River estuary was relatively low [17]. The dilution from the East China Sea may be one of the reasons for such low concentration [59]. In addition, Wang et al. [60] showed that the PAH composition in the Yangtze River estuary, which was dominated by four-ring to six-ring PAHs, was mainly caused by petroleum combustion, vehicle emission, and biomass combustion (mainly coal) in the nearshore area, whereas PAHs composition, which was dominated by two-ring to three-ring PAHs, in the farther shore zone originated from petroleum combustion of shipping processes and shoreside discharges.

3.5 Correlation between bio-TEQs and chemical analyses

According to Bols et al. [32], chem-TEQs were calculated by multiplying the specific REP factors for RTL-W1 cells with corresponding compound concentrations. The total biological response in the EROD assay (Bio-TEQs) and chem-TEQs of PAHs are shown in Figure 5. For sites Y1, Y3, Y6, and Y9, less than 5% of the induction could be explained by known priority PAHs, which were expressed as chem-TEQs. For sediment extracts from sites Y4 and Y7, 10% and 8% of the induction could be attributed to the presence of analyzed PAHs, respectively.

Several studies have shown that halogenated aromatic hydrocarbons and PAHs are chemicals that are usually related to dioxin-like activity, and can act as AhR agonists [11,61]. Generally, chemical analysis focused on substances that were considered to be a priority or are more relevant, and other non-priority substances were ignored. However, bio-TEQs are not explained by the chem-TEQs calculated from the concentrations of the analyzed priority compounds. In this study, less than 10% of the bio-TEQs could be attributed to the EPA-PAHs, whereas non-priority substances were shown as high inducers.

Concentrations of pollutants at different matrixes were based on reported results, and differences between the analytical methods have been neglected. According to Keiter et al. [7], 55% to 88% of AhR-mediated responses observed in upper Danube River sediments were not due to the measured priority PAHs, PCBs, and PCDDs/PCDFs. The highest bio-TEQ of sediments core layers (2.26×10^5 pg/g) was sampled at the historically contaminated dumping site in the Baltic Sea. This value was much higher than that obtained in this study, but it was lower than expected from the PAH concentration in the samples from the dumping site [2]. Special interactions of some contaminants or of contaminants with sediment components might be responsible for such low value. In contrast, the majority of bio-TEQ (approximately 58%) measured from the extracts of sediments from the Elbe River by EROD assays could be due to the priority PAHs [41]. In a number of studies involving sediments collected from the Elbe, Tietê, and Danube Rivers, the chemical measurements of priority PAHs or other persistent organic pollutants, such as PCBs and PCDDs/PCDFs, could only account for a small portion of the AhR-mediated potency [7,11,34]. The results from these above mentioned studies were consistent with those obtained in the present study. Kaisarevic et al. [12] showed that only minor portions of biologically derived TCDD-TEQs from waste water canal sediments could be due to the monitored PAHs with known relative potencies. In a previous study, PCBs and PCDDs/PCDFs were responsible for a minor portion of the total AhR-mediated activities of SPM in two floods [29]. The non-priority pollutants mainly mediated the high induction rates. Therefore, the main induction in the biotest systems was caused by non-priority pollutants [11,62]. Some studies showed that the non-priority substances might have caused AhR-mediated activity. Hinger et al. [50] showed that non-priority PAHs and NSO-Het are very potent AhR agonists. Heterocyclic polyaromatic compounds, including dinaphthofurans, 2-(2-naphthalenyl) benzothiophene, methylated chrysene, and benz[a]anthracene, were identified and confirmed as major cytochrome P4501A (CYP1A)-inducing compounds in a contaminated sediment of Bitterfeld (Germany) [63]. Furthermore, Brack et al. [11] demonstrated that PAHs, especially PAHs with a molecular weight between 228 and 252 g/mol, could explain the major dioxin-like potencies of sediment extracts from the Neckar river basin.

The potential contribution of non-priority pollutants to environmental hazards was indicated in this study. To better assess the environmental samples, a broader range of substances should be considered, and studies should not only focus on prioritized pollutants. Identification of unknown pollutants that cause the main AhR-mediated activity is recommended. EDA is a suitable tool for identifying the compounds.

Conclusion

The present study assessed the ecotoxicological hazard potential of the sediments from the Yangtze River estuary using NR retention and EROD induction assays with RTL-W1 cells. The results showed that cytotoxicity and AhR-mediated activity of sediment from Yangtze River estuary were at a minor to medium level when compared with those from other river systems. Concentrations of organic pollutants may be lower in the Yangtze River than in the Rhine River because of the comparably higher mass transport of water and particulate matter in the former, but such mass transport still results in comparably larger amounts of organic pollutants that end up in the Yangtze River's receiving

Table 3. Concentrations of the 16 US EPA-polycyclic aromatic hydrocarbons (PAHs; ng/g dw) in sediment samples from the Yangtze River estuary.

Sampling site	Y1	Y2	Y3	Y4	Y5	Y6	Y7	Y8	Y9
Naphthalene	6.0	47.0	11.0	22.0	7.0	4.0	26.0	7.0	8.0
Acenaphthylene	n.d.	4.0	n.d.	1.0	2.0	1.0	2.0	2.0	2.0
Acenaphthene	n.d.	2.0	n.d.	2.0	2.0	n.d.	n.d.	1.0	2.0
Fluorene	1.0	5.0	1.0	3.0	6.0	2.0	3.0	4.0	4.0
Phenanthrene	2.0	15.0	1.0	14.0	12.0	5.0	6.0	11.0	9.0
Anthracene	2.0	4.0	1.0	3.0	4.0	1.0	2.0	3.0	2.0
Fluoranthene	2.0	17.0	1.0	20.0	13.0	4.0	6.0	10.0	7.0
Pyrene	1.0	8.5	0.5	10.0	7.0	2.5	3.0	5.5	4.5
Benzo[a]anthracene	1.0	13.0	1.0	14.0	10.0	3.0	5.0	8.0	6.0
Chrysene	2.0	14.0	1.0	13.0	11.0	3.0	5.0	9.0	6.0
Benzo[b]fluoranthene	2.0	19.0	1.0	16.0	18.0	6.0	9.0	14.0	12.0
Benzo[k]fluoranthene	1.0	5.0	n.d.	6.0	5.0	1.0	3.0	4.0	3.0
Benzo[a]pyrene	n.d.	14.0	1.0	13.0	10.0	n.d.	4.0	9.0	6.0
Indeno[1,2,3-cd]pyrene	1.0	10.0	1.0	9.0	8.0	2.0	4.0	7.0	5.0
Dibenz[a,h]anthracene	n.d.	3.0	n.d.	2.0	2.0	n.d.	1.0	2.0	1.0
Benzo[g,h,i]perylene	1.0	10.0	1.0	8.0	8.0	2.0	4.0	7.0	5.0
Sum of EPA-PAHs	**22.0**	**190.5**	**21.5**	**156.0**	**125.0**	**36.5**	**83.0**	**103.5**	**82.5**

Note: n.d. = not detectable or below the detection limit.

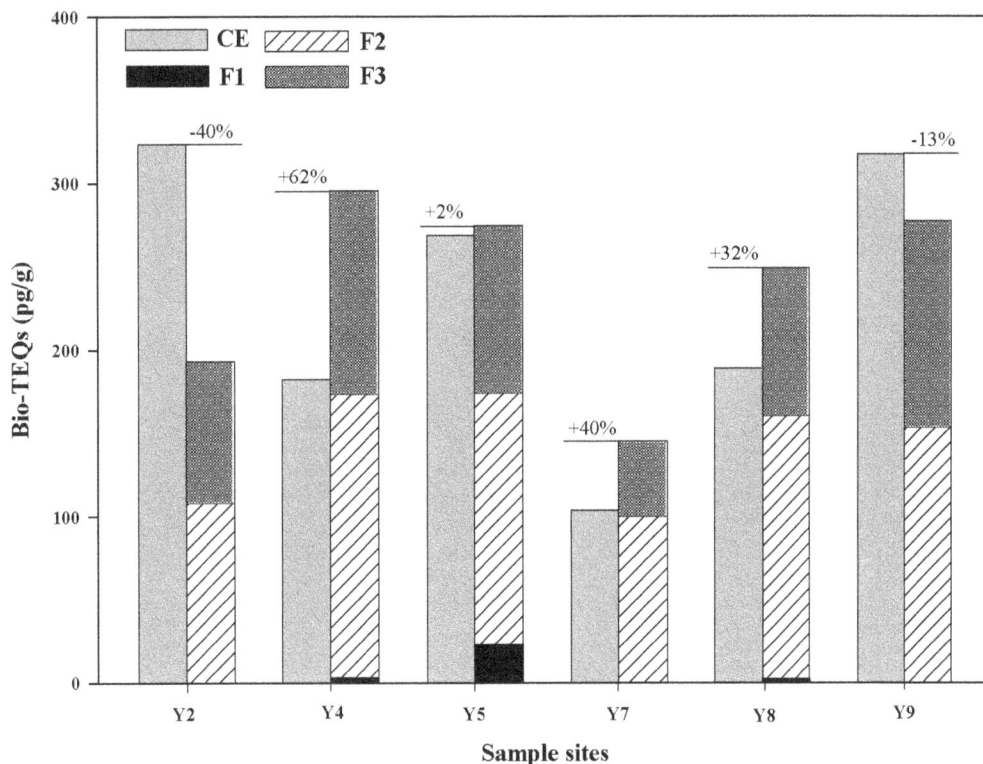

Figure 5. The contribution of Chem-TEQs of PAHs to Bio-TEQs. The total biological response of the crude extracts in the EROD assay with RTL-W1 cells are expressed as bio-TEQs. The chem-TEQs of the 16 measured EPA PAHs were calculated by multiplying compounds concentrations and relative equivalency potencies and are given in pg/g dw. Numbers in percent indicated the calculated contribution of these PAHs to the EROD induction in each crude extract.

water body. At the same contamination levels in both rivers, the 14-fold amount of toxic substances would still enter the East China Sea. Therefore, the effect of the Yangtze River on the East China Sea needs to be considered.

Results of the fractionation showed that weakly and moderately polar components and polar components showed the highest AhR-mediated activities and can maximum exceed about 60% of the crude extract. A combined analysis of chemical measurements of PAHs and the results from bioassays revealed that priority EPA-PAHs contributed only a minor portion of the determined AhR-mediated activities. Furthermore, identification of unknown pollutants causing the high biological AhR-mediated potency should be the focus of future research. EDA is a suitable tool for identifying the unknown pollutants and can be used in further studies to better protect the estuary and to serve as reference for environmental monitoring in this region. Moreover, for the protection of the Yangtze River estuary, some non-priority pollutants, which cause high ecotoxicity, should also be monitored.

Supporting Information

Table S1   Locations for the nine samples in this study.

Table S2   Cytotoxicity and dioxin-like activities of the crude sediment extracts in the RTL-W1 cells. NR_{50} values for sediment extracts are given in mg sediment equivalent per mL medium (mg/mL). Dioxin-like activity expressed as biological toxicity equivalents (Bio-TEQ) in pg/g dw.

Table S3   The dioxin-like activity of the multilayer fractions in the RTL-W1 cells and expressed as biological toxicity equivalents (Bio-TEQ) in pg/g dw.

Acknowledgments

The authors would like to express their thanks to Drs. Niels C. Bols and Lucy Lee (University of Waterloo, Canada) for providing RTL-W1 cells.

Author Contributions

Conceived and designed the experiments: HH LW. Performed the experiments: LL YS LZ YY. Analyzed the data: LL LC TF HX KE HH LW. Contributed reagents/materials/analysis tools: HH LW. Contributed to the writing of the manuscript: LL HH LW.

References

1. Hallare A, Seiler T-B, Hollert H (2011) The versatile, changing, and advancing roles of fish in sediment toxicity assessment–a review. Journal of Soils and Sediments 11: 141–173.

2. Wölz J, Borck D, Witt G, Hollert H (2009) Ecotoxicological characterization of sediment cores from the western Baltic Sea (Mecklenburg Bight) using GC–MS and in vitro biotests. Journal of Soils and Sediments 9: 400–410.

3. Hilscherova K, Machala M, Kannan K, Blankenship A, Giesy J (2000) Cell bioassays for detection of aryl hydrocarbon (AhR) and estrogen receptor (ER) mediated activity in environmental samples. Environmental Science and Pollution Research 7: 159–171.

4. Giltrap M, Macken A, McHugh B, McGovern E, Foley B, et al. (2011) In vitro screening of organotin compounds and sediment extracts for cytotoxicity to fish cells. Environmental Toxicology and Chemistry 30: 154–161.

5. Keiter S, Rastall A, Kosmehl T, Erdinger L, Braunbeck T, et al. (2006) Ecotoxicological Assessment of Sediment, Suspended Matter and Water Samples in the Upper Danube River. A pilot study in search for the causes for the decline of fish catches (12 pp). Environmental Science and Pollution Research 13: 308–319.

6. Lee IJ, Clemons J, Bechtel D, Caldwell S, Han K-B, et al. (1993) Development and characterization of a rainbow trout liver cell line expressing cytochrome P450-dependent monooxygenase activity. Cell Biology and Toxicology 9: 279–294.

7. Keiter S, Grund S, van Bavel B, Hagberg J, Engwall M, et al. (2008) Activities and identification of aryl hydrocarbon receptor agonists in sediments from the Danube river. Analytical and Bioanalytical Chemistry 390: 2009–2019.

8. Brack W (2003) Effect-directed analysis: a promising tool for the identification of organic toxicants in complex mixtures? Analytical and Bioanalytical Chemistry 377: 397–407.

9. Qu G, Shi J, Wang T, Fu J, Li Z, et al. (2011) Identification of tetrabromobisphenol A diallyl ether as an emerging neurotoxicant in environmental samples by bioassay-directed fractionation and HPLC-APCI-MS/MS. Environmental Science & Technology 45: 5009–5016.

10. Lübcke-von Varel U, Bataineh M, Lohrmann S, Löffler I, Schulze T, et al. (2012) Identification and quantitative confirmation of dinitropyrenes and 3-nitrobenzanthrone as major mutagens in contaminated sediments. Environment International 44: 31–39.

11. Brack W, Schirmer K, Erdinger L, Hollert H (2005) Effect-directed analysis of mutagens and ethoxyresorufin-O-deethylase inducers in aquatic sediments. Environmental Toxicology and Chemistry 24: 2445–2458.

12. Kaisarevic S, Varel UL-v, Orcic D, Streck G, Schulze T, et al. (2009) Effect-directed analysis of contaminated sediment from the wastewater canal in Pancevo industrial area, Serbia. Chemosphere 77: 907–913.

13. Chen G, White PA (2004) The mutagenic hazards of aquatic sediments: a review. Mutation Research/Reviews in Mutation Research 567: 151–225.

14. Hilscherova K, Kannan K, Kang Y-S, Holoubek I, Machala M, et al. (2001) Characterization of dioxin-like activity of sediments from a Czech River Basin. Environmental Toxicology and Chemistry 20: 2768–2777.

15. Liu LY, Wang JZ, Wei GL, Guan YF, Wong CS, et al. (2012) Sediment records of polycyclic aromatic hydrocarbons (PAHs) in the continental shelf of China: implications for evolving anthropogenic impacts. Environmental Science & Technology 46: 6497–6504.

16. Liu M, Hou L, Yang Y, Zou H, Lu J, et al. (2001) Distribution and sources of polycyclic aromatic hydrocarbons in intertidal flat surface sediments from the Yangtze estuary, China. Environmental Geology 41: 90–95.

17. Floehr T, Xiao H, Scholz-Starke B, Wu L, Hou J, et al. (2013) Solution by dilution?–A review on the pollution status of the Yangtze River. Environmental Science and Pollution Research: 1–38.

18. Müller M, Berg M, Yao ZP, Zhang XF, Wang D, et al. (2008) How polluted is the Yangtze river? Water quality downstream from the Three Gorges Dam. Science of The Total Environment 402: 232–247.

19. Wang C, Peng B (2002) analysis of micro organic compound pollution in major city river reaches of the main stem of the Changjiang river. Yangtze River 33: 4–9.

20. Sun YZ, Zhang B, Gao LR, Liu ZT, Zheng MH (2005) Polychlorinated Dibenzo-p-Dioxins and Dibenzofurans in Surface Sediments from the Estuary Area of Yangtze River, People's Republic of China. Bulletin of Environmental Contamination and Toxicology 75: 910–914.

21. Yang H, Zhuo S, Xue B, Zhang C, Liu W (2012) Distribution, historical trends and inventories of polychlorinated biphenyls in sediments from Yangtze River Estuary and adjacent East China Sea. Environmental Pollution 169: 20–26.

22. So M, Miyake Y, Yeung W, Ho Y, Taniyasu S, et al. (2007) Perfluorinated compounds in the Pearl River and Yangtze River of China. Chemosphere 68: 2085–2095.

23. Chen SJ, Gao XJ, Mai BX, Chen ZM, Luo XJ, et al. (2006) Polybrominated diphenyl ethers in surface sediments of the Yangtze River Delta: Levels, distribution and potential hydrodynamic influence. Environmental Pollution 144: 951–957.

24. Wu LL, Chen L, Hou JL, Zhang YL, Zhao JF, et al. (2010) Assessment of Sediment Quality of Yangtze River Estuary Using Zebrafish (Danio rerio) Embryos. Environmental Toxicology 25: 234–242.

25. Hollert H, Dürr M, Erdinger L, Braunbeck T (2000) Cytotoxicity of settling particulate matter and sediments of the Neckar River (Germany) during a winter flood. Environmental Toxicology and Chemistry 19: 528–534.

26. Luo J, Ma M, Liu C, Zha J, Wang Z (2009) Impacts of particulate organic carbon and dissolved organic carbon on removal of polycyclic aromatic hydrocarbons, organochlorine pesticides, and nonylphenols in a wetland. Journal of Soils and Sediments 9: 180–187.

27. Klee N, Gustavsson L, Kosmehl T, Engwall M, Erdinger L, et al. (2004) Changes in toxicity and genotoxicity of industrial sewage sludge samples containing nitro- and amino-aromatic compounds following treatment in bioreactors with different oxygen regimes. Environmental Science and Pollution Research 11: 313–320.

28. Borenfreund E, Puerner J (1985) A simple quantitative procedure using monolayer cultures for cytotoxicity assays (HTD/NR-90). Journal of tissue culture methods 9: 7–9.

29. Wölz J, Engwall M, Maletz S, Olsman Takner H, Bavel B, et al. (2008) Changes in toxicity and Ah receptor agonist activity of suspended particulate matter during flood events at the rivers Neckar and Rhine – a mass balance approach using in vitro methods and chemical analysis. Environmental Science and Pollution Research 15: 536–553.

30. Kennedy SW, Jones SP (1994) Simultaneous Measurement of Cytochrome P4501A Catalytic Activity and Total Protein Concentration with a Fluorescence Plate Reader. Analytical Biochemistry 222: 217–223.

31. Engwall M, Broman D, Brunström B, Ishaq R, Näf C, et al. (1996) Toxic potencies of lipophilic extracts from sediments and settling particulate matter (SPM) collected in a PCB-contaminated river system. Environmental Toxicology and Chemistry 15: 213–222.

32. Bols NC, Schirmer K, Joyce EM, Dixon DG, Greenberg BM, et al. (1999) Ability of Polycyclic Aromatic Hydrocarbons to Induce 7-Ethoxyresorufin-o-deethylase Activity in a Trout Liver Cell Line. Ecotoxicology and Environmental Safety 44: 118–128.

33. Keiter S, Braunbeck T, Heise S, Pudenz S, Manz W, et al. (2009) A fuzzy logic-classification of sediments based on data from in vitro biotests. Journal of Soils and Sediments 9: 168–179.

34. Rocha PS, Azab E, Schmidt B, Storch V, Hollert H, et al. (2010) Changes in toxicity and dioxin-like activity of sediments from the Tietê River (São Paulo, Brazil). Ecotoxicology and Environmental Safety 73: 550–558.

35. Yang G, Weng L, Li L (2008) Yangtze conservation and development report 2007: Science Press.

36. Liu M, Baugh PJ, Hutchinson SM, Yu L, Xu S (2000) Historical record and sources of polycyclic aromatic hydrocarbons in core sediments from the Yangtze Estuary, China. Environmental Pollution 110: 357–365.

37. Wen S, Hui Y, Yang F, Liu Z, Xu Y (2008) Polychlorinated dibenzo-p-dioxins (PCDDs) and dibenzofurans (PCDFs) in surface sediment and bivalve from the Changjiang Estuary, China. Chinese Journal of Oceanology and Limnology 26: 35.

38. Bian H, Li Z, Liu P, Pan J (2010) Spatial distribution and deposition history of nonylphenol and bisphenol A in sediments from the Changjiang River (Yangtze River) Estuary and its adjacent East China Sea. Acta Oceanologica Sinica 29: 44–51.

39. Schirmer K, Dixon DG, Greenberg BM, Bols NC (1998) Ability of 16 priority PAHs to be directly cytotoxic to a cell line from the rainbow trout gill. Toxicology 127: 129–141.

40. Fent K, Hunn J (1996) Cytotoxicity of organic environmental chemicals to fish liver cells (PLHC-1). Marine Environmental Research 42: 377–382.

41. Otte JC, Keiter S, Faßbender C, Higley EB, Rocha PS, et al. (2013) Contribution of Priority PAHs and POPs to Ah Receptor-Mediated Activities in Sediment Samples from the River Elbe Estuary, Germany. Plos One 8: e75596.

42. Giesy JP, Jude DJ, Tillitt DE, Gale RW, Meadows JC, et al. (1997) Polychlorinated dibenzo-p-dioxins, dibenzofurans, biphenyls and 2,3,7,8-tetrachlorodibenzo-p-dioxin equivalents in fishes from Saginaw Bay, Michigan. Environmental Toxicology and Chemistry 16: 713–724.

43. Yoo H, Khim JS, Giesy JP (2006) Receptor-mediated in vitro bioassay for characterization of Ah-R-active compounds and activities in sediment from Korea. Chemosphere 62: 1261–1271.

44. Xia J, Su G, Zhang X, Shi W, Giesy JP, et al. (2014) Dioxin-like activity in sediments from Tai Lake, China determined by use of the H4IIE-luc bioassay and quantification of individual AhR agonists. Environmental Science and Pollution Research 21: 1480–1488.

45. Hong S, Khim JS, Naile JE, Park J, Kwon BO, et al. (2012) AhR-mediated potency of sediments and soils in estuarine and coastal areas of the Yellow Sea region: A comparison between Korea and China. Environmental Pollution 171: 216–225.

46. Song M, Jiang Q, Xu Y, Liu H, Lam PK, et al. (2006) AhR-active compounds in sediments of the Haihe and Dagu Rivers, China. Chemosphere 63: 1222–1230.

47. Huisman P, De Jong J, Wieriks K (2000) Transboundary cooperation in shared river basins: experiences from the Rhine, Meuse and North Sea. Water Policy 2: 83–97.

48. Machala M, Vondráček J, Bláha L, Ciganek M, Neča J (2001) Aryl hydrocarbon receptor-mediated activity of mutagenic polycyclic aromatic hydrocarbons determined using in vitro reporter gene assay. Mutation Research/Genetic Toxicology and Environmental Mutagenesis 497: 49–62.

49. Clemons J, Myers C, Lee L, Dixon D, Bols N (1998) Induction of cytochrome P4501A by binary mixtures of polychlorinated biphenyls (PCBs) and 2, 3, 7, 8-tetrachlorodibenzo-p-dioxin (TCDD) in liver cell lines from rat and trout. Aquatic Toxicology 43: 179–194.

50. Hinger G, Brinkmann M, Bluhm K, Sagner A, Takner H, et al. (2011) Some heterocyclic aromatic compounds are Ah receptor agonists in the DR-CALUX assay and the EROD assay with RTL-W1 cells. Environmental Science and Pollution Research 18: 1297–1304.

51. Wölz J, Brack W, Moehlenkamp C, Claus E, Braunbeck T, et al. (2010) Effect-directed analysis of Ah receptor-mediated activities caused by PAHs in suspended particulate matter sampled in flood events. Science of The Total Environment 408: 3327–3333.

52. Engwall M, Broman D, Dencker L, Näf C, Zebühr Y, et al. (1997) Toxic potencies of extracts of sediment and settling particulate matter collected in the recipient of a bleached pulp mill effluent before and after abandoning chlorine bleaching. Environmental Toxicology and Chemistry 16: 1187–1194.

53. Brack W, Segner H, Möder M, Schüürmann G (2000) Fixed-effect-level toxicity equivalents–a suitable parameter for assessing ethoxyresorufin-O-deethylase induction potency in complex environmental samples. Environmental Toxicology and Chemistry 19: 2493–2501.

54. Gustavsson LK, Klee N, Olsman H, Hollert H, Engwall M (2004) Fate of Ah receptor agonists during biological treatment of an industrial sludge containing explosives and pharmaceutical residues. Environmental Science and Pollution Research 11: 379–387.

55. Grung M, Lichtenthaler R, Ahel M, Tollefsen K-E, Langford K, et al. (2007) Effects-directed analysis of organic toxicants in wastewater effluent from Zagreb, Croatia. Chemosphere 67: 108–120.

56. Thomas K, Langford K, Petersen K, Smith A, Tollefsen K (2009) Effect-directed identification of naphthenic acids as important in vitro xeno-estrogens and anti-androgens in North Sea offshore produced water discharges. Environmental Science & Technology 43: 8066–8071.

57. Tang Z, Liao H, Zhang L, Guo J, Wu F, et al. (2011) Distribution, source and risk assessment of polycyclic aromatic hydrocarbons in river sediment of Cheng-Yu economic zone. Huanjing Kexue (Environmental Science) 32: 2639–2644.

58. Feng C, Xia X, Shen Z, Zhou Z (2007) Distribution and sources of polycyclic aromatic hydrocarbons in Wuhan section of the Yangtze River, China. Environmental Monitoring and Assessment 133: 447–458.

59. Bouloubassi I, Fillaux J, Saliot A (2001) Hydrocarbons in Surface Sediments from the Changjiang (Yangtze River) Estuary, East China Sea. Marine Pollution Bulletin 42: 1335–1346.

60. Wang Y, Li X, Li BH, Shen ZY, Feng CH, et al. (2012) Characterization, sources, and potential risk assessment of PAHs in surface sediments from nearshore and farther shore zones of the Yangtze estuary, China. Environmental Science and Pollution Research 19: 4148–4158.

61. Whyte JJ, Jung RE, Schmitt CJ, Tillitt DE (2000) Ethoxyresorufin-O-deethylase (EROD) Activity in Fish as a Biomarker of Chemical Exposure. Critical Reviews in Toxicology 30: 347–570.

62. Hollert H, Dürr M, Olsman H, Halldin K, van Bavel B, et al. (2002) Biological and Chemical Determination of Dioxin-like Compounds in Sediments by Means of a Sediment Triad Approach in the Catchment Area of the River Neckar. Ecotoxicology 11: 323–336.

63. Brack W, Schirmer K (2003) Effect-directed identification of oxygen and sulfur heterocycles as major polycyclic aromatic cytochrome P4501A-inducers in a contaminated sediment. Environmental Science & Technology 37: 3062–3070.

16

Distance to High-Voltage Power Lines and Risk of Childhood Leukemia – an Analysis of Confounding by and Interaction with Other Potential Risk Factors

Camilla Pedersen[1]*, Elvira V. Bräuner[1,2], Naja H. Rod[3], Vanna Albieri[1], Claus E. Andersen[4], Kaare Ulbak[5], Ole Hertel[6,7], Christoffer Johansen[1,8], Joachim Schüz[9], Ole Raaschou-Nielsen[1]

1 Danish Cancer Society Research Center, Copenhagen Ø, Denmark, 2 Danish Building Research Institute, Aalborg University, Construction and Health, Copenhagen SV, Denmark, 3 Social Medicine Section, Department of Public Health, University of Copenhagen, Copenhagen K, Denmark, 4 Risø National Laboratory for Sustainable Energy, Radiation Research Division, Technical University of Denmark, Roskilde, Denmark, 5 National Institute of Radiation Protection, Herlev, Denmark, 6 Department of Environmental Science, Aarhus University, Roskilde, Denmark, 7 Department for Environmental, Social and Spatial Change (ENSPAC), Roskilde University, Roskilde, Denmark, 8 Oncology Clinic, Finsen Centre, Rigshospitalet 5073, University of Copenhagen, Copenhagen Ø, Denmark, 9 International Agency for Research on Cancer (IARC), Section of Environment and Radiation, Lyon, France

Abstract

We investigated whether there is an interaction between distance from residence at birth to nearest power line and domestic radon and traffic-related air pollution, respectively, in relation to childhood leukemia risk. Further, we investigated whether adjusting for potential confounders alters the association between distance to nearest power line and childhood leukemia. We included 1024 cases aged <15, diagnosed with leukemia during 1968–1991, from the Danish Cancer Registry and 2048 controls randomly selected from the Danish childhood population and individually matched by gender and year of birth. We used geographical information systems to determine the distance between residence at birth and the nearest 132–400 kV overhead power line. Concentrations of domestic radon and traffic-related air pollution (NO_x at the front door) were estimated using validated models. We found a statistically significant interaction between distance to nearest power line and domestic radon regarding risk of childhood leukemia (p = 0.01) when using the median radon level as cut-off point but not when using the 75th percentile (p = 0.90). We found no evidence of an interaction between distance to nearest power line and traffic-related air pollution (p = 0.73). We found almost no change in the estimated association between distance to power line and risk of childhood leukemia when adjusting for socioeconomic status of the municipality, urbanization, maternal age, birth order, domestic radon and traffic-related air pollution. The statistically significant interaction between distance to nearest power line and domestic radon was based on few exposed cases and controls and sensitive to the choice of exposure categorization and might, therefore, be due to chance.

Editor: Ramiro Garzon, The Ohio State University, United States of America

Funding: This work was supported by a grant by the foundation Children with Cancer UK (formerly Children with Leukaemia) http://www.childrenwithcancer.org.uk/, by Danish Energy Association for provision on data on power lines, and by the Danish Cancer Society. The funders had no role in study design and analysis, decision to publish, or preparation of the manuscript. Danish Energy have provided data on power lines.

Competing Interests: Regarding the financial disclosure statement the authors can confirm that Danish Energy is the Danish Energy Association, which is a commercial and professional organization for Danish energy companies. The Danish Energy Association has not supported the research project financially but the transmission companies have provided data on power lines for the project. They had no role in study design and analysis, decision to publish, or preparation of the manuscript.

* Email: camped@cancer.dk

Introduction

In 2001 a working group at the International Agency for Research on Cancer (IARC) classified exposure to extremely low-frequency magnetic fields (ELF-MF) as 'possibly carcinogenic to humans' and exposure to extremely low-frequency electric fields were grouped as 'not classifiable as to its carcinogenicity to humans' [1]. The classification of ELF-MF was primarily based on epidemiological findings, showing an association between residential exposure to ELF-MF and childhood leukemia [2]. There is no known biological explanation for this association and the epidemiological findings have not been supported by animal studies [3]. Therefore, it is not known whether the observed association reflects a causal relationship or is due to bias, confounding or chance [2,3].

In 2005 a large-scale case-control study from Great Britain [4] showed an association between proximity of residence at birth to high-voltage power lines and the risk of childhood leukemia. The association extended beyond distances where the 'power line'-induced ELF-MF exceed background levels, which suggests that the association was not explained by the magnetic field but perhaps by some other risk factor. Several studies have looked for potential confounders which could explain the observed association between ELF-MF and childhood leukemia [5–18] including socioeconomic status, residential mobility, residence type, viral contacts, environmental tobacco smoke, dietary agents, and traffic density; but none of them appear to explain the association [19]. Little is known about the etiology of childhood leukemia and there are only few established causes of childhood leukemia including exposure to ionizing radiation (X-rays and gamma rays) and certain genetic diseases such as Down's syndrome [20,21]. The limited knowledge of the etiology of childhood leukemia makes it difficult to exclude the possibility of some yet unknown risk factor or of the combination of a number of risk factors, which could confound the analysis between ELF-MF and childhood leukemia [19].

The lack of an accepted biological explanation for the observed association between ELF-MF and childhood leukemia have raised doubt about the causality of the association. Several mechanisms of how extremely low-frequency electric and magnetic fields can cause cancer have, however, been proposed [3]. One hypothesis is that the electric field from power lines interacts with airborne pollutant particles and thereby increases the harmful effect of these particles [22–25]. Airborne particles such as tobacco smoke, radon decay products, chemical pollutants, spores, bacteria and viruses might all affect health. These particles can be deposited on the skin or in the airways by inhalation [3,25]. In 1999 Fews et al. [23] found a higher deposition of pollutant aerosols on the body under high-voltage power lines, which they argued was due to the oscillation of charged particles in the electric field from the power line. Further, Fews et al. [24] suggested that the electric field from high-voltage power lines can cause electrical breakdown of the air resulting in emission of clouds of positive and negative ions, which can charge particles that pass through them. Since charged particles are more likely than uncharged particles to be deposited when close to the walls of the respiratory airways or to the skin, this could increase the exposure and thereby the adverse health effect of such particles.

The major objective of our study was, therefore, to investigate possible interactions between distance from residence to nearest power line and domestic radon and traffic-related air pollution, respectively, in relation to childhood leukemia risk; to our knowledge this has not previously been investigated. Additionally, we investigated whether adjusting for potential confounding factors alters the association between distance from residence at birth to nearest power line and childhood leukemia. We adjusted for factors observed to be associated with childhood leukemia risk in other studies, specifically socioeconomic status of the municipality, urbanization, maternal age, birth order, domestic radon and traffic-related air pollution [20,21,26,27], which might also be related to distance from residence to nearest power line. The main effect of distance to nearest power line and childhood leukemia risk has been reported in a previous paper [28].

Materials and Methods

Ethics statement

The Danish Data Protection Agency (2007-41-0239) approved the study. In accordance with Danish law written consent was not obtained as the study was entirely register-based and did not involve biological samples from, or contact with study participants.

Cases and controls

We identified all children in Denmark diagnosed with leukemia (all types) before the age of 15 years during the period 1968–1991 (inclusively) through the virtually complete nationwide Danish Cancer Registry [29]. Children with a previous cancer diagnosis were excluded. Two controls for each case were individually matched by gender and year of birth and were selected by incidence density sampling from the Danish Civil Registration System. All children born in Denmark, who were alive, without cancer and living in Denmark at the time of diagnosis of their matched case were eligible to become a control in the study.

We identified 1024 cases diagnosed with leukemia and selected 2048 matched controls, constituting a population of 3072 children (Figure S1).

Distance to nearest power line

The addresses of cases and controls at birth were obtained from the Danish Civil Registration System and we identified geographical coordinates by linkage to the Danish Address Database. The geographical coordinates refer to the front door of each residence (for apartments the main entrance door) and are precise within a few meters. We collected data on existing and historical 132–400 kV overhead power lines with alternating current from the seven Danish transmission companies. The mapped grid includes 4336 km of current and historical power lines (23.1% 400 kV, 0.9% 220 kV, 46.3% 150 kV, 29.8% 132 kV). In addition information on the date when the line was put into and out of operation was collected. For the lines for which we only had information on the year of operation (77.5%), we set the date of operation to 31 December and the date when the line was put out of operation was set to 1 January. This restrictive approach was applied to avoid assigning exposure to unexposed individuals. The distance from residence at birth to the nearest power line that existed at the date of birth was calculated in ArcGIS 9.3. We successfully calculated distance for 2797 (91.0%) of the addresses at birth. Distance could not be calculated due to missing information in the address for 100 (9.8%) cases and 175 (8.5%) controls (Figure S1). Distance was categorized into three groups: 0–199 meters, 200–599 meters and ≥600 meters, in accordance with the categorization used by Draper et al. [4].

Domestic radon

Domestic radon concentration at address at birth was estimated using a validated regression model constructed to predict radon concentrations in Danish dwellings on the basis of register data including geographical region, soil type, and house characteristics such as type of house, floor, basement, and building materials. Details on the model are given in previous papers [30,31].

Exposure to domestic radon at address at birth was successfully estimated for 2904 (94.5%) addresses. We could not estimate domestic radon exposure for 62 (6.1%) cases and 106 (5.2%) controls (Figure S1).

Traffic-related air pollution

We used the sum of nitrogen oxide gasses, NO_x (nitrogen monoxide (NO) + nitrogen dioxide (NO_2)), as an indicator for air pollution from traffic because NO_x correlates strongly with other traffic-related pollutants in Danish streets, especially ultrafine PM (Particulate matter): $r = 0.93$ for total particle number concentration (size, 10–700 nm) and $r = 0.70$ for PM_{10} (mass of particles with an aerodynamic diameter less than 10 μm) [32,33]. The average concentration of NO_x at the front door of the address at birth was estimated by use of the Operational Street Pollution Model taking into account both the street-level pollution as well as the background pollution [34,35,36,37]. In short the model was used to estimate the air pollution level from street information (such as width of the street, the height of the buildings, the distance between building and street sections with no buildings), the traffic density on the street, the proportion of vehicles weighing more than 3,500 kg, the average speed in combination with information on emission factors for the Danish car fleet, meteorological variables (wind speed, temperature and solar radiation) and background level of air pollution at the front door.

Exposure to traffic-related air pollution at the birth address was successfully estimated for 2942 (95.8%) addresses and could not be estimated for 45 (4.4%) cases and 85 (4.2%) controls (Figure S1).

Potential confounders

Socioeconomic status of the municipality (the average gross income of the municipality) has been described in detail previously [27], where it was estimated for children born in the period 1976–1991. Children born in the period 1953–1975 were assigned the municipality income level based on data in 1976. Definition of rural and urban area was based on the official Danish classification. Information on maternal age and birth order was provided by linkage with the Danish Civil Registration System using the personal identification number. We also adjusted the analyses for domestic radon and traffic-related air pollution.

We had no information on socio-economic status of the municipality for 6 cases and 12 controls, on urbanization for 100 cases and 175 controls, on maternal age for 43 cases and 77 controls and on birth order for 43 cases and 78 controls (Figure S1).

Statistical analysis

Because of the nested case-control design of the study, with equal follow-up time for a case and the matched controls, the rate ratios [38] (hereafter denoted relative risks (RR)) for childhood cancer and 95% confidence intervals (CIs) were estimated by conditional logistic regression models with the "PROC PHREG" procedure in SAS 9.2.

We analyzed the association between distance to nearest power line and risk of childhood leukemia both with and without adjustment for potential confounders. The potential confounders included socio-economic status of the municipality (categorical; < 10 pct., 10–90 pct., >90 pct.), urbanization (categorical; countryside, urban area), maternal age at time of birth (categorical; <30y, ≥30y), birth order (categorical; 1, ≥2), domestic radon (categorical; <50 pct., 50–90 pct., >90 pct.) and traffic-related air pollution (categorical; <50 pct., 50–90 pct., >90 pct.).

We analyzed possible interactions between distance to power lines and domestic radon and traffic-related air pollution, respectively. Since there is no known biological relevant cut-off point for radon or NO_x, we dichotomized domestic radon and traffic-related air pollution using the median value as cut-off; in sensitivity analyses we used the 75th percentile as cut-off point. In an additional sensitivity analysis we divided radon and NO_x into tertiles in order to investigate a potential dose-response association. All interaction analyses were adjusted for the potential confounders: socioeconomic status of the municipality, urbanization, maternal age at time of birth, birth order and domestic radon and traffic-related air pollution, respectively. From the former study on radon [31] we had information on radon for cases and controls until 1994 and therefore we conducted an additional sensitivity analysis (unadjusted) of the interaction between power lines and radon to maximize the power. Furthermore, the interaction was analyzed including radon as a continuous variable. The p-value for the interaction was obtained by comparing and testing the model with the interaction term against the model with only the main effect by means of the likelihood ratio test.

Due to small samples in some of the categories in the analysis of the interaction between distance to power line and radon, exact methods for logistic regression were also applied using LogXact. It was only possible to compute models without adjustment for confounding factors.

Results

We excluded 139 (13.6%) cases and 248 (12.1%) controls due to missing data on distance to power line or any of the potential confounding factors: domestic radon, traffic-related air pollution, socioeconomic status of the municipality, urbanization, maternal age or birth order. Further, we excluded 6 cases and 179 controls since they no longer had a matched control or case, leaving 879 (85.8%) cases and 1621 (79.2%) controls for the analyses (Figure S1).

In Table 1 the distribution of potential confounders by case-controls status and by distance to nearest power line is shown. Cases were more likely than controls to be living in a municipality with a lower socioeconomic status at time of birth, to have a mother at the age of 30 or older at time of birth, and to have at least one older sibling. Cases and controls were similar regarding exposure to domestic radon and traffic-related air pollution at address at birth and regarding whether they were living in a town or at the countryside at the time of birth. Children living close to power lines tended to live in the countryside, to be exposed to higher concentrations of domestic radon, and to be exposed to smaller concentrations of air pollution than children living further away. Children living close to power lines were similar to children living far away regarding socioeconomic status of the municipality, maternal age and birth order.

Table 2 presents the crude and adjusted associations between distance to power lines and childhood leukemia. Adjusting for potential confounders had virtually no effect on the estimated association between distance to nearest power line and childhood leukemia.

Table 3 shows the respective joint effects of distance to nearest power line and domestic radon and air pollution on leukemia. We found a statistically significant interaction between distance to nearest power line and domestic radon (p = 0.01). The RR for children living 0–199 meters of a power line and being exposed to domestic radon ≥42 Bq/m^3 was 2.88 (95% CI: 1.01–8.27) compared to children living ≥600 meters from a power line and exposed to domestic radon <42 Bq/m^3. Children living 200–599 meters from a power line and being exposed to domestic radon < 42 Bq/m^3 had a lower risk with an RR of 0.24 (95% CI: 0.07–0.83) compared to children living ≥600 meters from a power lines and exposed to domestic radon <42 Bq/m^3. There was no evidence of interaction between distance to nearest power line and traffic-related air pollution (p = 0.73). The sensitivity analysis including domestic radon and traffic-related air pollution with a cut-off at the 75th percentile showed no significant interaction with domestic radon (p = 0.90) or air pollution (p = 0.59) (Tables S1) and there was no dose-response association when radon and air pollution were divided into tertiles (Table S2). The interaction between power lines and radon remained statistically significant in the sensitivity analysis including radon until 1994, however, the RR for children exposed to radon concentrations of ≥42 Bq/m^3 and living within 200 meters of a power line was no longer statistically significant (RR = 2.62 (95% CI: 0.97–7.07)) (Table S3). When radon was included as a continuous variable the interaction between distance to power line and radon was not statistically significant (p = 0.37) (data not shown). The result of the interaction analysis using exact methods was similar to the result of the unadjusted analysis using asymptotic methods (both presented in Table S4). The RR for children living within 200 m of a power line and exposed to radon concentrations of ≥42 Bq/m^3 was still elevated but not statistically significant.

Table 1. Potential confounders by case-control status and distance to nearest power line.

	Case-control status				P-value (χ²-test)	Distance to nearest power line (meters)						P-value (χ²-test)
	Cases		Controls			0–199		200–599		≥600		
	N	(%)	N	(%)		N	(%)	N	(%)	N	(%)	
Socioeconomic status					0.02							0.15
10% most poor	46	(5.2)	52	(3.2)		0	(0.0)	1	(0.9)	97	(4.1)	
80% middle group	739	(84.1)	1362	(84.0)		19	(86.4)	86	(81.1)	1996	(84.2)	
10% most rich	94	(10.7)	207	(12.8)		3	(13.6)	19	(17.9)	279	(11.8)	
Urbanization					0.63							0.01
Town	639	(72.7)	1193	(73.6)		15	(68.2)	65	(61.3)	1752	(73.9)	
Country side	240	(27.3)	428	(26.4)		7	(31.8)	41	(38.7)	620	(26.1)	
Maternal age (years)					0.01							0.89
<30	644	(73.3)	1260	(77.7)		16	(72.7)	82	(77.4)	1806	(76.1)	
≥30	235	(26.7)	361	(22.3)		6	(27.3)	24	(22.6)	566	(23.9)	
Birth order					0.02							0.62
1	363	(41.3)	746	(46.0)		8	(36.4)	44	(41.5)	1057	(44.6)	
>1	516	(58.7)	875	(54.0)		14	(63.6)	62	(58.5)	1315	(55.4)	
Domestic radon (Bq/m³)[1]					0.77							<0.01
<42	432	(49.2)	818	(50.5)		7	(31.8)	26	(24.5)	1217	(51.3)	
42–101	360	(41.0)	640	(39.5)		13	(59.1)	62	(58.5)	925	(39.0)	
≥102	87	(9.9)	163	(10.1)		2	(9.1)	18	(17.0)	230	(9.7)	
NOx at the front door (ppb)[1]					0.21							0.04
<9	460	(52.3)	790	(48.7)		13	(59.1)	62	(58.5)	1175	(49.5)	
9–20	338	(38.5)	662	(40.8)		8	(36.4)	42	(39.6)	950	(40.1)	
≥21	81	(9.2)	169	(10.4)		1	(4.6)	2	(1.9)	247	(10.4)	
Total	879	(100.0)	1621	(100.0)		22	(100.0)	106	(100.0)	2372	(100.0)	

[1]Cut-point is the 50th and 90th percentile.

Table 2. Crude and adjusted RRs for leukemia in association with distance to nearest power line.

Distance to nearest power line (meters)	Cases N (%)	Controls N (%)	Total	Model 1[a] RR (CI)	P-value	Model 2[b] RR (CI)	P-value	Model 3[c] RR (CI)	P-value	Model 4[d] RR (CI)	P-value
0–199	10 (1.1)	12 (0.7)	22	1.65 (0.71–3.83)		1.77 (0.76–4.11)		1.73 (0.74–4.02)		1.68 (0.72–3.92)	
200–599	29 (3.3)	77 (4.8)	106	0.68 (0.44–1.05)		0.69 (0.45–1.07)		0.70 (0.45–1.09)		0.70 (0.45–1.09)	
≥600	840 (95.6)	1532 (94.5)	2372	1.00	0.11	1.00	0.11	1.00	0.13	1.00	0.13
Total	879 (100.0)	1621 (100.0)	2500								

[a] The crude model.
[b] Adjusted for socioeconomic status and urbanization.
[c] Adjusted for the same as model 2 and for maternal age and birth order.
[d] Adjusted for the same as model 3 and for domestic radon and air pollution.

Discussion

We found a statistically significant interaction between distance to nearest power line and domestic radon regarding risk of childhood leukemia when using the median radon level as cut-off point but not when using the 75th percentile of radon exposure as cut-off point. We found no evidence of an interaction between distance to nearest power line and traffic-related air pollution. We found almost no change in the estimated association between distance to power lines and risk of childhood leukemia when adjusting for socioeconomic status of the municipality, urbanization, maternal age, birth order, domestic radon and traffic-related air pollution.

Strengths and limitations

Our study was a register-based case-control study covering the entire population of Denmark. Cases were identified in a virtually complete national cancer registry, and the Central Population Registry provided an excellent basis for an unbiased sampling of controls. The potential for selection bias in our study is, therefore, minimal. We identified around 90% of current as well as historical power lines with ≥132 kV and according to two of the transmission companies (covering approximately 25% of all power lines) the data on power lines had a precision of 3–5 meters. The geographical coordinates used to identify the front door at the address at birth had a precision of few meters. For apartments the main entrance door was identified, which for the majority of apartments is within 5 meters, however, for some types of apartments a larger deviation may be expected.

Distance to nearest power line, domestic radon and air pollution was estimated for the address at birth, although exposure accumulated over all childhood addresses might be more relevant [31]. However, since interactions were in focus in the present study, we included the exposures at birth addresses only to ensure that the exposure to power lines (distance from residence) and exposure to domestic radon and traffic-related air pollution was present at the same address, which is essential for the analysis of interaction.

It is a limitation of the study that the analyses of interaction were based on few cases and controls for some of the cells of table 3. Data on air pollution was available from a previous study for children diagnosed during the period 1968–1991. It was not feasible to collect data for an extended period, though it would have contributed with more cases and controls and thereby more power in the analyses.

Exposure to domestic radon and traffic-related air pollution at the birth address were both modeled using successfully validated prediction models which have been applied in previous studies [31,39–41]. Radon values predicted by the model have previously been validated against measured values: Some non-differential misclassification occurs, but 80% of the lowest exposures (<50 Bq/m^3) were correctly classified and 60% of the highest exposures (>100 Bq/m^3) were correctly classified [31]. The internationally widely used [36] Operational Street Pollution Model used to predict the NO$_x$ concentration at the front door has been validated by comparing the calculated nitrogen dioxide (NO$_2$) values with measured values at 204 locations in Denmark with an R^2 value of 0.82 [42]. Even though more precise estimates for exposure may be obtained with measured values, modeled estimates make it possible to avoid participation bias and to obtain a larger amount of participants than would be possible with measurement in each house.

Table 3. The joint effects of distance to nearest power line and domestic radon and air pollution, respectively, on leukemia.

	Adjusted			P-value for interaction
	RR (95% CI)	(N cases; N controls)		
	Distance (meters)			
	0–199	200–599	≥600	
Domestic radon (Bq/m^3)[1, 2]				
<42	0.33 (0.04–2.80)	0.24 (0.07–0.83)	1.00	0.01
	(1; 6)	(3; 23)	(428; 789)	
≥42	2.88 (1.01–8.27)	0.85 (0.52–1.39)	0.95 (0.79–1.16)	
	(9; 6)	(26; 54)	(412; 743)	
NO$_x$ at the front door (ppb)[1, 3]				
<9	2.21 (0.73–6.64)	0.68 (0.38–1.21)	1.00	0.73
	(7; 6)	(17; 45)	(436; 739)	
≥9	1.02 (0.25–4.17)	0.67 (0.34–1.33)	0.92 (0.76–1.11)	
	(3; 6)	(12; 32)	(404; 793)	

[1]Cut-point is the median.
[2]The adjusted analysis includes following potential confounders: socioeconomic status, urbanization, maternal age, birth order and air pollution.
[3]The adjusted analysis includes following potential confounders: socioeconomic status, urbanization, maternal age, birth order and domestic radon.

Comparison with other studies and interpretation

In a previous study, where we included all children diagnosed with leukemia before the age of 15 years during the period 1968–2006, we report a OR of 0.76 (95% CI: 0.40–1.45) for children who lived 0–199 meters from the nearest power line compared with children who lived ≥600 meters away [28]. In the present study we report an RR of 1.68 (95% CI: 0.72–3.92) when living 0–199 meters from the nearest power line compared to those living ≥600 from the nearest power line. The 95% CIs of the two estimates widely overlap and none of the estimates are statistically significant, which indicate that the different estimates are due to chance and are two versions of the same null-result.

We found a statistically significant interaction between distance from residence to nearest power line and domestic radon but no interaction between distance and traffic-related air pollution. The interaction with radon showed a higher risk of childhood leukemia if living close to power lines and being exposed to high radon concentrations. This is in line with the hypothesis by Fews et al. [24] that ions of positive and negative charge emitted from power lines increases the proportion of charged radon particles and thereby increases the deposition of radon decay products on the skin or in the airways. Thus, the hypothesis by Fews et al. could explain the interaction observed in the present study. On the other hand the effect of the ions-clouds is expected to have the highest impact on particles arising outdoors whilst the exposure to radon is highest indoor where it can concentrate. According to an independent Advisory Group on Non-Ionising Radiation, advising The National Radiological Protection Board (NRPB) in UK, some of the ions are expected to enter the houses but the effect of the ion-clouds must be lower inside the house, since some of the ions would be deposited on the walls where entering the houses [3,25]. There could also be other explanations for an interaction between power lines and domestic radon in relation to childhood leukemia; for instance if the exposure to power lines makes the child more vulnerable to exposure to radon, or vice versa.

The interpretation of the observed interaction is not straight-forward since it is not only based on a statistically significant higher risk among those exposed to high radon and living close to a power line but also a statistically significant lower risk among those being exposed to low domestic radon and living 200–599 meters from a power line, which is not easy to explain. The observed interaction might be due to chance because of few cases and controls in several cells. The finding of no statistically significant interaction when the 75th percentile was used as cut-off also suggests that the interaction found when using the median as cut-off might be due to chance. Even if the interaction between power lines and radon in relation to childhood leukemia is true, the potential effect in terms of absolute risk would be very small.

We found almost no change in the estimated association between distance to power lines and risk of childhood leukemia when adjusting for socioeconomic status of the municipality, urbanization, maternal age, birth order, domestic radon and traffic-related air pollution. This is in line with other studies investigating the association between ELF-MF and childhood leukemia, where the inclusion of potential confounders had no substantial effect on the estimates [5–18].

Conclusion

The statistically significant interaction observed between power lines and radon might be due to chance since numbers were small and the interaction was no longer significant when changing the cut-point for radon exposure. We found no support for an interaction between power lines and traffic-related air pollution and no change in the estimated association between distance to power lines and the risk of childhood leukemia by adjustment for potential confounding factors.

Supporting Information

Figure S1   Number of cases and controls in the study.

Table S1   The joint effects of distance to nearest power line and domestic radon and air pollution, respectively, on leukemia.

Table S2   The joint effects of distance to nearest power line and domestic radon and air pollution, respectively, on leukemia.

Table S3   The joint effects of distance to nearest power line and domestic radon on leukemia for the period 1968–1994.

Table S4   The joint effects of distance to nearest power line and domestic radon on leukemia risk with exact methods and with asymptotic methods.

Author Contributions

Conceived and designed the experiments: ORN JS CJ CP. Performed the experiments: CP. Analyzed the data: CP EVB NHR VA ORN. Contributed reagents/materials/analysis tools: CEA KU OH. Wrote the paper: CP.

References

1. IARC Working Group on the Evaluation of Carcinogenic Risks to Humans (2002) Non-ionizing radiation, Part 1: Static and extremely low-frequency (ELF) electric and magnetic fields. In: IARC Monographs on the Evaluation of Carcinogenic Risks to Humans volume 80. Lyon: IARCPress.
2. Schüz J, Ahlbom A (2008) Exposure to electromagnetic fields and the risk of childhood leukaemia: a review. Radiat Prot Dosimetry 132: 202–211. doi:10.1093/rpd/ncn270.
3. WHO (2007) Extremely low frequency fields. Geneva: World Health Organization. (Environmental health criteria 238).
4. Draper G, Vincent T, Kroll ME, Swanson J (2005) Childhood cancer in relation to distance from high voltage power lines in England and Wales: a case-control study. BMJ 330: 1290–1292.
5. Wünsch-Filho V, Pelissari DM, Barbieri FE, Sant'Anna L, de Oliveira CT, et al. (2011) Exposure to magnetic fields and childhood acute lymphocytic leukemia in São Paulo, Brazil. Cancer Epidemiol 35: 534–539. doi:10.1016/j.canep.2011.05.008.
6. Kroll ME, Swanson J, Vincent TJ, Draper GJ (2010) Childhood cancer and magnetic fields from high-voltage power lines in England and Wales: a case-control study. Br J Cancer 103: 1122–1127. doi:10.1038/sj.bjc.6605795.
7. Schüz J, Grigat JP, Brinkmann K, Michaelis J (2001) Residential magnetic fields as a risk factor for childhood acute leukaemia: results from a German population-based case-control study. Int J Cancer 91: 728–735.
8. Ahlbom A, Day N, Feychting M, Roman E, Skinner J, et al. (2000) A pooled analysis of magnetic fields and childhood leukaemia. Br J Cancer 83: 692–698.
9. UK Childhood Cancer Study Investigators (2000) Childhood cancer and residential proximity to power lines. Br J Cancer 83: 1573–1580.
10. UK Childhood Cancer Study Investigators (1999) Exposure to power-frequency magnetic fields and the risk of childhood cancer. Lancet 354: 1925–1931.
11. McBride ML, Gallagher RP, Thériault G, Armstrong BG, Tamaro S, et al. (1999) Power-frequency electric and magnetic fields and risk of childhood leukemia in Canada. Am J Epidemiol 149: 831–842.
12. Green LM, Miller AB, Villeneuve PJ, Agnew DA, Greenberg ML, et al. (1999) A case-control study of childhood leukemia in southern Ontario, Canada, and exposure to magnetic fields in residences. Int J Cancer 82: 161–170.
13. Green LM, Miller AB, Agnew DA, Greenberg ML, Li J, et al. (1999) Childhood leukemia and personal monitoring of residential exposures to electric and magnetic fields in Ontario, Canada. Cancer Causes Control 10: 233–243.
14. Linet MS, Hatch EE, Kleinerman RA, Robison LL, Kaune WT, et al. (1997) Residential exposure to magnetic fields and acute lymphoblastic leukemia in children. N Engl J Med 337: 1–7.
15. Michaelis J, Schüz J, Meinert R, Menger M, Grigat JP, et al. (1997) Childhood leukemia and electromagnetic fields: results of a population-based case-control study in Germany. Cancer Causes Control 8: 167–174.
16. Petridou E, Trichopoulos D, Kravaritis A, Pourtsidis A, Dessypris N, et al. (1997) Electrical power lines and childhood leukemia: a study from Greece. Int J Cancer 73: 345–348.
17. London SJ, Thomas DC, Bowman JD, Sobel E, Cheng TC, et al. (1991) Exposure to residential electric and magnetic fields and risk of childhood leukemia. Am J Epidemiol 134: 923–937.
18. Savitz DA, Wachtel H, Barnes FA, John EM, Tvrdik JG (1988) Case-control study of childhood cancer and exposure to 60-Hz magnetic fields. Am J Epidemiol 128: 21–38.
19. Kheifets L, Shimkhada R (2005) Childhood leukemia and EMF: review of the epidemiologic evidence. Bioelectromagnetics 26(Suppl 7): S51–S59.
20. Little J (1999) Epidemiology of Childhood Cancer. IARC scientific publications 149. Lyon: International Agency for Research on Cancer.
21. Rossig C, Juergens H (2008) Aetiology of childhood acute leukaemias: current status of knowledge. Radiat Prot Dosimetry 132: 114–118. doi:10.1093/rpd/ncn269.
22. Henshaw DL, Ross AN, Fews AP, Preece AW (1996) Enhanced deposition of radon daughter nuclei in the vicinity of power frequency electromagnetic fields. Int J Radiat Biol 69: 25–38.
23. Fews AP, Henshaw DL, Keitch PA, Close JJ, Wilding RJ (1999) Increased exposure to pollutant aerosols under high voltage power lines. Int J Radiat Biol: 75: 1505–1521.
24. Fews AP, Henshaw DL, Wilding RJ, Keitch PA (1999) Corona ions from powerlines and increased exposure to pollutant aerosols. Int J Radiat Biol 75: 1523–1531.
25. AGNIR - Advisory Group on Non-Ionising Radiation (2004) Particle deposition in the vicinity of power lines and possible effects on health. Chilton: National Radiological Protection Board, (Documents of the NRPB, Vol. 15, No. 1).
26. Eden T (2010) Aetiology of childhood leukaemia. Cancer Treat Rev 36: 286–297. doi:10.1016/j.ctrv.2010.02.004.
27. Raaschou-Nielsen O, Obel J, Dalton S, Tjønneland A, Hansen J (2004) Socioeconomic status and risk of childhood leukaemia in Denmark. Scand J Public Health 32: 279–286.
28. Pedersen C, Raaschou-Nielsen O, Rod NH, Frei P, Poulsen AH, et al. (2013). Distance from residence to power line and risk of childhood leukemia: a population-based case-control study in Denmark. Cancer Causes Control. doi:10.1007/s10552-013-0319-5.
29. Gjerstorff ML (2011) The Danish cancer registry. Scand J Public Health 39(7 Suppl): 42–45. doi:10.1177/1403494810393562.
30. Andersen CE, Raaschou-Nielsen O, Andersen HP, Lind M, Gravesen P, et al. (2007) Prediction of 222Rn in Danish dwellings using geology and house construction information from central databases. Radiat Prot Dosimetry 123: 83–94.
31. Raaschou-Nielsen O, Andersen CE, Andersen HP, Gravesen P, Lind M, et al. (2008) Domestic radon and childhood cancer in Denmark. Epidemiology 19: 536–543. doi:10.1097/EDE.0b013e318176bfcd.
32. Hertel O, Solvang S, Andersen HV, Palmgren F, Wåhlin P, et al. (2001) Human exposure to traffic pollution. Experience from Danish studies. Pure Appl Chem 73: 137–145.
33. Ketzel M, Wåhlin P, Berkowicz R, Palmgren F (2003) Particle and trace gas emission factors under urban driving conditions in Copenhagen based on street and roof-level observations. Atmos Environ 37: 2735–2749.
34. Raaschou-Nielsen O, Hertel O, Vignati E, Berkowicz R, Jensen SS, et al. (2000) An air pollution model for use in epidemiological studies: evaluation with measured levels of nitrogen dioxide and benzene. J Expo Anal Environ Epidemiol 10: 4–14.
35. Berkowicz R (2000) OSPM - a parametrised street pollution model. Environ Monit Assess 65: 323–331.
36. Kakosimos KE, Hertel O, Ketzel M, Berkowicz R (2010) Operational Street Pollution Model (OSPM) – a review of performed application and validation studies, and future prospects. Environ Chem 7: 485–503. doi:10.1071/EN10070.
37. Jensen SS, Berkowicz R, Hansen HS, Hertel O (2001) A Danish decision-support GIS tool for management of urban air quality and human exposures. Transportation Res. Part D: Transport and Environment, 6: 229–241.
38. Prentice RL, Breslow NE (1978) Retrospective studies and failure time models. Biometrika 65: 153–158.
39. Bräuner EV, Andersen ZJ, Andersen CE, Pedersen C, Gravesen P, et al. (2013) Residential radon and brain tumour incidence in a Danish cohort. PLoS One 8: e74435. doi:10.1371/journal.pone.0074435.
40. Bräuner EV, Andersen CE, Sørensen M, Andersen ZJ, Gravesen P, et al. (2012) Residential radon and lung cancer incidence in a Danish cohort. Environ Res 118: 130–136. doi:10.1016/j.envres.2012.05.012.
41. Bräuner EV, Andersen CE, Andersen HP, Gravesen P, Lind M, et al. (2010) Is there any interaction between domestic radon exposure and air pollution from traffic in relation to childhood leukemia risk? Cancer Causes Control 21: 1961–1964. doi:10.1007/s10552-010-9608-4.
42. Ketzel M, Berkowicz R, Hvidberg M, Jensen SS (2011) Evaluation of AirGIS: a GIS-based air pollution and human exposure modelling system. Int J Environment and Pollution 47: 226–238.

17

A New Approach to Standardize Multicenter Studies: Mobile Lab Technology for the German Environmental Specimen Bank

Dominik Lermen[1], Daniel Schmitt[2], Martina Bartel-Steinbach[1], Christa Schröter-Kermani[3], Marike Kolossa-Gehring[3], Hagen von Briesen[1], Heiko Zimmermann[1,2,4]*

1 Department of Cell Biology & Applied Virology, Fraunhofer-Institute for Biomedical Engineering, St. Ingbert, Saarland, Germany, **2** Department of Laboratory & Information Technology, Fraunhofer-Institute for Biomedical Engineering, St. Ingbert, Saarland, Germany, **3** Federal Environment Agency (UBA), Berlin, Berlin, Germany, **4** Saarland University, Saarbruecken, Saarland, Germany

Abstract

Technical progress has simplified tasks in lab diagnosis and improved quality of test results. Errors occurring during the pre-analytical phase have more negative impact on the quality of test results than errors encountered during the total analytical process. Different infrastructures of sampling sites can highly influence the quality of samples and therewith of analytical results. Annually the German Environmental Specimen Bank (ESB) collects, characterizes, and stores blood, plasma, and urine samples of 120–150 volunteers each on four different sampling sites in Germany. Overarching goal is to investigate the exposure to environmental pollutants of non-occupational exposed young adults combining human biomonitoring with questionnaire data. We investigated the requirements of the study and the possibility to realize a highly standardized sampling procedure on a mobile platform in order to increase the required quality of the pre-analytical phase. The results lead to the development of a mobile epidemiologic laboratory (epiLab) in the project "Labor der Zukunft" (future's lab technology). This laboratory includes a 14.7 m^2 reception area to record medical history and exposure-relevant behavior, a 21.1 m^2 examination room to record dental fillings and for blood withdrawal, a 15.5 m^2 biological safety level 2 laboratory to process and analyze samples on site including a 2.8 m^2 personnel lock and a 3.6 m^2 cryofacility to immediately freeze samples. Frozen samples can be transferred to their final destination within the vehicle without breaking the cold chain. To our knowledge, we herewith describe for the first time the implementation of a biological safety laboratory (BSL) 2 lab and an epidemiologic unit on a single mobile platform. Since 2013 we have been collecting up to 15.000 individual human samples annually under highly standardized conditions using the mobile laboratory. Characterized and free of alterations they are kept ready for retrospective analyses in their final archive, the German ESB.

Editor: Clive M. Gray, University of Cape Town, South Africa

Funding: DL and HvB received funding for the sampling from the Federal Ministry for the Environment, Nature Conservation, Building and Nuclear Safety. Grant Number: 301 02 048 http://www.bmub.bund.de/en/. DS and HZ received funding from the government and the State Chancellery of the federal state Saarland (Germany). Grant Number: C/1-LdZ-2011 http://www.saarland.de/staatskanzlei.htm. The funders had no role in study design, data collection and analysis, decision to publish, or preparation of the manuscript.

Competing Interests: Dominik Lermen, Daniel Schmitt, Martina Bartel-Steinbach and Heiko Zimmermann are with Fraunhofer IBMT. Fraunhofer IBMT is a non-for-profit research institute of the Fraunhofer-Society in Germany. The mobile laboratory has been conceptually designed by Fraunhofer IBMT and was tailored to the needs and plans of Fraunhofer IBMT in the framework of a research project funded by German federal state Saarland.

* Email: heiko.zimmermann@ibmt.fraunhofer.de

Introduction

Multicenter sampling events are the cornerstone of a high diversity of epidemiologic studies, health related environmental monitoring, and human biomonitoring studies [1–8]. These studies have an increasing demand for standardized working conditions and a standardized pre-analytical phase since misleading results due to improper sampling conditions are currently more relevant than errors occurring during lab analysis [9–14]. Once a sample is collected in an improper way the best analytical tool will not be able to reveal its pristine information. Errors in the pre-analytical phase mostly occur at the time of specimen collection. According to Bonini and colleagues (2002) errors in the pre-analytical phase predominate in the laboratory, ranging from 31.6% to 75% compared to errors that may occur in the analytical or post-analytical phase [15–19].

To reduce errors in the pre-analytical phase of multicenter sampling events, mobile units can provide the required infrastructure on various sampling sites and therewith simplify standardization. Mobile units based on diverse vehicle platforms are known from both the medical field and environmental research. Medical applications are mainly dedicated to acute medicine, thus ambulances equipped with specific instruments for the investigation of diseases like myocardial infarction [20,21], lung cancer [22], and stroke [23,24]. Mobile units are favorably used in natural disaster scenarios [25,26], during the outbreak of severe infectious diseases [27], and in military campaigns [28]. However, the latter examples show that more advanced mobile medical units for interventions are rather based on containers than on vans or trailers. Medical laboratories "on wheels" are only reported for restricted analytical procedures [1] and in the field of human health monitoring programs [29]. Environmental research is utilizing mobile units for on-site analysis of water and soil [30]. Specific units have been used to determine atmospheric pollutants [31–34] and xenobiotics [35]. Interestingly, samples are mostly analyzed on-site and sample storage is of minor concern.

The German ESB, as one example of a multicenter study in Germany, is a central element of the German environmental monitoring system. It is an archive of samples from representative animals and plants, soil, suspended particles, and human samples like blood, plasma, 24 h-urine, hair, and saliva (http://www.umweltprobenbank.de/en/). The German ESB collects, cryopreserves, cryostores, and analyzes human samples since the early 1980s [36,37]. Every year a maximum of 15.000 human samples from up to 600 young adults at four sites in Germany (Muenster, Halle/Saale, Greifswald, Ulm) are being collected. The blood, plasma and 24 h-urine samples subsequently get analyzed on selected environmental pollutants and physiological parameters. Stored samples allow rapid retrospective monitoring of emerging contaminants whenever needed. Since it is not known which chemicals will be of interest in the future and which concentrations of these chemicals can be found in environmental or human samples, it is highly important to avoid contaminations during the sampling procedure and during the pre-analytical phase. During sampling dental status is recorded. Medical history, exposure-relevant behavior, dietary habits, and living conditions of each volunteer are documented, using a standardized and self-reported questionnaire. Analytical results and data resulting from the questionnaire of each volunteer are statistically evaluated and interpreted, and afterwards reported to the Federal Environment Agency (UBA) and the Federal Ministry for the Environment, Nature Conservation, Building and Nuclear Safety (BMUB) on an annual basis. Thereby, the German ESB generates important information on internal exposures of humans and provides a scientific basis to decide on the necessity of risk reduction measures to protect human health and the environment as well as to control their success. Recent examples from the German ESB human related work are retrospective analyses of heavy metals, Hexamoll, DINCH, Bisphenol A, phthalates and perfluorinated compounds in body fluids [6,38–42].

The sampling of human samples for the German ESB was conducted in the facilities of collaborative institutes and universities in 2012 and the years before. For the routine implementation of the sampling process, a specific infrastructure considering the requirements of acts and regulations on biological safety and quality assurance was required and set up at each site. At least five separated rooms, including a reception, two diagnostic areas, a BSL 2 laboratory, a waiting room, and a van for transportation of samples with a complete cryo-equipment and an oxygen monitoring system were needed. For all processes of the sampling, SOPs were established by the German ESB (http://www.umweltprobenbank.de/de/documents/10022). For each sampling site these SOPs had to be adapted to the specific conditions of the site. Besides, all members of staff had to be trained specifically according to those SOPs. Having this in mind, the question arose in how far it would be possible to integrate a BSL 2 laboratory and the required infrastructure on a single mobile platform. Hence, the goal of this study was to identify the requirements for the implementation of a mobile biosafety level 2 laboratory and to evaluate its feasibility in a routine sampling in 2013.

Materials and Methods

Ethical statement

The study protocol of the German ESB was approved by the ethics committee of the Medical Association Saarland, Germany. The positive vote was made available to the partners of the sampling areas Muenster, Halle/Saale, Greifswald, and Ulm for submission to the local ethics committees. All study participants gave written informed consent on standardized forms approved by the same ethics committee. The right to know or not to know was guaranteed and records on the investigation results were supplied to the participants immediately after the analyses of the samples were completed.

Process and infrastructure of the routine sampling in 2012

Acquisition of participants and the pre-sampling phase. At each sampling site a maximum of 75 female and 75 male healthy students between 20 and 29 years were recruited. Volunteers were asked to register via an online registration form and to choose a defined date for blood withdrawal (https://umweltprobenbank.fraunhofer.de). Subsequently, every volunteer received a parcel with a container for collecting the 24 h-urine and an information kit. This kit contained general information about the processes and goals of the German ESB, specific information about the types of samples, the sampling procedure, and an instruction on how to collect the 24 h-urine. Furthermore, a material transfer agreement, two copies of a consent form, information on medical confidentiality and protection of privacy, and a 10-page standardized self-reported questionnaire to record medical history, individual behavior, and potential exposures were included. Volunteers were asked to fill in the questionnaire, to collect their 24 h-urine, to sign the consent form and to bring all of these items to their chosen appointment.

Medical history, exposure-relevant behavior, socio-demographic data and incentives. At the sampling site the volunteers handed over the described documents. Every questionnaire was checked for completeness and plausibility to increase data quality. The urine samples were directly transferred into the laboratory for further processing. The questionnaire was linked to the volunteers sample via a respective anonymous number. An allowance was paid to every volunteer after the sampling was finished. Two staff members and two separated desks were required for the implementation of the reception of 75 volunteers per day. A room of at least 8 m^2 with two separate cabins was needed for this first stage of the sampling process.

Dental fillings. For the evaluation of the internal exposure to chemicals (such as mercury or bisphenol A) released from dental fillings, their number and size were recorded for every volunteer by a dentist and one assistant. Dental fillings were recorded following the respective SOP. A separate room of at least 6 m^2 was equipped with a treatment chair and a desk. Other necessary materials are tongue depressors (Assistant, 4365), stomatoscopes (Hager & Werken, Brillant No. 4, 605400), gloves, and lab coats.

Blood collection. Blood samples were taken under medical supervision. Two teams of one nurse and one assistant each were necessary to realize the blood withdrawal of 75 volunteers per day. Blood withdrawal was done following the respective SOP. Therefore, safety needles (Saarstedt, Myltifly Safety, 85.1637.205) and sterile 20 ml syringes (BD, Discardit, 300296) were used. All sample tubes were rinsed according to the respective SOP with 2% nitric acid (Merck, Emsure, 1.00456.2500), methanol (Merck, Emsure, 1.06009.2500), and 18 MΩ ultrapure water (Millipore, Milli-Q Integral 5,

Table 1. Overview of provided instruments.

Task	Instrument
Blood withdrawal	2×Laminar flow cabinet
	2×Treatment bed
Laboratory (Urine processing)	1×Aerometer
	1×Electronic balance
	1×Conductivity measurement device
	2×Pipets
Laboratory (Blood processing)	1×Laminar flow cabinet
	2×Pipets
	2×Centrifuge
Freezing and Storing	1×80°C Freezer
	1×LN tank (150 l)
	1×LN samples storage tank (420 l)

ZRXQ005T0) prior to the sampling to avoid any organic and inorganic contamination and were supplemented with heparin (Ratiopharm, Heparin-Natrium-25000, PZN-3029843). Finally 140 ml blood was collected in seven 20 ml syringes and 5 ml were dropped out of the safety needle into a sample tube. The transfer of blood from the syringe into the sampling tubes was done in two laminar flow safety cabinets (Aura Mini, BioAir, LV 30000). Afterwards, samples were immediately transferred into the laboratory for further processing. For safety reasons a set of medical devices, e.g., stethoscope, blood glucose meter, blood pressure meter, and medicine (e.g., physiologic salt solutions, drugs for improved circulation) was assembled in an emergency bag and kept aside ready to use.

The laboratory and the sample preparation process. According to international and national biological agents regulation untested human body fluids, such as blood, have to be considered as potentially infectious. In general, the ESB samples are collected from healthy volunteers, not representing any risk group. However, a maximum level of protection would be supplied by a BSL 2 laboratory. Such a laboratory did not exist at each sampling site. Therefore, the necessary infrastructure in the available laboratories was established by IBMT. Following the German regulations with regard to biosafety, general requirements on a biological safety level 2 laboratory are as follows:

Staff members have to wear protective clothing (lab coat and gloves). A personal lock must be provided that has to be equipped with a handsfree sink and disposable towels, an emergency shower, an eye shower, disinfectants and hygiene regulations. The laboratory itself must be a separate room with surfaces easy to clean and resistant against disinfectants. Doors and windows must be closed while working in the laboratory. The entrance door has to show clearly the biohazard sign and should have a window. All processes that may lead to the formation of aerosols must be done in a laminar flow safety cabinet or staff members must wear task specific protective clothing, e.g. goggles, surgical masks, gloves, and lab coats [43].

Blood and 24 h-urine samples were directly processed upon entering the laboratory following the respective SOPs. After the blood samples were transferred into the lab, four of the seven collected blood tubes were directly prepared and packed for freezing. The remaining three blood tubes, each containing 20 ml of blood, were centrifuged for plasma separation (Eppendorf, 5810 R, 5811 000.424). Separation of plasma from the remaining blood cells and portioning into sample tubes was done in a laminar flow safety cabinet (BioAir, EF/S4, H071001) using pipets (Eppendorf, Research, 3120 632.000) with 5 ml pipet tips (Eppendorf, epTIPS, 0030 073.169).

Weight, density, and conductivity of the urine samples were determined in the lab. Therefore, the lab was equipped with an electronic balance (Mettler-Toledo, 11124926), an aerometer (Assisstent, 60008), and a conductivity measurement device (Mettler-Toledo, 51302936). Urine samples were portioned into 9×13 ml (10 ml urine) and 3×30 ml (20 ml urine) cleaned tubes and prepared for freezing. Immediately after sample processing

Table 2. Overview of required members of staff and their tasks.

Task	Members of stuff
Reception and data control	2×Scientists
Dental fillings	1×Dentist, 1×Assistant
Blood withdrawal	1×Medical Doctor, 3×Assistant
Laboratory (Urine processing)	2×Medical Technical Assistant
Laboratory (Blood processing)	1×Medical Technical Assistant
Freezing and Storing	1×Medical Technical Assistant
	Total: 12

Table 3. Sample types per Volunteer.

Sample type	Size	Cryo-Repository	Real-Time-Monitoring
Blood	20 ml	3	
Blood	5 ml		1
Plasma	3 ml	7	2
24 h-Urine	10 ml	7	2
24 hUrine	20 ml	3	
	Total	**25**	

tubes were transferred into a –80°C freezer (Heraeus, HFC586 PLUS-V14, 77710200) and kept there overnight to avoid disruption of the sampling tubes due to a rapid expansion of the freezing liquid. On the following day, samples were transferred into the cryogenic storage vessel (Cryotherm, Bio Safe 420), pre-cooled down to –160°C using liquid nitrogen.

Infrastructure, logistics and stuff members. Several instruments were required to conduct the sampling with regard to biosafety regulations and quality assurance. Except of the cryogenic storage vessel for samples and the cryogenic vessel for liquid nitrogen supply which were installed in the transporter, all devices (see table 1) were provided by IBMT and were delivered by a forwarder to each sampling site one day before sampling. At the same day the infrastructure was installed in the rented laboratories of the cooperating institute. All instruments and analytical tools were re-validated and the revalidation was documented in the frame of a quality management system according the GCLP. At least three additional assistants were needed to set up the required infrastructure and to remove it one day after sampling.

The realization of the sampling event required at least 12 members of staff. Table 2 gives an overview of the required members of staff and their related tasks.

Generated samples and subsequent analysis. During the sampling procedure different samples were generated depending on their use. Table 3 gives an overview of the generated samples per person. In general, a subset of samples was directly separated for real-time monitoring (RTM) of selected chemicals. After the sampling procedure in 2012 a part of these samples was transferred to the IBMT laboratory to investigate clinical chemical parameters. 24 h-urine samples were thawed and creatinine was measured. Furthermore, creatinine, triclycerides, total protein, and cholesterol were measured in thawed plasma samples. Clinical chemistry of both body fluids was measured using the cobas c111 analyzer (Roche, 04 777 433 001).

The other part of the RTM samples was transferred to a respective laboratory that analyzed metal compounds in blood and urine. The results of these studies were published in an annual report to the UBA, brochures and on the homepage of the German ESB (www.umweltprobenbank.de). Each volunteer received a personalized letter with his own results after these RTM analyses were completed. Thereafter, personal data were eliminated and samples were made anonymous according to the German data privacy act for their further storage in the German ESB.

In total, 20 samples per volunteer were cryopreserved and long-time stored immediately after processing and portioning in the laboratory. Considering a maximum of 600 participating volunteers per year, a total of 15.000 subsamples could have been

Table 4. Result of requirement analysis for rooms.

Room/compartment	Instruments/furniture	Floor space [m²]
Reception	Table, chairs (4), kitchenette	10.6
Privacy compartment (2 times)	Table, chairs (2)	2×1.5
Waiting area	Chairs (3)	2.5
Dental status	Dentist's chair, chairs	4.6
Blood withdrawal and initial processing (2 times)	Chair for blood withdrawal, laminar flow cabinet, chairs	2×7.0
BSL 2 laboratory		
- personnel locker room	Handsfree sink, hand desinfection	2.6
- main room	Laminar flow cabinet, centrifuge, cobas C111, pipets, conductivity meter, balance	7.9
- storage room	LN tank (150 l), storage tank (420 l)	3.4
Toilet		1.1
Technical compartment		2.9

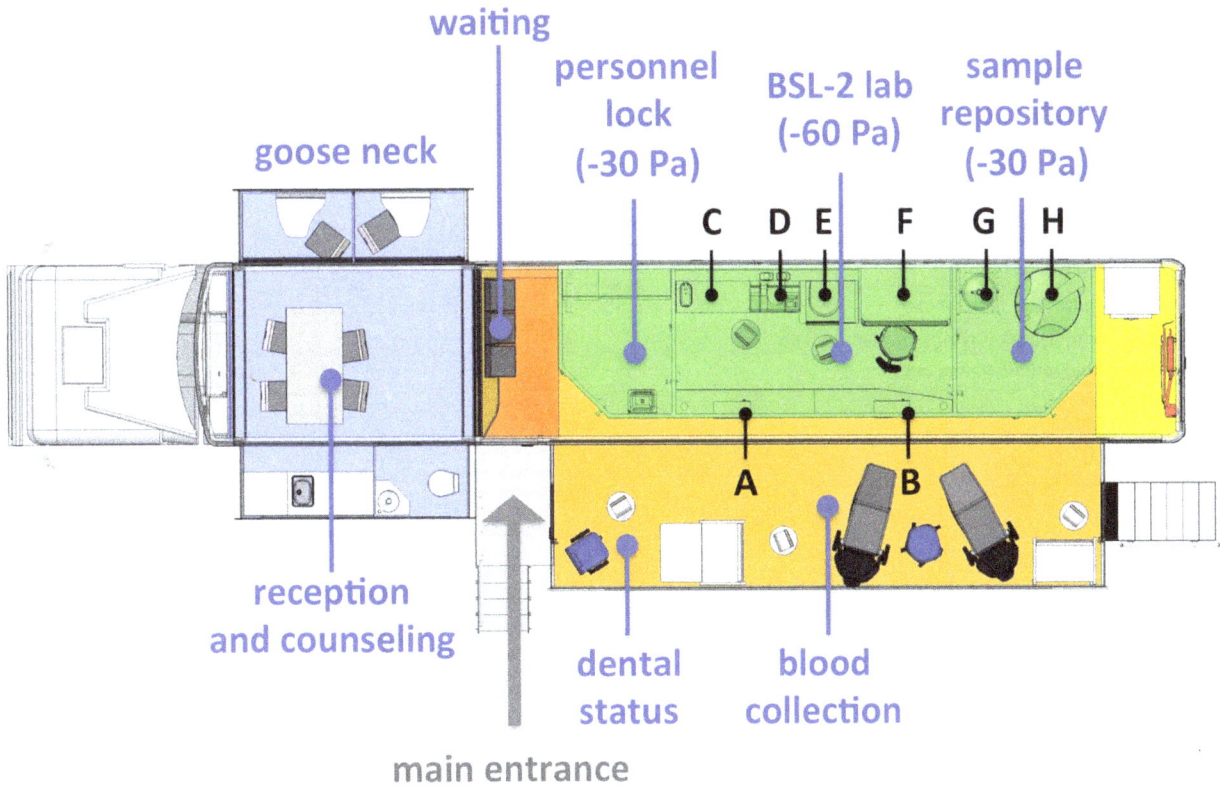

Figure 1. General layout and concept of the mobile epidemiologic laboratory (epiLab). Three motoric driven expansion units provide a reception and counseling area (14.5 m²) and an examination room (21.1 m²). The laboratory block (15.5 m²) has three rooms (personnel lock, BSL-2 lab and sample repository) with individual pressure levels for secure operation.

Figure 2. Mobile epiLab at first mission. A: The expandable unit. B: View to the examination room. C: BSL-2 area in the mobile epiLab with urine processing area (background) and blood processing area (foreground).

Table 5. Different workplaces and instruments in the laboratory block.

Workspace/Instrument	Letter
Material lock - Urine	A
Material lock - Blood	B
Urine/aerometer, balance, conductivity measurement device	C
Clinical Chemistry/Roche - Cobas c111 analyzer	D
Centrifugation/centrifuge	E
Blood processing/laminar flow cabinet	F
LN Supply/LN storage tank (Cryotherm - Saturn 300)	G
Sample Cryo-preservation/storage tank	H

generated. About 3.000 of them got used for real-time monitoring and approximately 12.000 got stored in the German ESB.

Results and Discussion

Requirements review of the sampling in 2012

The samplings in 2012 in Muenster (Westphalia), Halle (Saale), Greifswald and Ulm were conducted according to the procedures stated above. The establishment of a quality management system (QMS) following the GCLP has been started by IBMT under these

conditions. It has turned out that the varying locations with different standards with respect to laboratory regulations have large implications on the QMS. According to the analysis of process requirements, a mobile laboratory was developed to further optimize, standardize and harmonize these multicenter sampling events and to realize the sampling in 2013 for the first time with a mobile BSL 2 facility.

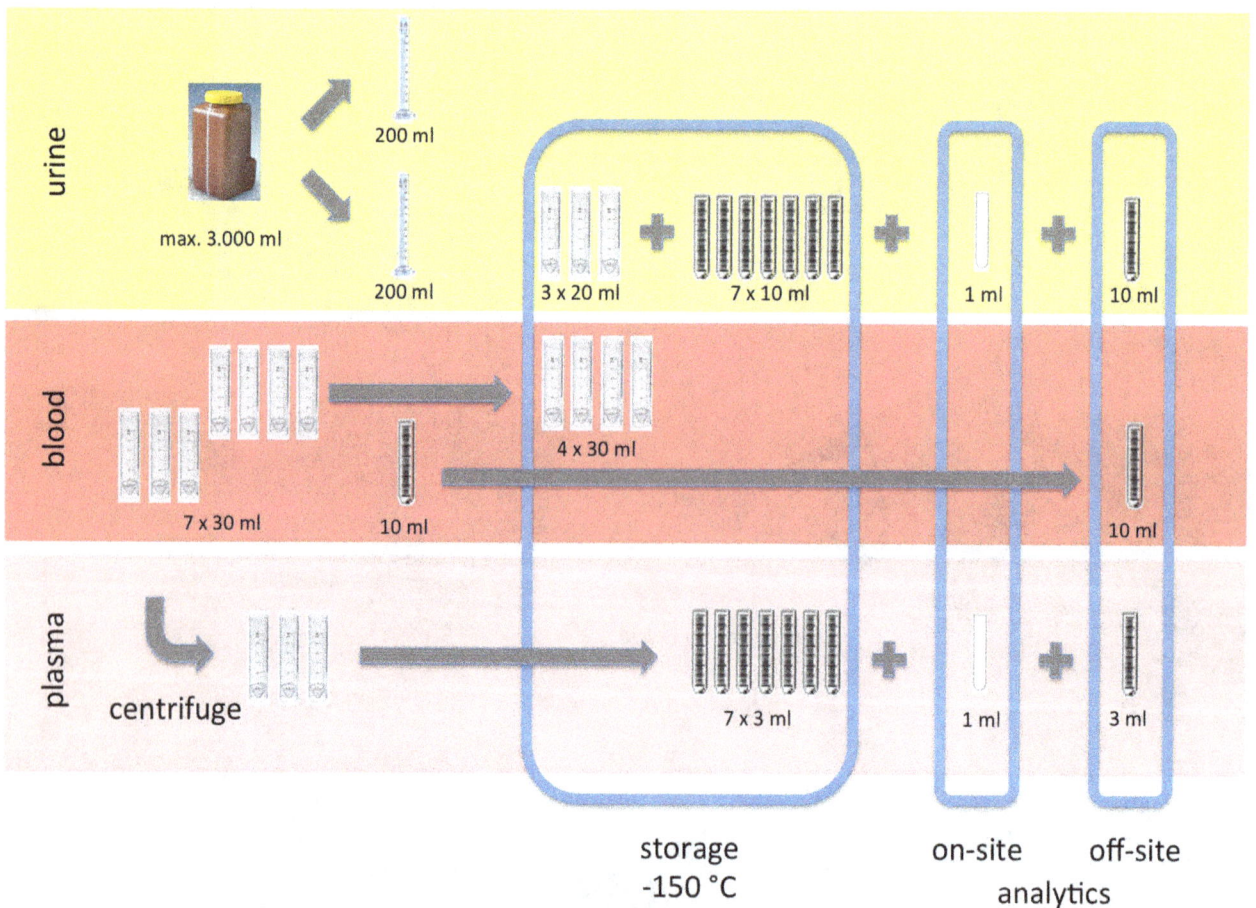

Figure 3. Sample workflow and pre-analytics in the mobile epiLab. Urine and blood samples are processed at the same time and aliquots are stored in the mobile cryo-repository and prepared for off-site analytics. Urine and plasma samples are also analyzed on-site.

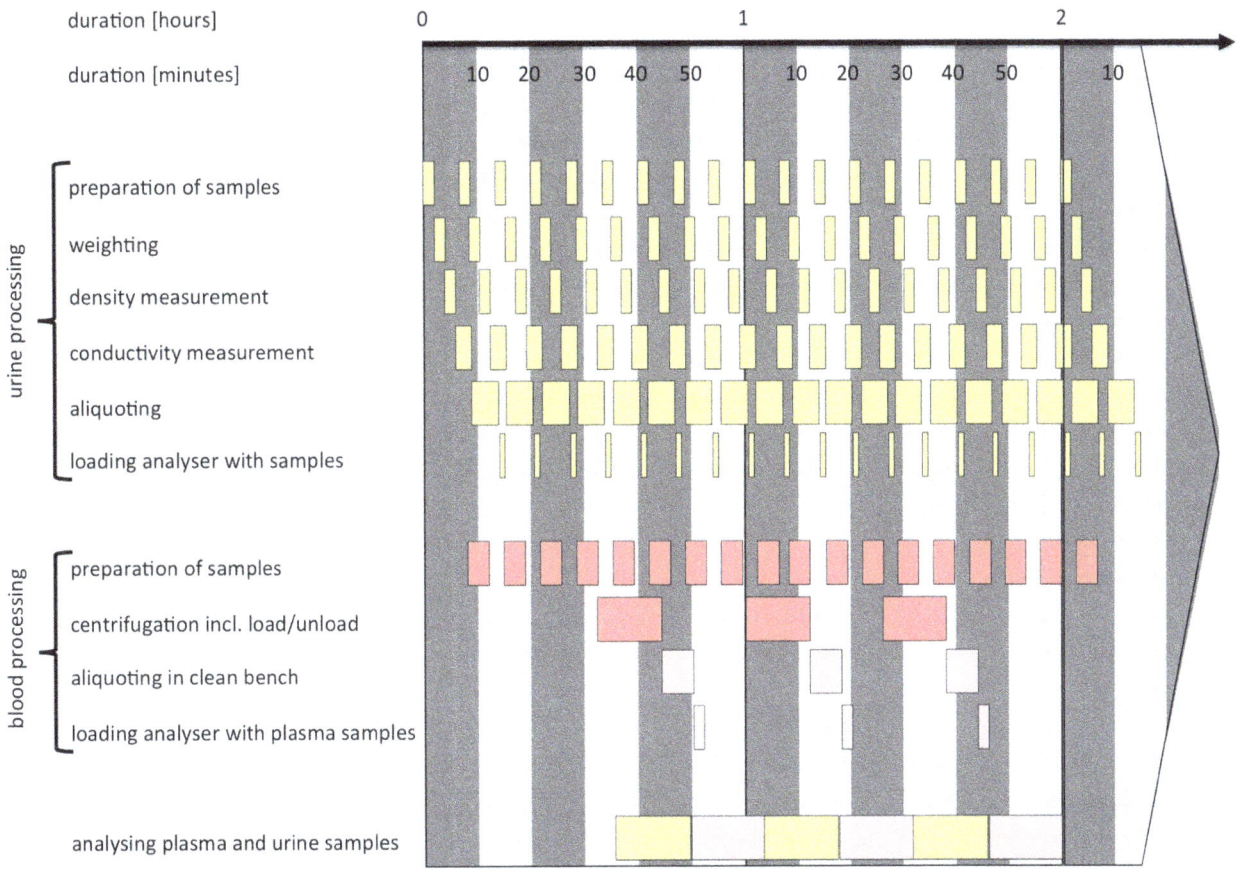

Figure 4. Schematic workflow for the processing of urine (yellow), blood (red) and plasma (pink) samples in the mobile epiLab. Three volunteers per 20 min are assumed for the optimum throughput shown here. The starting delay for the blood processing is caused by the on-site blood draw.

Optimization of room layout and workflow

A detailed analysis of the requirements for floor space has been carried out based on the experience of the sampling procedures in 2012. A minimum total floor space of 50 m² is required in a mobile platform to carry out all processes of the workflow (see Table 4).

A semitrailer with a maximum allowed width and length of 2.5 m by 14 m respectively can thus provide the largest connected floor space (35 m²) according to the German road traffic act (special transports with specific allowance and under police control are not considered). Experiences with laboratory vehicles developed in the project "Labor der Zukunft" (www.labor-der-zukunft. com) showed that 32 m² can be realized including all required technical rooms for heating, ventilation and air conditioning. Consequently pull-out extensions have been considered and investigated. A central pull-out compartment of approximately 8 m length was proposed to provide additional space (16 m²). Two smaller pull-out compartments (approximately 3 m² each) are located on both sides at the bow of the semitrailer (goose neck).

Urine and blood processing can be carried out in one single laboratory room, but a dedicated workflow has to be implemented to prevent cross interference. In order to separate and localize the sample entrance into the main laboratory, two different material locks were implemented. One lock was specifically designed for six urine containers. The location and use of this lock was designed to enable single-handed delivery of the urine sample into the laboratory.

Figure 1 gives an overview of the developed mobile epidemiologic laboratory (epiLab). As shown, two motoric driven expansible units in the goose neck offer 14.5 m² of room. This part of the trailer is equipped with a sanitary facility, a small kitchen and two small cabins with chairs and folding tables. It is used as reception and counseling unit and allows controlling the self-reported questionnaire of each participant in a private atmosphere. A large motoric driven expansible unit reveals the diagnostic area (21.1 m²) and the actual biosafety level 2 laboratory block (15.5 m²). The diagnostic area is split into three parts. One part close to the entrance is used as a waiting room, the next part is for the dentist to ascertain dental status (Figure 1, marked *dental status*), and the largest part in the back that is equipped with two laminar flow cabinets and two treatment chairs for blood collection (Figure 1, marked *blood collection*). Figure 1 shows the two material locks in the front of the laboratory block, which allow a direct transfer of urine samples (Figure 1, A) and blood samples (Figure 1, B) into the laboratory. The actual laboratory is organized into three categories: A personnel lock, the main laboratory and the sample repository. A cascade of low pressure with steps of −30 Pa (lock and repository) and −60 Pa respectively (laboratory) is realized with a ventilation system using 100% fresh filtrated air and allowing air renewal rates up to 14 times per hour. Table 5 gives an overview of the different workplaces and instruments that are used during the sampling and which are shown in Figure 1. Figure 2 shows the mobile epiLab at its first mission in Muenster in January 2013.

Table 6. Overview of acquired and actual participants in 2012 and 2013.

Year	Sampling Site	Date of sampling	Total number of acquired volunteers	Number of volunteers that participated	Number of female volunteers	Number of male volunteers	Number of former participations
2012	Muenster	20.–21.01.	155	133	63	70	21
	Halle/Saale	26.–27.03.	136	129	64	65	29
	Greifswald	14.–15.04.	135	124	64	60	56
	Ulm	05.–06.06.	140	120	59	61	23
	Total		**566**	**506**	**250**	**256**	**129**
2013	Muenster	25.–26.01.	153	131	69	62	37
	Halle/Saale	19.–20.03.	141	124	67	57	12
	Greifswald	20.–21.04.	145	131	68	63	53
	Ulm	04.–05.05.	139	124	68	56	17
	Total		**578**	**510**	**272**	**238**	**119**

Organization of the general sample procedure and sample processing in the mobile epiLab

In general, after medical history and exposure relevant behavior was recorded and the questionnaire was controlled, the volunteer was asked to place his urine sample into the first material lock (Figure 1, A) by himself. Weight, conductivity and density were measured immediately after the sample entered the lab following the respective SOP. After recording the dental status of the volunteer by a dentist and one assistant, blood was withdrawn in the respective area of the examination room from a medical doctor and one assistant (see Figure 1). Blood samples were immediately placed into the second lock by the assistant (Figure 1, B) and directly processed and prepared for clinical chemistry and cryopreservation. All results were documented following the respective SOP in standardized documentation forms. The procedure applied in 2012 required cooling down samples from body or room temperature to –80°C and later deep freezing to below –140°C in order to prevent the plastic tubes from thermal cracking. With regard to simplifying this process in 2013 a storage vessel with liquid nitrogen cooling was adapted to provide different temperature zones. So, samples can be deep frozen in a gentle two-step process. The Flow chart in Figure 3 describes the sample processing procedure for urine, blood, and plasma samples that is specifically designed for the mobile lab unit and to prevent cross interference.

A specific time schedule was set up to manage the required up to 75 volunteers a day and the respective amount of samples and to avoid bottlenecks and waiting periods. A timeframe of 15 to 20 minutes was defined for the realization of all sampling tasks for three volunteers. Figure 4 shows the schedule for each sample type.

Integration of clinical chemistry

In 2012 clinical chemistry of the body fluids was measured not on site, but in a laboratory at the IBMT headquarter. Therefore, immediately after the sampling, samples were frozen as described in "Materials and Methods" and transferred to the laboratory. Using the mobile laboratory in 2013, for the first time it was possible to integrate the analysis of the described clinical chemical parameters into the sampling process and thus avoiding freezing and transportation of the samples. Hence, blood samples were centrifuged as described. Plasma was separated in the laminar flow safety cabinet and subsequently 1 ml of plasma was pipetted into a respective tube for analysis of clinical chemical parameters using the cobas c111 analyzer (Roche). Urine was homogenized via soft panning and 1 ml of urine was directly pipetted into a respective cobas tube. Thus, results of the analysis for all participating volunteers of each day were available only one hour after the sampling.

Participants and number of samples in 2012 and 2013

An overview of the acquired volunteers and the actual number of volunteers that participated in 2012 and 2013 for each of the four sampling sites is given in Table 6. As table 6 reveals, using the mobile laboratory and the described workflow, in 2013 the same amount of volunteers and even more were managed within the same time period. Finally, according to the number of volunteers and the number of subsamples described in Table 3, in 2012 a total of 12.650 subsamples from 566 volunteers were generated during the samplings at the four different sampling sites, whereas 12.750 subsamples from 578 volunteers were generated during the samplings in 2013 using the mobile epiLab.

Table 7. Distances of mission locations from headquarter (Sulzbach/Saar) and repository (Muenster).

Mission location	Distance to headquarter [km]	Distance to repository [km]
Muenster (Westphalia)	400	0
Halle (Saale)	600	400
Greifswald	1000	600
Ulm	300	600

Sample transport

All samples have to be brought to the central repository in Muenster without breaking the cooling chain. Table 7 gives an overview of the distances between the sampling sites and the IBMT headquarter and between the sampling sites and the central repository in Muenster. In 2012, this transport was carried out with a fixed integrated cryotank in a transfer vehicle. Consequently, each sample had to be handled twice for each sampling. In order to avoid these handling processes, in 2013 a removable cryogenic storage vessel has been established in the mobile sampling unit of the epiLab. This tank can be brought, together with all samples, to the central archive in Muenster, unloaded via an included tail lift and connected to the liquid nitrogen supply in the repository.

The Quality Management System

The German ESB already exists since 1985. Since that time standard operating procedures (SOPs) were established for all relevant steps and were released by the UBA and the German ESB consortium [36]. These SOPs were used to document all the processes in an internal quality management system but not by national or international guidelines.

Experiences made during the sampling events in 2012 were used to define the requirements for a QMS. In a first attempt the existing QMS was converted into a QMS according to GCLP. A QM handbook was written describing all relevant SOPs. It was clearly recognized that the four different sampling sites provide different infrastructures and laboratory settings. Therefore, workflows and processes had to be reorganized for each of the sampling sites preventing a strict standardization of these pre-analytical processes. A QMS according to GCLP was already implemented by Fraunhofer IBMT for their HIV Specimen Cryorepository (HSC) funded by the Bill & Melinda Gates Foundation [44]. These specific guidelines are only internal guidelines for networks funded by the Division of AIDS (DAIDS) of the National Institute for Health (NIH) and do not constitute an international accepted regulation. For this reason, the IBMT is now adapting the QM system to the international DIN EN ISO/IEC 17025 regulation [45,46] and is aiming an accreditation of the sampling by Germany's National Accreditation Body (DAkkS). In 2013, with the availability of the mobile laboratory, all processes and workflows related to the sampling event were adapted to the new conditions and standardized. Following the described requirements all SOPs were adapted to the available rooms, workflows and analyses. Finally, the implementation of the mobile lab guarantees the same facility and infrastructure specifically adapted to the need of the sampling procedure of the German ESB. Thus guaranteeing identical treatment of all samples at all sites.

Stand-alone capabilities

In order to ensure stand-alone operation of the mobile unit and to minimize the requirements for potential locations, the following requirements have been defined:

Fresh and waste water. The main requirements resulted from the collection of urine samples. 150 volunteers per sampling can lead to a maximum of 450 l (3 l per volunteer) to be discarded. Assuming a mean value of 1 l of fresh water used per urine sample for flushing, the following minimum requirements for fresh and waste water per mission can be projected: 150 l of fresh water and 600 l of waste water. Including some extra capacities for hand washing and toilet use, a 400 l fresh water tank and a 1.000 l waste water tank are implemented. Both tanks are made of stainless steel and are equipped with electrical heating for frost protection.

Heating, ventilation and air condition (HVAC). In order to cope with the different climate conditions during the mission, the large number of staff and volunteers on board and the regulatory requirements for the laboratory (constant temperature and air exchange), a complex and powerful HVAC system had to be designed and implemented. A water based floor heating driven by an engine block heater (10 kW Diesel) is used to establish basic heating during the winter. The laboratory is supplied with 100% fresh air at an exchange rate of up to 14 times per hour (laboratory only). This fresh air is filtered and conditioned by a central climate system. The target temperature for the incoming air can be set in the range of 20 to 24°C and is specified for ambient temperatures from –10 to 34°C. A low pressure cascade can be established in the laboratory on demand. All secondary rooms are equipped with individual HVAC units with heat pump function for heating.

Electrical supply. The entire trailer is designed to be operated with a diesel-based electric generator (34 kVA and 120 l tank). In normal mode, the system lasts for 10 h. An energy management system is implemented to cope with shortages by switching off energy of peripheral systems (lighting, kitchen). Although the generator is acoustically shielded, land line electric power is used for standard operation. A 32 A (CEE) landline is required as a minimum, 6 A (CEE) is recommended to prevent energy management cut-offs and to refresh all internal batteries. The critical laboratory instruments and part of the IT infrastructure are connected to an uninterrupted power supply unit (USV). Via this USV, switching from land line to generator can be done without shortage of electrical energy in the laboratory.

Conclusions

Since 2013 Fraunhofer IBMT collects up to 15.000 individual human subsamples annually under highly standardized conditions using the introduced mobile laboratory. Clinical chemical parameters like cholesterol, total protein, creatinine, and triglycerides are now analyzed directly on-site. At the same time samples are frozen in the adjacent cryorepository and stored at temperatures below –140°C. Frozen samples are transferred within the vehicle to its final destination without breaking the cooling chain. Since all processes are conducted following ESB specific SOPs according to the DIN EN ISO/IEC 17025 regulation, the quality of each sample is guaranteed. Finally, difficulties in the process of standardization resulting from different sampling sites are avoided and the collected samples are kept ready, characterized and free of change for retrospective analyses in their final archive, the German ESB.

Lab settings and analysis tools in the Fraunhofer IBMT mobile laboratory can be changed and modified. With an adapted QM system this mobile lab allows the standardization of every multicenter sampling event in the field of human related research, scientific surveys, and clinical trials.

Acknowledgments

The authors wish to thank the government and the State Chancellery of the federal state Saarland (Germany), the German Federal Ministry for the Environment, Nature Conservation, Building and Nuclear Safety, and the German Federal Environment Agency.

Author Contributions

Conceived and designed the experiments: HZ DS DL HvB. Performed the experiments: DL DS MB. Analyzed the data: HZ HvB DL DS CS MK. Contributed reagents/materials/analysis tools: HZ HvB. Contributed to the writing of the manuscript: DL DS MB CS MK HvB HZ. Obtained permission from ethics committee: DL HvB HZ.

References

1. Njanpop-Lafourcade B-M, Hugonnet S, Djogbe H, Kodjo A, N'douba AK, et al. (2013) Mobile Microbiological Laboratory Support for Evaluation of a Meningitis Epidemic in Northern Benin. PLoS ONE 8: e68401.
2. Lorente L, Martín MM, Abreu-González P, Domínguez-Rodríguez A, Labarta L, et al. (2013) Prognostic Value of Malondialdehyde Serum Levels in Severe Sepsis: A Multicenter Study. PLoS ONE 8: e53741.
3. Wiesmüller GA, Gies A (2011) Environmental Specimen Bank for Human Tissues. Encyclopedia of Environmental Health: 507–527.
4. Ozarda Y, Ichihara K, Barth Julian H, Klee G (2013) Protocol and standard operating procedures for common use in a worldwide multicenter study on reference values. Clinical Chemistry and Laboratory Medicine. 1027.
5. Kolossa-Gehring M, Becker K, Conrad A, Schröter-Kermani C, Schulz C, et al. (2012) Environmental surveys, specimen bank and health related environmental monitoring in Germany. International Journal of Hygiene and Environmental Health 215: 120–126.
6. Becker K, Schroeter-Kermani C, Seiwert M, Rüther M, Conrad A, et al. (2013) German health-related environmental monitoring: Assessing time trends of the general population's exposure to heavy metals. International Journal of Hygiene and Environmental Health 216: 250–254.
7. Schulz C, Seiwert M, Babisch W, Becker K, Conrad A, et al. (2012) Overview of the study design, participation and field work of the German Environmental Survey on Children 2003–2006 (GerES IV). International Journal of Hygiene and Environmental Health 215: 435–448.
8. Joas R, Casteleyn L, Biot P, Kolossa-Gehring M, Castano A, et al. (2012) Harmonised human biomonitoring in Europe: Activities towards an EU HBM framework. International Journal of Hygiene and Environmental Health 215: 172–175.
9. Lima-Oliveira G, Lippi G, Salvagno GL, Dima F, Brocco G, et al. (2013) Management of preanalytical phase for routine hematological testing: is the pneumatic tube system a source of laboratory variability or an important facility tool? International Journal of Laboratory Hematology 36: 37–40.
10. Romero A, Cobos A, Gómez J, Muñoz M (2012) Role of training activities for the reduction of pre-analytical errors in laboratory samples from primary care. Clinica Chimica Acta 413: 166–169.
11. Lippi G, Becan-McBride K, Behúlová D, Bowen Raffick A, Church S, et al. (2013) Preanalytical quality improvement: in quality we trust. Clinical Chemistry and Laboratory Medicine. 229.
12. Favaloro EJ, Funk DM, Lippi G (2012) Pre-analytical Variables in Coagulation Testing Associated With Diagnostic Errors in Hemostasis. Lab Medicine 43: 1–10.
13. Simundic AM, Cornes M, Grankvist K, Lippi G, Nybo M, et al. (2013) Survey of national guidelines, education and training on phlebotomy in 28 European countries: an original report by the European Federation of Clinical Chemistry and Laboratory Medicine (EFLM) working group for the preanalytical phase (WG-PA). Clin Chem Lab Med 51: 1585–1593.
14. Carraro P, Zago T, Plebani M (2012) Exploring the Initial Steps of the Testing Process: Frequency and Nature of Pre-Preanalytic Errors. Clinical Chemistry 58: 638–642.
15. Bonini P, Plebani M, Ceriotti F, Rubboli F (2002) Errors in Laboratory Medicine. Clinical Chemistry 48: 691–698.
16. Hammerling JA (2012) A Review of Medical Errors in Laboratory Diagnostics and Where We Are Today. Lab Medicine 43: 41–44.
17. Lazzari MA (2012) Errors in Specimen Processing and the Potential Misdiagnosis of Acute Renal Failure in a 12-year-old Girl. Lab Medicine 43: 9–11.
18. Hooijberg E, Leidinger E, Freeman KP (2012) An error management system in a veterinary clinical laboratory. Journal of Veterinary Diagnostic Investigation 24: 458–468.
19. Plebani M (2010) The detection and prevention of errors in laboratory medicine. Annals of Clinical Biochemistry 47: 101–110.
20. Walsh MJ, Shivalingappa G, Scaria K, Morrison C, Kumar B, et al. (1972) Mobile coronary care. Br Heart J 34: 701–704.
21. Wennerblom B, Holmberg S, Wedel H (1982) The effect of a mobile coronary care unit on mortality in patients with acute myocardial infarction or cardiac arrest outside hospital. European Heart Journal 3: 504–515.
22. Takizawa M, Sone S, Hanamura K, Asakura K (2001) Telemedicine system using computed tomography car of high-speed telecommunication vehicle. Information Technology in Biomedicine, IEEE Transactions on 5: 2–9.
23. Walter S, Kostopoulos P, Haass A, Keller I, Lesmeister M, et al. (2012) Diagnosis and treatment of patients with stroke in a mobile stroke unit versus in hospital: a randomised controlled trial. The Lancet Neurology 11: 397–404.
24. Walter S, Kostopoulos P, Haass A, Lesmeister M, Grasu M, et al. (2011) Point-of-care laboratory halves door-to-therapy-decision time in acute stroke. Annals of Neurology 69: 581–586.
25. Krol DM, Redlener M, Shapiro A, Wajnberg A (2007) A mobile medical care approach targeting underserved populations in post-Hurricane Katrina Mississippi. J Health Care Poor Underserved 18: 331–340.
26. Shapiro A, Seim L, Christensen RC, Dandekar A, Duffy MK, et al. (2006) Chronicles From Out-of-State Professionals: Providing Primary Care to Underserved Children After a Disaster: A National Organization Response. Pediatrics 117: 412–415.
27. Chui P, Chong P, Chong B, Wagener S (2007) Mobile Biosafety Level-4 Autopsy Facility-An Innovative Solution. Applied Biosafety 12.4: 238.
28. King B, Jatoi I (2005) The mobile Army surgical hospital (MASH): a military and surgical legacy. J Natl Med Assoc 97: 648–656.
29. Wahner HW, Looker A, Dunn WL, Walters LC, Hauser MF, et al. (1994) Quality control of bone densitometry in a national health survey (NHANES III) using three mobile examination centers. Journal of Bone and Mineral Research 9: 951–960.
30. Arvela H, Markkanen M, Lemmelä H (1990) Mobile Survey of Environmental Gamma Radiation and Fall-Out Levels in Finland After the Chernobyl Accident. Radiation Protection Dosimetry 32: 177–184.
31. Kolb CE, Herndon SC, McManus JB, Shorter JH, Zahniser MS, et al. (2004) Mobile Laboratory with Rapid Response Instruments for Real-Time Measurements of Urban and Regional Trace Gas and Particulate Distributions and Emission Source Characteristics. Environmental Science & Technology 38: 5694–5703.
32. Zavala M, Herndon SC, Slott RS, Dunlea EJ, Marr LC, et al. (2006) Characterization of on-road vehicle emissions in the Mexico City Metropolitan Area using a mobile laboratory in chase and fleet average measurement modes during the MCMA-2003 field campaign. Atmos Chem Phys 6: 5129–5142.
33. Wang M, Zhu T, Zheng J, Zhang RY, Zhang SQ, et al. (2009) Use of a mobile laboratory to evaluate changes in on-road air pollutants during the Beijing 2008 Summer Olympics. Atmos Chem Phys 9: 8247–8263.
34. Weibring P, Edner H, Svanberg S (2003) Versatile Mobile Lidar System for Environmental Monitoring. Applied Optics 42: 3583–3594.
35. Prohl A, Boge KP, Alsen-Hinrichs C (1997) Activities of an Environmental Analysis Van in the German Federal State Schleswig-Holstein. Environ Health Perspect 105: 844–849.
36. Wiesmüller GA, Eckard R, Dobler L, Günsel A, Oganowski M, et al. (2007) The Environmental Specimen Bank for Human Tissues as part of the German Environmental Specimen Bank. International Journal of Hygiene and Environmental Health 210: 299–305.
37. Emons H, Schladot JD, Schwuger MJ (1997) Environmental specimen banking in Germany – present state and further challenges. Chemosphere 34: 1875–1888.
38. Kolossa-Gehring M, Becker K, Conrad A, Schröter-Kermani C, Schulz C, et al. (2012) Environmental surveys, specimen bank and health related environmental monitoring in Germany. International Journal of Hygiene and Environmental Health 215: 120–126.
39. Koch HM, Kolossa-Gehring M, Schroter-Kermani C, Angerer J, Bruning T (2012) Bisphenol A in 24[thinsp]h urine and plasma samples of the German Environmental Specimen Bank from 1995 to 2009: A retrospective exposure evaluation. J Expos Sci Environ Epidemiol 22: 610–616.

40. Wilhelm M, Hölzer J, Dobler L, Rauchfuss K, Midasch O, et al. (2009) Preliminary observations on perfluorinated compounds in plasma samples (1977–2004) of young German adults from an area with perfluorooctanoate-contaminated drinking water. International Journal of Hygiene and Environmental Health 212: 142–145.

41. Schröter-Kermani C, Müller J, Jürling H, Conrad A, Schulte C (2013) Retrospective monitoring of perfluorocarboxylates and perfluorosulfonates in human plasma archived by the German Environmental Specimen Bank. International Journal of Hygiene and Environmental Health 216: 633–640.

42. Schütze A, Kolossa-Gehring M, Apel P, Brüning T, Koch HM (2013) Entering markets and bodies: increasing levels of the novel plasticizer Hexamoll DINCH

in 24 hr urine samples from the German Environmental Specimen Bank. International Journal of Hygiene and Environmental Health. 217.2: 421–426.

43. Ausschuss für Biologische Arbeitsstoffe (2013) Technische Regel für Biologische Arbeitsstoffe 100: Schutzmaßnahmen für Tätigkeiten mit biologischen Arbeitsstoffen in Laboratorien Gemeinsames Ministerialblatt 51/52: 1010–1042.

44. Germann A, Durst CHP, Ihmig FR, Shirley SG, Schön U, et al. (2009) Global HIV Vaccine Research Cryorepository-GHRC. Procedia in Vaccinology 1: 49–62.

45. Honsa JD, McIntyre DA (2003) ISO 17025: Practical Benefits of Implementing a Quality System. Journal of AOAC International 86: 1038–1044.

46. Kohl H (1998) The new ISO 17025– basic idea. Accreditation and Quality Assurance 3: 422–425.

18

Study on the Traffic Air Pollution inside and outside a Road Tunnel in Shanghai, China

Rui Zhou[1], Shanshan Wang[2]*, Chanzhen Shi[1], Wenxin Wang[1], Heng Zhao[1], Rui Liu[1], Limin Chen[1], Bin Zhou[1,3]*

1 Shanghai Key Laboratory of Atmospheric Particle Pollution and Prevention (LAP3), Department of Environmental Science & Engineering, Fudan University, Shanghai 200433, China, 2 School of Environment and Architecture, University of Shanghai for Science and Technology, Shanghai 200093, China, 3 Fudan Tyndall Centre, Fudan University, Shanghai 200433, China

Abstract

To investigate the vehicle induced air pollution situations both inside and outside the tunnel, the field measurement of the pollutants concentrations and its diurnal variations was performed inside and outside the Xiangyin tunnel in Shanghai from 13:00 on April 24th to 13:00 on April 25th, 2013. The highest hourly average concentrations of pollutants were quantified that CO, NO, NO_2 and NO_X inside the tunnel were 13.223 mg/m^3, 1.829 mg/m^3, 0.291 mg/m^3 and 3.029 mg/m^3, respectively, while the lowest ones were 3.086 mg/m^3, 0.344 mg/m^3, 0.080 mg/m^3 and 0.619 mg/m^3. Moreover, the concentrations of pollutants were higher during the daytime, and lower at night, which is relevant to the traffic conditions inside the tunnel. Pollutants concentrations inside the tunnel were much higher than those outside the tunnel. Then in a case of slow wind, the effect of wind is much smaller than the impact of pollution sources. Additionally, the $PM_{2.5}$ concentrations climbed to the peak sharply (468.45 $\mu g/m^3$) during the morning rush hours. The concentrations of organic carbon (OC) and elemental carbon (EC) in $PM_{2.5}$ inside the tunnel were 37.09–99.06 $\mu g/m^3$ and 22.69–137.99 $\mu g/m^3$, respectively. Besides, the OC/EC ratio ranged from 0.72 to 2.19 with an average value of 1.34. Compared with the results of other tunnel experiments in Guangzhou and Shenzhen, China, it could be inferred that the proportion of HDVs through the Xiangyin tunnel is relatively lower.

Editor: Yinping Zhang, Tsinghua University, China

Funding: This work was partially supported by the National Natural Science Foundation of China under grant No. 21277029, 40975076, 41365010, Science and Technology Commission of Shanghai Municipality (Grant: 12231201001, 12DJ1400102), and China Meteorological Administration (Grant: GYHY201106045-8). The authors are thankful for the great help from Shanghai Environmental Protection Bureau (Grant: 2013-76, 2013-03) and Design Institute of Urban Transport and Underground Space. The funders had no role in study design, data collection and analysis, decision to publish, or preparation of the manuscript.

Competing Interests: The authors have declared that no competing interests exist.

* Email: binzhou@fudan.edu.cn (BZ); shanshan.wang@usst.edu.cn (SSW)

Introduction

The primary air pollutants emitted by motor vehicles are carbon monoxide (CO), nitrogen oxides (NO_X, including NO and NO_2), hydrocarbons (HCs) and particulate matter (PM). Vehicular exhaust has become a main source of air pollution. Among these emissions, CO and NO_X emissions are accounting for more than 80% and 40% of the total urban emissions in many big cities like Beijing, Shanghai and Guangzhou in China [1–3]. The particles from vehicular exhaust are mainly composed of organic carbon (OC) and elemental carbon (EC); and the combinational effect of OC and EC can reduce visibility, accounting for 30% to 40% of the total extinction [4]. Moreover, the HC and NO_X are precursors to secondary ozone formation and aerosols. These products and the vehicle-induced pollutants can cause damages to human health, including the harms to respiratory system, cardiovascular system etc. [5,6].

With the increasing recognition of the adverse effects of vehicle-induced air pollution and the relationship between the pollutants and diseases, some solutions have been considered [7]. For example, constructing tunnels to redirect traffic away from surface roads not only alleviates the increasing serious problem of traffic congestion, but also largely improves ambient air quality [8]. However, it has also been found that inadequate ventilation in the tunnels combined with high traffic volume can result in elevated concentrations of vehicle-induced air pollutants. He [9] characterized comprehensively the $PM_{2.5}$ emissions inside the Zhujiang Tunnel in the Pearl River Delta region of China. It was found that the organic compounds in vehicular $PM_{2.5}$ emissions were less influenced by the geographic area and fleet composition. Thereby it is more suitable for across extensive regions to use in aerosol source apportionment modeling. The diffusion of exhaust gas jetted out from city traffic tunnel under different conditions of the structure size of wind tower, jet velocity of exhaust gas and relevant ambient meteorological parameters, and the decay rates of exhaust gas were discussed by Tian [10]. Compared with the environmental effect of shaft discharge with cavern mouth diffusion by the Gauss model and TOP model theory, it was suggested that the environmental pollution contribution of CO is remarkably weaker than that of NO_2 under the same conditions. Besides, the environmental effect of cavern mouth diffusion, which is affected by air direction evidently, is more serious than that of shaft discharge [11].

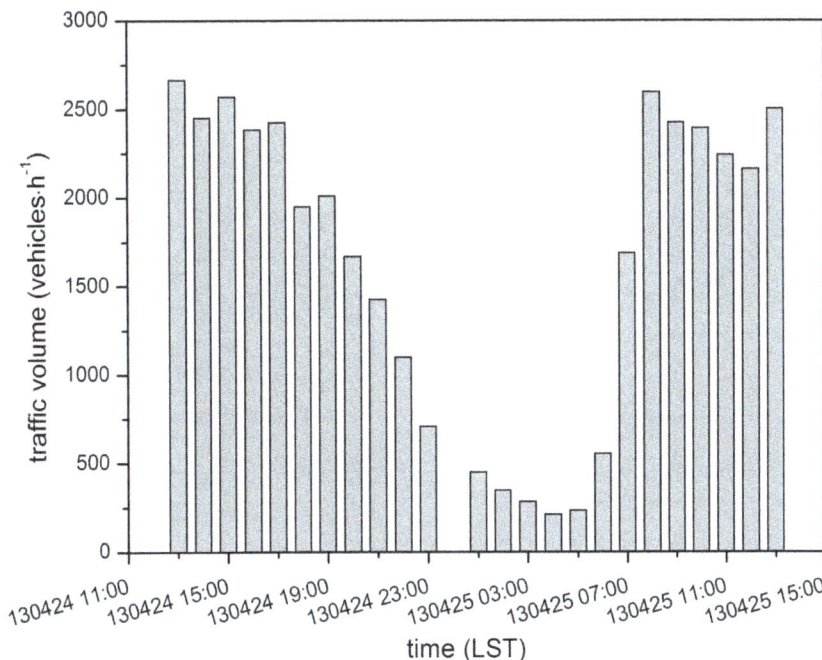

Figure 1. Traffic volume from 13:00 on April 24th to 13:00 on April 25th in the Xiangyin tunnel (south bore).

The aim of this study is to figure out the characteristics of the concentrations of vehicle-induced air pollutants inside and outside the tunnel (the Xiangyin tunnel, Shanghai) over a 24-hours period, as well as the relationship between these variables. Furthermore, the carbonaceous substances from $PM_{2.5}$ in the tunnel, namely OC and EC, along with their ratio were discussed.

Methodology

Experimental location

The Xiangyin tunnel is located in the northeast of Shanghai, which is crossing the Huangpu River. It is an urban two-bore tunnel (north and south bores) with two lanes of traffic per bore (without walkways). The tunnel has a length of 2.6 km with designed speed capacity of 80 km/h, and the ventilation mode is longitudinal. Two exhausting towers are located on the east side of the Huangpu River (about 120 m away from the exit), and another same set on the west side, while two horizontal axial exhausting fans are set to centralize and diffuse exhausts. The volume flow rate, power and pressure of the fans are 125 m^3/s, 250 kW and 1000 Pa, respectively. There were no specific permissions required for this experiment location, and the experiment study did not involve endangered or protected species.

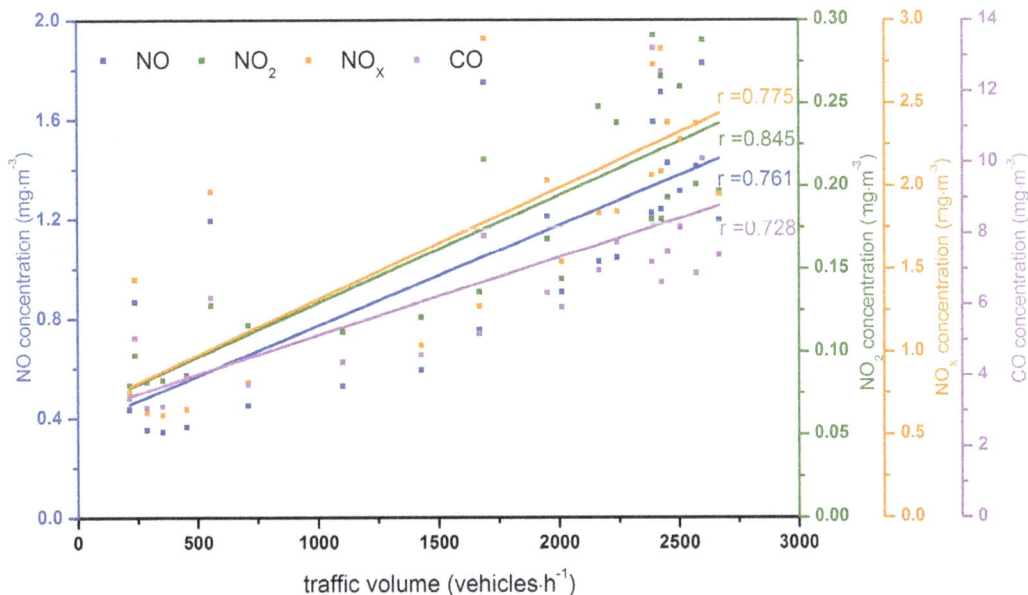

Figure 2. The correlation between the traffic volume and pollutants (NO, NO_2, NO_X, CO) concentrations inside the tunnel.

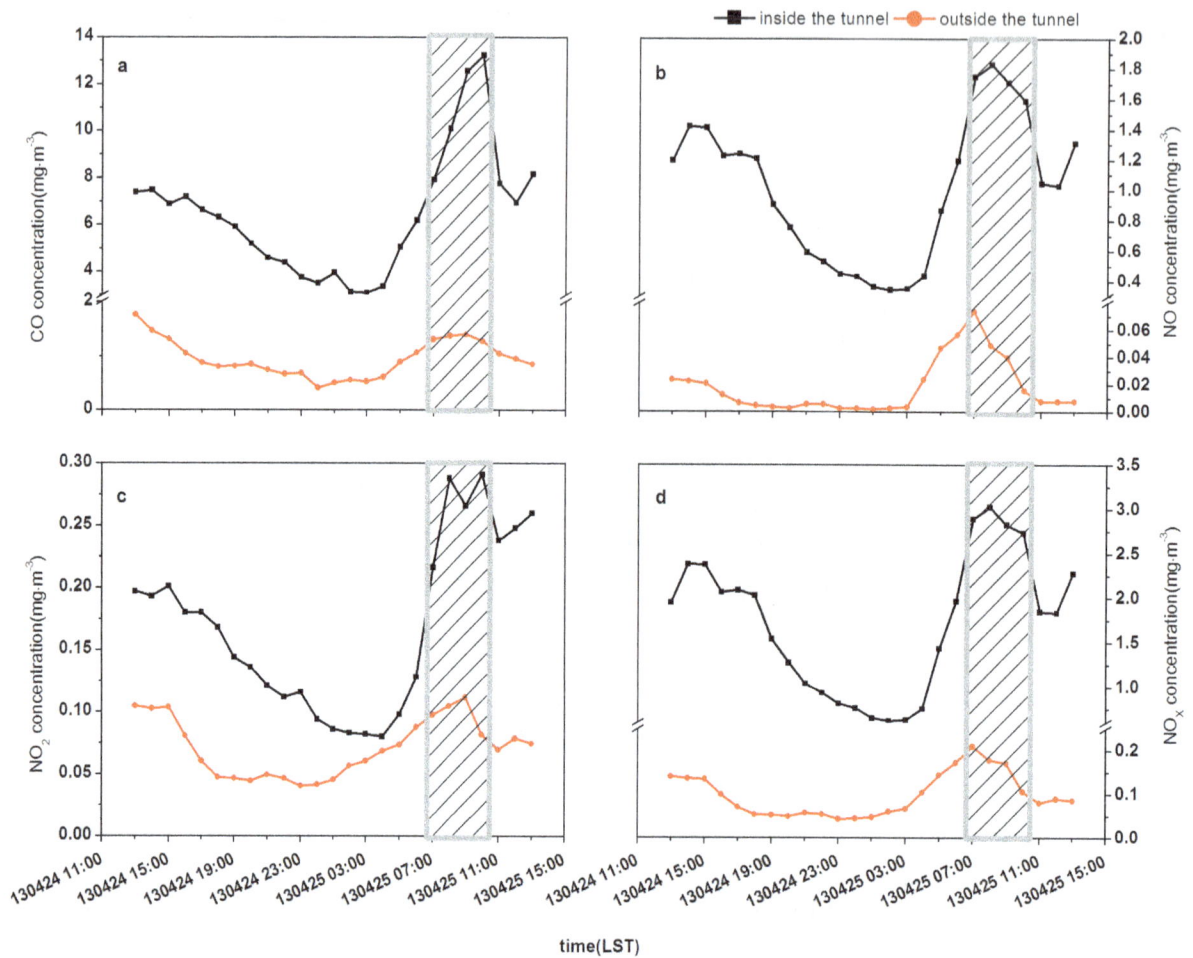

Figure 3. Comparisons of pollutants concentrations inside and outside the tunnel (a: CO, b: NO, c: NO$_2$, d: NO$_X$; shaded areas: the period of the fan working).

Monitoring sites and instruments

The monitoring sites (121.58°E,31.30°N) were positioned on the east side of the exhausting tower, both inside and outside the tunnel. The monitoring time was from 13:00 on April 24th to 13:00 on April 25th, 2013, during which one horizontal axial exhausting fan was working from 7:00 to 10:00 on April 25th. The monitoring site outside the tunnel was to the west of the exhausting tower, which was located on the east side of the Huangpu River. And the online data about meteorology and pollutants outside the tunnel were gained with a monitoring van at the monitoring site. The meteorological data included temperature, relative humidity, atmospheric pressure, as well as wind speed and direction. The pollutants were mainly carbon monoxide (CO), nitrogen oxides (NO$_X$), nitric oxide (NO) and nitrogen dioxide (NO$_2$).

The concentration of CO was measured by a carbon monoxide analyzer (Model 48i, Thermo scientific), which used gas filter correlation technology to detect the amount of CO. The concentration of NO-NO$_2$-NO$_X$ mixture was measured by a nitrogen oxides analyzer (Model 42i, Thermo scientific) using chemiluminescence.

Inside the tunnel, the monitoring site was located at the top of south bore of the tunnel, which was 10 m away from the exhausting fan. The monitoring instruments used inside the tunnel

were same as those used outside the tunnel to measure the concentrations of the same pollutants (CO, NO, NO$_2$, NO$_X$).

Sampling and instruments

For the PM$_{2.5}$ samples, the sampling site was located at the top of south bore of the tunnel, the same as the monitoring site inside the tunnel. The sampling time was from 13:00 on April 24th to 11:00 on April 25th with the temporal resolution of an hour. PM$_{2.5}$ concentrations were measured by the gravimetric method, while the glass fiber membrane filters and an intelligent medium volume PM$_{2.5}$ sampler (Model 2030, Laoying, Qingdao Laoshan Applied Technology Institute) were used. The flow rate of the sampler was set at 100 L/min, and the cumulative sample volume was saved automatically after each sampling. In order to determine the PM$_{2.5}$ mass concentrations, glass fiber membrane filters were pro- and post- weighted at least twice on an analytical balance, with a sensitivity of 0.1 mg. Before weighing, the membrane filters were equilibrated in a temperature humidity chamber at 20°C with a relative humidity of 50±5% for 24 hours.

To determine the mass concentrations of OC and EC, the samples were analyzed by a Thermal/Optical Carbon Analyzer (Model 2001A, Desert Research Institute). The technical method of the analyzer is thermo-optical reflection. The operation of the analyzer is based on the preferential oxidation of OC compounds and elemental EC at different temperatures. It relies on the fact that organic compounds can be volatilized from the sample deposit in a non-oxidizing helium (He) atmosphere, while elemental carbon must be combusted by an oxidizer. In this study only 7 samples (April 24th 16:00, 19:00, 22:00; April 25th 1:00, 4:00, 7:00 and 10:00) were actually analyzed.

Statistical analysis methods

Correlation analysis was used to illustrate the correlation of two variables, which is performed by OriginPro 8.1 software (OriginLab Cooperation). To test the significance of the relationship between variables, the hypothesis test was taken for the correlation coefficient. P-values [12,13] are often coupled to a significance or alpha (α) level, which is also set ahead of time, usually at 0.05 (5%). Other significance levels, such as 0.1 or 0.01, are also used, depending on the field of study; P-value was set as 0.01 in this study. Thus, if a P-value was found to be less than 0.01, then the result would be considered statistically significant and then the null hypothesis would be rejected. By Origin 8.1 software, correlation coefficient (r) was calculated, and based on the degree of freedom (ν), the corresponding r_α could be taken from the correlation coefficient r boundary value table (Table S1). If $r \geq r_\alpha$, then P<0.01, which means the correlation between the variables significant.

Results and Discussion

Traffic volume and pollutants concentrations inside the tunnel

Fig. 1 displays the traffic volume from west to east direction, i.e. south bore of the tunnel, during the monitoring period (the data for 0:00–1:00 on April 25th was missing). In the daytime (8:00–20:00), the traffic volume were around 2000 vehicles per hour and the maximum occurred between 13:00 and 14:00 on 24th with 2666 vehicles passed through the tunnel. In contrast, the traffic

volume declined during the night, with the lowest volume only of 214 vehicles during the 4:00–5:00 period. Due to the heavy congestion and consequent slow movement of traffic, in the morning rush hour the actual number of cars passing through the tunnel was not that large in spite of how busy the road appeared. However, the road appeared relatively quiet in the afternoon, as a similar number of vehicles passed through the tunnel quickly. Therefore, the hourly traffic volume did not differ greatly during the daytime.

Fig. 2 shows the correlations between the traffic volume and the concentrations of pollutants (NO, NO_2, NO_X, CO) inside the tunnel. The correlation coefficients (r) of the traffic volume and the concentrations of NO, NO_2, NO_X, CO were 0.761, 0.845, 0.775 and 0.728 respectively, from which it could be assumed that the concentrations of the pollutants were all associated with the traffic volume. Moreover, according to the statistics results, the degree of freedom (ν) was equal to 22 and the corresponding r_α was 0.515 referred to the correlation coefficient r boundary value table. It was obvious that the correlation coefficients were all greater than r_α. Consequently, P-Values were all smaller than 0.001 suggesting the relationships between the traffic volume and pollutant concentrations were significant. However, the traffic volume was not the only factor that determines the pollutants concentrations. Those concentrations were determined by many factors, such as vehicle speed, vehicle type, and fuel, etc. [14]. Due to the limitation of this experimental design, only the traffic volume was monitored and the other information about the traffic condition was unknown.

Comparison of pollutants concentrations inside and outside the tunnel

Comparisons of the hourly average concentrations of each pollutant inside and outside the tunnel, obtained using the real-time monitoring instruments, are shown in Fig. 3.

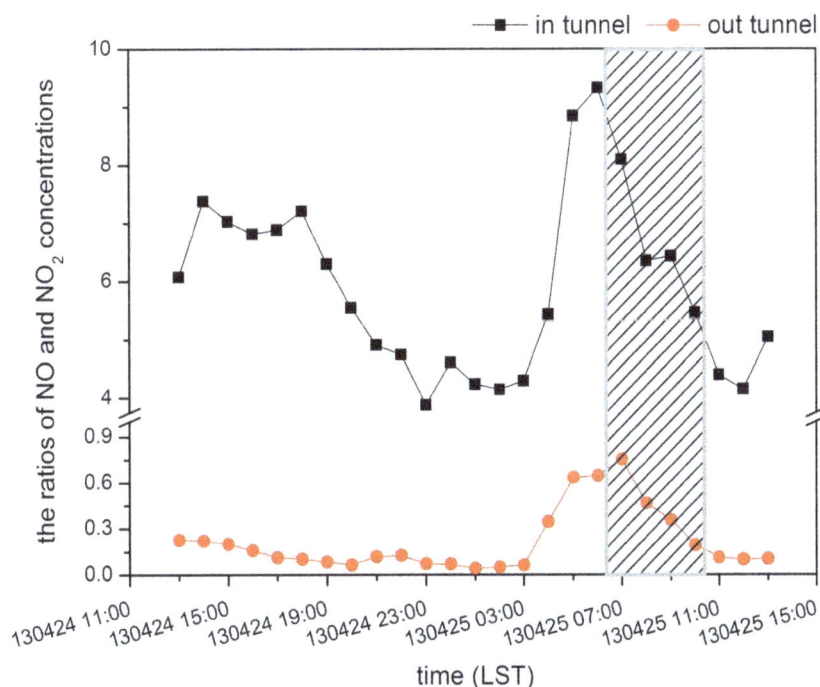

Figure 4. Ratios of NO and NO₂ concentrations inside and outside the tunnel (shaded area: the period of the fan working).

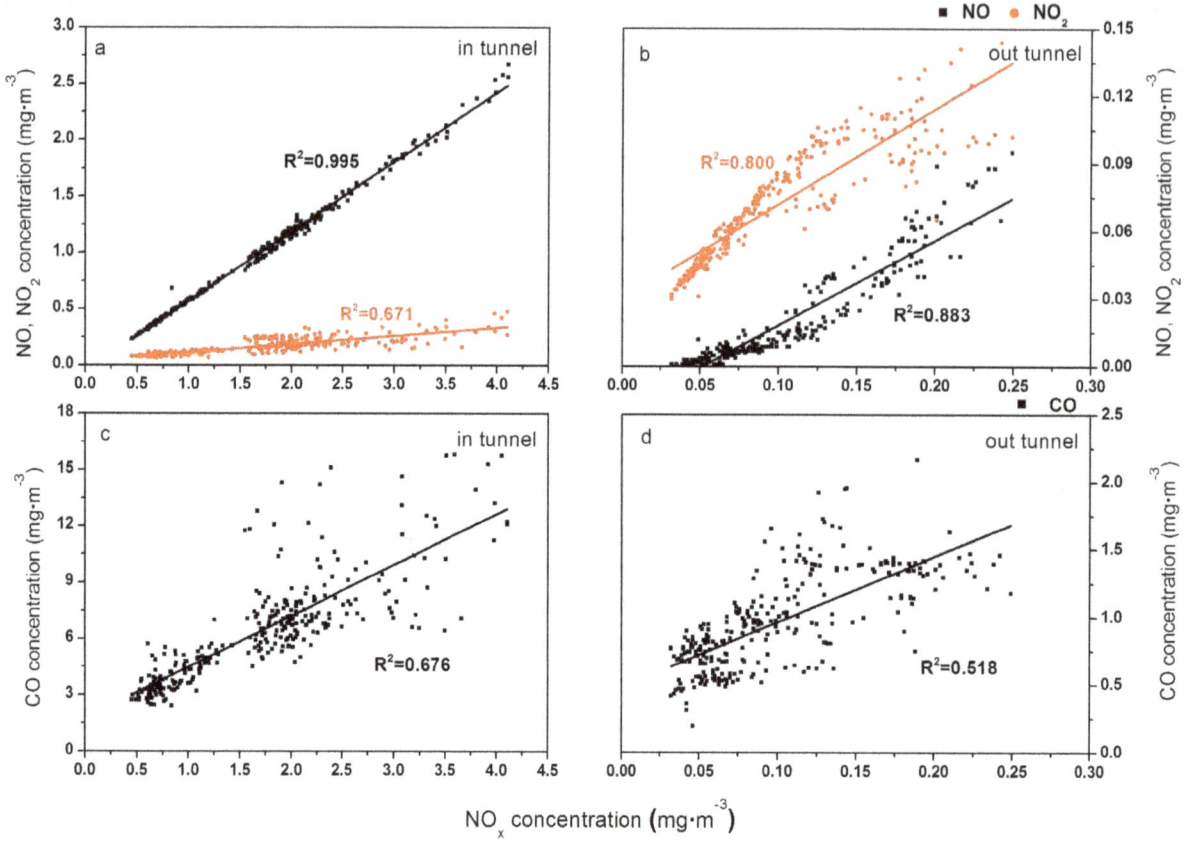

Figure 5. Fitting results of NO, NO₂, CO with NOₓ (a: NO, NO₂ inside the tunnel, b: NO, NO₂ outside the tunnel, c: CO inside the tunnel, d: CO outside the tunnel).

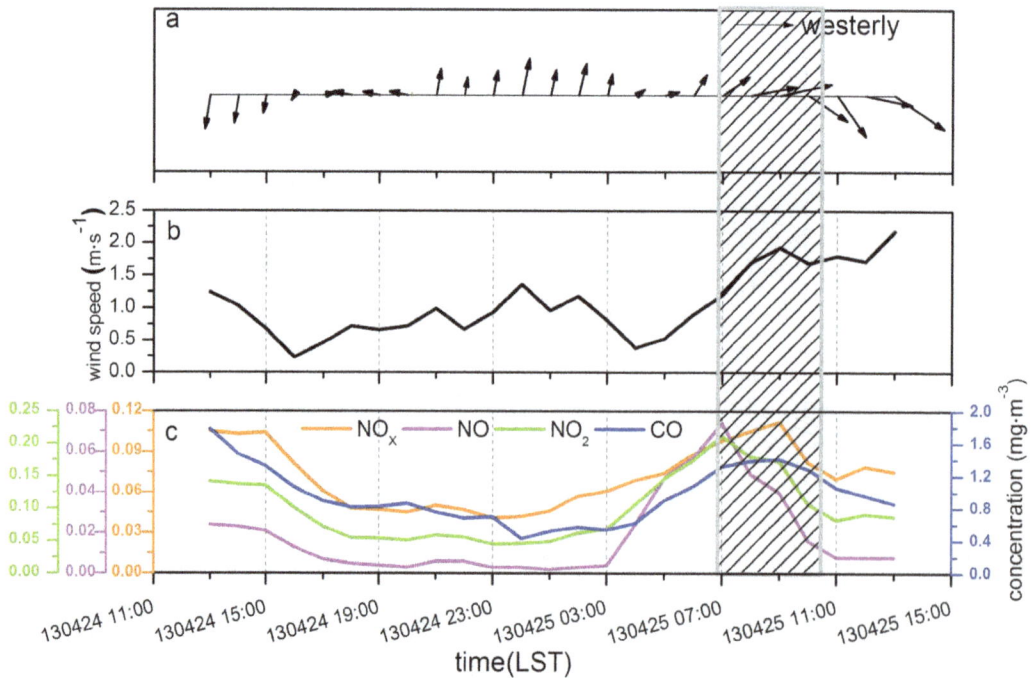

Figure 6. Hourly average meteorological parameters and pollutants concentrations measured by the monitoring van outside the tunnel (a: wind direction, b: wind speed, c: pollutants concentrations; shaded area: the period of the fan working).

The variations of hourly averaged CO concentrations inside and outside the tunnel were illustrated in Fig. 3.a. Higher concentrations appeared in the morning rush hours, while the values declined to the lowest at midnight. However, the concentrations of CO rose quickly to a peak level during the 8:00–10:00 period. This was associated with traffic congestion so the fact that a large amount of CO was accumulated without diffusion in time. The exhausting fan was working at this time so a portion of CO that accumulated in the tunnel was discharged, which would make the CO concentration of the monitoring site (at the top of the tunnel) increase. Besides, CO concentrations inside the tunnel were much higher than those outside the tunnel by 6 to 7 times on average. The main reasons were that the traffic volume on the surface road was less than that inside the tunnel. Moreover, compared to the relatively isolated environment in the tunnel, the surface road was in an open environment, which was more conducive to the diffusion of CO.

Comparing the NO concentrations inside the tunnel with those outside the tunnel in Fig. 3.b, it could be concluded that concentrations were higher in the daytime (especially in the morning) and lower at night. Furthermore, NO concentrations inside the tunnel were almost 100 times higher than that outside the tunnel. The trend of NO concentrations in the tunnel was mainly associated with the traffic situation and the effect of the fan since NO increased sharply when the fan was working. The situation was more complicated outside the tunnel since there were some important chemical reactions involved. NO could react quickly with O_3 to form NO_2, but NO_2 could also be decomposed back to NO in the condition of the light. Due to these chemical reactions, NO concentrations outside the tunnel were quite low at night, and then rose in the daytime.

In Fig. 3.c, the trends of NO_2 concentrations inside and outside the tunnel were not similar. The concentrations of NO_2 were much lower than the other pollutants inside the tunnel. However,

the trend of NO_2 concentrations was similar to those of CO and NO, for which the concentrations in the daytime were higher than at night, and the highest concentrations occurred during rush hours. The variation of NO_2 concentrations outside the tunnel was not significant, with a maximum value that was only twice the minimum, whereas the ratios of the maximum and minimum values of CO and NO were 4.15 and 37, respectively. NO_2 was not the main pollutant emitted directly by vehicles. In addition, the concentration of ozone was low inside the tunnel in the general situation, which means that the amount of NO_2 generated from the reaction with NO was low. Outside the tunnel, NO_2 could be generated quickly from NO, and with the presence of the light, NO_2 could revert back to NO. Therefore, the concentrations of NO_2 didn't change too much overall.

As the main components concentrations of NO_X, NO and NO_2, their ratios inside and outside the tunnel were shown in Fig. 4. The concentration of NO in the tunnel was much higher than NO_2 (about 6 times), implying NO_X in the tunnel was primarily composed of NO. But the situation outside the tunnel was almost the opposite. NO_2 was much higher than NO except for the early morning of April 25th when the concentration of NO climbed slightly to around two thirds of that of NO_2. The variation in NO_X concentrations can be found from Fig. 3.d and Fig. 5.a and b, which shows the results of fitting linear models to the NO, NO_2 and NO_X data for the inside and outside the tunnel, respectively. Fig. 5.a shows that the correlation coefficient (r) of NO and NO_X concentrations was 0.998, while that of NO_2 and NO_X concentrations was 0.820. Inside the tunnel the stronger correlation between NO_X and NO implies that NO was a more important component of NO_X than NO_2. As mentioned above, the NO concentration was much higher than NO_2. As a result NO_X concentrations were mainly determined by NO concentrations. Compared with the results inside the tunnel, there was little difference between the correlation coefficient (r) of NO and NO_X

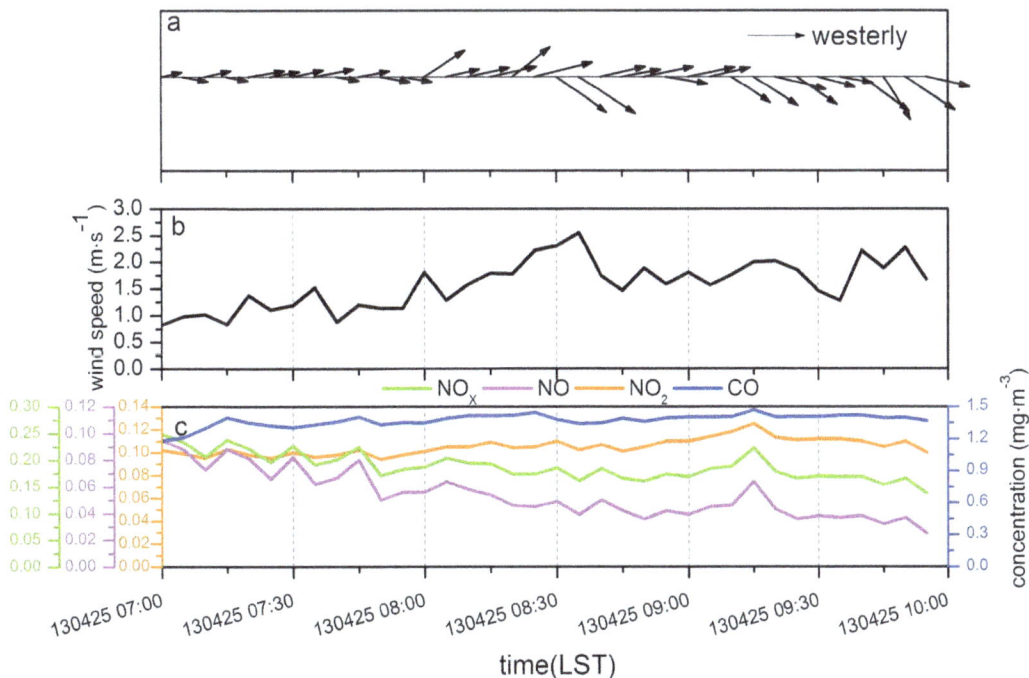

Figure 7. Average meteorological parameters and pollutants concentrations in 5 minutes intervals from 7:00 to 10:00 on April 25th (a: wind direction, b: wind speed, c: pollutants concentrations).

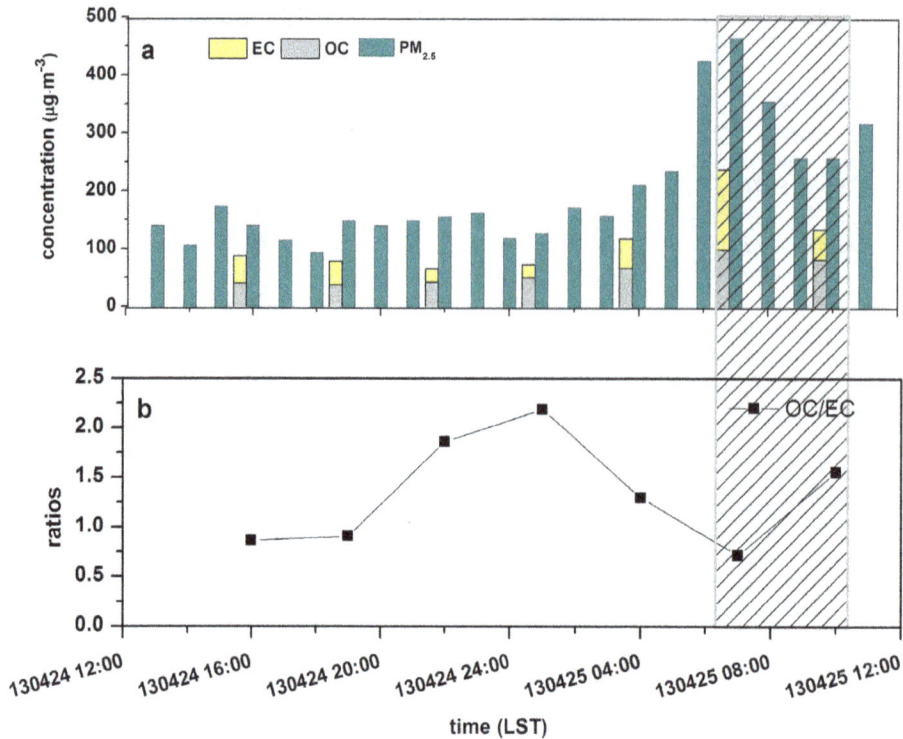

Figure 8. PM$_{2.5}$ concentrations and OC and EC concentrations in PM$_{2.5}$, as well as the OC/EC ratio inside the tunnel (shaded area: the period of the fan working).

concentrations and that of NO$_2$ and NO$_X$ concentrations outside the tunnel from Fig. 5.b. They were 0.895 and 0.940, respectively, which means NO$_X$ concentrations outside the tunnel depended on both NO and NO$_2$. Thus, the trend of NO$_X$ concentrations inside the tunnel was close to the trend of NO concentrations, while outside the tunnel the trend of NO$_X$ concentrations was between the trend lines of NO and NO$_2$ concentrations and relatively closer to that of NO (Fig. 3). Besides, comparing the correlation coefficients of NO$_X$ and CO concentrations (Fig. 5.c and d), we could further analyze the sources of NO$_X$ inside and outside the tunnel. The correlation coefficients were 0.823 and 0.721 inside

and outside the tunnel, respectively, and showing the NO$_X$ inside the tunnel comes from the same source as CO, i.e. traffic exhaust. On the other hand, the sources of NO$_X$ outside the tunnel included a small contribution from nearby industrial sources and the traffic exhaust.

Meteorological conditions and pollutant concentrations outside the tunnel

By analyzing the variation of the pollutants concentrations inside and outside the tunnel, the sources of pollution and some

Table 1. Mean OC/EC ratios in PM$_{2.5}$ of tunnel studies.

Tunnel	City	Mean OC/EC	Analytical method	Reference
Zhujiang tunnel	Guangzhou	0.49	Thermal optical transmittance	He et al. [9]
Tanglang Hill tunnel	Shenzhen	0.52	Thermal optical transmittance	Liu et al. [23]
Xueshan Tunnel	Taiwan	1.26	Thermal optical reflectance	Zhu et al. [25]
Mount Victoria tunnel	Wellington	1.00	Thermal optical reflectance	Ancelet et al. [17]
Janio Quadios tunnel	Sao Paulo	1.59	Thermal optical transmittance	Brito et al. [18]
Rodoanal Mario Covas tunnel	Sao Paulo	0.49	Thermal optical transmittance	Brito et al. [18]
Sepulveda tunnel	Los Angeles	0.76	Thermal optical reflectance	Gillies et al. [20]
Squirrel Hill tunnel	Pittsburgh	1.00	Thermal optical transmittance	Grieshop et al. [21]
Urban tunnel	Marseille	0.56	Thermal optical transmittance	Haddad et al. [19]
Kaisermühlen tunnel	Vienna	0.30	Thermal two step combustion	Handler et al. [22]
Marquês de Pombal tunnel	Lisbon	0.29~0.37	Thermal optical transmittance	Pio et al. [24]
Xiangyin tunnel	Shanghai	1.34	Thermal optical reflectance	this study

chemical reactions had a major influence on the pollutant concentrations outside the tunnel. Fig. 6 shows the meteorological parameters measured by the monitoring van during the monitoring period, including the hourly average wind speed and direction. It seems the influence of meteorological conditions was little.

According to Fig. 6.a, at the height of 7 m, the main wind direction was northeast in the afternoon of 24th, and then it changed to southeast from the evening of 24th until the early morning of 25th. The prevailing wind directions were southwest and northwest during the morning. The wind speed changed a little as shown in Fig. 6.b, from about 1 m/s since the afternoon of 24th till the early morning of 25th to about 1.5–2.5 m/s during the morning of 25th. Compared with the trends in pollutant concentrations in Fig. 6.c, the wind direction and speed had hardly any effect on the diurnal variation of pollutant concentrations. Diurnal variations of pollutants concentrations were mainly determined by the sources of pollution and some chemical reactions. For most of the monitoring period, the fan was not working; the sources of pollution were the traffic and industry nearby. However, when the fan was working during the morning rush hours (7:00–10:00 on 25th), the pollutants emitted through the wind tower, which might have been affected by ambient wind, could influence the pollutants concentrations measured by the monitoring van. Then this time period was chosen to analyze whether wind direction and speed affect pollutant concentrations.

Fig. 7 shows the average meteorological parameters and pollutants concentrations over 5 minutes intervals from 7:00 to 10:00 on April 25th. During these three hours, the dominant direction was westerly from Fig. 7.a. During the first one and a half hours the wind direction was mostly WSW, then it turned to WSW from WNW between 8:30 and 9:00. The prevailing wind direction was northwest in the last one hour. Also the wind speed of the northwest wind was greater than that of the southwest wind. Comparing with the trends of NO, NO_2 and NO_X concentrations, when the wind speed was low, the concentration was high. As a result, the speed of the wind might affect the pollutants concentrations to a certain extent, but the direction of the wind affected it little. That was mainly because the pollution sources of NO_X were not only the traffic exhaust, but also the exhaust of the industrial factories nearby and thus NO_X was produced all around. With respect to NO_X, the change of the CO concentration was affected by the wind direction. The monitoring van was at the west of the exhausting wind towers. Thus when the winds blew southwest or northwest, the van was located upwind. Thus, the monitoring results were influenced little by pollution from the tunnel, but mainly reflected the traffic pollution on the surface road. Some researchers have reported the similar result, which considered that the impact of pollution sources are much more significant than the effect of the wind. Chen [15] preliminary analyzed the influences of vehicular pollution sources and meteorological conditions for PM_{10} concentrations in the downtown streets in Beijing. The variation of concentrations in the streets (near the vehicular pollution sources) was affected by pollution sources greatly, and only at the conditions of unstable weather, the impact of pollution sources was little. Yang [16] calculated the contribution rates of the meteorological conditions and traffic control measurements for particle concentrations decreasing during the Beijing Olympics, the results showed that the traffic control measurements had a more apparent impact on decreasing the roadside $PM_{2.5}$ concentrations.

$PM_{2.5}$ and the OC, EC content inside the tunnel

The variation of $PM_{2.5}$ concentrations inside the tunnel was shown in Fig. 8. During the morning rush hours, the concentra-

tions climbed to the peak value of 468.45 $\mu g/m^3$ sharply. Fig. 8 also gives the contents of organic carbon and elemental carbon in $PM_{2.5}$ inside the tunnel as well as their ratios. OC concentrations varied from 37.09 $\mu g/m^3$ to 99.06 $\mu g/m^3$ while EC concentrations ranged from 22.69 $\mu g/m^3$ to 137.99 $\mu g/m^3$. Additionally, the contribution of OC and EC to $PM_{2.5}$ were 21.15%–38.45% and 14.33%–32.83%, respectively. Thus, TC contributed to $PM_{2.5}$ from 40.99% to 61.11%. From Fig. 8.a, it shows that high values of OC and EC appeared at the beginning of the morning rush hours, while low values of OC appeared at nightfall and low values of EC appeared at midnight. It could be noticed that the maximum of OC/EC ratio was 2.19, while the minimum was 0.72, with the mean value of 1.34. Comparing with other tunnel studies around the world, the mean OC/EC ratios were indicated in Table 1 [9,17–25]. It seems that EC concentrations were higher than OC in the tunnel in many cases, and the ratios between OC and EC were less than 2 even in the case that OC concentrations were higher than EC. In general, the OC/EC ratios are more than 2 in the urban ambient air, as OC can be produced by photochemical reactions; while inside the tunnel under the circumstance of no light, it was difficult to generate secondary chemical reactions to produce more OC, so the OC/EC ratios were relatively smaller. On the other hand, EC was primarily emitted by HDVs, and thus fewer HDVs, the OC/EC ratios are larger instead. In addition, compared with the results of tunnel experiments in Guangzhou [9] and Shenzhen [23], it could be found that the mean of OC/EC in this study is higher than that of those two tunnel experiments. Heavy duty vehicles were accounting for 18.4% to 35.1% of the total traffic volume passing through the Zhujiang tunnel in Guangzhou with the averaged OC/EC of 0.42. And for the Tanglang Hill tunnel in Shenzhen, the mean OC/EC of light duty vehicles was 1.2 while that of heavy duty vehicles was 0.42. Under the condition of the 33.8% proportion of HDVs, the OC/EC ratio was 0.48. Then it could be inferred that the proportion of HDVs passed through the Xiangyin tunnel is relatively lower.

Conclusions

Through this study, it has been found out that the highest hourly average concentrations of CO, NO, NO_2 and NO_X inside the tunnel were 13.223 mg/m^3, 1.829 mg/m^3, 0.291 mg/m^3 and 3.029 mg/m^3, respectively, while the lowest ones were 3.086 mg/m^3, 0.344 mg/m^3, 0.080 mg/m^3 and 0.619 mg/m^3. Concentrations of pollutants (CO, NO, NO_2, NO_X) were higher during the daytime (especially in the morning rush hours), and lower at night, which is related to the traffic condition in the tunnel. Additionally, the concentrations of pollutants (CO, NO, NO_2 and NO_X) inside the tunnel were higher than those outside the tunnel, about 7, 109, 2 and 18 times, respectively.

Referring to the impact of wind on the concentrations of the pollutants outside the tunnel, the effect of wind was not as significant as the impact of pollution sources in the case of slow wind.

$PM_{2.5}$ concentrations climbed to the peak value of 468.45 $\mu g/m^3$ during the morning rush hours. As for the contents of organic carbon and elemental carbon in $PM_{2.5}$ inside the tunnel, those were 21.15%–38.45% and 14.33%–32.83%, respectively. Moreover, the maximum of OC/EC ratio was about 2.19, and the minimum was 0.72, while the mean value was 1.34. The ratios suggested that the proportion of HDVs through the Xiangyin tunnel might be relatively lower.

Author Contributions

Conceived and designed the experiments: RZ SSW LMC BZ. Performed the experiments: RZ SSW CZS WXW HZ RL. Analyzed the data: RZ SSW. Contributed reagents/materials/analysis tools: RZ SSW. Wrote the paper: RZ SSW BZ.

References

1. Yang F, Yu L, Song GH (2004) Application of small sampling approach to estimating vehicle mileage accumulations for Beijing, China. Transport Res Rec 1880: 77–82.

2. Wang HK, Chen CH, Huang C, Fu LX (2008) On-road vehicle emission inventory and its uncertainty analysis for Shanghai, China. Sci Total Environ 398: 60–67.

3. Xue JP, Tian WL, Zhang QY (2010) Development of NO_X emission inventory from motor vehicles in Hangzhou and study on its influence on air quality. Research of Environmental Sciences 23(5): 613–618.

4. Zhao HJ, Che HZ, Zhang XY, Ma YJ, Wang YF, et al. (2013) Characteristics of visibility and particulate matter (PM) in an urban area of Northeast China. Atmospheric Pollution Research 4: 427–434.

5. Chen CH, Kan HD, Huang C, Li L, Zhang YH, et al. (2009) Impact of ambient air pollution on public health under various traffic policies in Shanghai, China. Biomed Environ Sci 22: 210–215.

6. Beelen R, Hoek G, van den Brandt PA, Goldbohm RA, Fischer P, et al. (2008) Long-term exposure to traffic-related air pollution and lung cancer risk. Epidemiology 19: 702–710.

7. Beevers SD, Carslaw DC (2005) The impact of congestion charging on vehicle emissions in London. Atmos Environ 39: 1–5.

8. Cowie CT, Rose N, Gillett R, Walter S, Marks GB (2012) Redistribution of traffic related air pollution associated with a new road tunnel. Environ Sci Technol 46: 2918–2927.

9. He LY, Hu M, Zhang YH, Huang XF, Yao TT (2008) Fine particle emissions from on-road vehicles in the Zhujiang Tunnel, China. Environ Sci Technol 12: 4461–4466.

10. Tian LW, Yu JH (2011) Study on the high altitude discharge of city traffic tunnel pollution. Contamination Control & Air-condition Technology 6(2): 30–34.

11. Wang J, Zhang X, Zhang RP (2009) Analysis of environmental effect of centralized discharge from the long tunnel in city. Chinese Journal of Underground Space and Engineering 5(1): 196–200.

12. Sandra S (2007) Elementary statistics using JMP (SAS Press) (PAP/CDR ed.). Cary, NC: SAS Institute. pp. 166–169.

13. Steve M (2006) "Probability helps you make a decision about your results". Statistics Explained: An Introductory Guide for Life Scientists (1st ed.). Cambridge, United Kingdom: Cambridge University Press. pp. 44–56.

14. Cheng Y, Lee SC, Ho KF, Louis PKK (2006) On-road particulate matter ($PM_{2.5}$) and gaseous emissions in the Shing Mun Tunnel, Hong Kong. Atmos Environ 40: 4235–4245.

15. Cheng XH, Xu XD, Ding GA, Chen ZY, Huang JH, et al. (2007) Diurnal variation characteristics of PM_{10} mass concentration from traffic exhaust emission and its influencing factors on street air quality in Beijing. Journal of Safety and Environment 7: 77–82.

16. Yang L (2011) Study on characteristics of roadside particulate matter with different patterns in traffic flow. Thesis, Tsinghua University. Available: http://cdmd.cnki.com.cn/Article/CDMD-10003-1013023952.htm. Accessed April 2011.

17. Ancelet T, Davy PK, Trompetter WJ, Markwitz A, Weatherburn DC (2011) Carbonaceous aerosols in an urban tunnel. Atmos Environ 45: 4463–4469.

18. Brito J, Rizzo LV, Herckes P, Vasconcellos PC, Caumo SES, et al. (2013) Physical-chemical characterization of the particulate matter inside two road tunnels in the São Paulo metropolitan area. Atmos Chem Phys Discussion 13: 20839–20883.

19. El Haddad I, Marchand N, Dron J, Temime-Roussel B, Quivet E, et al. (2009) Comprehensive primary particulate organic characterization of vehicular exhaust emissions in France. Atmos Environ 43: 6190–6198.

20. Gillies JA, Gertler AW, Sagebiel JC, Dippel WA (2001) On-road particulate matter ($PM_{2.5}$ and PM_{10}) emissions in the Sepulveda Tunnel, Los Angeles, California. Environ Sci Technol 35: 1054–1063.

21. Grieshop AP, Lipsky EM, Pekney NJ, Takahama S, Robinson AL (2006) Fine Particle Emission Factors from Vehicles in a Highway Tunnel: Effects of Fleet Composition and Season. Atmos Environ 40: S287–S298.

22. Handler M, Puls C, Zbiral J, Marr I, Puxbaum H, et al. (2008) Size and composition of particulate emissions from motor vehicles in the Kaisermuhlen-Tunnel, Vienna. Atmospheric Environment 42: 2173–2186.

23. Liu C, Huang XF, Lan ZJ, He LY (2012) A tunnel test for $PM_{2.5}$ emission factors of motor vehicles in Shenzhen. Chinese Journal of Environmental Science & Technology 35(12): 150–153.

24. Pio C, Cerqueira M, Harrison RM, Nunes T, Mirante F, et al. (2011) OC/EC ratio observations in Europe: Re-thinking the approach for apportionment between primary and secondary organic carbon. Atmos Environ 45: 6121–6132.

25. Zhu CS, Chen CC, Cao JJ, Tsai CJ, Chou CCK, et al. (2010) Characterization of carbon fractions for atmospheric fine particles and nanoparticles in a highway tunnel. Atmos Environ 44: 2668–2673.

19

Exposure to Low-Dose Bisphenol A Impairs Meiosis in the Rat Seminiferous Tubule Culture Model: A Physiotoxicogenomic Approach

Sazan Ali[1], Gérard Steinmetz[2], Guillaume Montillet[3], Marie-Hélène Perrard[3], Anderson Loundou[4], Philippe Durand[3¤], Marie-Roberte Guichaoua[1◗], Odette Prat[2*◗]

1 Institut Méditerranéen de Biodiversité et d'Ecologie marine et continentale (IMBE), Centre National de la Recherche Scientifique (CNRS) UMR 7263/ Institut de Recherche pour le Développement (IRD) 237, Faculté de Médecine, Aix-Marseille Université (AMU), Marseille, France, **2** Institute of Environmental Biology and Biotechnology (IBEB), Life Science division, French Alternative Energy and Atomic Energy Commission (CEA), Marcoule, Bagnols-sur-Cèze, France, **3** Institut de Génomique Fonctionnelle de Lyon (IGFL), Centre National de la Recherche Scientifique (CNRS) UMR 5242/ Institut National de la Recherche Agronomique (INRA), Ecole Normale Supérieure de Lyon (ENS), Lyon, France, **4** Unité d'Aide Méthodologique à la Recherche clinique, Faculté de Médecine, Aix-Marseille Université (AMU), Marseille, France

Abstract

Background: Bisphenol A (BPA) is one of the most widespread chemicals in the world and is suspected of being responsible for male reproductive impairments. Nevertheless, its molecular mode of action on spermatogenesis is unclear. This work combines physiology and toxicogenomics to identify mechanisms by which BPA affects the timing of meiosis and induces germ-cell abnormalities.

Methods: We used a rat seminiferous tubule culture model mimicking the *in vivo* adult rat situation. BPA (1 nM and 10 nM) was added to the culture medium. Transcriptomic and meiotic studies were performed on the same cultures at the same exposure times (days 8, 14, and 21). Transcriptomics was performed using pangenomic rat microarrays. Immunocytochemistry was conducted with an anti-SCP3 antibody.

Results: The gene expression analysis showed that the total number of differentially expressed transcripts was time but not dose dependent. We focused on 120 genes directly involved in the first meiotic prophase, sustaining immunocytochemistry. Sixty-two genes were directly involved in pairing and recombination, some of them with high fold changes. Immunocytochemistry indicated alteration of meiotic progression in the presence of BPA, with increased leptotene and decreased diplotene spermatocyte percentages and partial meiotic arrest at the pachytene checkpoint. Morphological abnormalities were observed at all stages of the meiotic prophase. The prevalent abnormalities were total asynapsis and apoptosis. Transcriptomic analysis sustained immunocytological observations.

Conclusion: We showed that low doses of BPA alter numerous genes expression, especially those involved in the reproductive system, and severely impair crucial events of the meiotic prophase leading to partial arrest of meiosis in rat seminiferous tubule cultures.

Editor: Xuejiang Guo, Nanjing Medical University, China

Funding: The Research Consortium ECCOREV n° 3098 (Ecosystemes Continentaux et Risques Environnementaux) CNRS/Aix-Marseille Université funded this study (AOI 2010, grant number 7). The funders had no role in study design, data collection and analysis, decision to publish, or preparation of the manuscript.

Competing Interests: P. Durand is currently affiliated with Kallistem but his contribution to this study was made when he was an employee of the IGFL, UMR 5242 CNRS INRA Ecole Normale Supérieure de Lyon 1, F-69342 Lyon France. The authors can affirm that no conflicting interest exists in that case.

* Email: odette.prat@cea.fr

◗ These authors contributed equally to this work.

¤ Current address: Kallistem SAS, ENS, Lyon, France

Introduction

Bisphenol A (4, 4′-isopropylidenediphenol), or BPA, is one of the world's most highly produced chemicals, used to manufacture epoxy resins and polycarbonate plastics. According to physiologically based pharmacokinetic studies, BPA is found in human serum, urine, milk and fat, with plasma levels ranging from 0.2 to 20 ng/mL (or 1 to 100 nM) [1,2,3,4,5,6]. This substance is mainly absorbed by the digestive tract. BPA is an endocrine-disrupting chemical (EDC), which could interact with both α- and β-estrogen receptors [7] and bind to androgen receptors [8]. Emerging evidence suggests that this molecule may influence multiple endocrine-related pathways [9]. Several controversies have divided scientific opinion regarding the adverse effects of BPA in the testis and reproductive organs [10,11]. Indeed, previous studies indicate that there are no reproductive effects of BPA [12,13,14,15]. Nevertheless, numerous findings suggest that BPA adversely affects the male reproductive system. Tohei et al [16] showed that bisphenol A inhibits testicular function in adult male rats. In mice, testicular hypotrophy and decreased daily sperm production were observed in the presence of BPA [17,18]. In humans, BPA also reduced sperm concentration, motility and morphology [19]. BPA exposure may also induce apoptosis in rat germ cells *in vivo* [20] and in cultured rat Sertoli cells [21], and has the potential to redistribute several known Sertoli cell junctional proteins [22,23]. Subsequent studies also demonstrated that BPA is genotoxic. The accumulation of DNA damage in germ cells was induced by BPA exposure via oxidative stress [24]. BPA causes meiotic abnormalities in oocytes [25,26,27,28,29,30] and in male germ cells of the adult rat [31]. Despite these numerous studies of the effects of BPA on sperm quality, few investigations have been conducted on the crucial meiotic step of spermatogenesis [24,32]. Thus, the molecular action of BPA on spermatogenesis remains largely unknown.

We conducted a fine analysis of the first meiotic prophase with low doses of BPA (1 nM and 10 nM), approximating levels in biological fluids [33]. Decreased efficiency of sperm production in mice appeared at a dose of 20 µg/kg/day (20 ng/g body weight/day) [34]. This study was performed using a validated rat seminiferous tubule culture model [35], able to reproduce spermatogenesis *ex vivo*. This model allows the analysis of cellular responses induced by exposure to low doses of toxic substances for three weeks. This period of time corresponds, in the rat, to the development of spermatogenesis and mimics puberty, a critical period of life with regard to endocrine disruptors [36,37].

It appears that a model "sensitive" to the possible adverse effects of chemicals is indeed of the highest importance for toxicological studies. These models must be able to respond to very low concentrations of toxicants. It must also be underlined that, in order to prevent "false-positive" results, particular attention must be paid to toxicant test concentrations, which must be realistic. Our intention was to investigate whether BPA, at the selected doses for three weeks, could alter the chronology of meiosis and induce morphological abnormalities, and to apprehend its mechanisms of action, combining toxicogenomic and physiological approaches.

Microarray-based transcriptional profiling is a powerful and ultrasensitive tool for monitoring altered cellular functions and pathways under the action of toxicants, providing a wealth of information for sketching the mode of action of toxic substances [38,39,40] or for finding new toxicity bioindicators [41]. However, to date very few transcriptome analyses have been conducted to comprehend the molecular action of BPA on spermatogenesis [42,43,44]. Using transcriptomics, we were able to detect changes in gene expression at biological doses, in the nanomolar range. We studied whether specific patterns of gene modulation could be associated with cytological changes of the first meiotic prophase, observed by immunostaining of the synaptonemal complexes (SC) with an anti-SCP3 antibody. This model allowed for immunocytochemistry and transcriptomic experiments using the same cultures at the same exposure times.

Material and Methods

Animals

The entire study was performed *ex vivo* using cultures of seminiferous tubules. For these cultures, Male 23-day-old Sprague Dawley rats from Charles River France Inc. (supplier: Janvier,

France), having undergone no treatment, were used. Animals were housed 3/ cage, at temperature $21 \pm 3°C$, light cycle 12–12 (6 pm-6 am), diet made of SDS VRF1(from Special Diets Services), water filtered 0.1 µm in bottle system, sawdust bedding, autoclaved. Rats were anesthetized with chloroform then decapitated. At the age of 22–23 days the most advanced germ cells are late spermatocytes [45] allowing to study the whole meiotic phase under our culture conditions [35].

In order to counterbalance interanimal variations, testes from eight rats were pooled in every culture, and used immediately, as previously described [35].

The same population was seeded for control cultures and cultures exposed to toxicant. Analyses were performed at days 8 (D8), 14 (D14) and 21 (D21) of the cultures. All procedures were approved by the Scientific Research Agency (approval number 69306) and conducted in accordance with the guidelines for care and use of laboratory animals. The experimental protocol was designed in compliance with recommendations of the European Economic Community (EEC) (86/ 609/EEC) for the care and use of laboratory animals.

Preparation and culture of seminiferous tubules

The technique of seminiferous tubule culture has been described previously [35]. Cultures were performed with and without BPA. When required, BPA was added beginning from day 2, at 1 nM or 10 nM (Sigma-Aldrich Corporation, St. Louis, USA) in the basal compartment of the bicameral culture chamber. BPA concentrations were selected on the basis of those found in human and rat plasma, i.e. 0.2 to 20 ng/mL, meaning 1 to 100 nM [1,2,3,4,6]. 0.3% DMSO was used as the BPA dilution vehicle; the same solvent concentration was introduced in control cultures.

RNA extraction, labelling and microarray experiments

Two different pools of seminiferous tubules were exposed to two concentrations of BPA (1 nM and 10 nM) or to complete medium with vehicle (control cells) for 8, 14, and 21 days. Total RNA was extracted using the RNeasy Mini kit (Qiagen). RNAs were quantified with the Nanodrop 1000 spectrophotometer; their qualities were assessed with the Agilent 2100 Bioanalyzer. RNA samples were amplified and labeled with the cyanine-3 fluorophore using a Low Input QuickAmp Labeling Kit (Agilent). Hybridization was performed using Agilent Oligo Microarrays (Rat V3 4×44K). Fluorescence was scanned and signal data were extracted with Feature Extraction Software (Agilent).

Cytological methods

Samples treatment. Spreading, and immunocytological localization of SC axial and lateral elements, were performed according to [46]. After spreading by cytocentrifugation at 30 g, slides were fixed in 2% paraformaldehyde (Merk Darmstad, Germany). A rabbit polyclonal anti-SCP3 antibody (Abcam, Cambridge, UK Ab 15093) was used at a 1:100 dilution, to reveal axial elements and lateral elements of the SC. Detection was performed with an FITC-conjugated anti-goat immunoglobulin G (Abcam, Cambridge, UK) at a dilution of 1:100. Slides were mounted in antifade medium (Vectashield, Vector Laboratories, Burlingame, USA).

Microscope analysis. A Zeiss Axioplan 2 Fluorescence Photomicroscope (Carl Zeiss, Oberkochen, Germany) was used to observe the spermatocyte nuclei. Primary spermatocytes, stained with the anti-SCP3 antibody, were selected to evaluate the respective percentages of leptotene, zygotene, pachytene and diplotene stages: 100 to 200 nuclei were analyzed for each culture, for control cultures, and for each time and BPA dose condition. We evaluated the percentages of the three pachytene substages, P1, P2 and P3, corresponding to early, mid and late pachytene substages, in the rat [37]. These substages were defined according to the condensation degree of the sex bivalent during the pachytene stage; 50 nuclei were analyzed for each condition. The pachytene index (PI) was evaluated for each culture and for each time and BPA dose. We defined the PI in rat by the ratio P3/P1+P2+P3 [37]. The percentages of nuclei showing SC abnormalities were quantified at each time point, in both control cultures and cultures exposed to 1 nM and 10 nM BPA. For each stage, and for each abnormality, we researched a possible dose-and-time variation.

Statistical analysis

Transcriptomic analysis. In this experimental design, six independent analyses were conducted versus each specific control for the considered time point: a) 1 nM BPA-exposed cells for 8 days, b) 1 nM BPA-exposed cells for 14 days, c) 1 nM BPA-exposed cells for 21 days, d) 10 nM BPA-exposed cells for 8 days, e) 10 nM BPA-exposed cells for 14 days, f) 10 nM BPA-exposed cells for 21 days. For each analysis, eight raw fluorescence data files (four controls and four tests) were submitted to GeneSpring Software GX11 (Agilent Technologies) using a widely used method for determining the significance change of gene expression [38,47]. The fold change cutoff between control and exposed samples was set to 1.5. Genes significantly up- or downregulated were determined by an unpaired t-test, with a p-value <0.05 and a Benjamini-Hochberg false discovery rate correction. We thus obtained probe sets that were significantly induced or repressed after exposure to BPA.

Immunocytochemistry (ICC). Statistical analysis was performed using PASW Statistics Version 17.0.2 (IBM SPSS Inc., Chicago, IL, USA). Continuous variables are expressed as means±SD. Comparisons of means between two groups were performed using a Student's t-test. All tests were two-sided. The statistical significance was defined as p<0.05. Three biological replicates were analyzed for D8 and D14, and two replicates for D21. Each experiment included controls (vehicle only) and tests (BPA). The total number of nuclei analyzed was 4630, combining all doses and time points.

Biological analysis

Lists of genes significantly induced or repressed after exposure to BPA were uploaded into Ingenuity Pathway Analysis Software (IPA, Ingenuity Systems, www.ingenuity.com) for biological analysis by comparison with the Ingenuity Knowledge Database. These lists of altered genes were then processed to investigate the functional distribution of these genes, as defined by Gene Ontology. Datasets and known canonical pathways associations were measured by IPA by using a ratio (R) of the number of genes from a dataset that map to a specific pathway divided by the total number of genes that map to this canonical pathway. A Fisher's exact test was used to determine a p-value representing the significance of these associations.

Quantitative RT-PCR

Total RNA was isolated according to the manufacturer's instructions using the RNeasy Kit (Qiagen), and treated with DNase. RNA purity and concentration were determined by UV on a Nanodrop Spectrophotometer and integrity was assessed on an Agilent 2100 Bioanalyzer (Agilent Technologies). All the samples used in this study showed 28S/18S ratio signing intact and pure RNA. Differential analysis of RNA from cells exposed to

NPs and from unexposed cells was performed by qRT-PCR with the Sybr Green PCR Master Mix (Finzyme) Kit according to the manufacturer's instructions, on Opticon II (Biorad). Primer (Sigma) sequences were, for *Stra8*: 5′ CAGCCTCAAAGTGG-CAGGTA 3′ (forward) and 5′ GGGAGAGGGAGTGGGACA-GAT 3′ (reverse); for *Mlh1*: 5′ CGCCATGCTGGCCTTA-GATA 3′ (forward) and 5′ CCTCCAAAGGCGGCACATA 3′ (reverse); for *Prdm9*: 5′ AGAATGAGAAAGCCAACAGCA 3′ (forward) and 5′ AGACTCCTTAGAAGTTTTAGCAGA 3′ (reverse); for *Sycp1*: 5′ GAGAGAAGACCGTTGGGCA 3′ (forward) and 5′ TCCATTGCAAGTAAAAGCAACA 3′ (reverse); for *Fpr3*: 5′ ACTGTGAGCCTGGCTAGGAA 3′ (forward) and 5′ CTCGTGAAGCACGGCTAGAA 3′ (reverse); for *Dmc1*: 5′ CTTTCCGTCCAGATCGCCTT 3′ (forward) and 5′ AAAATGCCGGCTTCTTCGTG 3′ (reverse); for *Card11*: 5′ CTCAGGCCCAGTTTCTCCAG 3′ (forward) and 5′ CTGTTGAGCTCTGTGGAGGG 3′ (reverse); for *Nfkb1*: 5′ GGAGATGGCCCACTGCTATC 3′ (forward) and 5′ TTCGGAAGGCCTCGAATGAC 3′ (reverse). For *Stra8, Mlh1, Prdm9, Sycp1, Fpr3, Dmc1, Card11, Nfkb1*, the amplicon sizes were 347, 200, 100, 286, 313, 160, 92, and 223 bp, respectively. The measurements were the means of six individual results and normalization was based on the total RNA mass quantified on the Nanodrop. Expression ratios were calculated according to Pfaffl et al. [48], where the relative expression ratio (R) of a target gene was calculated using PCR efficiency (E) and the CT (number of cycles at threshold) deviation of an unknown sample versus a control. The target gene fold change was then expressed as follows: $E^{(CT\ mean\ control\ -\ CT\ mean\ treatment)}$. Statistical significance was tested by Pair Wise Fixed Reallocation Randomization Test (REST software) where a p-value less than 0.05 was considered significant.

Results

Transcriptome analysis

Figure 1 indicates the number of significantly differentially expressed genes for each dose and time point. These figures encompass up- and downregulated transcripts. The number of genes affected by BPA increased markedly over the exposure time. At 1 nM BPA, this modification was time dependent. At 10 nM, there was, curiously, a decrease in the number of modulated genes at D14, but this number increased again at D 21. The entire list of significantly up- and downregulated genes for each dose and time point (FC>1.5 with p-value <0.05) is provided as supplementary material (Table S1).

Altered physiological functions and canonical pathways. We analyzed the distribution of altered genes per function, as defined in Gene Ontology (Figure 2), using Ingenuity Pathway Analysis. Radar plots helped to apprehend the complexity of toxicity, both in terms of amplitude and effect. This resulted in a specific pattern representing the toxicity for each individual dose and time point. Graphic overlay of all time points allows a visual comparison of the extension of adverse effects throughout the exposure time. The distribution of functions altered by BPA on the radar plot delineated similar patterns from D8 to D21, meaning that adverse effects were amplified with time but not with dose (Fig. 2A and 2B for 1 nM and 10 nM, respectively). Whatever the dose and time point the top three altered functions were cancer, cell death and cellular development, as shown in Fig. 2. For instance, at D21/1 nM BPA, the numbers of genes related to cancer, cell death and cellular development were 2927, 2422 and 1833, respectively. For reproductive system disease and DNA replication and repair, the numbers of genes were 746 and

Time and dose response relationship

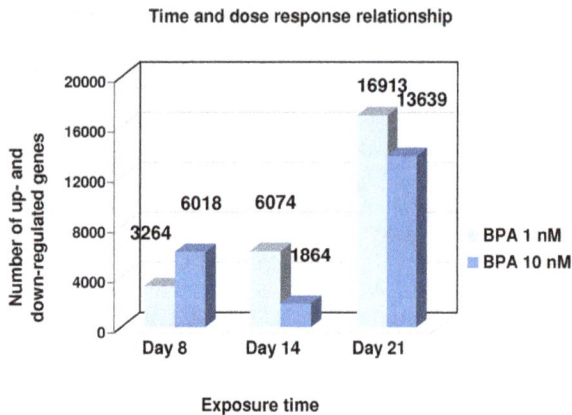

Figure 1. Total number of genes differentially up- or down-regulated by 1 nM and 10 nM BPA after 8, 14 and 21 days of exposure. Genes were selected with a fold change cutoff ≥1.5 (p-value <0.05). The global expression change compared with control cells was time but not dose dependent.

511, respectively. For other dose and time points, the numbers of genes per altered function are indicated in Figure 2.

Figure 3 shows the common canonical pathways disturbed by BPA (1 nM at D8, D14, and D21). Each canonical pathway is

constituted of a finite number of genes. For each time point, we calculated a ratio indicating the percentage of altered genes in our dataset belonging to a given canonical pathway (for precise calculation, see Material and Methods). We selected nine canonical pathways on the basis of the most significant p-values. These nine main canonical pathways were altered by BPA in a time-dependent manner. The following scores are given for D21/ 1 nM BPA, but all were altered early, at D8.

- *DNA double-strand break repair by homologous recombination,* R = 0.56, p-value 3.52×10^{-3}
- *LXR/RXR activation,* R = 0.68, p-value 4.26×10^{-3}
- *Aryl hydrocarbon receptor signaling,* R = 0.69, p-value 2.75×10^{-4}
- *NF-κB signaling,* R = 0.66, p-value 4.78×10^{-3}
- *NRF2-mediated oxidative stress response,* R = 0.74, p-value 1.29×10^{-8}
- *Xenobiotic metabolism signaling,* R = 0.66, p-value 6.76×10^{-4}
- *Apoptosis signaling,* R = 0.72, p-value 2.95×10^{-4}
- *Role of BRCA1 in DNA damage response,* R = 0.63, p-value 9.77×10^{-3}
- *Androgen signaling,* R = 0.56, p-value 2.59×10^{-2}

Transcription changes of genes expressed in meiotic and premeiotic cells. Of the 746 deregulated genes of the

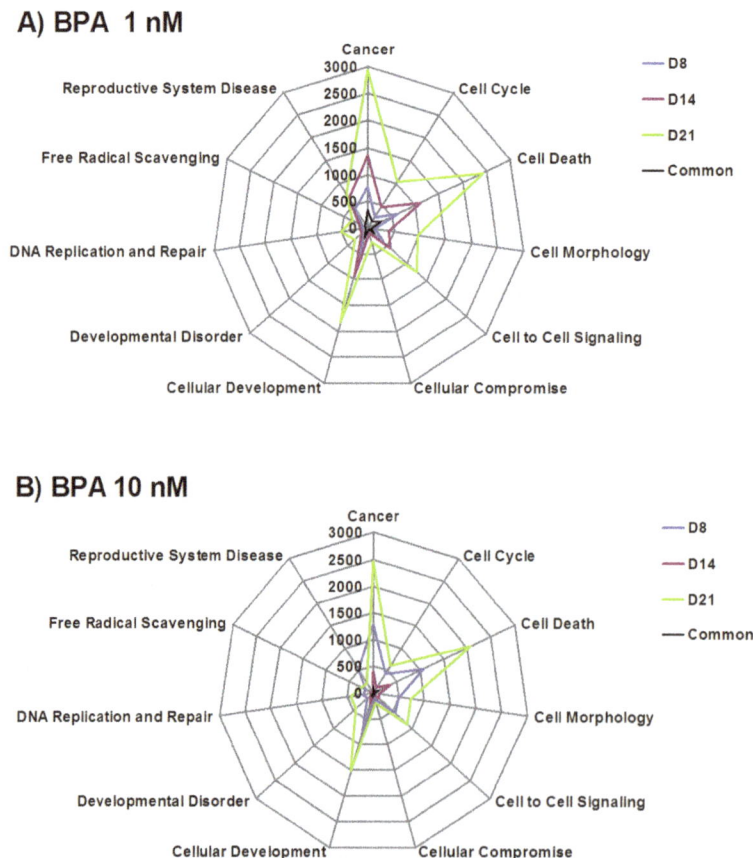

Figure 2. Distribution of differentially expressed genes per altered function. Genes significantly up- or downregulated after cell exposure to BPA for 8 days (blue), 14 days (red), or 21 days (green), and genes common to all time points (black) were examined and classified per function according to Gene Ontology. The results comprise a specific pattern of BPA toxicity, similar for all time points. The amplitude of the toxicity is given by the scale representing the number of modulated genes for each type of altered function. **A** – 1 nM BPA. **B** – 10 nM BPA.

Main altered canonical pathways

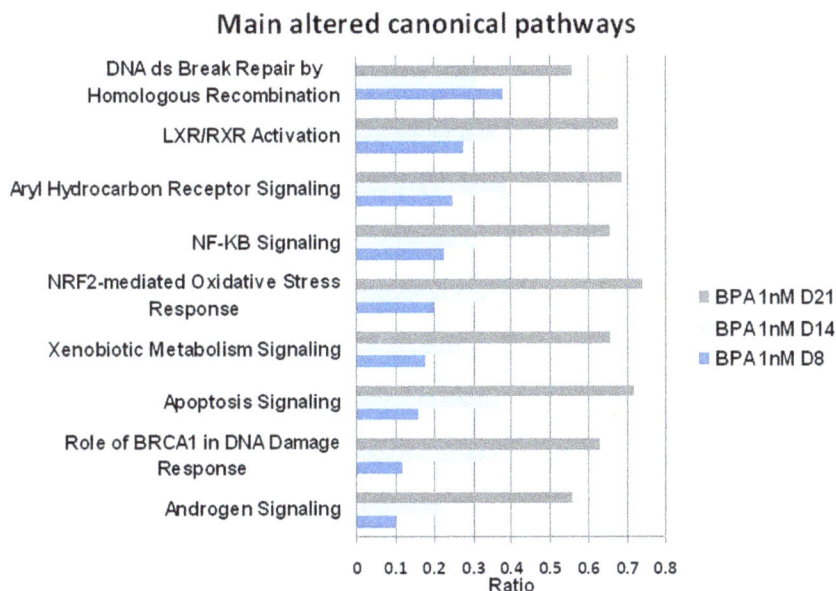

Figure 3. Comparative analysis of canonical pathways significantly altered by 1 nM BPA at D8, D14 and D21. The x-axis depicts gene ratios within a dataset mapping to the considered pathway (see Methods for calculation). A Fisher's exact test was used to determine a p-value representing the significance of these associations (p<0.01). In all cases, the ratios increased with exposure time.

reproductive system, we focused on 120 genes known to be involved in premeiotic steps or in the first meiotic prophase (Table S2). As for the total set of genes (Fig. 1), the number of BPA-affected genes increased markedly over the time of exposure, except for a decrease at D14/10 nM BPA. The highest fold changes were observed at D21/1 nM. Among these 120 genes, the number of downregulated genes (62.2%) widely exceeded the number of upregulated genes. The genes that had the greatest fold change were downregulated, except for Nos2, which was upregulated. The greatest fold change was observed for Stra8 (−37.83) which was deregulated in all conditions. Figure 4 shows that BPA deregulates genes involved in all the important processes of premeiotic steps and first meiotic prophase. The genes affected with the greatest fold change were mainly involved in meiotic initiation and recombination. Indeed, of the 120 genes, 62 were directly involved in these two functions. Some modified genes were involved in functions other than meiotic events but nevertheless essential to meiosis, such as transcriptional regulation, cell cycle, chromatin organization, protein stability, stress-induced responses and repression of retrotransposable elements. Most of this study's meiotic genes coded for nuclear proteins, some for cytoplasmic proteins, and rarely for plasma membrane proteins (Table S2). All of these genes are represented in Fig. 4. They are classified according to their respective functions in meiotic initiation, recombination and pairing. All have been shown to be interconnected in a network (Fig. 5) obtained by IPA.

qRT-PCR validation of the microarray data focused on meiosis. Quantitative RT-PCR was performed using the same batches of RNA as those evaluated by microarrays. Validation of the microarray data was investigated on five genes belonging to BPA-1 nM/D21, and on 7 genes belonging to BPA-10 nM/D21. Table 1 reports the compared fold changes obtained with microarray and qRT-PCR (p-value <0.05).

Immunocytological analysis of the meiotic prophase

Effect of BPA on the percentage of SCP3-stained meiotic stages. Each stage's normal morphological aspect of the first meiotic prophase – leptotene, zygotene, pachytene and diplotene – has been described previously [37]. The percentages of these four stages in the control cultures were of the same order of magnitude as those obtained previously. These percentages were evaluated from 100 to 200 nuclei for each culture, for every BPA dose and time condition, and for control cultures. We showed in the present study that BPA disrupted the progression of meiotic prophase in the cultures analyzed, for the two doses at the three time points.

The most obvious changes were observed at the leptotene and diplotene stages. In the treated cultures, the percentage of leptotene stage (Fig. 6A) increased for all days and concentrations compared with control cultures. This increase was at the limit of significance for 1 nM at D8 (11.4±2.8 versus 8.1±0.6, p = 0.06). The increase in leptotene stage was significant (p<0.05) for 10 nM at D8 (11.6±1.5 versus 8.1±0.6 in control), for 1 and 10 nM at D14 and D21 (D14: 13.8±1.1 and 11.7±2.3, respectively, versus 6.6±0.5 in control; D21: 13.1±0.2 and 17.6±3.8, respectively, versus 6.3±1.9 in control). In the same cultures, diplotene stage decreased for all days and concentrations compared with control cultures (Fig. 6B). The decrease in diplotene stage was not significant for 1 nM at D8. This decrease was significant (p< 0.05) for 10 nM at D8 (2.4±0.2 versus 5.8±1.5 in control), for 1 and 10 nM at D14 and D21 (D14: 3.2±1.5 and 4.2±1.5, respectively, versus 14.5±2.3 in control; D21: 2.1±1.2 and 3.2±1.5 versus 16.0±1.8 in control). Nevertheless, these changes of leptotene and diplotene stages were independent of the BPA concentration and of the exposure time.

Zygotene stage slightly decreased in the BPA-treated cultures, whereas pachytene stage slightly increased, but these variations were not significant, whatever the doses and time points.

Effect of BPA on the pachytene index. Although the percentage of pachytene stage did not vary significantly in this study for the majority of doses and time points, we observed a

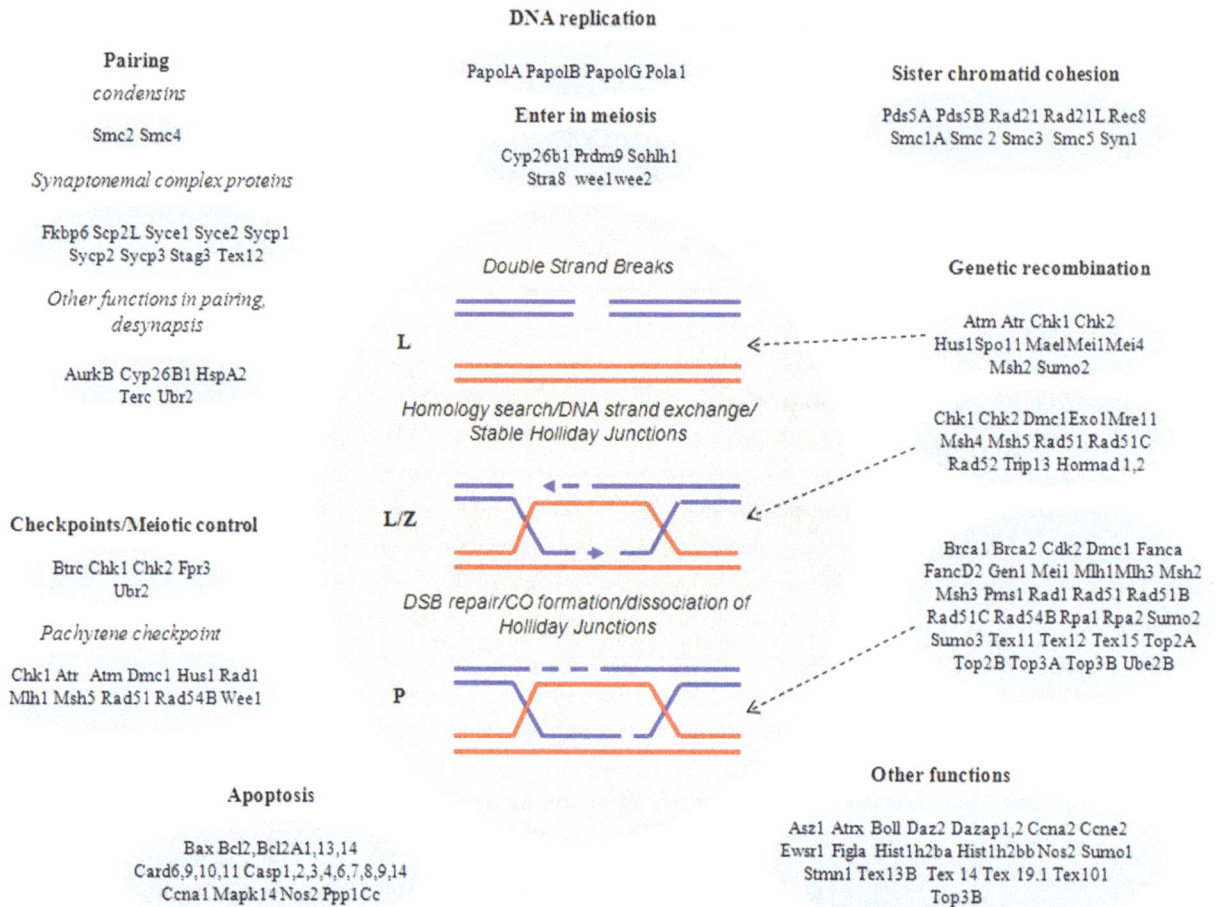

Pairing

condensins

Smc2 Smc4

Synaptonemal complex proteins

Fkbp6 Scp2L Syce1 Syce2 Sycp1
Sycp2 Sycp3 Stag3 Tex12

*Other functions in pairing,
desynapsis*

AurkB Cyp26B1 HspA2
Terc Ubr2

Checkpoints/Meiotic control

Btrc Chk1 Chk2 Fpr3
Ubr2

Pachytene checkpoint

Chk1 Atr Atm Dmc1 Hus1 Rad1
Mlh1 Msh5 Rad51 Rad54B Wee1

Apoptosis

Bax Bcl2,Bcl2A1,13,14
Card6,9,10,11 Casp1,2,3,4,6,7,8,9,14
Ccna1 Mapk14 Nos2 Ppp1Cc

DNA replication

PapolA PapolB PapolG Pola1

Enter in meiosis

Cyp26b1 Prdm9 Sohlh1
Stra8 wee1wee2

Double Strand Breaks

L

*Homology search/DNA strand exchange/
Stable Holliday Junctions*

L/Z

*DSB repair/CO formation/dissociation of
Holliday Junctions*

P

Sister chromatid cohesion

Pds5A Pds5B Rad21 Rad21L Rec8
Smc1A Smc 2 Smc3 Smc5 Syn1

Genetic recombination

Atm Atr Chk1 Chk2
Hus1 Spo11 Mael Mei1 Mei4
Msh2 Sumo2

Chk1 Chk2 Dmc1 Exo1 Mre11
Msh4 Msh5 Rad51 Rad51C
Rad52 Trip13 Hormad 1,2

Brca1 Brca2 Cdk2 Dmc1 Fanca
FancD2 Gen1 Mei1 Mlh1 Mlh3 Msh2
Msh3 Pms1 Rad1 Rad51 Rad51B
Rad51C Rad54B Rpa1 Rpa2 Sumo2
Sumo3 Tex11 Tex12 Tex15 Top2A
Top2B Top3A Top3B Ube2B

Other functions

Asz1 Atrx Boll Daz2 Dazap1,2 Ccna2 Ccne2
Ewsr1 Figla Hist1h2ba Hist1h2bb Nos2 Sumo1
Stmn1 Tex13B Tex 14 Tex 19.1 Tex101
Top3B

Figure 4. Diagram of the main stages of meiotic recombination and the corresponding stages of the meiotic prophase (L = leptotene, Z = zygotene and P = pachytene). The 120 genes showing a fold change ≥1.5 (p value ≤0.05) in BPA-treated cultures compared with controls, and involved in events of the meiotic prophase, were classified according to their function. This figure shows that the main functions of the first meiotic prophase are altered by BPA. Genes having several functions appear several times in this figure.

decrease of the P3 substage at all dose and time points ($p < 0.05$). For 1 nM and 10 nM, at D8: 13.0 ± 2.6 and 10.4 ± 2.3, respectively, versus 22.0 ± 2.6 in control cultures; at D14: 9.3 ± 3.1 and 7.3 ± 2.3, respectively, versus 24.7 ± 3.1 in control cultures; at D21: 8.3 ± 0.2 and 8.3 ± 1.7, respectively, versus 29.1 ± 2.3 in control cultures. Consequently, the PI also decreased at all doses and time points ($p < 0.05$) (Fig. 7). For 1 nM and 10 nM, at D8: 0.13 ± 0.03 and 0.10 ± 0.03, respectively, versus 0.22 ± 0.05 in control cultures; at D14: 0.06 ± 0.03 and 0.05 ± 0.02, respectively, versus 0.25 ± 0.03 in control cultures; at D21: 0.08 ± 0.007 and 0.08 ± 0.02, respectively, versus 0.29 ± 0.03 in control cultures.

BPA-induced axial element and SC abnormalities. SYCP3 revealed BPA-induced abnormalities of axial elements and SC at the leptotene, pachytene and diplotene stages.

– At *leptotene*, in the absence of BPA, thin and discontinuous axial cores held the nucleus area of the leptotene nuclei [36,37]. With BPA, abnormally long stretches of axial cores without indication of polarization appeared in these nuclei at both BPA concentrations and at three culture time points (Fig. 8A). For 1 nM and 10 nM at D8: 25.9 ± 1.2 and 30.2 ± 1.3, respectively; at D14: 22.4 ± 2.1 and 34.9 ± 2.1, respectively; at D21: 23.0 ± 6.7 and 26.5 ± 6.1, respectively.

– *At pachytene*, the prevalent abnormality observed in the presence of BPA was asynapsis, especially total asynapsis (Fig. 8B). The percentage of asynapsis increased significantly ($p < 0.05$) for all doses and time points with no dose or time dependency. For 1 and 10 nM, at D8: 24.0 ± 3.2 and 26.4 ± 2.8, respectively, versus 5.2 ± 2.2 in control cultures. At D14: 27.2 ± 2.9 and 24.6 ± 3.7, respectively, versus 9.1 ± 3.7 in control cultures. At D21: 44.8 ± 9.4 and 52.8 ± 7.8, respectively, versus 9.5 ± 1.4 in control cultures (Fig. 9A).

– The pulverized SC nuclei (Fig. 8C), proving apoptosis [49], significantly increased ($p < 0.05$) for all doses and time points (Fig. 9B). For 1 nM and 10 nM at D8: 6.2 ± 0.8 and 8.5 ± 3.0, respectively, versus 1.9 ± 0.5 in control cultures; D14: 11.3 ± 1.9 and 10.0 ± 1.0, respectively, versus 3.2 ± 0.6, in control cultures; D21: 42.5 ± 7.5 and 52.1 ± 11.5 versus 20.4 ± 1.6 for 10 nM).

– *At diplotene*, all spermatocytes contained univalents and fragmented lateral elements of SC (Fig. 8D).

Discussion

BPA effects were investigated on male meiosis, using a validated and reproducible seminiferous tubule culture model [35]. Under the present experimental ex-vivo conditions, testosterone, pro-

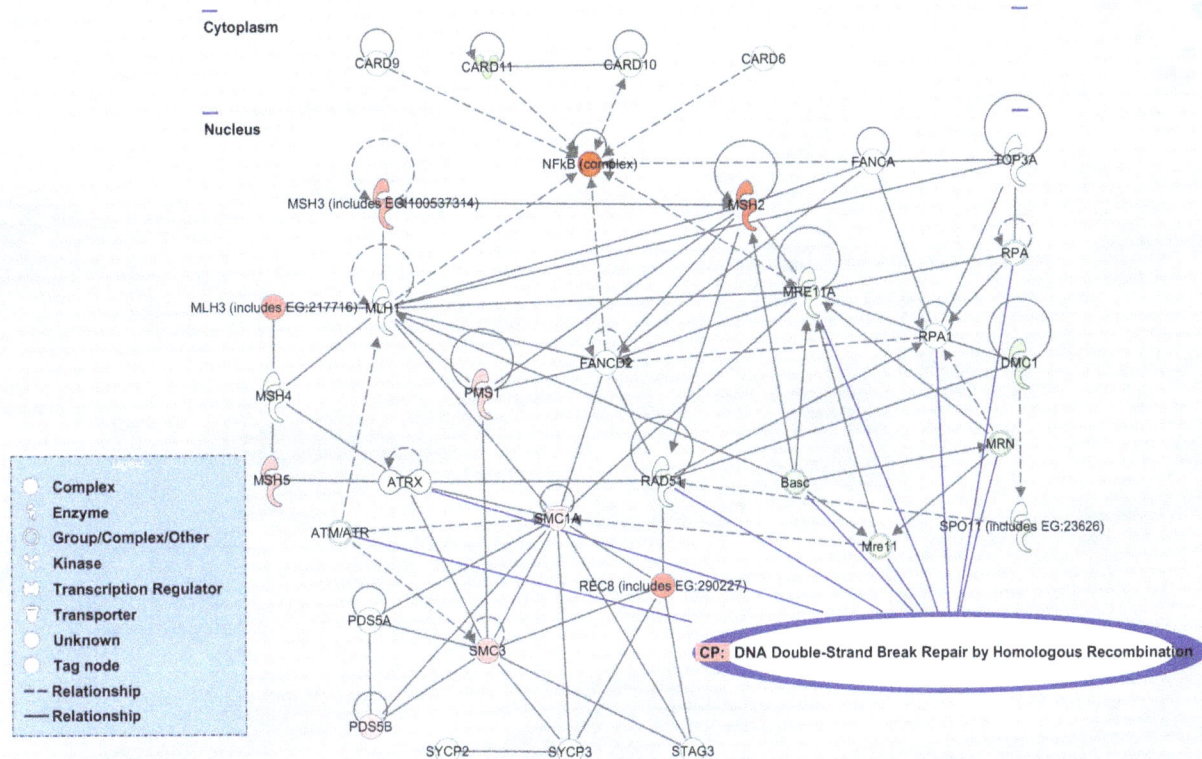

Figure 5. The top-ranked Ingenuity network identified within the group of 120 genes preceding meiotic divisions, involved in the first meiotic prophase, or essential to meiosis that were found to be differentially expressed in our datasets. This literature-based network shows the high level of connections between genes linked to DNA DSB repair in our datasets, considering all BPA doses and time points. The indicated values of fold change (induction or repression) are the maximal values found in the time course with 1 nM and 10 nM BPA. The red nodes are upregulated, the green nodes are downregulated.

duced in vivo by the Leydig cells and acting on the Sertoli cells, is added directly to the culture medium as described in Staub et al. (2002). As for the relationship between the Sertoli cells and the germ cells, we have shown previously that, under our experimental conditions, cellular junctions between the Sertoli cells and the germ cells, which are most important for germ cell differentiation, are maintained [50,51]. This model allows the analysis of induced responses by germ cell exposure to low doses of toxic substances for three weeks. This period of time corresponds, in the rat, to the development of spermatogenesis and mimics puberty, a critical period of life sensitive to endocrine disruptors. Indeed, a tiny concentration of endocrine disruptors can produce long-term adverse effects on the reproductive system [52]. Seminiferous tubules from 23-day-old Sprague-Dawley rats were used. At this age there are no round spermatids in the rat testes. Thus, we are sure that the round spermatids originate from the meiotic divisions which occurred in vitro [35]. Moreover, we previously showed similarities between the meiotic processes in vivo and ex vivo [36,37]. We performed transcriptomic analyses in BPA-treated cultures versus controls without any a priori concerning the results, varying the doses and time points. We performed germ cell ICC analyses on the same cultures, at the same doses and time points. We show that the transcriptomic results and morphological observations are consistent. The percentages of the four populations of spermatocytes under the control conditions, as well as the pachytene substages, were found in the present study to be very

similar to those described in our previous publications [36,37]. BPA concentrations were selected on the basis of the concentrations found in human and rat sera, i.e. 0.2 to 20 ng/mL, meaning 1 to 100 nM [1,2,3,4,6,33,53].

BPA alters important biological functions and canonical pathways in the seminiferous tubule cultures

The overall number of altered genes is a very good indicator of the level of cellular disturbance induced by a toxic compound [43,54,55]. Here, we observed no dose dependency in terms of number of significantly differentially expressed genes, but a time dependency (Fig. 1).

Radar plots showed that BPA alters important biological functions. We analyzed, for all doses and time points, the distribution of altered genes per function, as defined in Gene Ontology. The same functions were altered at each dose and time point, but the number of genes increased only with the exposure time (Fig. 2). Altered genes involved in cancer, cell death and cellular development were predominant. Notably, genes involved in disease of the reproductive system and in DNA replication and repair were also altered.

Although the functions described provide valuable information on the action of the involved genes, the canonical pathways help in understanding, in a faster and more drastic manner, the interactions between these genes themselves and the cellular mechanisms to which they belong. As shown in Fig. 3, the nine

Table 1. Microarray gene expression validation by qRT-PCR.

1 nM BPA, D21

Gene ID	Primers (5'-3')	Microarray fold change BPA/Ctrl	qRT-PCR fold change BPA/Ctrl
Stra8	F: CAGCCTCAAAGTGGCAGGTA	−37.8	−6.1
	R: GGGAGAGGAGTGGGACAGAT		
Mlh1	F: CGCCATGCTGGCCTTAGATA	−3.1	−1.1
	R: CCTCCAAAGGCGGCACATA		
Prdm9	F: AGAATGAGAAAGCCAACAGCA	−25.4	−5.6
	R: AGACTCCTTAGAAGTTTTAGCAGA		
Sycp1	F: GAGAGAAGACCGTTGGGCA	−9.1	−3.9
	R: TCCATTGCAAGTAAAAGCAACA		
Fpr3	F: ACTGTGAGCCTGGCTAGGAA	−25.1	−3.5
	R: CTCGTGAAGCACGGCTAGAA		

10 nM BPA, D21

Gene ID	Primers (5'-3')	Microarray fold-change BPA/Ctrl	qRT-PCR fold change BPA/Ctrl
Stra8	F: CAGCCTCAAAGTGGCAGGTA	−4.6	−6.9
	R: GGGAGAGGAGTGGGACAGAT		
Mlh1	F: CGCCATGCTGGCCTTAGATA	−2.3	−1.6
	R: CCTCCAAAGGCGGCACATA		
Prdm9	F: AGAATGAGAAAGCCAACAGCA	−3.9	−2.2
	R: AGACTCCTTAGAAGTTTTAGCAGA		
Frp3	F: ACTGTGAGCCTGGCTAGGAA	−2.6	−2.7
	R: CTCGTGAAGCACGGCTAGAA		
Dmc1	F: CTTTCCGTCCAGATCGCCTT	3.9	2.5
	R: AAAATGCCGGCTTCTTCGTG		
Card11	F: CTCAGGCCCAGTTTCTCCAG	−3.9	−11
	R: CTGTTGAGCTCTGTGGAGGG		
Nfkb1	F: GGAGATGGCCCACTGCTATC	3	5.9
	R: TTCGGAAGGCCTCGAATGAC		

Changes in the mRNA expression measured by transcriptomics and by quantitative real-time PCR. The fold changes in cells treated with BPA at 1 nM and 10 nM are expressed versus untreated cells, at D21. For qRT-PCR, the measurements were the means of six measurements (triplicates of two independent experiments) and normalization was based on the total RNA mass quantified on Nanodrop spectrophotometer. All expression levels in treated cells were significantly different from controls ($p < 0.05$). F = forward; R = reverse.

main disturbed canonical pathways in our cultures support the literature findings. The disturbance of these canonical pathways was time dependent. Apart from these observations, analyzing the direction of expression (induction or repression) of each specific transcript is difficult in the context of current knowledge. Since the mRNAs are produced in an oscillatory manner (in bursts) [56], their production is likely to obey extremely finely tuned processes, depending on many features, and should not be overinterpreted at this stage.

The *aryl hydrocarbon receptor signaling* pathway was altered as expected, as BPA is an aromatic substance. The *xenobiotic metabolism signaling* pathway by cytochrome P450 was highlighted, as expected, including P450-family genes. Tumor necrosis factor, Tnf, was strongly increased, up to 39 times at D14/1 nM BPA. Several genes of the Nfkb family were upregulated at all doses and time points. Seventy genes encoding heat-shock proteins were mobilized. Their role is to prevent protein aggregation under stress conditions.

The strong activation of *the NRF2-mediated oxidative stress response* pathway induced the overexpression of genes encoding Phase I and II metabolizing enzymes, as well as antioxidant proteins.

The *LXR/RXR activation* pathway was highly disturbed, meaning an effect of BPA on lipid metabolism, inflammation and cholesterol metabolism through the low-density lipoprotein receptor (LdlR). Genes regulated by LXR included ATP-binding cassette transporter A1 (Abca1), which was upregulated. This observation confirms previous descriptions of the implication of BPA in lipid metabolism [57,58].

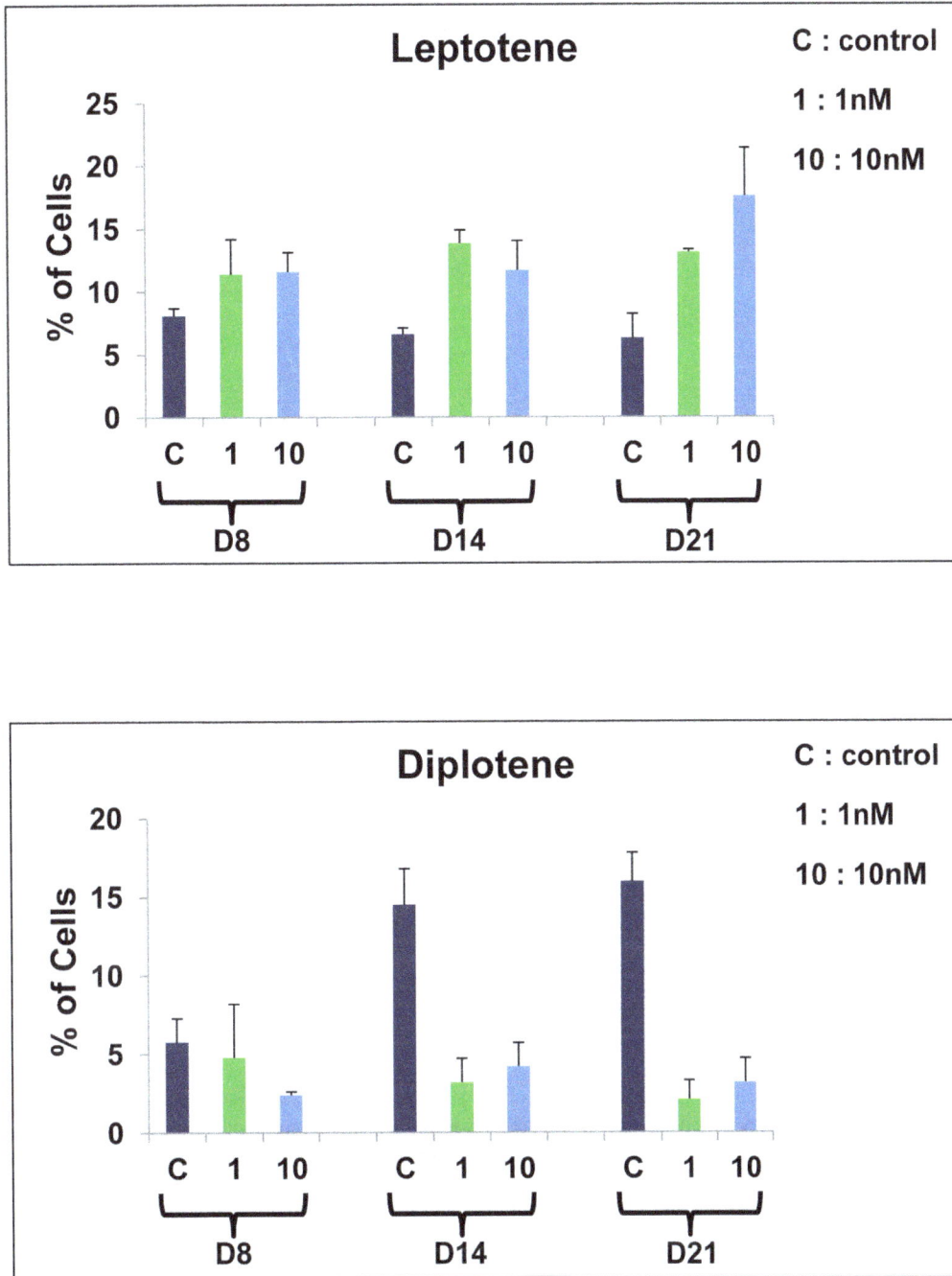

Figure 6. Effects of BPA on the percentages of leptotene and diplotene stages. A – percentage of leptotene stage increased at all doses and time points. The increase was at the limit of significance for 1 nM at D8 (p = 0.06) and significant for D14 and D21 at the two doses (p<0.05). **B** – diplotene stage percentage decreased at all doses and time points. The increase was not significant for 1 nM at D8 and was significant for D14 and D21 at the two doses (p<0.05). Each bar represents the mean ±SD (n = 3 cultures for D8 and D14, and 2 cultures for D21) versus control. C = control culture, 1 and 10 = 1 nM and 10 nM BPA, respectively.

An interesting finding was the alteration of the *Androgen signaling* pathway as proof of the role of BPA as an endocrine disruptor. At D21/1 nM BPA, the androgen receptor (AR) was downregulated (−2.7), and the androgen signaling pathway included 76 altered genes. BPA has also been described as an estrogen-like substance [9,11]. Indeed, in our hands, BPA leaves the marks of endocrine disruption. Numerous prolactin family genes were repressed, especially at 21 days of culture and more intensely at 1 nM than 10 nM. BPA altered the expression of tens of GPCR. At D21/1 nM BPA, the estrogen receptors, Esr1 and Esr2, were also significantly downregulated (−15.3 and −14.3, respectively). Members of the aldo-keto reductase family (Akr1c), and especially Akr1c12/C13 and Akr1c3, were downregulated. This late gene encodes an enzyme essential for testosterone synthesis and could explain the decreased testosterone synthesis already described with BPA *in vivo* [52].

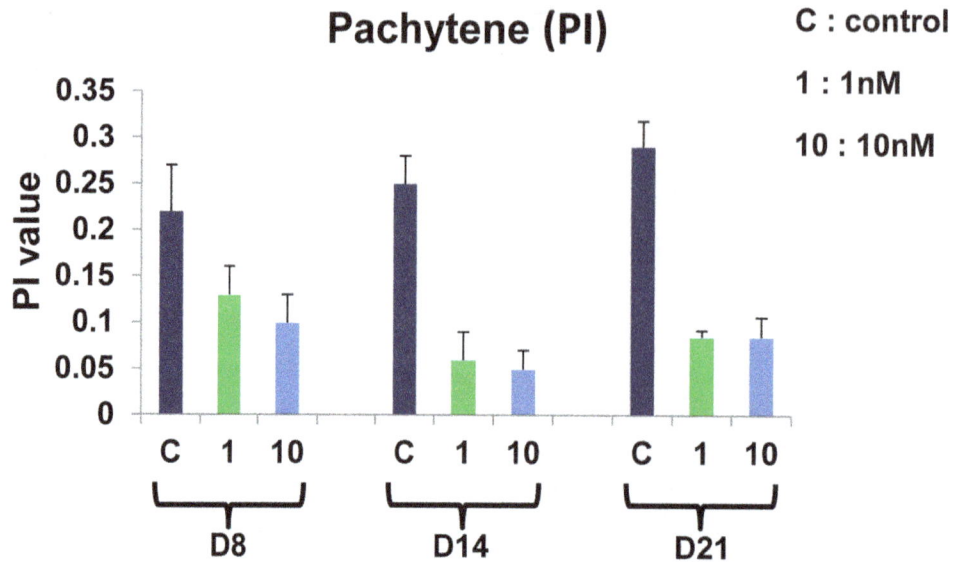

Figure 7. Effect of BPA on the pachytene stage progression. The pachytene index (PI) decreased at both dose and time points (p<0.05). C = control culture, 1 and 10 = 1 nM and 10 nM BPA, respectively. Each bar represents the mean ±SD (n = 3 cultures for D8 and D14, and 2 cultures for D21) versus control.

One can, logically, wonder why these features are visible at 1 nM and not at 10 nM. A lack of dose dependency is often a mark of endocrine disruptors, which induce nonmonotonic dose-response curves [5]. An endocrine disruptor can be equally as potent as endogenous hormones in some systems, causing biological effects at levels as low as picomolar [59]. One of the possible explanations for the absence of dose dependency might be that, after the receptor is bound by BPA and transcription of target genes has occurred, the reaction eventually must reach a plateau until the bound receptor must be inactivated [60]. As an antiandrogen, BPA can also trigger other mechanisms of action that are more complex or even antagonist to the previous one [61].

Many genes were implicated in genetic recombination, in particular the process of *DNA double-strand break repair by homologous recombination*, and are discussed below. The *role of BRCA1 in DNA damage response* is also a major altered pathway and is consistent with the many genes playing a role in cancer functions. Atm, Atr and Chk2, regulators of the DNA damage response, were downregulated in several experimental conditions (Table S1). Thus, our transcriptomic analysis on cultured seminiferous tubules supports current knowledge regarding BPA action.

Transcriptomic analysis sustains immunocytological observations

We showed in this study that ICC analysis revealed concomitant modifications of meiotic prophase chronology, defects of chromosome pairing, and induction of germ-cell apoptosis in cultures exposed to BPA. Combined transcriptomic analysis showed that BPA deregulates numerous genes involved in premeiotic steps and in the first meiotic prophase: DNA replication and sister chromatid cohesion, initiation of meiosis, pairing and genetic recombination, meiotic control and checkpoints, germ-cell apoptosis, and genes implicated in several other cellular processes essential to the life of the germ cells (Fig. 4). In the scope of this article, to explain the abnormalities revealed by ICC, we chose to focus in particular on 120 genes, preceding meiotic divisions, involved in the first meiotic prophase, or essential to meiosis (Table

S2). This table shows that differentially expressed genes are almost the same at both doses of BPA for a same time of exposure. This observation and the fact that most of these genes/proteins interact with each other as shown in Figure 5 could explain why the phenotype observed by ICC is the same for 1 and 10 nM.

Quantitative analysis of meiotic prophase revealed that BPA increases the percentage of leptotene nuclei, one-third of these showing abnormally long stretches of axial core. Abnormalities of the leptotene stage could be consistent with an alteration of cells to progress towards the zygotene stage. According to Zickler and Kleckner [62], the leptotene/zygotene transition appears to be an unusually complex and critical transition. Indeed, during this transient period the polarization of meiotic chromosome telomeres (bouquet stage) occurs, which is essentially concomitant with the onset of synaptonemal complex (SC) formation. This stage is also concomitant with the progression between DSBs and stable strand-exchange intermediates (double Holliday junctions). These observations are consistent with the microarray analysis, showing transcriptional changes of the genes involved in the early steps of genetic recombination as a homology search, DNA strand exchange and stable Holliday junction formation (Fig. 4). Thus, it is impossible for most germ cells to pass through the leptotene-zygotene transition. The most deregulated gene, Stra8 (Table S2), participates as a fundamentally positive regulator in the commitment of spermatocytes to meiosis, and regulates progression through the early stages of meiotic prophase [63]. In the study of Mark et al. [63], a small percentage of *Stra8-/-* spermatocytes progressed into the later stages of meiosis and showed a prolonged bouquet stage configuration. Leptotene nuclei with abnormally long stretches of axial core were also reported in Dmc1-/- mice [64]. In our study, Dmc1 was strongly repressed at D21/1 nM BPA and upregulated at D21/10 nM. The regulation of some key genes of meiosis was validated by qRT-PCR (Table 1).

Morphological analysis of pachytene spermatocytes indicated the prevalence of asynapsis – almost exclusively extended asynapsis. Homologous pairing failure has also been described in pachytene oocytes of C.elegans and mice exposed to BPA [25,30,65]. This abnormality is obviously related to the changes

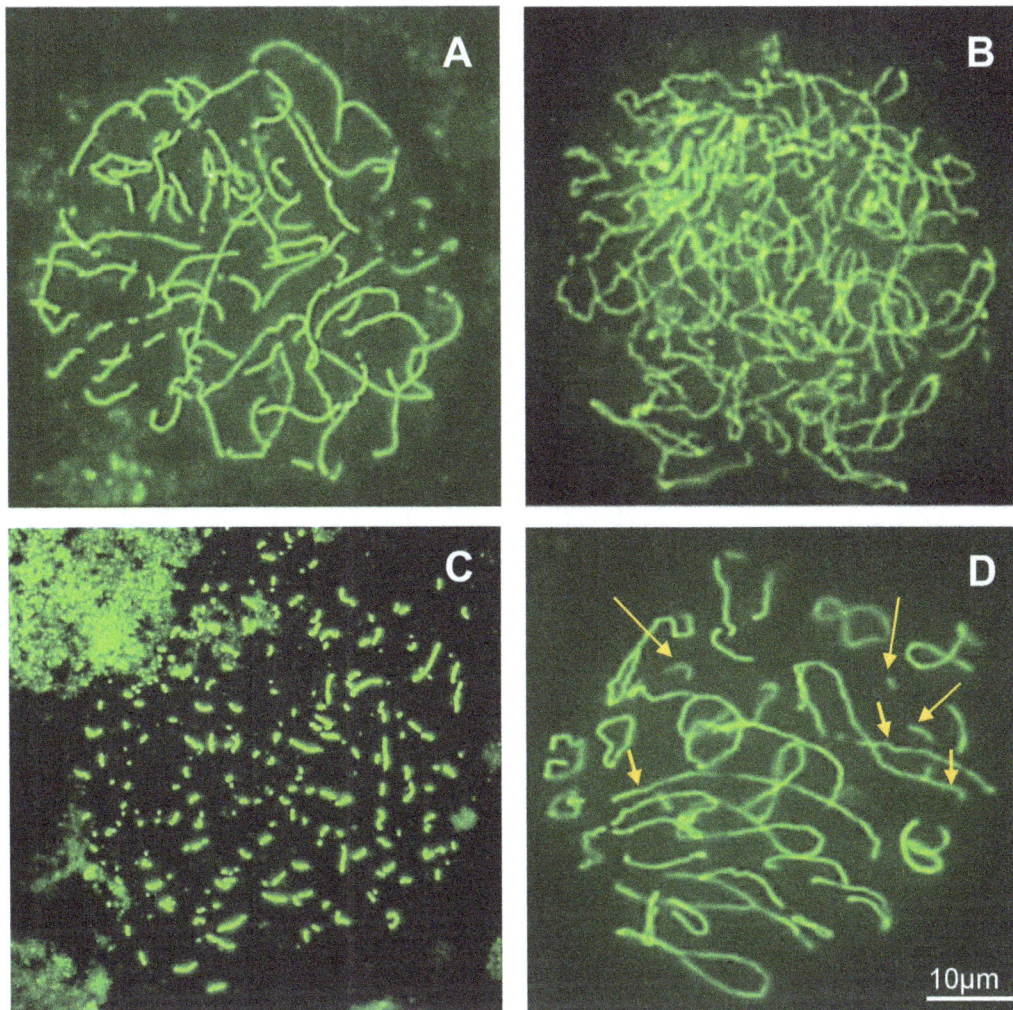

Figure 8. Pictures of abnormal spermatocyte nuclei cultured in the presence of BPA and stained with the anti-SCP3 antibody.
A – leptotene nucleus with abnormally long stretches of axial cores without indication of polarization; **B** – pachytene nucleus with total asynapsis; **C** – pachytene nucleus with pulverized synaptonemal complexes, proving apoptosis; **D** – diplotene nucleus with univalents (short arrows) and fragments (long arrows). Bar = 10 μm.

we observed in the transcription of many genes involved in homologous pairing and recombination. We previously reported [66] that high levels of extended asynapsis could arise from defects affecting these two crucial steps of meiosis. In the particular case of BPA, Allard and Colaiacovo [65] and Liu et al. [31] postulated that the DSB's repair machinery is impaired. In the present study, we showed that several genes involved in this process were deregulated by BPA (Table S2, Fig. 4). Several members of the Rec/Rad-51 family, which is known to play an important role in DNA repair by homologous recombination, were up- or down-regulated, with a high fold change for Rad51C. Genes involved in the DNA damage response, such as Brca1, Atm, Atr and Chk2, were deregulated. Nevertheless, other impaired processes may be involved in asynapsis formation. We showed that several genes involved in DSB generation (Spo11), in homology search and DNA strand exchange, and in the formation of Holliday junctions (Exo1, Hormad2, Msh4, Rad51C, Rad52, Tex15) were strongly deregulated. The gene, Fpr3, which functions in a checkpoint-like manner to ensure that chromosome synapsis is contingent on the initiation of recombination, was strongly underexpressed. Of the

proteins of genetic recombination, the SC proteins can be directly involved in the mechanisms of asynapsis. Indeed, the corresponding genes, Sycp1, Sycp2, Sycp2L and Sycp3, were clearly downregulated. It has also been reported that Stra8 plays a direct role in SC assembly [63]. The above discussion about asynapsed pachytene could explain the presence of univalents at the diplotene stage. Deregulation of genes which stabilize the Holliday junction (Msh4, Msh5) or maintain genomic integrity during DNA replication and recombination (Mlh1, Mlh3) could be involved in diplotene abnormalities. We emphasize that all these genes interact with each other. Their interconnections were visualized in a literature-based network showing the high level of connections between differentially expressed genes linked to DNA DSB repair in our datasets (Fig. 5). Thus, the phenotypes observed with ICC could result from the action of several of these interconnected genes and probably other genes expressed in germ cells and whose functions are not yet clearly defined.

Expression changes of genes revealed in our study could also explain some published data. We saw above that the increased number of Rad51 foci observed by Allard and Colaiacovo [65]

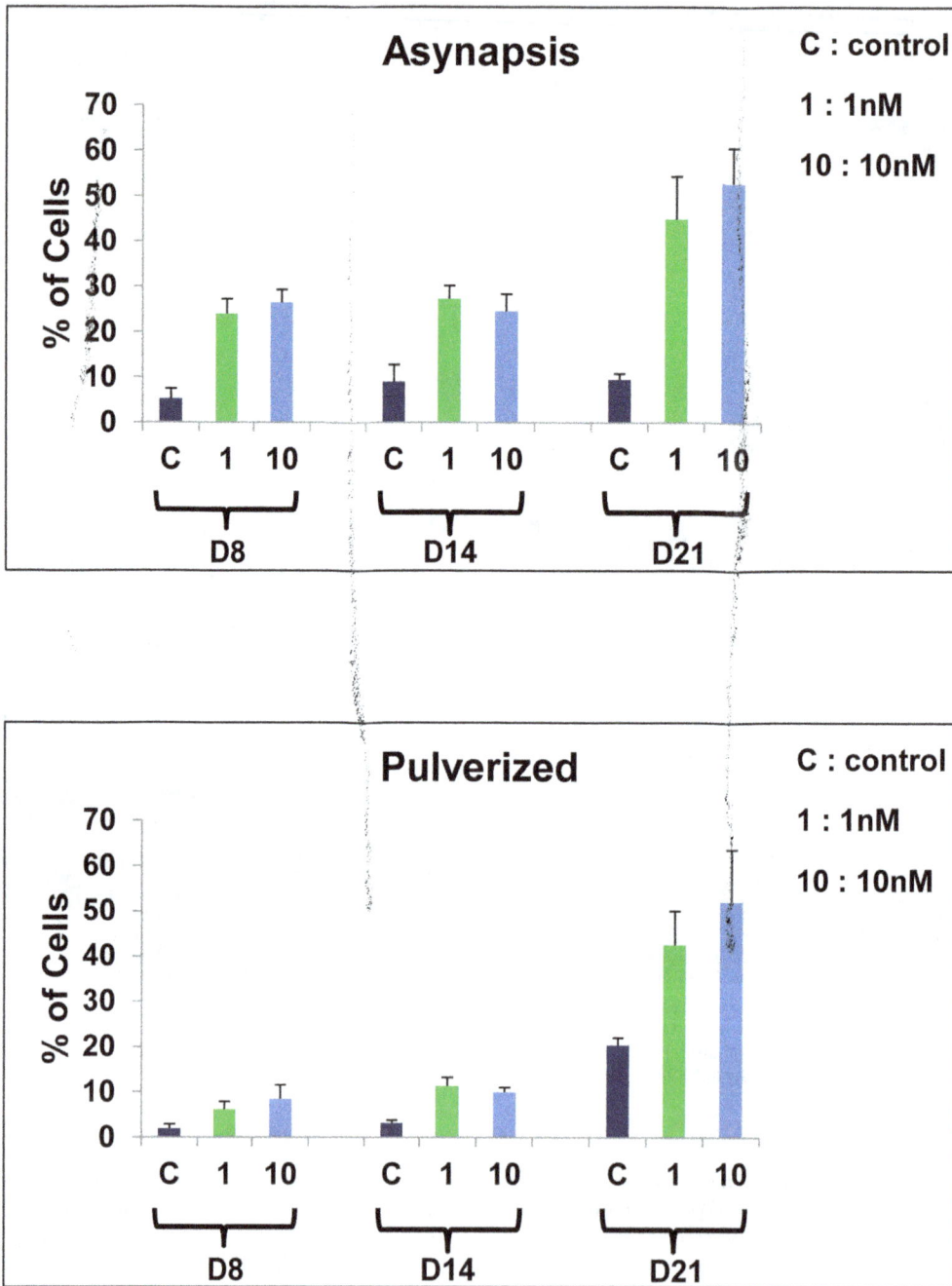

Figure 9. Changes of synaptonemal complex (SC) abnormality percentages during the pachytene stage in BPA-cultured spermatocytes. A – the percentage of nuclei with asynapsed SC increases at each dose and time point (p<0.05). **B** – the percentage of nuclei with pulverized SC (apoptotic nuclei) increases at each dose and time point (p<0.05). Each bar represents the mean ±SD (n = 3 cultures for D8 and D14, and 2 cultures for D21) versus control. **C** = control culture, 1 and 10 = 1 nM and 10 nM BPA, respectively.

might be consistent with transcriptional changes of the Rec/Rad51 family. The increased number and modified distribution of Mlh1 foci reported in pachytene oocytes following BPA exposure [25,30] might be consistent with the strong downregulation of Prdm9, a major player in hotspot specification [67].

Morphological analysis revealed the presence of apoptotic (pulverized) spermatocytes (Fig. 8C) whose percentage significantly increased at D21. Early features of apoptosis were also observed with the transcriptomic analysis. Radar plots (Fig. 2) show that cell

death is one of the most represented functions within the deregulated genes. Among the BPA-responsive apoptosis genes, we identified genes from the Card and Casp families. The antiapoptotic gene, Bcl2, was downregulated whereas the proapoptotic gene, Bax, was upregulated (Table S2). In addition Nfκb, which modulates the differentially expressed Card-family genes, was strongly induced.

Apoptosis might be induced in cells that fail to recombine and/or pair their homologous chromosomes [68]. Several checkpoints that monitor the progression of meiotic recombination are activated in response to the unrepaired meiotic DSBs. For example, Atm and Atr are checkpoint kinases that regulate meiotic DSB repair; both were downregulated in our study. Thus, the accumulation of apoptotic spermatocytes coupled with a decrease of the pachytene index is an indication of pachytene checkpoint activation by BPA. We have localized this checkpoint at the end of the P2 pachytene substage in humans [66] and rats [37]. The majority of asynapsed spermatocytes could, thus, be eliminated in this way. This could explain the decreasing percentage of diplotene nuclei in our BPA-exposed cultures.

These results demonstrate the interest of the analysis of SC in toxicological studies because they underline the specificities of each toxic substance. SC analysis is a highly sensitive indicator of potentially heritable effects of genotoxic agents [69]. All agents tested by Allen et al. [70,71] and Backer et al. [69] caused dose-dependent SC damage, which varied with the chemicals. We also previously showed with the same culture model that Cr (VI) treatment led to SC fragmentation whereas Cd treatment induced moth-eaten SC [36,37]. Presently, we show that BPA alters meiotic cell progression and increases asynapsis, without dose dependency.

Does BPA produce aneuploid gametes?

The pachytene checkpoint prevents chromosome missegregation by eliminating asynapsed pachytene spermatocytes that would lead to the production of aneuploid gametes [68]. Nevertheless, as shown in Fig. 4, BPA elicited expression changes of 11 genes implicated in pachytene checkpoint function, leading to a failure of checkpoint activation that could alleviate the meiotic arrest at this point.

Moreover, we previously showed that the pachytene checkpoint was not an absolute barrier [66], abnormal meiotic cells being able to complete spermatogenesis. In the present study, the presence of diplotene spermatocytes with univalents leads to the assumption that the existence of aneuploid metaphases II cannot be excluded. If so, low-level BPA exposure could induce errors in chromosome segregation and could produce aneuploid germ cells. Consistent with this hypothesis, studies of oocyte meiosis from female mice exposed to BPA indicate that BPA can affect chromosome segregation by disturbing synapsis and recombination [28,30]. A second mechanism involving the cell division machinery was suspected to explain a potential aneugenic effect of BPA [26]. It was demonstrated that BPA alters the centrosome dynamic and increases the number of mitotic and meiotic spindles with unaligned chromosomes. Nevertheless, according to Eichenlaub-Ritter et al. [27,72], low-level chronic BPA exposure does not appear to pose a risk for the induction of errors in chromosome segregation at first meiosis in mouse oocytes. These authors preferentially suggested that BPA induced meiotic arrest. However-er, we did not find in the literature any sperm chromosomal analyses of BPA-exposed male rodents to demonstrate the existence of aneuploidy.

According to Hunt et al. [73], studies in rodents allow predictions about humans to be made regarding reproductive effects of EDCs. Chalmel et al. [74] reported a cross-species expression profile between rodents and humans. Thus, our cytological and transcriptional results could be a predictor of the deleterious effects of low-dose BPA on human spermatogenesis.

Another point is that, although *ex vivo* models might be questionable for their lack of biotransformation and clearance compared to *in vivo* models, they do nevertheless represent a good alternative to animal testing regarding the necessary reproductive toxicity assays of thousands of chemicals.

Conclusion

This study provides arguments for the deleterious effects of BPA at low doses on male germ cells, by combining transcriptomic analyses and immunocytochemistry in an *ex vivo* rat seminiferous tubule culture model. Transcriptomic analyses showed that BPA altered the expression of genes involved in events preceding meiosis, its initiation and progression. Of the numerous genes differentially expressed by low-dose BPA exposure, we focused on 120 premeiotic and meiotic genes; some showed very elevated fold changes. Nevertheless we did not observe any dose dependency between 1 nM and 10 nM with both the techniques used. Only the gene expression analysis underlined a time dependency between D8, D14 and D21. Immunocytochemistry showed that low-dose BPA had deleterious effects on meiotic progression, and that the main alterations induced by BPA are asynapsis and apoptosis. These results bring additional arguments for the hypothesis that BPA alters pairing and recombination. Moreover, many differentially expressed genes were also involved in other important physiological functions, corroborating published findings, such as the triggering of xenobiotic metabolism, the disturbance of lipid metabolism, and endocrine disruption. Further analysis of our transcriptomic data could help to provide candidates for predictive biomarkers of meiotic abnormalities related to the toxic effects of chemicals on spermatogenesis.

Supporting Information

Table S1   List of genes up- or down-regulated at 1 nM and 10 nM BPA (D8, D14 and D21).

Table S2   List of the 120 genes differentially expressed under BPA exposure involved in the premeiotic and first meiotic prophase. Fold change values and cellular localization (N nuclear, C cytoplasm, PM plasma membrane, Un under-termined), are reported for the two BPA concentrations (1 and 10 nM) and at the three time-points (D8, D14 and D21). Red = up-regulated genes; green = down-regulated genes. The number of deregulated genes is time dependent but not dose-dependent.

Acknowledgments

We thank the Research Consortium ECCOREV n° 3098 (Ecosystemes Continentaux et Risques Environnementaux) CNRS / Aix-Marseille Université who funded this study (AOI 2010, grant number 7). We acknowledge Marc Fraterno for his help in creation of figure 8 and Thierry Orsière for fruitful discussions.

Author Contributions

Conceived and designed the experiments: OP MG MHP PD. Performed the experiments: SA GS GM. Analyzed the data: OP MG. Contributed reagents/materials/analysis tools: OP AL. Contributed to the writing of the manuscript: OP MG PD.

References

1. Ikezuki Y, Tsutsumi O, Takai Y, Kamei Y, Taketani Y (2002) Determination of bisphenol A concentrations in human biological fluids reveals significant early prenatal exposure. Hum Reprod 17: 2839–2841.

2. Takeuchi T, Tsutsumi O (2002) Serum bisphenol a concentrations showed gender differences, possibly linked to androgen levels. Biochem Biophys Res Commun 291: 76–78.

3. Takeuchi T, Tsutsumi O, Ikezuki Y, Takai Y, Taketani Y (2004) Positive relationship between androgen and the endocrine disruptor, bisphenol A, in normal women and women with ovarian dysfunction. Endocr J 51: 165–169.

4. Takeuchi T, Tsutsumi O, Nakamura N, Ikezuki Y, Takai Y, et al. (2004) Gender difference in serum bisphenol A levels may be caused by liver UDP-glucuronosyltransferase activity in rats. Biochem Biophys Res Commun 325: 549–554.

5. Vandenberg LN, Chahoud I, Heindel JJ, Padmanabhan V, Paumgartten FJ, et al. (2012) Urinary, circulating, and tissue biomonitoring studies indicate widespread exposure to bisphenol A. Cien Saude Colet 17: 407–434.

6. Zhang HQ, Zhang XF, Zhang LJ, Chao HH, Pan B, et al. (2012) Fetal exposure to bisphenol A affects the primordial follicle formation by inhibiting the meiotic progression of oocytes. MolBiolRep 39: 5651–5657.

7. Kuiper GG, Lemmen JG, Carlsson B, Corton JC, Safe SH, et al. (1998) Interaction of estrogenic chemicals and phytoestrogens with estrogen receptor beta. Endocrinology 139: 4252–4263.

8. Sohoni P, Sumpter JP (1998) Several environmental oestrogens are also anti-androgens. JEndocrinol 158: 327–339.

9. Rubin BS (2011) Bisphenol A: an endocrine disruptor with widespread exposure and multiple effects. J Steroid Biochem Mol Biol 127: 27–34.

10. Vandenberg LN, Hunt PA, Myers JP, Vom Saal FS (2013) Human exposures to bisphenol A: mismatches between data and assumptions. Rev Environ Health 28: 37–58.

11. Vandenberg LN, Maffini MV, Sonnenschein C, Rubin BS, Soto AM (2009) Bisphenol-A and the great divide: a review of controversies in the field of endocrine disruption. Endocr Rev 30: 75–95.

12. Ashby J, Tinwell H, Lefevre P, Joiner R, Haseman J (2003) The effect on sperm production in adult Sprague-Dawley rats exposed by gavage to bisphenol A between postnatal days 91-97. Toxicol Sci 2003 74(1): 129–138.

13. Howdeshell KL, Furr J, Lambright CR, Wilson VS, Ryan BC, et al. (2008) Gestational and lactational exposure to ethinyl estradiol, but not bisphenol A, decreases androgen-dependent reproductive organ weights and epididymal sperm abundance in the male long evans hooded rat. Toxicol Sci 102: 371–382.

14. LaRocca J, Boyajian A, Brown C, Smith SD, Hixon M (2011) Effects of in utero exposure to Bisphenol A or diethylstilbestrol on the adult male reproductive system. Birth Defects Res B Dev Reprod Toxicol 92: 526–533.

15. Tyl R, Myers C, Marr M, Thomas B, Keimowitz A, et al. (2002) Three-generation reproductive toxicity study of dietary bisphenol A in CD Sprague-Dawley rats. Toxicol Sci 68(1): 121–146.

16. Tohei A, Suda S, Taya K, Hashimoto T, Kogo H (2001) Bisphenol A inhibits testicular functions and increases luteinizing hormone secretion in adult male rats. Exp Biol Med (Maywood) 226: 216–221.

17. Aikawa H, Koyama S, Matsuda M, Nakahashi K, Akazome Y, et al. (2004) Relief effect of vitamin A on the decreased motility of sperm and the increased incidence of malformed sperm in mice exposed neonatally to bisphenol A. Cell Tissue Res 315: 119–124.

18. Al-Hiyasat AS, Darmani H, Elbetieha AM (2002) Effects of bisphenol A on adult male mouse fertility. EurJOral Sci 110: 163–167.

19. Meeker JD, Ehrlich S, Toth TL, Wright DL, Calafat AM, et al. (2010) Semen quality and sperm DNA damage in relation to urinary bisphenol A among men from an infertility clinic. Reprod Toxicol 30: 532–539.

20. Qiu LL, Wang X, Zhang XH, Zhang Z, Gu J, et al. (2013) Decreased androgen receptor expression may contribute to spermatogenesis failure in rats exposed to low concentration of bisphenol A. Toxicol Lett 219: 116–124.

21. Iida H, Maehara K, Doiguchi M, Mori T, Yamada F (2003) Bisphenol A-induced apoptosis of cultured rat Sertoli cells. Reprod Toxicol 17: 457–464.

22. Fiorini C, Tilloy-Ellul A, Chevalier S, Charuel C, Pointis G (2004) Sertoli cell junctional proteins as early targets for different classes of reproductive toxicants. ReprodToxicol 18: 413–421.

23. Li MW, Mruk DD, Lee WM, Cheng CY (2010) Connexin 43 is critical to maintain the homeostasis of the blood-testis barrier via its effects on tight junction reassembly. ProcNatlAcadSciUSA 107: 17998–18003.

24. Wu HJ, Liu C, Duan WX, Xu SC, He MD, et al. (2013) Melatonin ameliorates bisphenol A-induced DNA damage in the germ cells of adult male rats. Mutat Res 752: 57–67.

25. Brieno-Enriquez MA, Robles P, Camats-Tarruella N, Garcia-Cruz R, Roig I, et al. (2011) Human meiotic progression and recombination are affected by Bisphenol A exposure during in vitro human oocyte development. HumReprod 26: 2807–2818.

26. Can A, Semiz O, Cinar O (2005) Bisphenol-A induces cell cycle delay and alters centrosome and spindle microtubular organization in oocytes during meiosis. Molecular Human Reproduction 11: 389–396.

27. Eichenlaub-Ritter U, Vogt E, Cukurcam S, Sun F, Pacchierotti F, et al. (2008) Exposure of mouse oocytes to bisphenol A causes meiotic arrest but not aneuploidy. Mutat Res 651: 82–92.

28. Hunt PA, Koehler KE, Susiarjo M, Hodges CA, Ilagan A, et al. (2003) Bisphenol a exposure causes meiotic aneuploidy in the female mouse. CurrBiol 13: 546–553.

29. Lenie S, Cortvrindt R, Eichenlaub-Ritter U, Smitz J (2008) Continuous exposure to bisphenol A during in vitro follicular development induces meiotic abnormalities. Mutat Res 651: 71–81.

30. Susiarjo M, Hassold TJ, Freeman E, Hunt PA (2007) Bisphenol A exposure in utero disrupts early oogenesis in the mouse. PLoSGenet 3: e5.

31. Liu C, Duan W, Li R, Xu S, Zhang L, et al. (2013) Exposure to bisphenol A disrupts meiotic progression during spermatogenesis in adult rats through estrogen-like activity. Cell Death Dis 4: e676.

32. Liu C, Duan W, Zhang L, Xu S, Li R, et al. (2014) Bisphenol A exposure at an environmentally relevant dose induces meiotic abnormalities in adult male rats. Cell Tissue Res 355(1): 223–232.

33. Vandenberg LN, Chahoud I, Heindel JJ, Padmanabhan V, Paumgartten FJ, et al. (2010) Urinary, circulating, and tissue biomonitoring studies indicate widespread exposure to bisphenol A. Environ Health Perspect 118: 1055–1070.

34. vom Saal FS, Cooke PS, Buchanan DL, Palanza P, Thayer KA, et al. (1998) A physiologically based approach to the study of bisphenol A and other estrogenic chemicals on the size of reproductive organs, daily sperm production, and behavior. ToxicolIndHealth 14: 239–260.

35. Staub C, Hue D, Nicolle JC, Perrard-Sapori MH, Segretain D, et al. (2000) The whole meiotic process can occur in vitro in untransformed rat spermatogenic cells. ExpCell Res 260: 85–95.

36. Geoffroy-Siraudin C, Perrard MH, Chaspoul F, Lanteaume A, Gallice P, et al. (2010) Validation of a rat seminiferous tubule culture model as a suitable system for studying toxicant impact on meiosis effect of hexavalent chromium. ToxicolSci 116: 286–296.

37. Geoffroy-Siraudin C, Perrard MH, Ghalamoun-Slaimi R, Ali S, Chaspoul F, et al. (2012) Ex-vivo assessment of chronic toxicity of low levels of cadmium on testicular meiotic cells. ToxicolApplPharmacol 262: 238–246.

38. Fisichella M, Berenguer F, Steinmetz G, Auffan M, Rose J, et al. (2012) Intestinal toxicity evaluation of TiO2 degraded surface-treated nanoparticles: a combined physico-chemical and toxicogenomics approach in caco-2 cells. Part Fibre Toxicol 9: 1743–8977.

39. Guyton K, Kyle A, Aubrecht J, Cogliano V, Eastmond D, et al. (2009) Improving prediction of chemical carcinogenicity by considering multiple mechanisms and applying toxicogenomic approaches. Mutat Res 681(2–3): 230–240.

40. Hartung T, McBride M (2011) Food for Thought … on mapping the human toxome. ALTEX 28(2): 83–93.

41. Prat O, Berenguer F, Steinmetz G, Ruat S, Sage N, et al. (2010) Alterations in gene expression in cultured human cells after acute exposure to uranium salt: Involvement of a mineralization regulator. Toxicol In Vitro 24: 160–168.

42. Lopez-Casas PP, Mizrak SC, Lopez-Fernandez LA, Paz M, de Rooij DG, et al. (2012) The effects of different endocrine disruptors defining compound-specific alterations of gene expression profiles in the developing testis. Reprod Toxicol 33: 106–115.

43. Naciff JM, Hess KA, Overmann GJ, Torontali SM, Carr GJ, et al. (2005) Gene expression changes induced in the testis by transplacental exposure to high and low doses of 17{alpha}-ethynyl estradiol, genistein, or bisphenol A. Toxicol Sci 86: 396–416.

44. Tainaka H, Takahashi H, Umezawa M, Tanaka H, Nishimune Y, et al. (2012) Evaluation of the testicular toxicity of prenatal exposure to bisphenol A based on microarray analysis combined with MeSH annotation. J Toxicol Sci 37: 539–548.

45. Clermont Y (1972) Kinetics of spermatogenesis in mammals: seminiferous epithelium cycle and spermatogonial renewal. Physiol Rev 52: 198–236.

46. Metzler-Guillemain C, Guichaoua MR (2000) A simple and reliable method for meiotic studies on testicular samples used for intracytoplasmic sperm injection. FertilSteril 74: 916–919.

47. Wright WR, Parzych K, Crawford D, Mein C, Mitchell JA, et al. (2012) Inflammatory transcriptome profiling of human monocytes exposed acutely to cigarette smoke. PLoS One 7: 17.

48. Pfaffl MW, Horgan GW, Dempfle L (2002) Relative expression software tool (REST) for group-wise comparison and statistical analysis of relative expression results in real-time PCR. Nucleic Acids Res 30: e36.

49. Longepied G, Saut N, Aknin-Seifer I, Levy R, Frances AM, et al. (2010) Complete deletion of the AZFb interval from the Y chromosome in an oligozoospermic man. Hum Reprod 25: 2655–2663.

50. Gilleron J, Carette D, Durand P, Pointis G, Segretain D (2009) Connexin 43 a potential regulator of cell proliferation and apoptosis within the seminiferous epithelium. Int J Biochem Cell Biol 41(6): 1381–1390.

51. Godet M, Sabido O, Gilleron J, Durand P (2008) Meiotic progression of rat spermatocytes requires mitogen-activated protein kinases of Sertoli cells and close contacts between the germ cells and the Sertoli cells. Dev Biol 315 (1): 173–188.

52. Della Seta D, Minder I, Belloni V, Aloisi A, Dessi-Fulgheri F, et al. (2006) Pubertal exposure to estrogenic chemicals affects behavior in juvenile and adult male rats. Horm Behav 50(2): 301–307.

53. Shin BS, Kim CH, Jun YS, Kim DH, Lee BM, et al. (2004) Physiologically based pharmacokinetics of bisphenol A. J Toxicol Environ Health A 67: 1971–1985.

54. Lobenhofer EK, Cui X, Bennett L, Cable PL, Merrick BA, et al. (2004) Exploration of low-dose estrogen effects: identification of No Observed Transcriptional Effect Level (NOTEL). Toxicol Pathol 32: 482–492.

55. Ludwig S, Tinwell H, Schorsch F, Cavaille C, Pallardy M, et al. (2011) A molecular and phenotypic integrative approach to identify a no-effect dose level for antiandrogen-induced testicular toxicity. Toxicol Sci 122: 52–63.

56. Raj A, Peskin CS, Tranchina D, Vargas DY, Tyagi S (2006) Stochastic mRNA synthesis in mammalian cells. PLoS Biol 4: e309.

57. Miyawaki J, Sakayama K, Kato H, Yamamoto H, Masuno H (2007) Perinatal and postnatal exposure to bisphenol a increases adipose tissue mass and serum cholesterol level in mice. J Atheroscler Thromb 14: 245–252.

58. Rubin BS, Soto AM (2009) Bisphenol A: Perinatal exposure and body weight. Mol Cell Endocrinol 304: 55–62.

59. Welshons WV, Thayer KA, Judy BM, Taylor JA, Curran EM, et al. (2003) Large effects from small exposures. I. Mechanisms for endocrine-disrupting chemicals with estrogenic activity. Environ Health Perspect 111: 994–1006.

60. Vandenberg LN, Colborn T, Hayes TB, Heindel JJ, Jacobs DR, Jr., et al. (2012) Hormones and endocrine-disrupting chemicals: low-dose effects and nonmonotonic dose responses. Endocr Rev 33: 378–455.

61. Kortenkamp A, Faust O, Evans R, McKinlay R, Orton F, et al. (2011) State of the Art Assessment of Endocrine Disrupters. European Commission document Final report.

62. Zickler D, Kleckner N (1998) The leptotene-zygotene transition of meiosis. AnnuRevGenet 32: 619–697.

63. Mark M, Jacobs H, Oulad-Abdelghani M, Dennefeld C, Feret B, et al. (2008) STRA8-deficient spermatocytes initiate, but fail to complete, meiosis and undergo premature chromosome condensation. J Cell Sci 121: 3233–3242.

64. Pittman DL, Cobb J, Schimenti KJ, Wilson LA, Cooper DM, et al. (1998) Meiotic prophase arrest with failure of chromosome synapsis in mice deficient for Dmc1, a germline-specific RecA homolog. MolCell 1: 697–705.

65. Allard P, Colaiacovo MP (2010) Bisphenol A impairs the double-strand break repair machinery in the germline and causes chromosome abnormalities. ProcNatlAcadSciUSA 107: 20405–20410.

66. Guichaoua MR, Perrin J, Metzler-Guillemain C, Saias-Magnan J, Giorgi R, et al. (2005) Meiotic anomalies in infertile men with severe spermatogenic defects. HumReprod 20: 1897–1902.

67. Baudat F, Buard J, Grey C, de MB (2010) Prdm9, a key control of mammalian recombination hotspots. MedSci(Paris) 26: 468–470.

68. Roeder GS, Bailis JM (2000) The pachytene checkpoint. Trends Genet 16: 395–403.

69. Backer L, Gibson J, Moses M, Allen J (1988) Synaptonemal complex damage in relation to meiotic chromosome aberrations after exposure of male mice to cyclophosphamide. Mutat Res 203(4): 317–330.

70. Allen J, Gibson J, Poorman P, Backer L, Moses M (1988) Synaptonemal complex damage induced by clastogenic and anti-mitotic chemicals: implications for non-disjunction and aneuploidy. Mutat Res 201(2): 313–324.

71. Allen JW, Poorman P, Backer L, Gibson J, Westbrook-Collins B, et al. (1988) Synaptonemal complex damage as a measure of genotoxicity at meiosis. Cell Biol Toxicol 4(4): 487–494.

72. Eichenlaub-Ritter U, Vogt E, Cukurcam S, Sun F, Pacchierotti F, et al. (2008) Evaluation of aneugenic effects of bisphenol A in somatic and germ cells of the mouse Exposure of mouse oocytes to bisphenol A causes meiotic arrest but not aneuploidy. MutatRes 651: 64–70.

73. Hunt PA, Susiarjo M, Rubio C, Hassold TJ (2009) The bisphenol A experience: a primer for the analysis of environmental effects on mammalian reproduction. BiolReprod 81: 807–813.

74. Chalmel F, Rolland AD, Niederhauser-Wiederkehr C, Chung SS, Demougin P, et al. (2007) The conserved transcriptome in human and rodent male gametogenesis. Proc Natl Acad Sci USA 104: 8346–8351.

20

PCB-153 Shows Different Dynamics of Mobilisation from Differentiated Rat Adipocytes during Lipolysis in Comparison with PCB-28 and PCB-118

Caroline Louis[1], Gilles Tinant[1], Eric Mignolet[1], Jean-Pierre Thomé[2], Cathy Debier[1]*

1 Institut des Sciences de la Vie, Université catholique de Louvain, Louvain-la-Neuve, Belgium, **2** Laboratoire d'Ecologie animale et d'Ecotoxicologie, Université de Liège, Liège, Belgium

Abstract

Background: Polychlorinated biphenyls (PCBs) are persistent organic pollutants. Due to their lipophilic character, they are preferentially stored within the adipose tissue. During the mobilisation of lipids, PCBs might be released from adipocytes into the bloodstream. However, the mechanisms associated with the release of PCBs have been poorly studied. Several *in vivo* studies followed their dynamics of release but the complexity of the *in vivo* situation, which is characterised by a large range of pollutants, does not allow understanding precisely the behaviour of individual congeners. The present *in vitro* experiment studied the impact of (*i*) the number and position of chlorine atoms of PCBs on their release from adipocytes and (*ii*) the presence of other PCB congeners on the mobilisation rate of such molecules.

Methodology/Principal Findings: Differentiated rat adipocytes were used to compare the behaviour of PCB-28, -118 and -153. Cells were contaminated with the three congeners, alone or in cocktail, and a lipolysis was then induced with isoproterenol during 12 hours. Our data indicate that the three congeners were efficiently released from adipocytes and accumulated in the medium during the lipolysis. Interestingly, for a same level of cell lipids, PCB-153, a hexa-CB with two chlorine atoms in *ortho*-position, was mobilised slower than PCB-28, a tri-CB, and PCB-118, a penta-CB, which are both characterised by one chlorine atom in *ortho*-position. It suggests an impact of the chemical properties of pollutants on their mobilisation during periods of negative energy balance. Moreover, the mobilisation of PCB congeners, taken individually, did not seem to be influenced by the presence of other congeners within adipocytes.

Conclusion/Significance: These results not only highlight the obvious mobilisation of PCBs from adipocytes during lipolysis, in parallel to lipids, but also demonstrate that the structure of congeners defines their rate of release from adipocytes.

Editor: Arun Rishi, Wayne State University, United States of America

Funding: The authors have no support or funding to report.

Competing Interests: The authors have declared that no competing interests exist.

* Email: cathy.debier@uclouvain.be

Introduction

Polychlorinated biphenyls (PCBs) are a class of environmentally persistent pollutants that biomagnify throughout food chains. Ingestion of contaminated food, and especially fat-rich animal products, represents 90% of the mean uptake of humans [1]. Adipose tissue is then the main reservoir for the storage of these highly lipophilic molecules [2]. The cytoplasm of adipocytes is almost exclusively composed of lipid droplets (LDs) [3], which appear to be the principal targets for PCBs [4]. These cells therefore have an enormous capacity to accumulate lipophilic pollutants [5].

During periods of weight loss in humans, lipids from adipose tissue are mobilised, leading to an increase of PCB concentrations in this tissue [6,7]. Evidence from wildlife indicates same trends [8–14]. This phenomenon suggests that PCBs are less efficiently mobilised from adipocytes than fatty acids. However, a release of PCBs in the blood circulation does occur during such periods of weight loss and appears to become more important when the adipose stores are already significantly reduced [2,6,15–18]. In addition to being a reservoir, the adipose tissue can thus also be an internal source of lipophilic pollutants for the rest of the body [2]. Once in the bloodstream, pollutants are able to contaminate other tissues or be transferred in maternal milk. The exposure to PCBs is associated to adverse effects on human and animal health [19,20]. Among others, PCBs are involved in endocrine disruption, immuno- and neuro-toxicity as well as in the development of cardiovascular diseases and type-2 diabetes [21–25]. A correlation between the rise of persistent organic pollutants (POPs), such as PCBs, in the serum and the alterations of skeletal muscle oxidative capacity has been suggested in humans [16]. Furthermore, individuals who underwent bariatric surgery exhibited a positive association between POP serum levels and a diminished improvement of lipid values and liver markers [6].

Even if several *in vivo* studies report a release of PCBs from adipose tissue during lipolytic process [6,10,13–15], little is known concerning the chemical and biochemical factors that govern their mobilisation and transfer into the circulation. *In vivo* studies on long-term fasting wild animals report an impact of the fasting stage as well as the degree of lipophilicity of PCB congeners on their dynamics of mobilisation from the adipose tissue [8,10,13,14]. The transfer from adipose tissue into the blood circulation appears to be selective and strongly dependent on the log K_{ow} value of the compounds, with less lipophilic PCBs being more efficiently released. *In vivo* models being usually complex, the *in vitro* cultures of adipocytes would be useful to precisely understand the mobilisation of PCB congeners as a function of their chemical structure. A recent study from our group investigated the dynamics of accumulation of three PCB congeners, differing in the position and number of their chlorine atoms (PCB-28, log $K_{ow} = 5.71$; PCB-118, log $K_{ow} = 6.57$ and PCB-153, log $K_{ow} = 6.80$) in cultured adipocytes [26]. The accumulation profile revealed significant differences between PCB congeners. Their release during lipolysis was however not investigated.

In this study, we followed and compared the dynamics of mobilisation of PCB-28, PCB-118 and PCB-153 from *in vitro* differentiated rat adipocytes. Cells were contaminated by the three congeners, added individually or in cocktail, at the same concentrations in the culture medium. Lipolysis was then triggered over 12 hours with a lipolytic medium supplemented with isoproterenol, a well-known synthetic catecholamine [27,28]. The levels of PCBs in the extracellular medium and adipocytes were regularly assessed. The present experiment allowed (*i*) to estimate the impact of the number and position of chlorine atoms of PCBs on their release from adipocytes and (*ii*) to assess the impact of the presence of other PCB congeners on the mobilisation dynamics of such molecules.

Experimental Procedures

Primary cultures of rat adipocytes

Differentiated rat adipocytes were obtained and cultured as described previously [5,28]. Experimental procedures in animals were approved by the Animal Care and Use Committee of the Université catholique de Louvain (#103201) and were performed in accordance with the "Principles of Laboratory Animal Care" (NIH Publication 85–23). Two-month-old male Wistar rats (Centre d'Elevage Janvier, Le Genest Saint Isle, France) were sacrificed by decapitation. The fat tissue of the stromal-vascular fraction was sampled and then digested in a solution of collagenase (1250 U/ml type II; Sigma-Aldrich, Bornem, Belgium). The digested tissue underwent three filtrations and three centrifugations in order to obtain a final pellet of stromal-vascular cells, which was suspended in a medium composed of Dulbecco's Modified Eagle Medium (DMEM, 4.5 g/l glucose, Gibco-Invitrogen, Merelbeke, Belgium), 10% (*v:v*) heat-inactivated foetal bovine serum (FBS, PAA, A&E Scientific, Marcq, Belgium) and an antibiotic and antifungal mixture. Cells were seeded at a mean density of 18,000 cells per cm^2 on 6-well plates (Corning CellBIND Surface, Corning, Elscolab, Kruibeke, Belgium) (day 0) and incubated at 37°C in a humidified atmosphere containing 10% CO_2 in air for 24 hours to allow cell sedimentation and adhesion. Twenty-four hours after the isolation of progenitor cells (day 1), the medium was replaced by a differentiation medium composed of DMEM (4.5 g/l glucose), 10% (*v:v*) heat-inactivated FBS, 100 U/ml penicillin – 100 U/ml streptomycin – 250 ng/ml amphotericin B mixture (Lonza, Verviers, Belgium), 10 nM dexamethasone (Sigma-Aldrich), 10 μM ciglitizone (Sigma-Al-

drich) and 5 μg/ml insulin (Sigma-Aldrich). This medium was renewed every 48 hours until day 10 in order to obtain differentiated adipocytes.

Cell treatment

At day 10, adipocytes were incubated with a medium supplemented with PCB congeners (Dr. Ehrenstorfer GmbH, Ausburg, DE) during 12 hours (37°C – 10% CO_2 in air). The PCBs were added to the culture medium as ethanolic solution and four conditions of PCB contamination were tested: (*i*) 2,4,4'-trichlorobiphenyl (PCB-28); (*ii*) 2,3',4,4',5-pentachlorobiphenyl (PCB-118); (*iii*) 2,2',4,4',5,5'-hexachlorobiphenyl (PCB-153); (*iv*) an equimolar mixture of three PCB congeners (PCB-28, -118 and -153), also called cocktail of PCBs. In all conditions of contamination, each PCB was added to the culture medium at a concentration of 300 nM, which is within the range of concentrations found in *in vivo* and *in vitro* studies [4,29,30]. Impact of the ethanol vehicle was tested earlier [4].

Lipolysis experiment

At day 11, lipolytic process was induced to differentiated adipocytes as previously described in Louis et al. [28]. The differentiation medium in contact with adipocytes was removed and replaced by a lipolytic medium composed of DMEM (1.0 g/l glucose, Gibco–Invitrogen), 5% (*v:v*) heat-inactivated FBS, 2% (*w:v*) bovine albumin (Sigma-Aldrich) and 1 μM isoproterenol (Sigma-Aldrich). The lipolytic medium was renewed every 3 hours and the process was carried out for 12 hours. In the same way as in Louis et al. [28], cells from one plate (*i.e.* cells coming from the same PCB contamination) were collected every 3 hours and pooled in order to assess the cellular PCB and protein contents as well as the levels of fatty acids of cellular neutral lipids (NLs). Likewise, free fatty acids (FFAs), glycerol and PCBs released in the extracellular medium were quantified every 3 hours in all conditions.

Cellular protein assessment

Every 3 hours, cells were washed with phosphate-buffer saline (Sigma-Aldrich) at 37°C and then collected in an aqueous solution composed of 35 mM sodium dodecyl sulfate (Merck, Darmstadt, Germany), 60 mM Tris buffer (Merck) and 10 mM ethylenediaminetetraacetic acid (Sigma-Aldrich). After homogenisation, the cellular protein content was quantified by using the Bicinchoninic Acid Protein Assay kit (Sigma-Aldrich) with bovine serum albumin (Sigma-Aldrich) as calibration curve [28].

Cellular neutral lipid assessment

Cells were collected as described for the determination of protein content. The method used for the extraction and the isolation of the NL fraction (i.e. triglycerides (TGs), diglycerides, monoglycerides and cholesterol esters) from cell lysates is described in details in Louis et al. [28]. Briefly, the lipids were extracted with a mixture of chloroform/methanol/water (2:2:1, *v:v:v*) (Biosolve, Valkenswaard, The Netherlands) containing triheptadecanoin (Larodan, Malmö, Sweden) used as internal standard. After centrifugation, the supernatant was discarded; the chloroform phase was evaporated and samples were then suspended into 200 μl chloroform. In order to isolate NLs, samples were loaded on solid-phase extraction columns (Bond Elut NH$_2$, 200 mg, Varian, Middelburg, The Netherlands). The NL fraction collected with chloroform/2-propanol (2:1, *v:v*) (Biosolve) was evaporated and a methylation step was performed by adding 0.1 M KOH (Sigma-Aldrich) in methanol at 70°C for 1 hour and then, by adding 1.2 M HCl in methanol at 70°C for 15 min. The addition of hexane followed by deionized water allowed extracting the fatty acid methyl esters by a centrifugation step. Those were separated by gas chromatography [31]. Each peak was then identified and quantified by comparison with pure methyl ester standards (Larodan and Nu-Check Prep, Elysian, MN, USA). Data were processed with the ChromQuest 4.2 software (ThermoFinnigan, Milan, Italy). Thereafter, results were expressed by moles of fatty acids in cellular NLs. For the sake of simplicity, we refer to μmol NLs/mg protein in the text [28].

Extracellular free fatty acid assessment

FFA contents in the lipolytic medium were quantified with the Wako NEFA HR kit (Sopachem, Eke, Belgium) following the manufacturer's instructions [28].

Extracellular glycerol assessment

Glycerol released in the extracellular medium was measured with an *in vitro* enzymatic colorimetric test using glycerol-3-phosphate-oxidase (Diasys Free Glycerol FS kit, Sopachem) according to the manufacturer's instructions.

PCB assessment

At each studied time of the lipolysis, cells and extracellular medium were collected in EPA vials (Alltech, Lokeren, BE) with 5 ml of *n*-hexane (Biosolve) in order to perform a liquid-liquid extraction by a 10-min shaking. The hexane phase was transferred into a tube and PCB-112 (Dr. Ehrenstorfer GmbH) was added as internal standard. All samples were then purified by acid and Florisil clean-up steps as described in Debier et al. [8]. Purified samples were collected in *n*-hexane. Five μl of anhydrous nonane (Sigma-Aldrich) were added to samples and solvent was then evaporated under a gentle stream of nitrogen. The purified extracts were suspended into a hexane solution of Mirex (200 pg/μl) (Dr. Ehrenstorfer GmbH) used as internal standard for the correction of the extract volume injected in GC/MS. PCB congeners were separated and quantified with a gas chromatograph (GC Trace, ThermoFinnigan) equipped with an automatic split/splitless type injector (CTC Analytics, Zwingen, Switzerland), a fused silica capillary column (30 m × 0.25 mm internal diameter; 0.25 μm film) (Rxi-5ms, Restek, Bellefonte, PA, USA) and a mass spectrometer (Trace DSQ, ThermoFinnigan). The system used helium as carrier gas at a constant flow rate of 1.1 ml per minute. The temperature of injector was 230°C. The oven temperature program was as follows: 2 min at 60°C, gradual heating from 60 to 140°C at the rate of 20°C per minute, 1 minute at 140°C, gradual heating from 140 to 290°C at the rate of 2.5°C per minute, 10 minutes at 290°C and gradual cooling from 290°C to 60°C at the rate of 10°C per minute. Molecules were sent to mass spectrometer by the line transfer at 290°C. The ion source of the detector was kept at 230°C. PCBs were identified according to their retention time. Data were recorded using XCalibur 1.3 software (ThermoFinnigan). Quantification was performed by comparison to an external standard composed of 28 congeners (IUPAC numbers: PCB-8, -18, -28, -44, -52, -66, -77, -81, -101, -105, -114, -118, -123, -126, -128, -138, -153, -156, -157, -167, -169, -170, -180, -187, -189, -195, -206 and -209) in a certified calibration mixture (AccuStandard, New Haven, CT, USA). Five dilutions (concentration ranging from 25 to 500 pg/μl) were used in order to draw a linear calibration curve for each PCB.

Quality control

Blanks were run with sample series to control extraction and clean-up steps. The PCB recovery was calculated on the basis of the internal standard, PCB-112. Results were accepted only if the recoveries were between 70 and 130% according to EC [32]. All results were corrected to obtain 100% recovery [8]. The quality control was assessed through an interlaboratory comparison.

Cytotoxicity assessment

The potential cytotoxicities of the PCBs and the lipolytic treatment were assessed by measuring the release of lactate dehydrogenase (LDH) in the extracellular medium. The activity of LDH was determined using the cytotoxicity detection kit (Roche Diagnostics, Vilvoorde, Belgium) according to the manufacturer's instructions in (*i*) the differentiation medium after 12 hours of PCB exposure and (*ii*) the lipolytic media collected every 3 hours during the lipolytic process. Before the PCB exposure and the lipolysis, some cells were lysed with 1% Triton x-100 (Sigma-Aldrich) and were used as full toxicity control [28]. No treatments appeared toxic (< 5% of control) as compared to the full toxicity control (results not shown).

Statistical analysis

Data are presented as means ± SEM of three independent experiments for the conditions with PCB congeners alone and means ± SEM of five independent experiments for the condition with the cocktail of PCBs. The statistical analysis was performed by SAS 9.3 software (SAS Institute Inc., Cary, USA). Differences between treatments were assessed with mixed linear models and a Tukey's test [28]. Differences were deemed significant at *p*-values<0.05.

Results

1. Incorporation of PCBs in adipocytes

At day 10, differentiated rat adipocytes were exposed to four different treatments of PCBs during a 12-hour period. Three PCB congeners (PCB-28, -118 and -153) were added to the culture medium at a concentration of 300 nM, alone or in cocktail. Assessment of PCBs within adipocytes was carried out before the induction of lipolysis. The concentration of each congener in adipocytes, expressed as nmol per unit of cellular protein, was statistically similar, whatever the kind of contamination (alone or in cocktail) ($0.328 < p < 1.000$) (Table 1). Same conclusions were drawn when the results were expressed per unit of NLs ($0.207 < p < 1.000$). Since the dynamics of accumulation of PCBs in cells vary with cellular lipid content [26], the levels of cellular NLs were quantified and no statistical difference was noted ($0.457 < p < 0.978$) (Table 1).

2. The time-course of lipolysis

At day 11, adipocytes contaminated with PCBs (alone or in cocktail) were incubated with a lipolytic medium supplemented with 1 μM isoproterenol, which was replaced every 3 hours for 12 hours. Since the aim of this work was to study the kinetics of PCB release during the mobilisation of lipids in adipocytes, we firstly ensured of the efficiency of lipolysis. Cellular NLs as well as FFAs and glycerol released in the culture medium were quantified over 12 hours (Figure 1). The FFA and glycerol initially present in the lipolytic medium, before being in contact with adipocytes, were subtracted from each result, leading to a value of 0 at 0 hour (Figure 1B–C). The total FFAs and glycerol released by the adipocytes throughout a given period were calculated by adding the quantities measured at each period of 3 hours. For example,

Table 1. Efficient accumulation of PCBs in adipocytes during a 12-hour period.

	Contamination by a PCB congener alone:			Contamination by a cocktail of three PCB congeners:		
	PCB-28	PCB-118	PCB-153	PCB-28	PCB-118	PCB-153
PCBs (nmol/mg protein)	1.7±0.1	2.0±0.1	1.8±0.1	1.8±0.1	2.1±0.1	1.8±0.1
PCBs (nmol/mg NLs)	0.47±0.08	0.62±0.09	0.48±0.08	0.61±0.08	0.70±0.08	0.60±0.08
NLs (μmol/mg protein)	13.6±1.1	13.1±1.3	15.6±1.1	11.4±1.1	11.4±1.1	11.4±1.1

Quantities of PCBs accumulated in differentiated rat adipocytes, expressed in nmol per unit of cellular protein and per unit of neutral lipids (NLs), after a dose of PCBs was added during 12 hours in the culture medium. Quantities of NLs, expressed in μmol per unit of cellular protein are also presented. Data represent the means of (*i*) three independent experiments ± SEM for conditions with one PCB congener alone, (*ii*) five independent experiments ± SEM for conditions with the cocktail of PCBs. There was no significant difference of PCB and NL concentrations between the three PCB congeners, whatever the kind of contamination, alone or in cocktail ($p>0.05$).

the amount of total FFAs or glycerol released after 6 hours of lipolysis corresponds to the sum of FFAs or glycerol released between 0 and 3 hours and between 3 and 6 hours [28].

A significant decrease of cellular NLs was observed between early and late lipolysis for the conditions with the PCB-28 ($p = 0.005$) and cocktail of PCBs ($p = 0.029$). The cell NLs slightly decreased over 12 hours within adipocytes contaminated with PCB-118 ($p = 0.298$) and PCB-153 ($p = 0.097$). For all conditions, the greatest reduction occurred during the first 3 hours of the experiment (Figure 1A). Indeed, a mean loss of $57±3\%$ initial NLs was noted between 0 and 12 hours, whereas after 3 hours of lipolysis, adipocytes already lost $33±6\%$ of NLs on average, compared to the initial levels ($0.372<p<0.721$ for all conditions of contamination between 0 and 3 hours). The comparison of NL levels between the four PCB treatments at a given period did not highlight any difference ($0.255<p<1.000$).

The decrease of cellular lipid contents was accompanied by a significant increase of total FFAs in the lipolytic medium over the 12-hour period ($p<0.001$) (Figure 1B). The increase of total FFAs was more pronounced during the first 3 hours of lipolysis, during which an average release of $63±4\%$ of total FFAs occurred ($p<0.001$). FFA concentrations then continued to increase slightly, but not significantly, between the subsequent consecutive periods ($0.491<p<0.985$ for all conditions of contamination between 3 and 6 hours, 6 and 9 hours, 9 and 12 hours). No difference in the release of FFAs was noted between the four conditions of contamination at a given time of the lipolytic process ($0.273<p<1.000$). As for total FFAs, contaminated adipocytes released a significant amount of total glycerol throughout the lipolytic process ($p<0.001$ between 0 and 12 hours) (Figure 1C). The major part of the release ($49±3\%$) also occurred during the first 3 hours of lipolysis for all conditions ($p<0.001$) (Figure 1C). Here again, no difference in the release of glycerol was noted between the four conditions of PCB contamination over the 12-hour period ($0.119<p<0.999$). Taking all conditions of PCB contamination together, the FFA/glycerol ratios lied between 1.4 and 1.8.

3. Comparative dynamics of mobilisation of PCB-28, -118 and -153 present in cocktail in adipocytes

In the present section, we compared the dynamics of mobilisation of PCB-28, -118 and -153 (present in cocktail within the cells) from the same adipocytes undergoing a lipolytic process over a 12-hour period. In order to strictly compare the different mobilisations between congeners, we expressed the results as percentages of the amounts initially present in adipocytes (i.e. at 0 hour). The potential presence of PCBs in the lipolytic medium, before being in contact with the cells, has been tested and was negligible (result not shown). The proportions of PCBs in the

lipolytic medium at 0 hour were thus set at 0%. Regarding the subsequent studied times of the lipolysis (3, 6, 9 and 12 hours), the PCB levels in the medium were calculated by adding the quantities evaluated at each period of 3 hours. For example, the amounts of each PCB congener released after 6 hours of lipolytic treatment correspond to the sum of PCB released between 0 and 3 hours and between 3 and 6 hours.

In parallel to the lipid mobilisation, there was a release of PCBs from adipocytes into the culture medium (Figure 2). An important drop of the PCB cellular content occurred during the first half of the lipolytic process ($p<0.001$ between 0 and 6 hours for the three PCBs). It was followed by a slight, but still significant reduction of PCB-28, -118 and -153 in adipocytes during the second part of the lipolysis ($p<0.001$ between 6 and 12 hours for the three PCBs). Accordingly, a reversed trend was noted in the lipolytic medium: the proportions of PCB-28, -118 and -153 accrued during the first 6 hours ($p<0.001$ between 0 and 6 hours for the three PCBs). The PCB accumulation in the culture medium was more moderate, but still significant ($0.001<p<0.006$), during the second part of the lipolytic process.

Even if all congeners dropped in the cells and increased in the culture medium, the dynamics of release of PCB-153 somewhat differed from those of PCB-28 and -118. Indeed, despite the fact that the amounts of the three congeners within adipocytes were statistically similar before the beginning of the lipolysis (Table 1), the percentages of PCB-153 in adipocytes remained higher than the ones of PCB-28 and PCB-118 at 3, 6 and 9 hours of lipolysis ($0.001<p<0.032$ for PCB-28; $0.031<p<0.037$ for PCB-118). The difference disappeared at 12 hours between PCB-153 and PCB-28 ($p = 0.139$), but remained significant between PCB-153 and PCB-118 ($p = 0.027$). On the other hand, the cellular percentage of PCB-28 and PCB-118 did not differ between each other throughout the lipolytic process ($0.373<p<1.000$). The slower mobilisation of PCB-153 from adipocytes was reflected by a slower increase in the culture medium as compared to PCB-28 and PCB-118 after 3 hours of lipolysis ($p<0.001$ for the both comparisons). After 6 hours, the accumulation in the medium was weaker for PCB-153 than PCB-28 ($p = 0.027$) and similar between PCB-153 and PCB-118 ($p = 0.178$). The proportions of PCB-153 were then similar to those of PCB-28 and -118 for the subsequent hours ($0.275<p<0.986$ at 9 and 12 hours). On the other hand, the percentages of PCB-28 and -118 in the extracellular medium were similar to each other over the lipolysis ($0.429<p<1.000$).

4. Impact of the presence of other PCB congeners on the dynamics of mobilisation of PCB-28, -118 and -153

We investigated if the dynamics of mobilisation of one PCB congener was influenced by the presence of the other congeners.

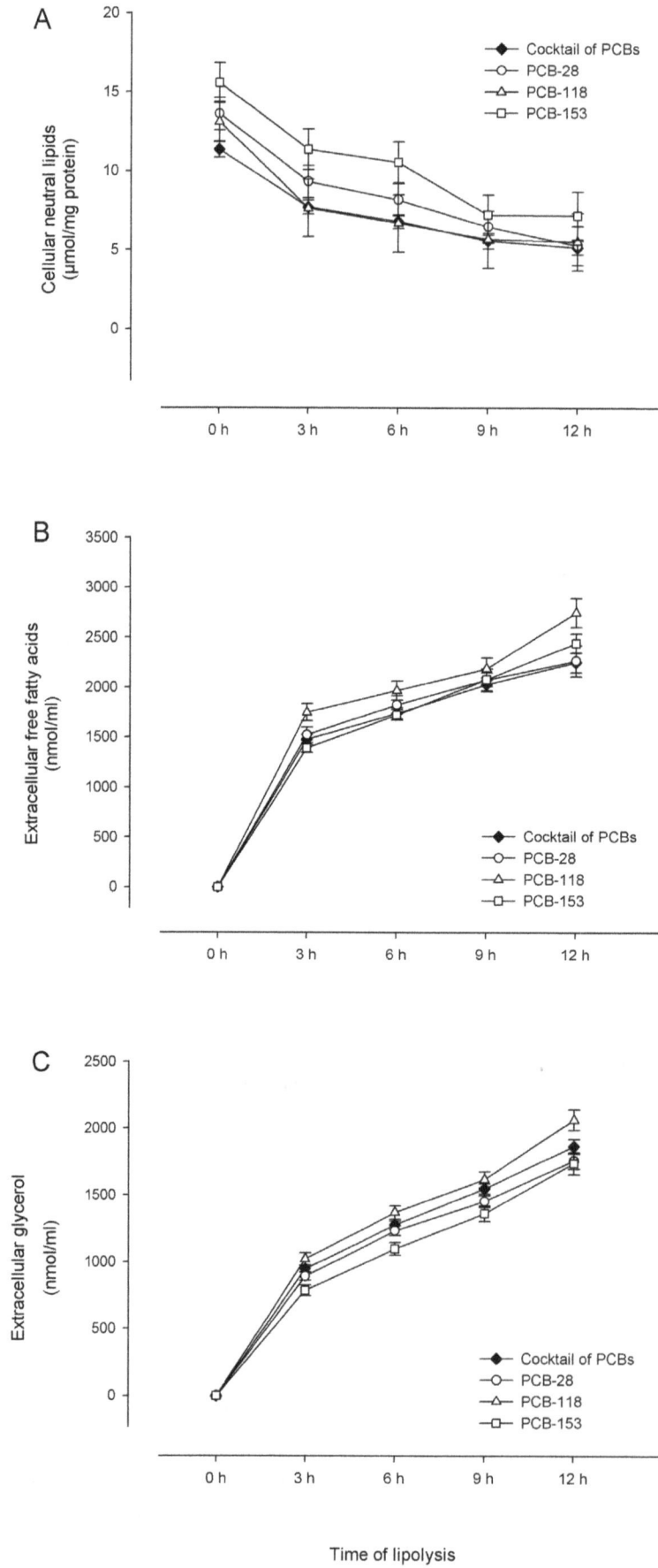

Time of lipolysis

Figure 1. Lipolytic treatment decreased cellular neutral lipids and increased extracellular fatty acids and glycerol. At day 11, differentiated rat adipocytes, which were previously contaminated with PCBs, were incubated with a lipolytic medium supplemented with 1 µM isoproterenol. We renewed the lipolytic medium every 3 hours for 12 hours. Cellular neutral lipids (corresponding to µmol of fatty acids in cellular neutral lipids) were expressed per mg of total cell protein (A). A significant decrease of cellular lipid contents was noted throughout the lipolytic process for the conditions with PCB-28 ($p = 0.005$) and the cocktail of PCBs ($p = 0.029$). The decrease was slighter for the conditions with PCB-118 ($p = 0.298$) and PCB-153 ($p = 0.097$). Total extracellular free fatty acids (B) and total extracellular glycerol (C) were expressed per ml of medium. Quantities of total free fatty acids and glycerol in the medium were obtained by adding the quantities released during the periods of 3 hours (e.g. total free fatty acids at 6 hours correspond to the sum of total free fatty acids released between 0 and 3 hours and between 3 and 6 hours). A significant increase of total free FAs and total glycerol was observed over the 12-hour lipolytic treatment ($p<0.001$ for all conditions). Data represent the means of (i) three independent experiments ± SEM for conditions with one PCB congener alone, (ii) five independent experiments ± SEM for conditions with the cocktail of PCBs.

To achieve this goal, we contaminated adipocytes either with one of the three PCB congeners (PCB-28, -118 or -153) or with a cocktail of the three congeners. Here also, the results are expressed as percentages of the amounts initially present in adipocytes. As previously described, the proportions of PCBs just before the lipolytic process (i.e. 0 hour) was set to 0% since no PCB congener was quantified in the initial lipolytic medium. Here also, the total PCB congeners released by the adipocytes throughout a given period were calculated by adding the quantities measured at each period of 3 hours.

At each studied time throughout the 12-hour period, similar proportions of PCB-28 (Figure 3A), PCB-118 (Figure 3B) and PCB-153 (Figure 3C) within adipocytes were quantified between both conditions of contamination (i.e. congeners alone and in cocktail) ($0.134<p<0.934$). Accordingly, no difference was noted between the percentages of PCBs released in the lipolytic medium in contact with adipocytes in both conditions of contamination ($0.122<p<0.916$), except after 3 hours of lipolysis, where the

proportion of PCB-153 measured in the medium were higher when it was added in cocktail than when it was added alone ($p = 0.046$).

Discussion

Considerable accumulation of PCBs within adipocytes

Differentiated rat adipocytes were exposed to three targeted PCB congeners (PCB-28, -118 and -153), which differ by the number and the position of the chlorine atoms [19,33]. After 12 hours, similar concentrations of PCBs were stored within adipocytes. Same observations were already drawn after 4 hours of exposure for the same congeners [26]. The different molecular structures of PCBs did thus not seem to influence their accretion within adipocytes on long term. Previous studies from our group highlighted the importance of the amount of cellular NLs, acting as a trap for the accumulation of PCBs within adipocytes [5,26]. The fact that cellular NL levels were similar between our

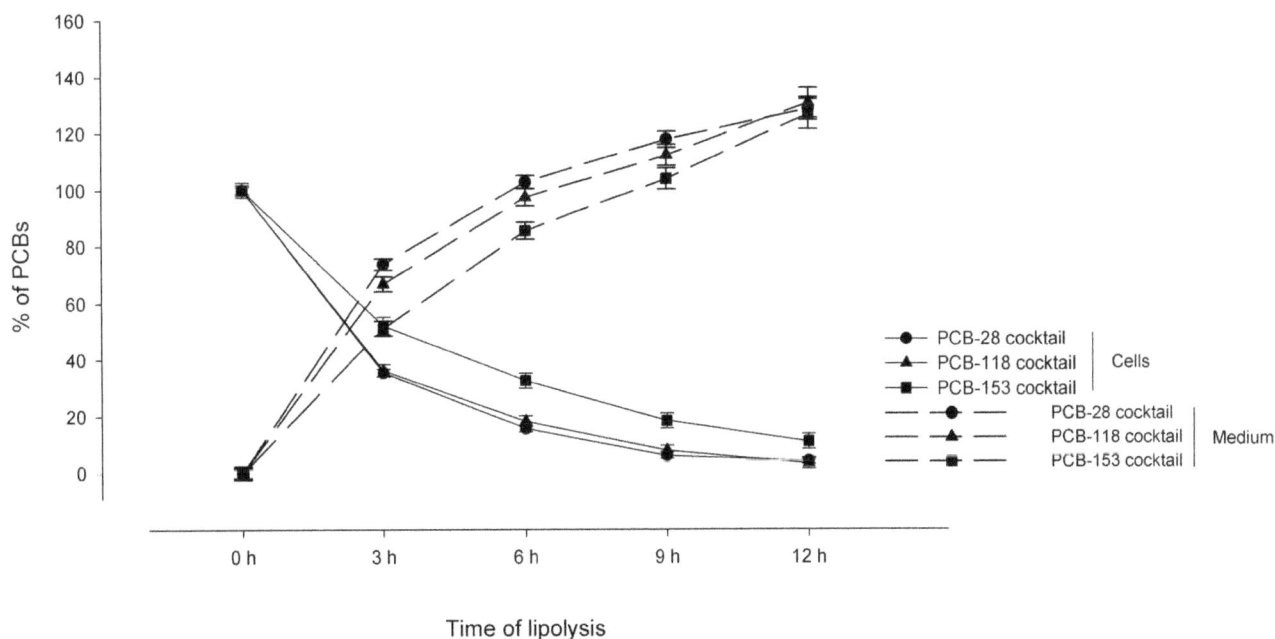

Figure 2. Lipolytic treatment decontaminated the adipocytes, inducing an accumulation of PCB congeners in the extracellular medium. At day 11, contaminated adipocytes underwent a lipolytic process. The PCB contents as well as the proportions of PCBs released in the extracellular medium were assessed every 3 hours. For each PCB congener, all results were expressed in percentage of the amounts initially present in the cells. Proportions of one PCB congener in the medium were obtained by adding the amounts released during periods of 3 hours (e.g. proportion of one congener at 6 hours corresponds to the sum of this congener released between 0 and 3 hours and between 3 and 6 hours). An important drop of cell PCB proportions was observed during the first 6 hours of lipolysis ($p<0.001$) and was followed by a more moderate decrease of the cell PCB percentages during the last 6 hours ($p<0.001$). In parallel, an important increase of PCBs was noted in the extracellular medium during the first half of lipolysis ($p<0.001$) and was followed by a slower accumulation during the second half of lipolysis ($0.001<p<0.006$). Data represent the means of five independent experiments ± SEM.

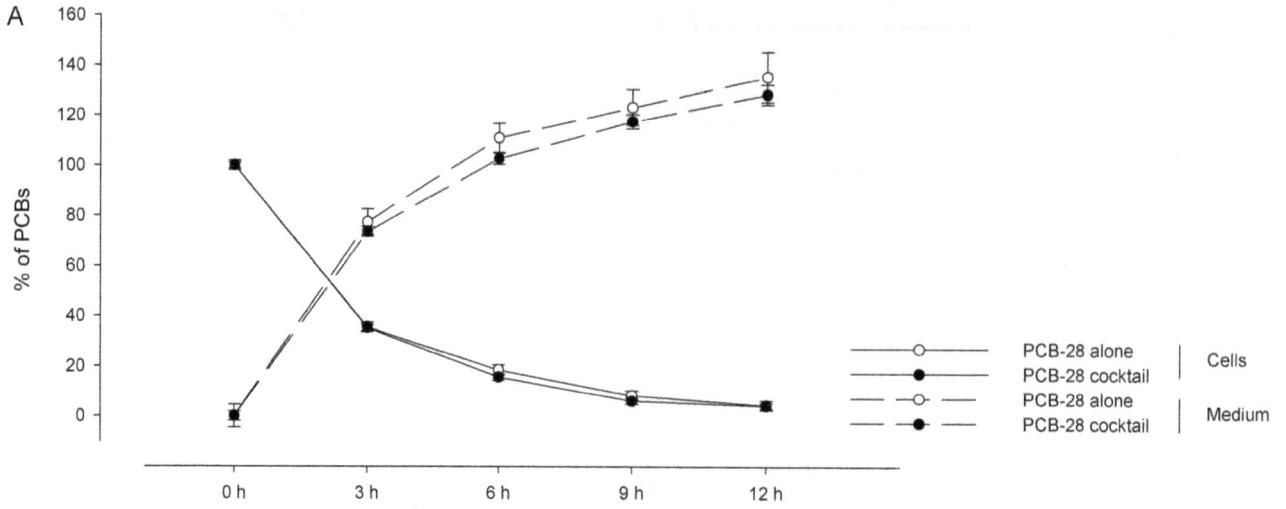

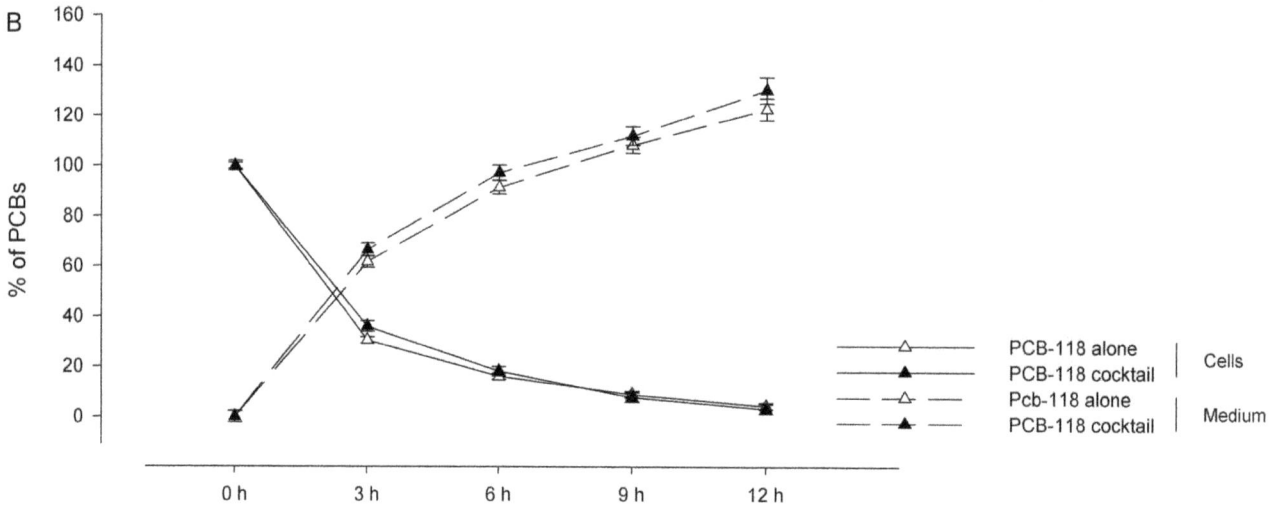

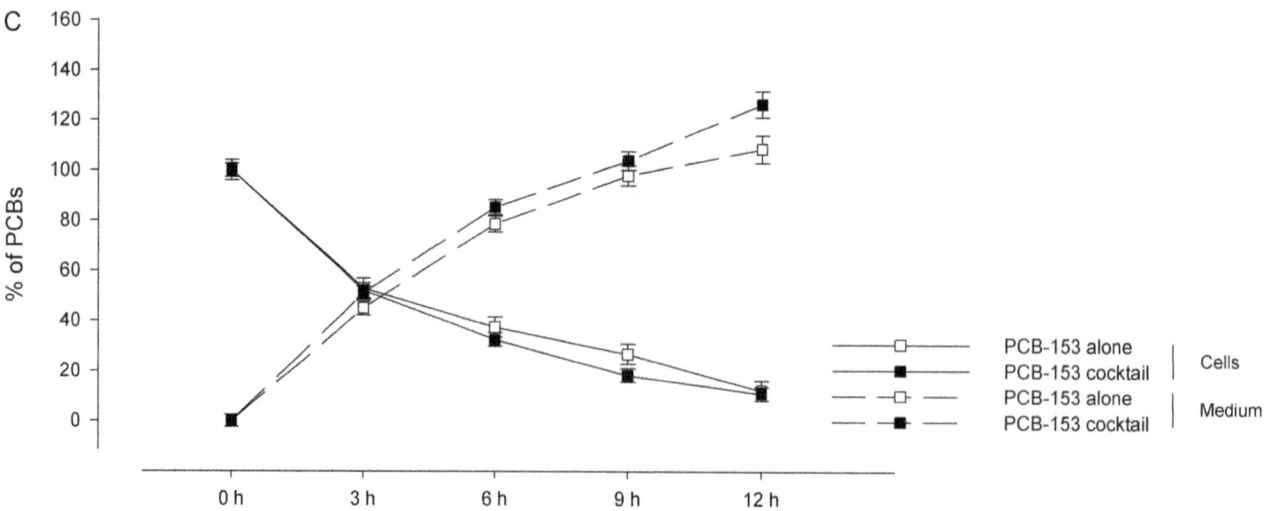

Time of lipolysis

Figure 3. Presence of other congeners did not influence the dynamic of PCB mobilisation. At day 11, differentiated rat adipocytes, which were previously contaminated with either individual PCB congeners or with a cocktail of PCBs, underwent a lipolytic process. The cellular levels of PCBs before the lipolytic process were quantified and set at 100%. During 12-hour period of lipolysis, the contents of PCB congeners within adipocytes and in the extracellular medium were assessed every 3 hours. The results for PCB-28 (A), PCB-118 (B) and PCB-153 (C) were expressed by the percentage of initial amounts of each congener. Within a condition of contamination, proportions of one PCB congener in the medium were obtained by adding the quantities released during the periods of 3 hours (e.g. proportion of one congener at 6 hours corresponds to the sum of this congener released between 0 and 3 hours and between 3 and 6 hours). At each given time of lipolytic treatment, no differences were noted between the proportions of each PCB (i.e. PCB-28, -118 and -153) in both conditions of contamination (i.e. congeners alone or in cocktail), either in the cells or in the lipolytic medium ($0.122 < p < 0.916$). Only the percentage of PCB-153 in the lipolytic medium was lower when taken alone as compared to the condition in cocktail after 3 hours of lipolysis ($p = 0.046$). Data represent the means of (*i*) three independent experiments ± SEM for conditions with one PCB congener alone, (*ii*) five independent experiments ± SEM for conditions with the cocktail of PCBs.

experimental conditions is most probably at the origin of the identical accumulation of PCBs within adipocytes. While PCB concentrations in the culture medium were in the same range than those measured in the human serum [29,30], the PCB levels found in cultured adipocytes after 12 hours of incubation (data from Table 1 are equivalent ~175 ng PCB congeners per mg NLs) were much higher than those measured *in vivo*, in the human adipose tissue (from 0.02 to 0.66 ng total PCBs per mg lipids) [34–38]. It reflects a high propensity of differentiated rat adipocytes to store PCBs [4]. Such differences have already been noticed previously and the reasons are discussed in details elsewhere [5]. Briefly, the higher *in vitro* concentrations of PCBs within adipocytes could result from the extended contact between the cells and the contaminated culture medium (12 hours). On the contrary, in the *in vivo* situation, the PCB congeners, transported in the circulation by lipoproteins and plasma albumin [33,39,40], are in continual movement thanks to the blood flow. In addition, the culture medium contains only a low concentration of serum (10%) [4], and therefore very low levels of lipoproteins and albumin, which could contribute to a smaller retention of PCBs in this hydrophilic compartment and a higher storage in the lipophilic compartment represented by adipocytes. The differentiated rat adipocytes are also organised as a monolayer whereas *in vivo* adipose tissue shows a complex 3D-structure. Furthermore, lipolysis, which occurs regularly *in vivo*, may lead to the mobilisation of PCBs from adipocytes. Finally, the circulating PCBs may be taken up by other tissues such as the liver.

Similar dynamics of mobilisation between cellular lipids and PCBs

Once the lipolytic pathway was induced, adipocytes started to mobilise their lipid content. A decrease of the cellular NLs could be observed throughout the 12-hour experiment, with a more pronounced lipolytic action during the first 3 hours. This sharper decrease of NL content at early lipolysis was in accordance with our previous study [28]. As a result of the mobilisation of cellular NLs, FFAs and glycerol were released in the extracellular medium. The lipolytic treatment also led to the release of PCBs from adipocytes to the extracellular medium. The dynamics of mobilisation of PCBs exhibited some parallelism with those of cellular lipids, as the major part of PCB discharge occurred during the first hours of lipolysis as well. Previously, it was shown that *in vitro* epididymal adipocytes isolated from rats also unloaded PCB-153 during a lipolytic treatment of 50 min with 0.8 µM isoproterenol [41]. The release of PCBs might accompany the mobilisation of cell lipids, which agrees with previous studies on the behaviour of dioxins [42,43]. In addition, cellular TG content is an important parameter governing the accumulation of PCBs in adipocytes [5,26]. PCBs are stored almost exclusively within the LDs [4]. As this lipophilic pool is reduced during lipolysis [28], the capacity of storage is thus also lessened, promoting the release of PCBs in the extracellular medium, where they could be tightly

associated with diverse lipoproteins (present in the 5% serum) and bovine albumin (2%).

Although an obvious release of PCBs occurred from adipocytes during the lipolytic experiment, some molecules of PCBs could be taken up again by the cells as previously suggested in *in vivo* studies [13,14]. This phenomenon is well known for FFAs, which are partly reabsorbed by adipocytes and re-esterified into newly synthesized TGs [28,44]. A complete hydrolysis of one mole of TGs leads to the release of three moles of free FAs and one mole of glycerol. This could be translated by a free FA/glycerol ratio equivalent to 3.0. However, free FA/glycerol ratios were lower than 3.0 in our experiments, which likely reflects a reuptake of free FAs by the cells. Nevertheless, this phenomenon might have been somewhat limited in our experimental conditions, because of the renewal of the lipolytic medium every 3 hours [28].

Studies investigating the release of POPs from adipose tissue during periods of weight loss in animals and humans usually report an increase of the concentrations of PCBs and related compounds in adipose tissue, despite their significant discharge in the blood circulation [6,14,15,18]. This increase suggests a less efficient mobilisation of PCBs from this tissue than lipids and a concentration of these lipophilic pollutants in the remaining amount of fat cells. The adipose tissue is a macroscopic structure that is irrigated by blood vessels. During adipose tissue lipolysis, it is possible that PCBs are transferred to adipocytes that still contain significant amounts of lipids in their LDs instead of being all released into the circulation. It is also possible that PCBs are released in the bloodstream together with the lipids and then reabsorbed by the adipose tissue as a result of their higher affinity for the remaining lipids present in the cells [13]. Our *in vitro* model differs from the *in vivo* situation among others by the fact that it is characterised by only one layer of cells, which is in direct contact with the extracellular medium that is regularly renewed. A reuptake of PCBs by the cells and/or a migration of PCBs to deeper adipocyte layers that are still filled with fat are thus not possible. Moreover, the high PCB concentrations, which were found in the cultivated adipocytes before the lipolytic induction, might also promote the massive release of congeners in the extracellular medium.

Differences of release according to the kind of PCB congener

When the three congeners were added in cocktail to the culture medium, we could observe that PCB-153 was less efficiently mobilised from adipocytes than PCB-28 and PCB-118 during the first part of the lipolytic process. This difference however disappeared at 12 hours of lipolysis. The slower mobilisation of PCB-153 from adipocytes reflects the fact that, besides the cellular lipid content, the rate of release is also governed by the physico-chemical properties of the congeners, which are defined by the number and the position of chlorine atoms on the biphenyl core [2]. If we consider the electrostatic potentials of PCBs [26], PCB-

153 exhibits a large electron-deficient zone. This characteristic makes this congener rather lipophilic, which is reflected by the higher partition coefficient n-octanol/water (log K_{ow} = 6.80). On the other hand, PCB-28 and PCB-118 have a reduced electron-deficient zone, translated by lower log K_{ow} (PCB-28: log K_{ow} = 5.71 and PCB-118: log K_{ow} = 6.57). PCB-153 could thus be more trapped within LDs than PCB-28 and PCB-118 and as a consequence, be released more slowly.

In addition, it was previously observed that a small proportion of PCB-153 was sequestrated in the cell membranes when isolated primary adipocytes absorbed the PCB congeners present in the culture medium for 2 hours [4]. It was not the case for PCB-28 and -118. Likewise, several studies showed that PCB-52 and -153, two di-*ortho*-substituted PCBs, intercalate between membrane phospholipids similarly to cholesterol and have an impact on the membrane fluidity in fish, rodent and chicken cells [45–49]. In the present study, the release of PCB-153 from adipocytes could thus be slowed down by its association with cell membranes, as compared to PCB-28 and PCB-118, two mono-*ortho*-substituted congeners. The fact that PCB-153 has two chlorine atoms in the *ortho* position on the biphenyl core induces a more perpendicular layout of the phenyl rings. It means that PCB-153 occupies a larger bulk than PCB-28 and PCB-118, which could be involved in the sequestration of PCB-153 within membranes and its slower mobilisation from adipocytes.

Previous findings from our group, investigating the uptake of PCBs by differentiated adipocytes, highlighted that PCB-28 enters the cells more rapidly than PCB-118 and PCB-153 [4]. If, during lipolysis, a reuptake of PCBs by the cells occurs, PCB-28 might thus be taken up more rapidly than the other congeners, which could lead to an underestimation of the differences of mobilisation kinetics during the lipolytic treatment.

Similar rate of release when PCBs are present alone or in cocktail

In the *in vivo* situation, tissues are exposed to a cocktail of contaminants that might interact with each other's, regarding either the toxicokinetics or the toxicodynamics of the molecules. Here, we investigated the effect of a simple combination of PCBs (three congeners) on their release by adipocytes during lipolysis. To do this, the discharge of PCB-28, -118 and -153 from

adipocytes was followed either alone, or in cocktail (i.e. with two other congeners). In the two conditions of contamination, the dynamics of PCB release were similar, meaning that the mobilisation of PCB congeners was not influenced by the presence of other congeners within adipocytes in these experimental conditions. As noted above, PCB-153 influences the properties of cell membranes. One could thus have expected that this congener could influence the dynamics of release of other PCBs from the cells.

Conclusion

Our results showed an efficient accumulation of PCB-28, -118 and -153 in adipocytes. Once lipolysis was induced, the congeners were massively mobilised from cells into the culture medium, in parallel with the release of lipids. The dynamics of discharge however differed between the three investigated congeners. The release of PCB-153 was slightly but significantly slower than the ones of PCB-28 and -118. The phenomenon might be explained by the fact that PCB-153 is more lipophilic than the two other congeners and could thus be more trapped in LDs. In addition, PCB-153 being a di-*ortho*-substituted congener, it is more bulky, which could be involved in its partial sequestration within cell membrane [5] and its slower mobilisation from adipocytes. On the other hand, the dynamics of mobilisation was not influenced by the presence of the other two congeners.

Acknowledgments

The authors are very grateful to Coralie Piget and Marie-Thérèse Ahn from "Institut des Sciences de la Vie" (ISV), UCLouvain, for technical assistance. Members of "Support en méthodologie et calcul statistique" (Institut multidisciplinaire pour la modélisation et l'analyse quantitative, UCLouvain, Belgium) are gratefully acknowledged for the collaboration in the statistical analyses. We also greatly appreciated the help and advice of Guillaume Bernard for picture processing.

Author Contributions

Conceived and designed the experiments: CL GT. Performed the experiments: GT. Analyzed the data: CL GT. Contributed reagents/materials/analysis tools: EM JPT CD. Contributed to the writing of the manuscript: CL. Engaged in active discussions: CL GT EM JPT CD.

References

1. Djien Liem AK, Furst P, Rappe C (2000) Exposure of populations to dioxins and related compounds. Food Additives & Contaminants 17: 241–259.
2. La Merrill M, Emond C, Kim MJ, Antignac JP, Le Bizec B, et al. (2013) Toxicological function of adipose tissue: focus on persistent organic pollutants. Environmental Health Perspectives 121: 162–169.
3. Sbarbati A, Accorsi D, Benati D, Marchetti L, Orsini G, et al. (2010) Subcutaneous adipose tissue classification. European Journal of Histochemistry 54: e48.
4. Bourez S, Le Lay S, Van den Daelen C, Louis C, Larondelle Y, et al. (2012) Accumulation of polychlorinated biphenyls in adipocytes: Selective targeting to lipid droplets and role of caveolin-1. PLoS ONE 7: e31834.
5. Bourez S, Joly A, Covaci A, Remacle C, Larondelle Y, et al. (2012) Accumulation capacity of primary cultures of adipocytes for PCB-126: Influence of cell differentiation stage and triglyceride levels. Toxicology Letters 214: 243–250.
6. Kim MJ, Marchand P, Henegar C, Antignac JP, Alili R, et al. (2011) Fate and complex pathogenic effects of dioxins and polychlorinated biphenyls in obese subjects before and after drastic weight loss. Environmental Health Perspectives 119: 377–383.
7. Chevrier J, Dewailly E, Ayotte P, Mauriege P, Despres JP, et al. (2000) Body weight loss increases plasma and adipose tissue concentrations of potentially toxic pollutants in obese individuals. International Journal of Obesity and Related Metabolic Disorders 24: 1272–1278.
8. Debier C, Pomeroy PP, Dupont C, Joiris C, Comblin V, et al. (2003) Quantitative dynamics of PCB transfer from mother to pup during lactation in

UK grey seals *Halichoerus grypus*. Marine Ecology Progress Series 247: 237–248.
9. Debier C, Pomeroy PP, Thomé JP, Mignolet E, de Tillesse T, et al. (2004) An unexpected parallelism between Vitamin A and PCBs in seal milk. Aquatic Toxicology 68: 179–183.
10. Debier C, Chalon C, Le Bœuf BJ, de Tillesse T, Larondelle Y, et al. (2006) Mobilization of PCBs from blubber to blood in northern elephant seals (*Mirounga angustirostris*) during the post-weaning fast. Aquatic Toxicology 80: 149–157.
11. Vanden Berghe M, Mat A, Arriola A, Polain S, Stekke V, et al. (2010) Relationships between vitamin A and PCBs in grey seal mothers and pups during lactation. Environmental Pollution 158: 1570–1575.
12. Debier C, Crocker DE, Houser DS, Vanden Berghe M, Fowler M, et al. (2012) Differential changes of fat-soluble vitamins and pollutants during lactation in northern elephant seal mother–pup pairs. Comparative Biochemistry and Physiology Part A 162: 323–330.
13. Vanden Berghe M, Weijs L, Habran S, Das K, Bugli C, et al. (2012) Selective transfer of persistent organic pollutants and their metabolites in grey seals during lactation. Environment International 46: 6–15.
14. Louis C, Dirtu AC, Stas M, Guiot Y, Malarvannan G, et al. (2014) Mobilisation of lipophilic pollutants from blubber in northern elephant seal pups (*Mirounga angustirostris*) during the post-weaning fast. Environmental Research 132: 438–448.
15. Hue O, Marcotte J, Berrigan F, Simoneau M, Doré J, et al. (2006) Increased plasma levels of toxic pollutants accompanying weight loss induced by hypocaloric diet or by bariatric surgery. Obesity Surgery 16: 1145–1154.

16. Imbeault P, Tremblay A, Simoneau J-A, Joanisse DR (2002) Weight loss-induced rise in plasma pollutant is associated with reduced skeletal muscle oxidative capacity. American Journal of Physiology 282: E574–E579.

17. Debier C, Le Boeuf BJ, Ikonomou MG, de Tillesse T, Larondelle Y, et al. (2005) Polychlorinated biphenyls, dioxins, and furans in weaned, free-ranging northern elephant seal pups from central California, USA. Environmental Toxicology and Chemistry 24: 629–633.

18. Dirtu AC, Dirinck E, Malarvannan G, Neels H, Van Gaal L, et al. (2013) Dynamics of organohalogenated contaminants in human serum from obese individuals during one year of weight loss treatment. Environmental Science & Technology 47: 12441–12449.

19. Carpenter DO (2006) Polychlorinated biphenyls (PCBs): routes of exposure and effects on human health. Reviews on Environmental Health 21: 1–23.

20. Robertson L, Hansen L (2001) PCBs: Recent advances in environmental toxicology and health effects. The University Press of Kentucky, Lexington, Kentucky.

21. Kester MH, Bulduk S, Tibboel D, Meinl W, Glatt H, et al. (2000) Potent inhibition of estrogen sulfotransferase by hydroxylated PCB metabolites: A novel pathway explaining the estrogenic activity of PCBs. Endocrinology 141: 1897–1900.

22. Gupta C (2000) Reproductive malformation of the male offspring following maternal exposure to estrogenic chemicals. Proceedings of the Society for Experimental Biology and Medicine 224: 61–68.

23. Schantz SL, Widholm JJ, Rice DC (2003) Effects of PCB exposure on neuropsychological function in children. Environmental Health Perspectives 111: 357–576.

24. Lee DH, Steffes MW, Sjödin A, Jones RS, Needham LL, et al. (2011) Low dose organochlorine pesticides and polychlorinated biphenyls predict obesity, dyslipidemia, and insulin resistance among people free of diabetes. PLoS ONE 6: e15977.

25. Dirinck E, Jorens PG, Covaci A, Geens T, Roosens L, et al. (2011) Obesity and persistent organic pollutants: Possible obesogenic effect of organochlorine pesticides and polychlorinated biphenyls. Obesity 19: 709–714.

26. Bourez S, Van den Daelen C, Le Lay S, Poupaert J, Larondelle Y, et al. (2013) The dynamics of accumulation of PCBs in cultured adipocytes vary with the cell lipid content and the lipophilicity of the congener. Toxicology Letters 216: 40–46.

27. Zhou L, Wang X, Yang Y, Wu L, Li F, et al. (2011) Berberine attenuates cAMP-induced lipolysis via reducing the inhibition of phosphodiesterase in 3T3-L1 adipocytes. Biochimica et Biophysica Acta (BBA) 1812: 527–535.

28. Louis C, Van den Daelen C, Bourez S, Donnay I, Larondelle Y, et al. (2014) Efficient in vitro adipocyte model of long-term lipolysis: A tool to study the behaviour of lipophilic compounds. In Vitro Cellular & Developmental Biology - Animal 50: 507–518.

29. Wassermann M, Wassermann D, Cucos S, Miller HJ (1979) World PCBs map: Storage and effects in man and his biologic environment in the 1970s. Annals of the New York Academy of Sciences 320: 69–124.

30. Meeker JD, Maity A, Missmer SA, Williams PL, Mahalingaiah S, et al. (2011) Serum concentrations of polychlorinated biphenyls in relation to in vitro fertilization outcomes. Environmental Health Perspectives 119: 1010–1016.

31. Dang Van QC, Focant M, Mignolet E, Turu C, Froidmont E, et al. (2011) Influence of the diet structure on ruminal biohydrogenation and milk fatty acid composition of cows fed extruded linseed. Animal Feed Science and Technology 169: 1–10.

32. EC (2002) European Commission. Council L221/8. Official Journal of the European Communities 262: 8–36.

33. Matthews HB, Surles JR, Carver JG, Anderson MW (1984) Halogenated biphenyl transport by blood components. Fundamental and Applied Toxicology 4: 420–428.

34. Wang N, Kong D, Cai D, Shi L, Cao Y, et al. (2010) Levels of polychlorinated biphenyls in human adipose tissue samples from Southeast China. Environmental Science & Technology 44: 4334–4340.

35. De Saeger S, Sergeant H, Piette M, Bruneel N, Van de Voorde W, et al. (2005) Monitoring of polychlorinated biphenyls in Belgian human adipose tissue samples. Chemosphere 58: 953–960.

36. Moon HB, Lee DH, Lee Y, Choi M, Choi HG, et al. (2012) Polybrominated diphenyl ethers, polychlorinated biphenyls, and organochlorine pesticides in adipose tissues of Korean women. Archives of Environmental Contamination and Toxicology 62: 176–184.

37. Arrebola JP, Cuellar M, Claure E, Quevedo M, Antelo SR, et al. (2012) Concentrations of organochlorine pesticides and polychlorinated biphenyls in human serum and adipose tissue from Bolivia. Environmental Research 112: 40–47.

38. Malarvannan G, Dirinck E, Dirtu AC, Pereira-Fernandes A, Neels H, et al. (2013) Distribution of persistent organic pollutants in two different fat compartments from obese individuals. Environment International 55: 33–42.

39. Becker MM, Gamble W (1982) Determination of the binding of 2,4,5,2',4',5'-hexachlorobiphenyl by low density lipoprotein and bovine serum albumin. Journal of Toxicology and Environmental Health 9: 225–234.

40. Spindler-Vomachka M, Vodicnik MJ, Lech JJ (1984) Transport of 2,4,5,2',4',5'-hexachlorobiphenyl by lipoproteins in vivo. Toxicology and Applied Pharmacology 74: 70–77.

41. Gallenberg LA, Ring BJ, Vodicnik MJ (1987) Influence of lipolysis on the mobilization of 2,4,5,2',4',5'-hexachlorobiphenyl from adipocytes in vitro. Journal of Toxicology and Environmental Health 20: 163–171.

42. Koppe JG (1995) Nutrition and breast-feeding. European Journal of Obstetrics & Gynecology and Reproductive Biology 61: 73–78.

43. Irigaray P, Mejean L, Laurent F (2005) Behaviour of dioxin in pig adipocytes. Food and Chemical Toxicology 43: 457–460.

44. Edens NK, Leibel RL, Hirsch J (1990) Mechanism of free fatty acid re-esterification in human adipocytes in vitro. Journal of Lipid Research 31: 1423–1431.

45. López-Aparicio P, Merino MJ, Sánchez E, Recio MN, Pérez-Albarsanz MA (1997) Effect of Aroclor 1248 and two pure PCB congeners upon the membrane fluidity of rat renal tubular cell cultures. Pesticide Biochemistry and Physiology 57: 54–62.

46. Gonzalez A, Odjélé A, Weber JM (2013) PCB-153 and temperature cause restructuring of goldfish membranes: Homeoviscous response to a chemical fluidiser. Aquatic Toxicology 144–145: 11–18.

47. Yilmaz B, Sandal S, Chen CH, Carpenter DO (2006) Effects of PCB 52 and PCB 77 on cell viability, $[Ca^{2+}]_i$ levels and membrane fluidity in mouse thymocytes. Toxicology 217: 184–193.

48. Campbell AS, Yu Y, Granick S, Gewirth AA (2008) PCB association with model phospholipid bilayers. Environmental Science & Technology 42: 7496–7501.

49. Katynski AL, Vijayan MM, Kennedy SW, Moon TW (2004) 3,3',4,4',5-pentachlorobiphenyl (PCB 126) impacts hepatic lipid peroxidation, membrane fluidity and β-adrenoceptor kinetics in chick embryos. Comparative Biochemistry and Physiology Part C 137: 81–93.

21

Soil Mineral Composition Matters: Response of Microbial Communities to Phenanthrene and Plant Litter Addition in Long-Term Matured Artificial Soils

Doreen Babin[1¤]**, Cordula Vogel**[2]**, Sebastian Zühlke**[3]**, Michael Schloter**[4]**, Geertje Johanna Pronk**[2,5]**, Katja Heister**[2]**, Michael Spiteller**[3]**, Ingrid Kögel-Knabner**[2,5]**, Kornelia Smalla**[1]*

1 Institute for Epidemiology and Pathogen Diagnostics, Julius Kühn-Institut - Federal Research Centre for Cultivated Plants (JKI), Braunschweig, Germany, **2** Lehrstuhl für Bodenkunde, Technische Universität München, Freising-Weihenstephan, Germany, **3** Institut für Umweltforschung (INFU), Lehrstuhl für Umweltchemie und Analytische Chemie der Fakultät für Chemie und Chemische Biologie, Technische Universität Dortmund, Dortmund, Germany, **4** Research Unit for Environmental Genomics, Helmholtz Zentrum München, German Research Center for Environmental Health, Neuherberg, Germany, **5** Institute for Advanced Study, Technische Universität München, Garching, Germany

Abstract

The fate of polycyclic aromatic hydrocarbons (PAHs) in soil is determined by a suite of biotic and abiotic factors, and disentangling their role in the complex soil interaction network remains challenging. Here, we investigate the influence of soil composition on the microbial community structure and its response to the spiked model PAH compound phenanthrene and plant litter. We used long-term matured artificial soils differing in type of clay mineral (illite, montmorillonite) and presence of charcoal or ferrihydrite. The soils received an identical soil microbial fraction and were incubated for more than two years with two sterile manure additions. The matured artificial soils and a natural soil were subjected to the following spiking treatments: (I) phenanthrene, (II) litter, (III) litter + phenanthrene, (IV) unspiked control. Total community DNA was extracted from soil sampled on the day of spiking, 7, 21, and 63 days after spiking. Bacterial 16S rRNA gene and fungal internal transcribed spacer amplicons were quantified by qPCR and subjected to denaturing gradient gel electrophoresis (DGGE). DGGE analysis revealed that the bacterial community composition, which was strongly shaped by clay minerals after more than two years of incubation, changed in response to spiked phenanthrene and added litter. DGGE and qPCR showed that soil composition significantly influenced the microbial response to spiking. While fungal communities responded only in presence of litter to phenanthrene spiking, the response of the bacterial communities to phenanthrene was less pronounced when litter was present. Interestingly, microbial communities in all artificial soils were more strongly affected by spiking than in the natural soil, which might indicate the importance of higher microbial diversity to compensate perturbations. This study showed the influence of soil composition on the microbiota and their response to phenanthrene and litter, which may increase our understanding of complex interactions in soils for bioremediation applications.

Editor: Andrew C. Singer, NERC Centre for Ecology & Hydrology, United Kingdom

Funding: This study was funded by Deutsche Forschungsgemeinschaft within the framework of the priority program SPP1315 "Biogeochemical Interfaces in Soil" (SM59/8-2) (www.dfg.de). The funders had no role in study design, data collection and analysis, decision to publish, or preparation of the manuscript.

Competing Interests: The authors have declared that no competing interests exist.

* Email: kornelia.smalla@jki.bund.de

¤ Current address: Lehrstuhl für Hydrogeologie, Institut für Geowissenschaften, Friedrich-Schiller-Universität Jena, Jena, Germany

Introduction

Bioremediation refers to different cleanup strategies using living organisms for the removal of environmental pollutants, such as polycyclic aromatic hydrocarbons (PAHs), from soils contaminated by anthropogenic activities. Different parameters can severely affect the efficiency of the methods, as for instance the exposure time of the contaminant in soil [1], soil structure [2], pH [3], temperature [3], sorptive interfaces [4], and the organic carbon content [5]. In addition, the accessibility to oxygen may be of importance, since the fastest and most often used microbial pathway of PAH degradation involves the oxidation of the ring-structure via dioxygenases [6]. The concentration and availability of nutrients, such as nitrogen and phosphorus, represent another factor influencing the rate of biodegradation in contaminated soils [3,6]. To enhance the bioremediation efficiency, a current method is the addition of straw, compost, or manure to polluted soils which improves soil structure, oxygen transfer and provides energy sources for the soil microbiota [7].

Apart from those single factors and existing strategies to tackle them, the efficient application of biodegradation in the environment is still challenged by two scientific frontiers as reviewed recently by Jeon & Madsen [8]: On the one hand, although many studies exist on reactions of pollutants with clays and soil organic matter (OM), e.g. [2,5,9,10,11], we lack a comprehensive understanding of these abiotic soil interaction processes. On the other hand, there is restricted knowledge of the interaction of pollutants with native soil microbial communities within the natural soil system. So far, only a few studies have been carried out to compare the consequences of contamination on the microbiota in different soils. For instance, Bundy et al. [12] and Ding et al. [13] showed previously that each soil type has its individual microbial response to pollutants. In this respect, it is necessary to consider the role of the interplay between microbes and their physical soil environment [14]. This is of particular interest since the different organic, inorganic and biological soil components are in close contact and form complex, so-called biogeochemical interfaces where important ecosystem processes take place [15].

The influence of the soil mineral composition on the response of microbial communities to pollutants and the resulting effect on the biodegradative potential has been poorly addressed so far. It is known that different soil components, such as minerals and charcoal, can influence metabolic activity, the establishment and the microenvironment of the soil microbiota [16,17,18,19]; but might the soil composition interfere also with the microbial response to and biodegradation of pollutants? In order to answer this question, artificial soils are a good tool to simulate natural soil environments [20,21,22,23]. The design of artificial soils with defined compositions makes it possible to disentangle the influence of certain factors (e.g., soil components, soil OM, microbiota, pollutants) on a particular soil process. Based on this, we recently investigated the effect of the soil mineral composition and the presence of charcoal on the early soil interface development. We observed the fast formation of organo-mineral associations [22] and the dynamic effects of charcoal, clay minerals and metal oxides on the microbial community establishment [13,24]. After spiking one year-incubated artificial soils with the model PAH compound phenanthrene, bacteria from different artificial soils showed similar and soil composition-dependent responses to the pollutant [24]. Recently, Vogel et al. [25] reported on the maturation of artificial soils over a long incubation period of 842 days. The authors observed that clay minerals are the major long-term driver of bacterial communities and that artificial soils developed to soil-like systems [25]. We hypothesized that the establishment of these soil composition-driven microbial communities and interfaces over such a long period causes different responses of the microbiota to spiked phenanthrene. Furthermore, we assumed that plant litter addition stimulates the established microbial communities and their response to phenanthrene.

We therefore conducted a spiking experiment on artificial soils incubated for 842 days with four different treatments: phenanthrene (+P), litter (+L), litter and phenanthrene (+L+P), unspiked control (Figure 1). Four soil compositions were used based on the results of the aforementioned studies [19,24,25]: two soils differed in the type of clay mineral (montmorillonite [M] or illite [I]), one soil contained charcoal (C) and one soil had the metal oxide ferrihydrite (F) as additional soil component. The microbiota in differently treated artificial soils was compared to a natural soil which was spiked similarly. We sampled and extracted total community DNA (TC-DNA) on the day of spiking, 7 days after spiking (DAS), 21 DAS, and 63 DAS. The responses of the microbial communities to the spiking were studied by denaturing gradient gel electrophoresis (DGGE) fingerprints and quantitative real-time PCR (qPCR) of 16S rRNA gene and internal transcribed spacer (ITS) fragments. This study shows for the first time the effect of spiked phenanthrene and litter amendment on microbial communities which established over a long-term period as a function of the soil mineral composition and presence of charcoal.

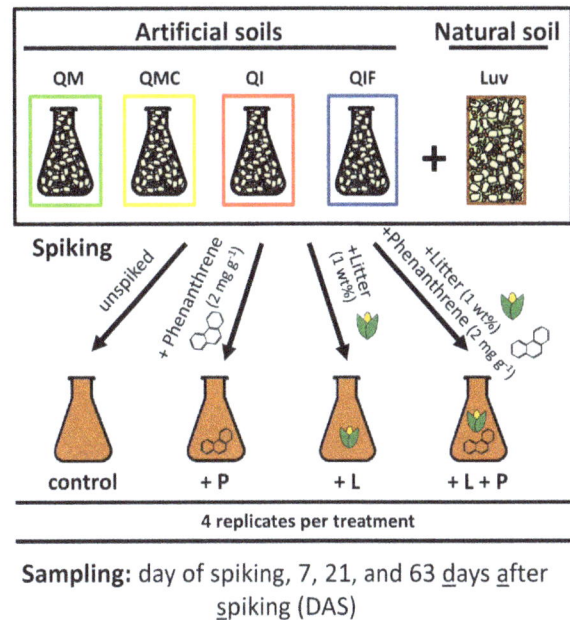

Figure 1. Experimental design. Long-term matured artificial soils QM, QMC, QI, QIF and the natural soil were subjected to following treatments: unspiked control, phenanthrene (+P), plant litter (+L), litter and phenanthrene (+L+P). Each treatment consisted of four replicates. Sampling was carried out on the day of spiking, 7, 21 and 63 days after spiking (DAS). Q-quartz, M-montmorillonite, C-charcoal, I-illite, F-ferrihydrite, Luv-natural soil (Luvisol).
doi:10.1371/journal.pone.0106865.g001

Materials and Methods

1. Soils used in this study

Design and model materials of the artificial soil compositions were previously described in detail by Pronk et al. [22] and Vogel et al. [25]. The following four artificial soil compositions were used in this experiment: QM (94% quartz +6% montmorillonite), QI (92% quartz +8% illite), QMC (94% quartz +4% montmorillonite +2% charcoal), and QIF (92% quartz +7% illite +1% ferrihydrite). Sand- and silt-sized quartz was used as structure material similarly for all artificial soil compositions. Sterile manure (4.5 wt%) was used as initial OM input. The inoculant was derived by water extraction from a topsoil of an agricultural site in southern Germany (Scheyern; 48°N, 11°E) which has been classified as a Luvisol. The inoculant was added to the soil compositions after an abiotic conditioning time of three days with 0.01 M $CaCl_2$ as artificial soil solution. Soils were incubated under constant environmental conditions (20°C, constant water holding capacity of 60%, in the dark) and gently mixed weekly for homogenization. After 562 days of incubation, the addition of sterile manure (4.5 wt%) was repeated to provide a fresh OM source, since previous experiments suggested the decrease of the microbial activity and macro-aggregation due to limited nutrients [22]. Soils were incubated up to 842 days and a detailed characterization of soil parameters as well as of microbial communities established during that incubation time can be found in Vogel et al. [25]. The natural soil used in this study was an arable topsoil (Ap) from a Luvisol (Luv) and was collected from the same site as the soil used for the extraction of the inoculant (Scheyern, Germany). The natural soil was characterized by an organic C content of 16 ± 0.8 mg g^{-1}, N content of 1.76 ± 0.1 mg g^{-1}, a C/N-ratio of

9.0 ± 0.4 and a pH value ($CaCl_2$) of 6.6 ± 0.1. Prior to spiking, all soils were homogenized and sieved to <2 mm.

The natural soil used for extraction of the inoculant and for the spiking experiment originated from a German research farm (Scheyern; 48°N, 11°E) and no specific permissions were required for collecting. The study did not involve endangered or protected species.

2. Spiking experiment

Artificial soils and the natural soil were subjected to the following four spiking treatments: +P, +L, +L+P, unspiked control (Figure 1). Each treatment consisted of four independent replicates. Briefly, phenanthrene spiking was done as follows: a seeding soil (20 mg phenanthrene per gram of soil) was prepared for each artificial soil and the natural soil by adding phenanthrene ($\geq97\%$, Merck Schuchardt, Hohenbrunn, Germany) dissolved in acetone (100 mg ml^{-1}) to 40 g of soil. Acetone was allowed to evaporate over night to avoid severe changes of the total microbial community. Then, the homogenized seeding soil was mixed to each soil replicate to reach a final phenanthrene concentration of 2 mg g^{-1} of soil [26]. Phenanthrene-unspiked soils received a seeding soil spiked with pure acetone solution. Dried plant litter consisting of maize and potato leaves (1:1) grown under controlled conditions in the greenhouse was used to conduct the litter amendments on soils (1 wt%). After spiking, the water content was adjusted to a water holding capacity of 60% for each soil composition and treatment. Soils were carefully homogenized in order to save the structure and formed interfaces of matured artificial soils. Incubation was carried out at 14°C, in the dark and at constant water content. Samples were taken on the day of spiking, 7 DAS, 21 DAS, and 63 DAS and subsequently stored at -20°C.

3. Extraction and purification of TC-DNA

Extraction of TC-DNA from 0.5 g soil (wet weight) was carried out according to the manufacturer's protocol (FastDNA spin Kit for soil) using FastPrep FP24 bead-beating system (MP Biomedicals, Santa Ana, California). DNA solutions were purified by GeneClean Spin Kit (Qbiogene, Inc., Carlsbad, California). Both steps were followed by controlling the DNA yield on an agarose gel. TC-DNA was kept at -20°C.

4. Amplification of bacterial 16S rRNA gene and ITS fragments

All PCRs were performed in a 25 µl reaction with purified and 1:10 diluted TC-DNA. Total bacterial community was assessed by primers F984GC/R1378 according to Gomes et al. [27]. On samples taken 21 DAS and 63 DAS, nested PCR approaches were carried out for important soil bacterial groups: *Alphaproteobacteria*, *Betaproteobacteria* as described by Gomes et al. [27], *Actinobacteria* [28], and *Pseudomonas* [29]. Diluted amplicons of group-specific PCRs were used as target in PCR with primers F984GC/R1378 to generate the GC-clamp for subsequent DGGE analysis. Fungal communities in artificial soils were studied based on ITS fragments amplified in a nested PCR with primers ITS1F/ITS4 and ITS1F-GC/ITS2 as described by Weinert et al. [30]. Modifications of the conditions of the above PCRs can be found in Babin et al. [24]. Primers with respective references are listed in Table S1.

5. DGGE analysis

The microbial community structure was studied by DGGE of amplified 16S rRNA gene or ITS fragments performed in an Ingeny PhorU system (Ingeny, Goes, The Netherlands). Gradient concentrations and DGGE conditions were specified earlier by Weinert et al. [30]. Gels were silver-stained according to Heuer et al. [31].

Extraction of treatment-specific DGGE bands and cloning were carried out as described by Babin et al. [24] and Smalla et al. [32]. Several clones with identical electrophoretic mobility for each band were subjected to sequencing using standard primer SP6 (Macrogen, Amsterdam, The Netherlands). Sequences were cleaned, compared and checked for similarity hits with 16S rRNA gene database entries using Nucleotide BLAST (http://blast.ncbi.nlm.nih.gov/Blast.cgi). Sequences can be retrieved in GenBank under accession numbers KJ145751-KJ145753.

6. Quantification of 16S rRNA gene and ITS fragments by qPCR

16S rRNA gene copy numbers were determined by qPCR 5'-nuclease assay with primer and TaqMan probe previously described by Suzuki et al. [33] in a 50 µl reaction volume. Concentrations of reagents and the thermal program can be found in Vogel et al. [25].

Quantification of the fungal ITS fragment was carried out according to the protocol established by Gschwendtner et al. [34] with modifications listed in Vogel et al. [25]. Both quantifications for 16S rRNA gene and ITS fragments were performed in a CFX96 Real-Time System (Biorad, München, Germany). Information on primers and TaqMan probe can be found in Table S1.

7. Phenanthrene analysis

The phenanthrene concentration was measured in three replicates per soil and treatment of samples taken on the day of spiking and 21 DAS. Quantification was done according to existing methods [26,35] with slight modifications as follows: One gram soil was extracted using 5 ml acetone and 5 ml cyclohexane in an ultrasonic bath and an overhead shaker. The final extract was separated with an acetonitrile-water gradient on a C18 column (Luna C18, 100 A, 150×2.0 mm, 3 µm; Phenomenex, Aschaffenburg, Germany). Phenanthrene was detected at 254 nm (UVD 340 S UV detector; Dionex, Sunnyvale, California). External quantification was done with multicomponent standard solution SRM 1647D (National Institute of Standards and Technology, Promochem, Wesel, Germany) at concentrations ranging from 1 to 10 µg ml^{-1}. Linearity was excellent ($R^2 = 0.996$).

8. Statistical analysis

DGGE profiles were compared pairwise using software GelCompar II 6.5 (Applied Maths, Sint-Martens-Latem, Belgium). Pairwise Pearson similarity coefficients were used for the construction of dendrograms based on the unweighted pair group method with arithmetic mean (UPGMA) cluster algorithm and to perform permutation tests for significant differences (d-values) between soil compositions and treatments according to Kropf et al. [36]. The qPCR data and phenanthrene concentrations were subjected to analysis of variance in conjunction with Tukey's HSD test with software R 3.0.2 using package *agricolae*. Multiple factor ANOVA was used to test for significant effects of soil, phenanthrene, and litter. Bacterial 16S rRNA gene and fungal ITS fragment copy numbers were log transformed prior to statistical analysis.

Results

1. The bacterial community structure in long-term matured artificial soils

By DGGE fingerprints, a high similarity of bacterial communities between the four independent replicates of each artificial soil was observed in samples taken on the day of spiking from the control treatment (Figure 2a). Several soil composition-dependent populations were observed which were marked in Figure 2a by arrows. A strong influence of the type of clay mineral, montmorillonite (bands A) and illite (bands C), on the bacterial community was visible as well as several populations with increased or decreased relative abundance due to the presence of charcoal (bands B) or ferrihydrite (bands D), respectively. UPGMA cluster analysis of DGGE fingerprints revealed a clustering according to the type of clay minerals and distinct sub-clusters for artificial soils containing charcoal (QMC) and ferrihydrite (QIF) (Figure 2b). Permutation tests based on Pearson similarity coefficients showed the highest difference between bacterial communities from artificial soils with different clay minerals present (QM vs. QI 30%). Smaller but also significant values were found for the effect of charcoal (QM vs. QMC 18%) and ferrihydrite (QI vs. QIF 8%).

2. Response of total bacteria to spiking in long-term matured artificial soils

The effects of +P, +L, +L+P treatments on the total bacterial community structure of the artificial soils were assessed over time. The differences (d-values) between communities in the treatment of interest vs. the control, i.e. without the respective spike, are reported in Table 1.

In samples taken 7 DAS, populations with increased relative abundance as inferred from band intensity due to the litter

addition were observed. Some of these litter responders were found in all artificial soils, but others depended on the soil composition (Figure S1, Figure S2). Bacteria in litter-amended artificial soils differed significantly from the controls. Smaller d-values for the litter effect were found in presence of phenanthrene (+L+P vs. +P) compared to the litter effect in phenanthrene-unspiked soils (+L vs. control, Table 1). Overall, lowest d-values for the litter effect were found in soils containing montmorillonite (QM, QMC). Compared to the litter effect, the response of bacteria to phenanthrene was low 7 DAS (Table 1).

Twenty-one DAS, remarkable shifts due to phenanthrene spiking were observed in bacterial communities of all artificial soils (white arrows, Figure 3, Figure S3). In QM+P soils, only one strong responder to phenanthrene was found, which showed 99% similarity to *Arthrobacter crystallopoietes* (band 1 marked by arrow, Figure 3; Table 2). Corresponding bands with similar electrophoretic mobility were detected in QI+P and QIF+P but not in QMC+P soils (DGGE not shown). In QMC+P, another phenanthrene-enhanced population was observed (Figure 3). More shifts in relative abundance in the bacterial community structure due to spiked phenanthrene were found in the QI+P and QIF+P fingerprints (Figure S3). Only a few ferrihydrite-specific responses to phenanthrene were observed (Figure S3). Permutation tests revealed that the response to phenanthrene was as strong as the effect of the litter amendment 21 DAS (d-values ca. 50%) except for QMC soil, in which the difference between the unspiked and phenanthrene-spiked treatment were only 17% (Table 1).

A band with similar electrophoretic mobility as the identified phenanthrene-responding population *Arthrobacter crystallopoietes* (band 1, Figure 3) was also detected in the presence of litter in QM+P+L, QI+P+L, and QIF+P+L soils. Several other populations were found which either increased or decreased in relative abundance due to phenanthrene only in the presence of litter (grey

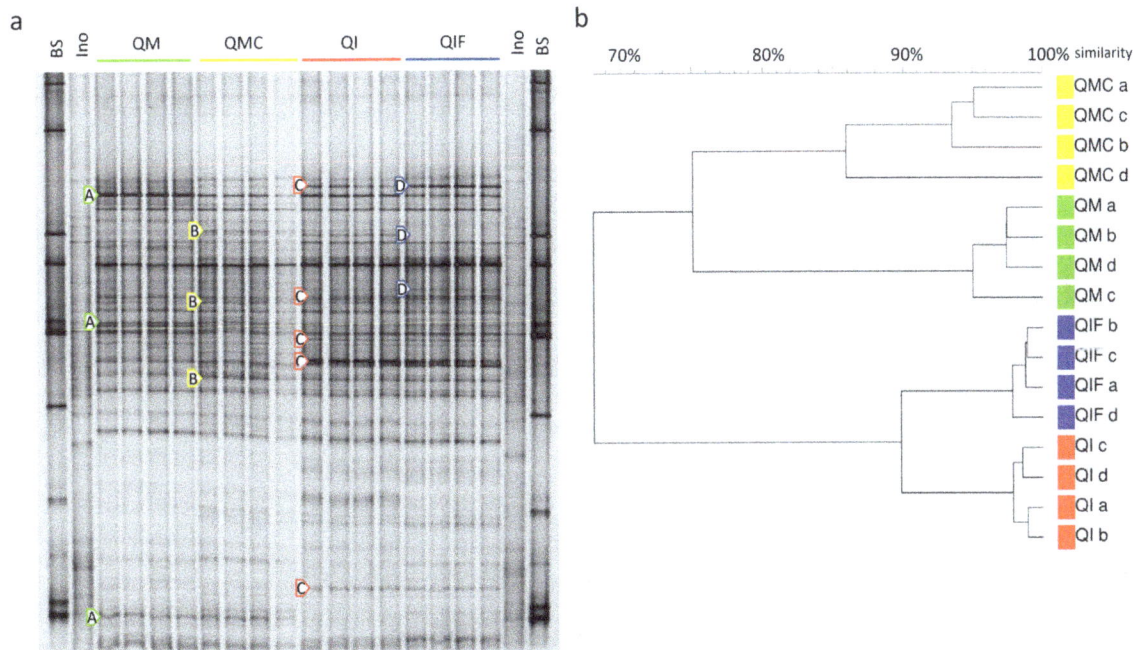

Figure 2. Bacterial communities of long-term matured artificial soils. a) Total bacterial DGGE fingerprints of unspiked control artificial soils (four replicates) sampled on the day of spiking. Arrows mark soil composition-specific populations in QM (bands A), QMC (B), QI (C), and QIF (D). BS-bacterial DGGE standard. Ino-Luvisol inoculant added to soil mixtures at the beginning of incubation. Q-quartz, M-montmorillonite, C-charcoal, I-illite, F-ferrihydrite. b) Corresponding UPGMA cluster analysis.

Table 1. Percent difference (d-values) between microbial communities of different spiking treatments for different taxa and different sampling times (21 and 63 days after spiking [DAS]) per soil.

% difference	Bacteria			Betaproteobacteria		Actinobacteria		Fungi	
	7 DAS	21 DAS	63 DAS	21 DAS	63 DAS	21 DAS	63 DAS	21 DAS	63 DAS
QM vs. QM+P	9	45	10	1	14	24	14	1	2
QM+L vs. QM+L+P	2	16	10	6	8	13	12	3	7
QM vs. QM+L	27	54	18	21	15	43	37	14	17
QM+P vs. QM+L+P	14	33	19	29	15	19	32	19	12
QMC vs. QMC+P	13	17	23	2	19	21	23	2	6
QMC+L vs. QMC+L+P	1	14	7	20	26	0	14	5	16
QMC vs. QMC+L	31	37	28	36	12	28	35	18	21
QMC+P vs. QMC+L+P	16	38	16	20	22	9	40	10	14
QI vs. QI+P	14	51	27	3	27	55	24	1	4
QI+L vs. QI+L+P	2	19	21	17	17	47	18	13	26
QI vs. QI+L	53	54	30	34	52	30	29	19	13
QI+P vs. QI+L+P	34	33	29	45	36	32	24	12	5
QIF vs. QIF+P	22	42	29	12	36	33	12	1	3
QIF+L vs. QIF+L+P	3	17	13	4	7	20	18	8	20
QIF vs. QIF+L	47	50	37	33	34	38	42	16	21
QIF+P vs. QIF+L+P	26	35	33	22	21	24	45	22	40
Luv vs. Luv+P	8	1	27	18	19	7	5	23	1
Luv+L vs. Luv+L+P	8	11	20	8	24	10	8	5	12
Luv vs. Luv+L	24	10	23	9	7	32	15	70	28
Luv+P vs. Luv+L+P	29	19	18	19	7	20	16	40	24

D-values were calculated based on pairwise Pearson similarity coefficients for the effect of phenanthrene (+P) in non-litter or litter-amended soils and the effect of litter (+L) in the absence or presence of phenanthrene. Bold numbers indicate significant differences ($p < 0.05$). Q-quartz, M-montmorillonite, C-charcoal, I-illite, F-ferrihydrite, Luv-natural soil (Luvisol).

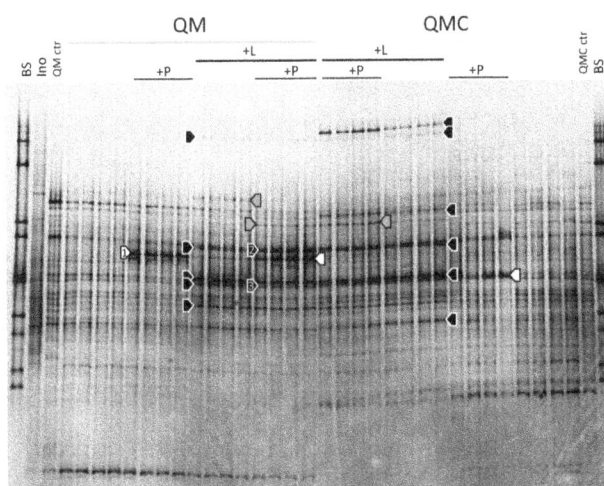

Figure 3. Response of bacterial communities to spiking in QM and QMC soils. DGGE fingerprints of bacterial communities in spiked QM and QMC soils sampled 21 days after spiking (control, phenanthrene (+P), litter (+L), litter and phenanthrene [+L+P]). Arrows mark populations responding to litter (black), phenanthrene (white), litter and phenanthrene (grey). Numbers indicate excised and cloned bands (Table 2). BS-bacterial DGGE standard. Ino-Luvisol inoculant added to soil mixtures at the beginning of incubation. QM/QMC ctr-bacterial community in QM or QMC, respectively, before spiking. Q-quartz, M-montmorillonite, C-charcoal.

bands, Figure 3, Figure S3). In the presence of litter, the bacterial community shifts in response to phenanthrene were less pronounced compared to the soils without litter amendment (Table 1).

At the sampling time 21 DAS, a remarkable effect of added litter was still observed in all soils in presence and absence of phenanthrene (Table 1). Two populations, which were enhanced in relative abundance in the presence of litter, were excised and cloned from QM+P+L fingerprints (Figure 3). Sequences were affiliated to *Stenotrophomonas maltophilia* (band 2) and *Arthrobacter* sp. (band 3; Table 2). By comparison of bacterial fingerprints of different soils, these two species were identified as responders in all artificial soils to the +L and +L+P treatment except for band 3 in QI+L+P and QIF+L+P soils. Populations with similar electrophoretic mobility compared to band 3

decreased in relative abundance in QI+L+P and QIF+L+P soils compared to QI+L and QIF+L (DGGE not shown).

In order to compare bacterial communities between artificial soils, 16S rRNA gene amplicons of all artificial soils from the same treatment were loaded on one DGGE gel. In samples taken 21 DAS, soil composition-dependent differences between bacteria among soils of the control treatment were observed (Table S2). These differences between artificial soils increased by the +P spiking compared to the control treatment (Table S2). High d-values were found for comparisons between bacteria in QM+P and other artificial soil compositions. The effect of ferrihydrite in litter-amended soils (QI+L vs. QIF+L) was comparable to the control treatment. Spiking of +P, +L, +L+P resulted in higher dissimilarities between bacteria in QM vs. QMC soils than in the control treatment.

In QI, QIF, and QM soils sampled 63 DAS, shifts in bacterial communities caused by spiking were similar to the shifts in samples of the respective artificial soil taken 21 DAS. In contrast, bacteria in QMC soils showed 63 DAS a stronger response to phenanthrene by pronounced shifts which were similar in their electrophoretic mobility to phenanthrene-responding populations in QM soils (Figure S4, Table 1).

3. Response of bacterial groups to spiking in long-term matured artificial soils

Cluster analysis of *Alphaproteobacteria* and *Pseudomonas* communities in samples taken 21 DAS showed effects of the soil composition and the litter addition, while the phenanthrene influence on these taxa was small (Figure S5). *Betaproteobacteria* in soils sampled 21 DAS were affected by the soil composition and litter amendment. A response of *Betaproteobacteria* to phenanthrene was only seen in the +L+P treatment (except for QIF; Table 1). Sixty-three DAS, more changes in betaproteobacterial fingerprints were observed due to phenanthrene compared to 21 DAS especially in QMC+P, QI+P, and QIF+P soils (Table 1). In QMC soils sampled 21 DAS, different shifts due to spiked phenanthrene were observed in actinobacterial fingerprints compared to QM soils (white arrows, Figure S6). *Actinobacteria* in QI and QIF soils responded in a similar way to phenanthrene, but shifts were more pronounced in QI soils (Table 1). A few soil composition and litter responders were found among the *Actinobacteria* (Figure S6, Table 1). The QMC soil exhibited clearly more treatment-responding actinobacterial populations than the QM soils at sampling time 63 DAS compared to 21 DAS (Table 1).

Table 2. Tentative phylogenetic affiliation of clones derived from bands excised from bacterial DGGE fingerprints in Figure 3.

Band	Clone (GenBank deposit)	Closest BLAST hits	% Identity	Accession number
1	QMP_1 (KJ145751)	*Arthrobacter crystallopoietes DSM 20117*	99	NR_026189
		Arthrobacter ramosus DSM 20546	98	NR_026193
		Arthrobacter pascens DSM 20545	98	NR_026191
2	QMPL_2 (KJ145752)	*Stenotrophomonas maltophilia R551-3*	99	NR_074875
		Stenotrophomonas humi R-32729	99	NR_042568
		Stenotrophomosa maltophilia IAM 12423	99	NR_041577
3	QMPL_3 (KJ145753)	*Arthrobacter sp. FB24*	99	NR_074590.1
		Arthrobacter humicola KV-653	99	NR_041546.1
		Arthrobacter oryzae KV-651	99	NR_041545.1

GenBank 14-March-2014.

4. Response of fungal communities to spiking in long-term matured artificial soils

At sampling time 21 DAS, the difference of fungal DGGE fingerprints between QI and QIF control soils and between QM and QMC control soils accounted for 9% and 20%, respectively. In all artificial soil samples taken 21 DAS and 63 DAS, weak fungal responses to +L addition were observed (black arrows, Figure 4, Figure S7), whereas no effects of +P were detected (Table 1). A profound effect on fungi was observed in all artificial soils for the +L+P spiking in samples taken 21 DAS, which even increased up to sampling time 63 DAS (Table 1). The +L+P treatment caused the detection of a few additional bands but also the disappearance of bands that were strongly abundant in the corresponding control. The shifts in fungal communities due to +L+P were similar in QI+L+P and QIF+L+P soils (grey bands, Figure 4). As for bacteria, the presence of charcoal also influenced the fungal response to +L+P treatment in QMC soils compared to QM by the appearance of additional responder bands (Figure S7).

5. Response of microbial communities to spiking in the natural soil

The bacterial community of the natural soil (Luv) revealed a different and more complex pattern in DGGE analysis than the artificial soils (Figure S8). Bacterial communities in the natural soil responded also to the different treatments, but the shifts were less pronounced, as indicated by lower d-values for the effect of spiking compared to the treatment effects in artificial soils (Figure S8, Table 1). The common responders of artificial soils (bands 1–3, Figure 3) were not found in the spiked natural soil. Twenty-one DAS, there was mainly a response of bacteria to the litter addition detectable in the natural soil. The effect of phenanthrene was retarded and only clearly visible in samples taken 63 DAS (Figure S8, Table 1). Taxon-specific DGGE analysis revealed that typical phenanthrene responders in the natural soil belonged to the *Betaproteobacteria* group (Table 1).

Fungal communities in the natural soil were also more complex and more heterogeneous than fungi in artificial soils. A significant response of fungi to +L was observed, especially 21 DAS, whereas shifts due to +L+P were not as pronounced as found in artificial soils (Table 1).

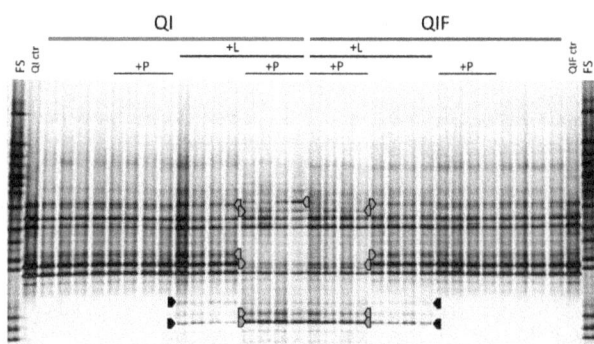

Figure 4. Response of fungal communities to spiking in QI and QIF soils. DGGE fingerprints of fungal communities in spiked QI and QIF soils sampled 63 days after spiking (control, phenanthrene (+P), litter (+L), litter and phenanthrene [+L+P]). Arrows mark populations responding to litter (black), litter and phenanthrene (grey). FS-fungal DGGE standard. QI/QIF ctr-fungal community in QI or QIF, respectively, before spiking. Q-quartz, I-illite, F-ferrihydrite.

6. Quantification of bacterial 16S rRNA gene and fungal ITS fragments

The amount of 16S rRNA genes was determined in artificial soils and the natural soil sampled 21 DAS (Figure 5a). In the control treatment, gene copy numbers among artificial soils were similar and ranged from 1.3×10^{10} to 1.6×10^{10}. A higher abundance was detected in the natural soil (Luv; 3.2×10^{10}). Phenanthrene caused a slight increase of 16S rRNA gene copies in QI+P, QMC+P, and QIF+P. Except for QI+L soils, litter addition resulted in clearly enhanced copy numbers. The highest numbers were found for all soils in the +L+P treatment. 16S rRNA gene copy numbers in the natural soil were relatively stable and were only enhanced in +L+P treatment. Analysis of variance revealed highly significant effects for litter (p<0.001), phenanthrene (p<0.001), and soil (p<0.001).

Fungal ITS fragments were quantified in differently treated artificial soils and the natural soil sampled 63 DAS (Figure 5b), since most pronounced shifts in the DGGE fingerprints were observed at this sampling time. Abundance of ITS fragments in artificial soils was in the range of the natural soil (Luv) in all treatments (3.2×10^7 to 2.1×10^8), except for soil QIF which showed significantly lower numbers (8.4×10^6 to 3×10^7). In all artificial soils, ITS copy numbers increased by spiking treatments compared to the control. The abundance of ITS fragments in the natural soil (Luv) was relatively stable over all treatments. Spiking (p<0.001), soil (p<0.001), and the interaction of both factors (p = 0.01) did significantly influence the ITS copy numbers.

7. Phenanthrene concentrations in spiked soils

Phenanthrene concentrations in spiked soils sampled on the day of spiking and 21 DAS are presented in Table 3. In unspiked control treatments, negligible amounts of phenanthrene were detected (<0.01 mg g^{-1} of soil; data not shown). Except for soil QM+P, slightly less phenanthrene than the spiked amount of 2 mg g^{-1} was recovered from all soils without litter amendment sampled on the day of spiking. In two replicates of soil QM+P, instead, a considerably lower concentration of phenanthrene (0.71 and 0.66 mg g^{-1}) was detected. Also in all +L+P soils sampled on the day of spiking, significantly lower concentrations were recovered compared to soils without litter (p<0.001). Twenty-one DAS, the phenanthrene concentrations in soils without the litter amendment decreased except for the two replicates of soil QM+P and one replicate of the natural soil (Luv+P) which showed higher amounts compared to the day of spiking. The lowest phenanthrene concentrations 21 DAS were found in QI+P and QIF+P soils. However, no significant differences were found between soils. The amount of phenanthrene in +L+P soils sampled 21 DAS was slightly lower compared to the concentrations measured on the day of spiking in +L+P soils. Soils containing illite had also the lowest amount of phenanthrene in +L+P treatment 21 DAS.

Discussion

1. Artificial soils exhibited soil composition-driven microbial communities

Amplicon-based fingerprinting analysis of bacteria and fungi in unspiked control artificial soils confirmed that the structure of the microbial communities was driven by soil composition. In agreement with Vogel et al. [25], who described the microbiota establishment in artificial soils incubated up to 842 days, we observed that clay minerals strongly influenced bacterial communities. Charcoal and ferrihydrite shaped bacterial communities to a lesser extent. However, Vogel et al. [25] did not find an effect of soil components on fungi in long-term incubated artificial soils as

a

b

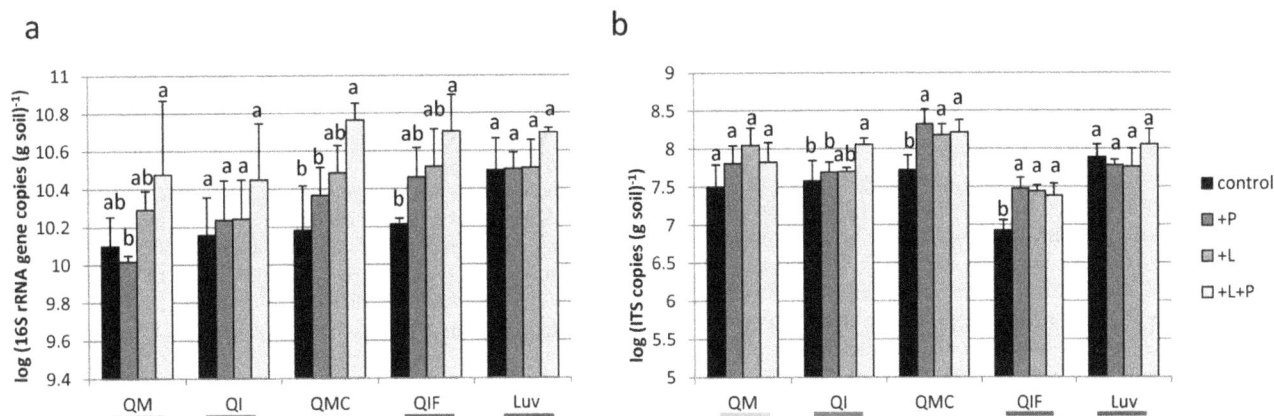

Figure 5. Quantification of 16S rRNA gene and ITS fragments in spiked soils. The qPCR analysis of (a) bacterial 16S rRNA gene copy numbers in samples taken 21 days after spiking (DAS) and (b) of fungal ITS fragment copy numbers in samples taken 63 DAS in the following treatments: unspiked (control), phenanthrene (+P), litter (+L), litter and phenanthrene (+L+P). Different letters within one soil composition mark a significant difference between treatments (Tukey test, $p < 0.05$). Bars indicate standard deviation of four replicates. Q-quartz, M-montmorillonite, C-charcoal, I-illite, F-ferrihydrite, Luv-natural soil (Luvisol).

observed in the present study. It is likely that the preparation of soils for the spiking treatments destroyed existing hyphae and aggregates so that soil components could re-induce their effects on fungi. In conclusion, microbial communities differed significantly among soil compositions incubated for 842 days.

2. Spiking changed bacterial communities in artificial soils

The litter addition resulted in rapid changes in the bacterial community composition in artificial soils as indicated by DGGE. The control fingerprint of 16S rRNA genes amplified from litter TC-DNA showed no similarities with bacteria in artificial soils

(Figures S2–S4). We therefore suggest that observed responders were not introduced by the litter amendment, but were due to a proliferation of artificial soil bacteria in response to the litter addition. *Stenotrophomonas maltophilia* and *Arthrobacter* were identified as potential litter responders that thrived equally well in all litter-amended artificial soils. Besides being a typical soil bacterium, *Arthrobacter* species are known for their tolerance to extreme conditions and their broad nutrient spectrum [37]. The trophic versatility of the *Stenotrophomonas* genus ranges from pathogenicity to humans, plant growth promoting and biocontrol properties, and the production of bioactive substances, to the ability to degrade a wide range of organic compounds, including

Table 3. Phenanthrene concentrations in mg g^{-1} soil determined in three replicates per soil composition in the phenanthrene (+P), litter and phenanthrene (+L+P) treatments on the day of spiking and 21 days after spiking (DAS).

P [mg g⁻¹]	+P		+L+P	
	day of spiking	21 DAS	day of spiking	21 DAS
QM	0.71	2.40	0.74	0.40
	1.76	1.08	0.49	0.68
	0.66	1.15	0.88	0.64
QMC	1.71	1.41	1.02	0.56
	1.63	1.53	0.75	0.54
	1.74	1.51	0.87	0.66
QI	1.36	0.89	0.92	0.26
	1.86	0.46	0.76	0.22
	1.78	0.81	0.66	0.30
QIF	1.14	0.33	0.85	0.27
	1.80	0.87	0.78	0.51
	1.74	1.33	0.60	0.45
Luv	1.45	0.77	0.56	0.73
	1.69	1.22	0.78	0.68
	1.68	1.72	0.90	0.77

Q-quartz, M-montmorillonite, C-charcoal, I-Illite, F-ferrihydrite, Luv-natural soil (Luvisol).

PAHs [38]. The latter fact might also have favored their establishment in +L+P treated soils. Furthermore, the litter addition caused a strong increase of bacterial 16S rRNA gene copy numbers in all soils (except QI+L). It is likely that only recalcitrant, polymerized organics remained of the two sterile manure inputs that artificial soils received during more than two years of maturation and that phenanthrene and litter thus represented easily available nutrient sources for the microbiota. The degradation of phenanthrene is supported by the lower phenanthrene concentrations measured, and the increase in 16S rRNA gene copy numbers found in all +P soils except QM+P (see section 4 for further discussion). Furthermore, phenanthrene resulted in several shifts in the bacterial community composition of samples taken 21 DAS and 63 DAS compared to the day of spiking. The high similarity of the sequenced DGGE band to *Arthrobacter crystallopoietes* provided further evidence for the proliferation of species potentially capable of degrading aromatic substances. *Arthrobacter crystallopoietes* has been reported to degrade aromatic compounds in soil [39,40] and to survive under long-term starvation conditions [41], which is consistent with our assumption of nutrient-limiting conditions in artificial soils [25]. Group-specific DGGE analyses suggested that *Actinobacteria* responded most to phenanthrene in all artificial soils which is in accordance with the phenanthrene response in one-year matured artificial soils [24]. However, in contrast to Babin et al. [24], we could not detect an increase in the dioxygenase gene abundance in +P and +L+P treated long-term matured artificial soils (data not shown). To conclude, spiking of long-term matured artificial soils with phenanthrene and litter resulted in significant shifts in the bacterial community structure.

3. Litter addition did not stimulate the response of bacterial communities to phenanthrene in artificial soils

We observed that the +L+P spiking to artificial soils resulted in clearly increased 16S rRNA gene copy numbers indicating that bacterial activity was triggered by this treatment. Several studies have demonstrated positive effects of compost, fertilizer, manure, and poultry litter addition on the degradation of contaminants in polluted soils [42,43,44,45]. We observed that bacterial community composition of artificial soils was significantly influenced by the +L+P spiking but to a lesser extent than by the +P spiking. Some litter-induced phenanthrene responders were observed, e.g. *Betaproteobacteria*, but, unfortunately, firm conclusions on the stimulatory effect of litter on degradation of phenanthrene cannot be drawn due to the low recovery of phenanthrene in presence of litter. In contrast to other studies, we added phenanthrene and litter simultaneously and the ground litter likely offered a high sorption surface. PAHs were previously shown to have a high affinity to soil OM [11]. We assume that phenanthrene was likely less accessible to bacteria in the +L+P than in the +P treatment. This assumption is supported by the lack of enhanced abundance of IncP-9 plasmids in +L+P spiked artificial soils (data not shown), even though nutrient sources in soils are generally considered to be microbial hotspots conducive to horizontal gene transfer events. Furthermore, we observed major shifts in *Pseudomonas* and *Alphaproteobacteria* fingerprints due to litter. It is possible that copiotrophic bacteria in artificial soils outcompeted phenanthrene degraders in +L+P treatment [46]. Hence, to get more insights into the influence of the +L+P spiking, further analyses of the microbial community changes (e.g. by pyrosequencing) and the metabolic activity are needed.

4. Soil composition controlled bacterial response to spiking

The spiking of +P, +L, +L+P to long-term matured artificial soils enabled us to study the response of the microbiota in a soil system established as a function of the soil composition. We observed similar and completely soil composition-specific responses to phenanthrene spiking confirming the results of our one year artificial soil study [24]. Similar responses among different bacterial communities indicate that comparable conditions in different artificial soils allowed the establishment of common taxa. However, it was more striking that phenanthrene even strengthened the difference between the microbiota of artificial soils, especially for soils containing different clay minerals. Correspondingly, other authors [12,13,26] have reported that contamination did not result in converging of microbial communities in different soil types. The different responses observed in the present study can be caused by the soil-specific initial microbial communities present in artificial soils before spiking treatments allowing or preventing the establishment of certain populations. These different initial microbial communities are caused by an establishment in an environment solely driven by the soil composition. That is why we conclude that the soil composition controlled the bacterial response to spiking. However, further analyses are needed to determine whether and to what extent indirect effects, i.e. the different interfaces established during incubation of artificial soils or a complex interaction between all soil components might have contributed to the observed differences.

Our results consistently showed that bacteria in QM soils responded less to phenanthrene compared to soils containing illite (QI, QIF). In a previous study of these artificial soils it was shown that the organic carbon concentration in QM soils was slightly higher compared to other soils [25] which might have resulted in sequestration of phenanthrene to the soil OM and its reduced bioavailability [5,9]. Furthermore, Vogel et al. [25] reported previously that QM and QMC soils incubated over a long-term period contained more aggregates than soils with illite. It is known that the biodegradation of PAHs is restricted in aggregates [2,47,48,49]. Montmorillonite is an expandable clay mineral with a high specific surface area which makes possible that phenanthrene was separated from bacterial cells by surface sorption or entrapment in small pores [10,50]. A lower degradation rate of phenanthrene in the presence of expandable clays has been shown in previous studies [51,52]. However, clay minerals also have previously been reported to have a positive influence on the biodegradation efficiency, e.g. by enabling the degradation of the bound pollutant by adhesion of microbial cells [51,53], facilitating transformation of chemical compounds [54], and stimulating microbial growth [55]. Our data suggest that soils containing illite might favor microbial degradation in this respect compared to montmorillonite. Unfortunately, this assumption cannot be proven, as the phenanthrene concentrations differed strongly among QM replicates. It is possible that mixing was not sufficient to prevent the formation of phenanthrene hotspots.

Besides clay minerals, we observed an effect of the charcoal component on the microbial response to spiking. Bacterial communities in QMC soils showed a delayed response to phenanthrene. Charcoal and soil OM in general are known to be the principal sorption sites for PAHs [11]. Along with the slightly higher phenanthrene concentrations in QMC compared to QI and QIF soils sampled 21 DAS, it can be assumed that only a slow desorption of phenanthrene from charcoal took place, which would be only degraded over a longer term than the incubation time with spikes used in this study [10].

5. Combined spiking of phenanthrene and litter increased effects on fungal communities

Fungal communities were almost insusceptible to the +P or +L spiking, whereas the combined +L+P spiking strongly changed the fungal community composition. Ligninolytic fungi are known to express enzymes with low substrate specificity that also makes them capable of degrading PAHs [7,56]. We suggest that some of the responsive fungi observed in artificial soils were favored by co-metabolism of phenanthrene, after enzyme expression was induced under nutrient-rich conditions as previously shown for several fungal strains [57,58].

6. Natural soil microbiota less responsive to spiking

Artificial soils incubated for more than two years were previously shown to develop to soil-like systems [25]. In the present study, we subjected these matured artificial soils and a natural soil to similar spiking treatments in order to compare the response of the microbiota. Since we were aware of several missing processes when incubating soils under laboratory conditions (e.g. freezing, thawing, transport, OM input), we were not surprised to observe a number of differences between the microbial responses. For instance, the lower amount of OM in artificial soils compared to the natural soil might have likely been an important factor resulting in higher bioavailability of phenanthrene. Furthermore, we suggest that the less pronounced response of natural soil microbial communities to spiking compared to artificial soils is caused by the higher diversity in the natural soil exhibiting a greater potential to compensate perturbations [59,60].

Conclusion

In the present study, we showed that soil composition and presence of charcoal matter by influencing the composition and the response of microbial communities to spiked phenanthrene and litter. We suggest that the soil mineral composition affects the formation of biogeochemical interfaces and of a soil-specific microbiota that control the response to introduced organic substrates. The artificial soil approach may therefore represent a valuable basis for understanding and integrating additional parameters to unravel the complex soil interaction network, and to improve current bioremediation strategies. However, further research is needed to disentangle the single role of the microbiota and the physical soil environment for the fate of pollutants.

Supporting Information

Figure S1   Response of bacterial communities to spiking in QM and QMC soils 7 days after spiking. DGGE fingerprints of bacterial communities in spiked QM and QMC soils sampled 7 days after spiking (control, phenanthrene (+P), litter (+L), litter and phenanthrene [+L+P]). Black arrows mark populations responding to litter. BS-bacterial DGGE standard. Q-quartz, M-montmorillonite, C-charcoal.

Figure S2   Response of bacterial communities to spiking in QI and QIF soils 7 days after spiking. DGGE fingerprints of bacterial communities in spiked QI and QIF soils sampled 7 days after spiking (control, phenanthrene (+P), litter (+ L), litter and phenanthrene [+L+P]). Black arrows mark populations responding to litter. BS-bacterial DGGE standard. Litter ctr-control fingerprint of litter used for amendments. Q-quartz, I-illite, F-ferrihydrite.

Figure S3   Response of bacterial communities to spiking in QI and QIF soils 21 days after spiking. DGGE fingerprints of bacterial communities in spiked QI and QIF soils sampled 21 days after spiking (control, phenanthrene (+P), litter (+ L), litter and phenanthrene [+L+P]). Arrows mark populations responding to litter (black), phenanthrene (white), litter and phenanthrene (grey). BS-bacterial DGGE standard. Ino-Luvisol inoculant added to soil mixtures at the beginning of incubation. QI/QIF ctr-bacterial community in QI or QIF, respectively, before spiking. Litter ctr-control fingerprint of litter used for amendments. Q-quartz, I-illite, F-ferrihydrite.

Figure S4   Response of bacterial communities in QM and QMC soils 63 days after spiking. DGGE fingerprints of bacterial communities in spiked QM and QMC soils sampled 63 days after spiking (control, phenanthrene (+P), litter (+L), litter and phenanthrene [+L+P]). Arrows mark populations responding to litter (black), phenanthrene (white), litter and phenanthrene (grey). BS-bacterial DGGE standard. QM/QMC ctr-bacterial community in QM or QMC, respectively, before spiking. Litter ctr-control fingerprint of litter used for amendments. Q-quartz, M-montmorillonite, C-charcoal.

Figure S5   Dendrograms of *Alphaproteobacteria* and *Pseudomonas* communities in artificial soils. UPGMA cluster analysis of DGGE fingerprints of taxon-specific 16S rRNA gene amplicons from artificial soils a) QM and QMC b) QI and QIF sampled 21 days after spiking (control, phenanthrene (+P), litter (+L), litter and phenanthrene [+L+P]). Q-quartz, M-montmorillonite, C-charcoal, I-illite, F-ferrihydrite.

Figure S6   Response of *Actinobacteria* in QM and QMC soils to spiking. DGGE fingerprints of actinobacterial communities in spiked QM and QMC soils sampled 21 days after spiking (control, phenanthrene (+P), litter (+L), litter and phenanthrene [+ L+P]). Arrows mark populations responding to litter (black), phenanthrene (white). BS-bacterial DGGE standard. Ino-Luvisol inoculant added to soil mixtures at the beginning of incubation. Q-quartz, M-montmorillonite, C-charcoal.

Figure S7   Response of fungal communities to spiking in QM and QMC soils. DGGE fingerprints of fungal communities in spiked QM and QMC soils sampled 63 days after spiking (control, phenanthrene (+P), litter (+L), litter and phenanthrene [+ L+P]). Arrows mark populations responding to litter (black), litter and phenanthrene (grey). FS-fungal DGGE standard. QM/QMC ctr-fungal community in QM or QMC, respectively, before spiking. Q-quartz, M-montmorillonite, C-charcoal.

Figure S8   Response of bacterial communities to spiking in the natural soil. DGGE fingerprints of bacterial communities in the spiked natural Luvisol soil (Luv) sampled 63 days after spiking (control, phenanthrene (+P), litter (+L), litter and phenanthrene [+L+P]). Arrows mark populations responding to litter (black), phenanthrene (white). BS-bacterial DGGE standard. Luv ctr-bacterial community in the natural soil before spiking.

Table S1   Primers and probe used in the study.

Table S2   Percent difference (d-values) between bacterial communities of different artificial soils per treat-

ment (control, phenanthrene (+P), litter (+L), litter and phenanthrene [+L+P]) 21 days after spiking. D-values were calculated based on pairwise Pearson similarity coefficients and were significant for all comparisons ($p < 0.05$). Q-quartz, M-montmorillonite, C-charcoal, I-illite, F-ferrihydrite.

Acknowledgments

The authors thank Ute Zimmerling for technical assistance and the reviewers for their valuable comments.

Author Contributions

Conceived and designed the experiments: M. Schloter GJP KH M. Spiteller IKK KS. Performed the experiments: DB CV. Analyzed the data: DB CV SZ. Contributed to the writing of the manuscript: DB KS.

References

1. Hatzinger PB, Alexander M (1995) Effect of aging of chemicals in soil on their biodegradability and extractability. Environ Sci Technol 29: 537–545.
2. Amellal N, Portal JM, Vogel T, Berthelin J (2001) Distribution and location of polycyclic aromatic hydrocarbons (PAHs) and PAH-degrading bacteria within polluted soil aggregates. Biodegradation 12: 49–57.
3. Alexander M (1999) Biodegradation and Bioremediation. San Diego: Academic Press. 453 p.
4. Müller S, Totsche KU, Kögel-Knabner I (2007) Sorption of polycyclic aromatic hydrocarbons to mineral surfaces. Eur J Soil Sci 58: 918–931.
5. Weissenfels WD, Klewer HJ, Langhoff J (1992) Adsorption of polycyclic aromatic hydrocarbons (PAHs) by soil particles: influence on biodegradability and biotoxicity. Appl Microbiol Biotechnol 36: 689–696.
6. Zhang XX, Cheng SP, Zhu CJ, Sun SL (2006) Microbial PAH-degradation in soil: Degradation pathways and contributing factors. Pedosphere 16: 555–565.
7. Haritash AK, Kaushik CP (2009) Biodegradation aspects of Polycyclic Aromatic Hydrocarbons (PAHs): A review. J Hazard Mater 169: 1–15.
8. Jeon CO, Madsen EL (2013) In situ microbial metabolism of aromatic-hydrocarbon environmental pollutants. Curr Opin Biotech 24: 474–481.
9. Nam K, Chung N, Alexander M (1998) Relationship between organic matter content of soil and the sequestration of phenanthrene. Environ Sci Technol 32: 3785–3788.
10. Lahlou M, Ortega-Calvo JJ (1999) Bioavailability of labile and desorption-resistant phenanthrene sorbed to montmorillonite clay containing humic fractions. Environ Toxicol Chem 18: 2729–2735.
11. Cornelissen G, Gustafsson Ö, Bucheli TD, Jonker MTO, Koelmans AA, et al. (2005) Extensive sorption of organic compounds to black carbon, coal, and kerogen in sediments and soils: Mechanisms and consequences for distribution, bioaccumulation, and biodegradation. Environ Sci Technol 39: 6881–6895.
12. Bundy JG, Paton GI, Campbell CD (2002) Microbial communities in different soil types do not converge after diesel contamination. J Appl Microbiol 92: 276–288.
13. Ding GC, Heuer H, Smalla K (2012) Dynamics of bacterial communities in two unpolluted soils after spiking with phenanthrene: soil type specific and common responders. Front Microbiol 3: 290.
14. Holden PA, Firestone MK (1997) Soil microorganisms in soil cleanup: How can we improve our understanding? J Environ Qual 26: 32–40.
15. Totsche KU, Rennert T, Gerzabek MH, Kögel-Knabner I, Smalla K, et al. (2010) Biogeochemical interfaces in soil: The interdisciplinary challenge for soil science. J Plant Nutr Soil Sci 173: 88–99.
16. Stotzky G (1986) Influence of soil mineral colloids on metabolic processes, growth, adhesion and ecology of microbes and viruses. In: Huang PM, Schmitzer M, editors.Interactions of soil minerals with natural organics and microbes: Special Publication No 17.Madison, WI: Soil Science Society of America. pp. 305–428.
17. Chenu C, Stotzky G (2002) Interactions between Microorganisms and Soil Particles: An Overview. In: Huang PM, Bollag JM, Senesi N, editors.Interac-Interactions between Soil Particles and Microorganisms. IUPAC series of Applied Chemistry ed.West Sussex: John Wiley & Sons. pp. 3–40.
18. Carson JK, Campbell L, Rooney D, Clipson N, Gleeson DB (2009) Minerals in soil select distinct bacterial communities in their microhabitats. Fems Microbiol Ecol 67: 381–388.
19. Ding GC, Pronk GJ, Babin D, Heuer H, Heister K, et al. (2013) Mineral composition and charcoal determine the bacterial community structure in artificial soils. Fems Microbiol Ecol 86: 15–25.
20. Madhok MR (1937) Synthetic Soil As A Medium for the Study of Certain Microbiological Processes. Soil Sci 44: 319–322.
21. Guenet B, Leloup J, Hartmann C, Barot S, Abbadie L (2011) A new protocol for an artificial soil to analyse soil microbiological processes. Appl Soil Ecol 48: 243–246.
22. Pronk GJ, Heister K, Ding G-C, Smalla K, Kögel-Knabner I (2012) Development of biogeochemical interfaces in an artificial soil incubation experiment; aggregation and formation of organo-mineral associations. Geoderma 189–190: 585–594.
23. OECD (1984) Earthworm, Acute Toxicity Tests. In: OECD, editor. 207. Guidelines for the Testing of Chemicals ed: OECD Publishing. pp. 9.
24. Babin D, Ding GC, Pronk GJ, Heister K, Kögel-Knabner I, et al. (2013) Metal oxides, clay minerals and charcoal determine the composition of microbial communities in matured artificial soils and their response to phenanthrene. Fems Microbiol Ecol 86: 3–14.
25. Vogel C, Babin D, Pronk GJ, Heister K, Smalla K, et al. (2014) Establishment of macro-aggregates and organic matter turnover by microbial communities in long-term incubated artificial soils. Soil Biol Biochem 10.1016/j.soilbio.2014.07.012. In press.
26. Ding GC, Heuer H, Zühlke S, Spiteller M, Pronk GJ, et al. (2010) Soil type-dependent responses to phenanthrene as revealed by determining the diversity and abundance of polycyclic aromatic hydrocarbon ring-hydroxylating dioxygenase genes by using a novel PCR detection system. Appl Environ Microbiol 76: 4765–4771.
27. Gomes NCM, Heuer H, Schönfeld J, Costa R, Mendonça-Hagler L, et al. (2001) Bacterial diversity of the rhizosphere of maize (*Zea mays*) grown in tropical soil studied by temperature gradient gel electrophoresis. Plant Soil 232: 167–180.
28. Heuer H, Krsek M, Baker P, Smalla K, Wellington EM (1997) Analysis of actinomycete communities by specific amplification of genes encoding 16S rRNA and gel-electrophoretic separation in denaturing gradients. Appl Environ Microbiol 63: 3233–3241.
29. Milling A, Smalla K, Maidl FX, Schloter M, Munch JC (2004) Effects of transgenic potatoes with an altered starch composition on the diversity of soil and rhizosphere bacteria and fungi. Plant Soil 266: 23–39.
30. Weinert N, Meincke R, Gottwald C, Heuer H, Gomes NC, et al. (2009) Rhizosphere communities of genetically modified zeaxanthin-accumulating potato plants and their parent cultivar differ less than those of different potato cultivars. Appl Environ Microbiol 75: 3859–3865.
31. Heuer H, Wieland G, Schönfeld J, Schönwälder A, Gomes NCM, et al. (2001) Bacterial community profiling using DGGE or TGGE analysis. In: Rochelle PA, editor.Environmental molecular microbiology: Protocols and applications.New York: Horizon Press Inc. pp. 177–190.
32. Smalla K, Wieland G, Buchner A, Zock A, Parzy J, et al. (2001) Bulk and rhizosphere soil bacterial communities studied by denaturing gradient gel electrophoresis: plant-dependent enrichment and seasonal shifts revealed. Appl Environ Microbiol 67: 4742–4751.
33. Suzuki MT, Taylor LT, DeLong EF (2000) Quantitative analysis of small-subunit rRNA genes in mixed microbial populations via 5′-nuclease assays. Appl Environ Microbiol 66: 4605–4614.
34. Gschwendtner S, Reichmann M, Müller M, Radl V, Munch JC, et al. (2010) Effects of genetically modified amylopectin-accumulating potato plants on the abundance of beneficial and pathogenic microorganisms in the rhizosphere. Plant Soil 335: 413–422.
35. Baran S, Oleszczuk P (2002) Chromatographic determination of polycyclic aromatic hydrocarbons (PAH) in sewage sludge, soil, and sewage sludge-amended soils. Pol J Environ Stud 11: 609–615.
36. Kropf S, Heuer H, Grüning M, Smalla K (2004) Significance test for comparing complex microbial community fingerprints using pairwise similarity measures. J Microbiol Methods 57: 187–195.
37. Jones D, Keddie R (2006) The Genus *Arthrobacter*. In: Dworkin M, Falkow S, Rosenberg E, Schleifer K-H, Stackebrandt E, editors.The Prokaryotes: Springer New York. pp. 945–960.
38. Ryan RP, Monchy S, Cardinale M, Taghavi S, Crossman L, et al. (2009) The versatility and adaptation of bacteria from the genus *Stenotrophomonas*. Nat Rev Microbiol 7: 514–525.
39. Hwang S, Tate RL (1997) Interactions of clay minerals with *Arthrobacter crystallopoietes*: Starvation, survival and 2-hydroxypyridine catabolism. Biol Fert Soils 24: 335–340.
40. Hwang S, Tate RL (1997) Humic acid effects on 2-hydroxypyridine metabolism by starving *Arthrobacter crystallopoietes* cells. Biol Fert Soils 25: 36–40.
41. Ensign JC (1970) Long-term starvation survival of rod and spherical cells of *Arthrobacter crystallopoietes*. J Bacteriol 103: 569–577.
42. Wischmann H, Steinhart H (1997) The formation of PAH oxidation products in soils and soil/compost mixtures. Chemosphere 35: 1681–1698.
43. Lindstrom JE, Prince RC, Clark JC, Grossman MJ, Yeager TR, et al. (1991) Microbial populations and hydrocarbon biodegradation potentials in fertilized shoreline sediments affected by the T/V Exxon Valdez oil spill. Appl Environ Microbiol 57: 2514–2522.
44. Wong JWC, Wan CK, Fang M (2002) Pig manure as a co-composting material for biodegradation of PAH-contaminated soil. Environ Technol 23: 15–26.
45. Williams CM, Grimes JL, Mikkelsen RL (1999) The use of poultry litter as co-substrate and source of inorganic nutrients and microorganisms for the ex situ biodegradation of petroleum compounds. Poult Sci 78: 956–964.

46. Fierer N, Bradford MA, Jackson RB (2007) Toward an ecological classification of soil bacteria. Ecology 88: 1354–1364.

47. Chang W, Akbari A, Snelgrove J, Frigon D, Ghoshal S (2013) Biodegradation of petroleum hydrocarbons in contaminated clayey soils from a sub-arctic site: the role of aggregate size and microstructure. Chemosphere 91: 1620–1626.

48. Nocentini M, Pinelli D (2001) Biodegradation of PAHs in aggregates of a low permeability soil. Soil Sediment Contam 10: 211–226.

49. Nam K, Kim JY, Oh DI (2003) Effect of soil aggregation on the biodegradation of phenanthrene aged in soil. Environ Pollut 121: 147–151.

50. Nam K, Alexander M (1998) Role of nanoporosity and hydrophobicity in sequestration and bioavailability: Tests with model solids. Environ Sci Technol 32: 71–74.

51. Hwang S, Cutright TJ (2003) Effect of expandable clays and cometabolism on PAH biodegradability. Environ Sci Pollut Res Int 10: 277–280.

52. Theng BKG, Aislabie J, Fraser R (2001) Bioavailability of phenanthrene intercalated into an alkylammonium-montmorillonite clay. Soil Biol Biochem 33: 845–848.

53. Warr LN, Perdrial JN, Lett MC, Heinrich-Salmeron A, Khodja M (2009) Clay mineral-enhanced bioremediation of marine oil pollution. Appl Clay Sci 46: 337–345.

54. Stotzky G, Rem LT (1966) Influence of Clay Minerals on Microorganisms. I. Montmorillonite and Kaolinite on Bacteria. Can J Microbiol 12: 547–563.

55. Chaerun SK, Tazaki K, Asada R, Kogure K (2005) Interaction between clay minerals and hydrocarbon-utilizing indigenous microorganisms in high concentrations of heavy oil: implications for bioremediation. Clay Miner 40: 105–114.

56. Sack U, Günther T (1993) Metabolism of PAH by fungi and correlation with extracellular enzymatic activities. J Basic Microbiol 33: 269–277.

57. Yateem A, Balba MT, Al-Awadhi N, El-Nawawy AS (1998) White rot fungi and their role in remediating oil-contaminated soil. Environ Int 24: 181–187.

58. Kaal EEJ, Field JA, Joyce TW (1995) Increasing Ligninolytic Enzyme-Activities in Several White-Rot Basidiomycetes by Nitrogen-Sufficient Media. Bioresource Technol 53: 133–139.

59. Griffiths BS, Ritz K, Bardgett RD, Cook R, Christensen S, et al. (2000) Ecosystem response of pasture soil communities to fumigation-induced microbial diversity reductions: an examination of the biodiversity-ecosystem function relationship. Oikos 90: 279–294.

60. van Bruggen AHC, Semenov AM (2000) In search of biological indicators for soil health and disease suppression. Appl Soil Ecol 15: 13–24.

Permissions

The contributors of this book come from diverse backgrounds, making this book a truly international effort. This book will bring forth new frontiers with its revolutionizing research information and detailed analysis of the nascent developments around the world.

We would like to thank all the contributing authors for lending their expertise to make the book truly unique. They have played a crucial role in the development of this book. Without their invaluable contributions this book wouldn't have been possible. They have made vital efforts to compile up to date information on the varied aspects of this subject to make this book a valuable addition to the collection of many professionals and students.

This book was conceptualized with the vision of imparting up-to-date information and advanced data in this field. To ensure the same, a matchless editorial board was set up. Every individual on the board went through rigorous rounds of assessment to prove their worth. After which they invested a large part of their time researching and compiling the most relevant data for our readers.

The editorial board has been involved in producing this book since its inception. They have spent rigorous hours researching and exploring the diverse topics which have resulted in the successful publishing of this book. They have passed on their knowledge of decades through this book. To expedite this challenging task, the publisher supported the team at every step. A small team of assistant editors was also appointed to further simplify the editing procedure and attain best results for the readers.

Apart from the editorial board, the designing team has also invested a significant amount of their time in understanding the subject and creating the most relevant covers. They scrutinized every image to scout for the most suitable representation of the subject and create an appropriate cover for the book.

The publishing team has been an ardent support to the editorial, designing and production team. Their endless efforts to recruit the best for this project, has resulted in the accomplishment of this book. They are a veteran in the field of academics and their pool of knowledge is as vast as their experience in printing. Their expertise and guidance has proved useful at every step. Their uncompromising quality standards have made this book an exceptional effort. Their encouragement from time to time has been an inspiration for everyone.

The publisher and the editorial board hope that this book will prove to be a valuable piece of knowledge for researchers, students, practitioners and scholars across the globe.

List of Contributors

Hui-Hsien Pan, Hai-Lun Sun, Min-Sho Ku, Ji-Nan Sheu and Ko-Huang Lue
Department of Pediatrics, Chung Shan Medical University Hospital, Taichung City, Taiwan R.O.C
School of Medicine, Chung Shan Medical University, Taichung City, Taiwan R.O.C

Pei-Fen Liao
Department of Pediatrics, Chung Shan Medical University Hospital, Taichung City, Taiwan R.O.C

Ko-Hsiu Lu
School of Medicine, Chung Shan Medical University, Taichung City, Taiwan R.O.C

Chun-Tzu Chen
Department of Pediatrics, Chung Shan Medical University Hospital, Taichung City, Taiwan R.O.C
Department of Health Policy and Management, Chung Shan Medical University, Taichung City, Taiwan R.O.C

Jar-Yuan Pai
Department of Health Policy and Management, Chung Shan Medical University, Taichung City, Taiwan R.O.C

Jing-Yang Huang
Institute of Public Health, Department of Public Health, Chung Shan Medical University, Taichung City, Taiwan R.O.C

Yuling Han., Lili Shi., Jing Meng, Hongbo Yu and Xiaoyu Zhang
College of Life Science and Technology, Huazhong University of Science and Technology, Wuhan, China

Ling Tong
School of Environmental Science and Engineering, Tianjin University, Tianjin, China
Key Laboratory of Pollution Process and Environmental Criteria (Ministry of Education), College of Environmental Science and Engineering, Nankai University, Tianjin, China

Kai Li
Department of Industrial Engineering, Nankai University, Tianjin, China

Qixing Zhou
Key Laboratory of Pollution Process and Environmental Criteria (Ministry of Education), College of Environmental Science and Engineering, Nankai University, Tianjin, China

Trong-Neng Wu and Chiu-Ying Chen
Department of Public Health, China Medical University, Taichung, Taiwan

Kuang-Hsi Chang
Department of Public Health, China Medical University, Taichung, Taiwan
Department of Medical Research, Taichung Veterans General Hospital, Taichung, Taiwan

Mei-Yin Chang
Department of Medical Laboratory Science and Biotechnology, School of Medical and Health Sciences, Fooyin University, Kaohsiung, Taiwan

Chih-Hsin Muo
Management Office for Health Data, China Medical University Hospital, Taichung, Taiwan

Chia-Hung Kao
Graduate Institute of Clinical Medical Science, College of Medicine, China Medical University, Taiwan
Department of Nuclear Medicine and PET Center, China Medical University Hospital, Taichung, Taiwan

Célina Roda, Ioannis Nicolis and Chantal Guihenneuc
Laboratoire Santé Publique et Environnement, EA 4064, Faculté de Pharmacie, Université Paris Descartes, Sorbonne Paris Cité, Paris, France

Isabelle Momas
Laboratoire Santé Publique et Environnement, EA 4064, Faculté de Pharmacie, Université Paris Descartes, Sorbonne Paris Cite´, Paris, France
Mairie de Paris, Direction de l9Action Sociale de l9Enfance et de la Santé, Cellule Cohorte, Paris, France

Hamed Layeghkhavidaki, Marie-Claire Lanhers, Samina Akbar, Lynn Gregory-Pauron, Thierry Oster, Cyril Feidt and Catherine Corbier
Unité de Recherche Animal et Fonctionnalités des Produits Animaux EA3998, Université de Lorraine, Vandoeuvre-lés-Nancy, France Institut National de Recherche Agronomique USC 0340, Vandoeuvre-lés-Nancy, France

Nathalie Grova and Brice Appenzeller
Laboratory of Analytical Human Biomonitoring, Centre de Recherche Public de la Santé, Luxembourg, Luxembourg

Jordane Jasniewski
Laboratoire d'Ingenérie des Biomolécules, Université de Lorraine, Vandoeuvre-lés-Nancy, France

Frances T. Yen
Unité de Recherche Animal et Fonctionnalités des Produits Animaux EA3998, Université de Lorraine Vandoeuvre-lés-Nancy, France
Institut National de Recherche Agronomique USC 0340, Vandoeuvre-lés-Nancy, France
Institut National de la Santé et de la Recherche Médicale, Vandoeuvre-lés-Nancy, France

Annette M. Hormann, Frederick S. vom Saal, Richard W. Stahlhut, Carol L. Moyer and Julia A. Taylor
Division of Biological Sciences, University of Missouri, Columbia, Missouri, United States of America

Susan C. Nagel
Department of Obstetrics, Gynecology and Women's Health, University of Missouri, Columbia, Missouri, United States of America

Mark R. Ellersieck
Department of Statistics, University of Missouri, Columbia, Missouri, United States of America

Wade V. Welshons
Department of Biomedical Sciences, University of Missouri, Columbia, Missouri, United States of America

Pierre-Louis Toutain
Université de Toulouse, INPT, ENVT, UPS, UMR1331, F- 31062 Toulouse, France
INRA, UMR1331, Toxalim, Research Centre in Food Toxicology, F-31027 Toulouse, France

Yunpeng Luo, Qiúan Zhu, Gang Yang, Yanzheng Yang and Yao Zhang
State Key Laboratory of Soil Erosion and Dryland Farming on the Loess Plateau, College of Forestry, Northwest A&F University, Yangling, Shaanxi, China

Huai Chen
State Key Laboratory of Soil Erosion and Dryland Farming on the Loess Plateau, College of Forestry, Northwest A&F University, Yangling, Shaanxi, China
Chengdu Institute of Biology, Chinese Academy of Sciences, Chengdu, China

Changhui Peng
State Key Laboratory of Soil Erosion and Dryland Farming on the Loess Plateau, College of Forestry, Northwest A&F University, Yangling, Shaanxi, China
Center of CEF/ESCER, Department of Biology Science, University of Quebec at Montreal, Montreal, Canada

Ina Ehlers, Tatiana R. Betson and Jürgen Schleucher
Department of Medical Biochemistry & Biophysics, Umeå University, Umeå, Sweden

Walter Vetter
Department of Food Chemistry, University of Hohenheim, Stuttgart, Germany

Zeyu Zhou, Hongtao Wang, Tan Chen and Wenjing Lu
Department of Environmental Science and Engineering, Tsinghua University, Beijing, P.R. China

Yaxin Zhang
College of Environmental Science and Engineering, Hunan University, Hunan, P.R. China

Elisabetta Ceretti, Donatella Feretti, Gaia C. V. Viola, Ilaria Zerbini, Rosa M. Limina, Claudia Zani, Francesco Donato and Umberto Gelatti
Unit of Hygiene, Epidemiology and Public Health, Department of Medical and Surgical Specialities, Radiological Sciences and Public Health, University of Brescia, Brescia, Italy

Michela Capelli and Rossella Lamera
Post-Graduate School of Public Health, University of Brescia, Brescia, Italy

Kyoung-Nam Kim and Sanghyuk Bae
Department of Preventive Medicine, Seoul National University College of Medicine, Seoul, Korea

Jin Hee Kim
Institute of Environmental Medicine, Medical Research Center, Seoul, Korea

Yun-Chul Hong
Department of Preventive Medicine, Seoul National University College of Medicine, Seoul, Korea
Institute of Environmental Medicine, Medical Research Center, Seoul, Korea

Ho-Jang Kwon
Department of Preventive Medicine and Public Health, Dankook University College of Medicine, Cheonan, Korea

Soo-Jong Hong
Department of Pediatrics, Childhood Asthma Atopy Center, Research Center for Standardization of Allergic Diseases, Asan Medical Center, University of Ulsan College of Medicine, Seoul, Korea

Byoung-Ju Kim
Department of Pediatrics, Haeundae Paik Hospital, Inje University College of Medicine, Busan, Korea

So-Yeon Lee
Department of Pediatrics, Hallym University Sacred Heart Hospital, Hallym University College of Medicine, Anyang, Korea

Hui Zeng, Wei-qun Shu, Ji-an Chen, Da-hua Wang, Wen-juan Fu, Ling-qiao Wang, Jiao-hua Luo, Liang Zhang, Yao Tan, Zhi-qun Qiu and Yu-jing Huang
Department of Environmental Hygiene, College of Preventive Medicine, Third Military Medical University, Chongqing, P. R. China

Lin Liu
The Lundberg-Kienlen Lung Biology and Toxicology Laboratory, Department of Physiological Sciences, Oklahoma State University, Stillwater, Oklahoma, United States of America

Peter Morfeld
Institute for Occupational Epidemiology and Risk Assessment (IERA) of Evonik Industries, Essen, Germany
Institute and Policlinic for Occupational Medicine, Environmental Medicine and Preventive Research, University of Cologne, Cologne, Germany

Michael F. Spallek
Institute of Occupational Medicine, Social Medicine and Environmental Medicine, Goethe-University, Frankfurt am Main, Germany
European Research Group on Environment and Health in the Transport Sector (EUGT), Berlin, Germany

David A. Groneberg
Institute of Occupational Medicine, Social Medicine and Environmental Medicine, Goethe-University, Frankfurt am Main, Germany

Li Liu, Ling Chen, Lili Zhang and Lingling Wu
Key Laboratory of Yangtze Water environment, Ministry of Education, Tongji University, Shanghai, China

Ying Shao
Key Laboratory of Yangtze Water environment, Ministry of Education, Tongji University, Shanghai, China
Department of Ecosystem Analysis, Institute for Environmental Research (Biology V), Aachen Biology and Biotechnology, RWTH Aachen University, Aachen, Germany

Tilman Floehr, Hongxia Xiao, Yan Yan and Kathrin Eichbaum
Department of Ecosystem Analysis, Institute for Environmental Research (Biology V), Aachen Biology and Biotechnology, RWTH Aachen University, Aachen, Germany

Henner Hollert
Key Laboratory of Yangtze Water environment, Ministry of Education, Tongji University, Shanghai, China
Department of Ecosystem Analysis, Institute for Environmental Research (Biology V), Aachen Biology and Biotechnology, RWTH Aachen University, Aachen, Germany
College of Resources and Environmental Science, Chongqing University, Chongqing, China
School of Environment, Nanjing University, Nanjing, China

Camilla Pedersen, Ole Raaschou-Nielsen and Vanna Albieri
Danish Cancer Society Research Center, Copenhagen Ø, Denmark

Elvira V. Bräuner
Danish Cancer Society Research Center, Copenhagen Ø, Denmark
Danish Building Research Institute, Aalborg University, Construction and Health, Copenhagen SV, Denmark

Naja H. Rod
Social Medicine Section, Department of Public Health, University of Copenhagen, Copenhagen K, Denmark

Claus E. Andersen
Risø National Laboratory for Sustainable Energy, Radiation Research Division, Technical University of Denmark, Roskilde, Denmark

Kaare Ulbak
National Institute of Radiation Protection, Herlev, Denmark

Ole Hertel
Department of Environmental Science, Aarhus University, Roskilde, Denmark
Department for Environmental, Social and Spatial Change (ENSPAC), Roskilde University, Roskilde, Denmark

Christoffer Johansen
Danish Cancer Society Research Center, Copenhagen Ø, Denmark
Oncology Clinic, Finsen Centre, Rigshospitalet 5073, University of Copenhagen, Copenhagen Ø, Denmark

Joachim Schüz
International Agency for Research on Cancer (IARC), Section of Environment and Radiation, Lyon, France

Dominik Lermen, Martina Bartel-Steinbach and Hagen von Briesen
Department of Cell Biology & Applied Virology, Fraunhofer-Institute for Biomedical Engineering, St. Ingbert, Saarland, Germany

Daniel Schmitt
Department of Laboratory & Information Technology, Fraunhofer-Institute for Biomedical Engineering, St. Ingbert, Saarland, Germany

Christa Schröter-Kermani and Marike Kolossa-Gehring
Federal Environment Agency (UBA), Berlin, Berlin, Germany

Heiko Zimmermann
Department of Cell Biology & Applied Virology, Fraunhofer-Institute for Biomedical Engineering, St. Ingbert, Saarland, Germany
Department of Laboratory & Information Technology, Fraunhofer-Institute for Biomedical Engineering, St. Ingbert, Saarland, Germany
Saarland University, Saarbruecken, Saarland, Germany

Rui Zhou, Chanzhen Shi, Wenxin Wang, Heng Zhao, Rui Liu and Limin Chen
Shanghai Key Laboratory of Atmospheric Particle Pollution and Prevention (LAP3), Department of Environmental Science & Engineering, Fudan University, Shanghai 200433, China

Shanshan Wang
School of Environment and Architecture, University of Shanghai for Science and Technology, Shanghai 200093, China

Bin Zhou
Shanghai Key Laboratory of Atmospheric Particle Pollution and Prevention (LAP3), Department of Environmental Science & Engineering, Fudan University, Shanghai 200433, China
Fudan Tyndall Centre, Fudan University, Shanghai 200433, China

Sazan Ali and Marie-Roberte Guichaoua
Institut Méditerranéen de Biodiversitéet d'Ecologie marine et continentale (IMBE), Centre National de la Recherche Scientifique (CNRS) UMR 7263/ Institut de Recherche pour le Développement (IRD) 237, Faculté de Médecine, Aix-Marseille Université (AMU), Marseille, France

Gérard Steinmetz and Odette Prat
Institute of Environmental Biology and Biotechnology (IBEB), Life Science division, French Alternative Energy and Atomic Energy Commission (CEA), Marcoule, Bagnols-sur-Céze, France

Guillaume Montillet, Marie-Hé léne Perrard and Philippe Durand
Institut de Génomique Fonctionnelle de Lyon (IGFL), Centre National de la Recherche Scientifique (CNRS) UMR 5242/ Institut National de la Recherche Agronomique (INRA), Ecole Normale Supérieure de Lyon (ENS),Lyon, France

Anderson Loundou
Unité d'Aide Méthodologique á la Recherche clinique, Faculté de Médecine, Aix-Marseille Université (AMU), Marseille, France

Caroline Louis, Gilles Tinant, Eric Mignolet and Cathy Debier
Institut des Sciences de la Vie, Université catholique de Louvain, Louvain-la-Neuve, Belgium

Jean-Pierre Thomé
Laboratoire d'Ecologie animale et d'Ecotoxicologie, Université de Liége, Liége, Belgium

Doreen Babin and Kornelia Smalla
Institute for Epidemiology and Pathogen Diagnostics, Julius Kühn-Institut - Federal Research Centre for Cultivated Plants (JKI), Braunschweig, Germany

Cordula Vogel and Katja Heister
Lehrstuhl für Bodenkunde, Technische Universität München, Freising-Weihenstephan, Germany

Sebastian Zühlke and Michael Spiteller
Institut für Umweltforschung (INFU), Lehrstuhl für Umweltchemie und Analytische Chemie der Fakultät für Chemie und Chemische Biologie, Technische Universität Dortmund, Dortmund, Germany

Michael Schloter
Research Unit for Environmental Genomics, Helmholtz Zentrum München, German Research Center for Environmental Health, Neuherberg, Germany

Geertje Johanna Pronk and Ingrid Kögel-Knabner
Lehrstuhl für Bodenkunde, Technische Universität München, Freising-Weihenstephan, Germany
Institute for Advanced Study, Technische Universität München, Garching, Germany

Index

www.ingramcontent.com/pod-product-compliance
Lightning Source LLC
Chambersburg PA
CBHW082050190326
41458CB00010B/3500